GLOBAL CLIMATE CHANGE, FOOD SECURITY, AND SUSTAINABLE DEVELOPMENT GOALS

Earth's temperature continues to rise due primarily to greenhouse gases (GHG) emitted by human activity. Corresponding environmental changes impact food production as farmlands are damaged or diminished. Food insecurity can lead to poverty, conflict, and mass migration as people look for better living conditions elsewhere.

In 2016, the United Nations adopted Sustainable Development Goals (SDGs) to tackle global issues like climate change and food security through 17 interconnected missions. SDGs challenge every country to promote better living conditions for all while protecting the planet.

Torrential rains from Hurricane Harvey flooded Houston in 2017. Global climate change may lead to more intense storms, threatening rural and urban places alike. *SDG11: Sustainable Cities and Communities* calls for more resilient, sustainable urban practices. (Ch. 3)

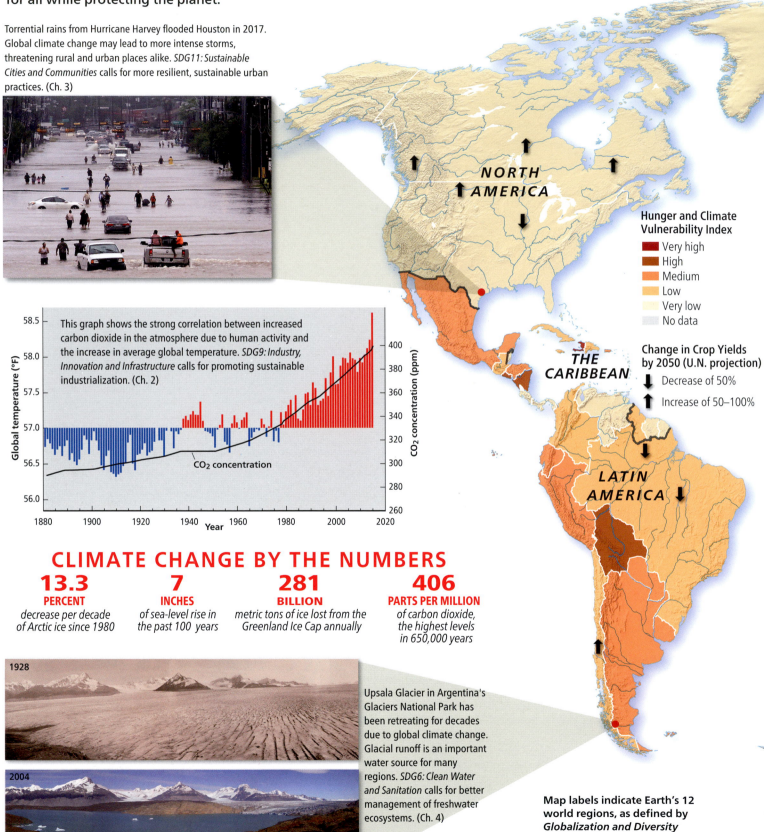

This graph shows the strong correlation between increased carbon dioxide in the atmosphere due to human activity and the increase in average global temperature. *SDG9: Industry, Innovation and Infrastructure* calls for promoting sustainable industrialization. (Ch. 2)

CO₂ concentration — CO_2 concentration

Global temperature (°F) / CO_2 concentration (ppm) / Year

Hunger and Climate Vulnerability Index
- Very high
- High
- Medium
- Low
- Very low
- No data

Change in Crop Yields by 2050 (U.N. projection)
- ↓ Decrease of 50%
- ↑ Increase of 50–100%

NORTH AMERICA

THE CARIBBEAN

LATIN AMERICA

CLIMATE CHANGE BY THE NUMBERS

13.3 PERCENT
decrease per decade of Arctic ice since 1980

7 INCHES
of sea-level rise in the past 100 years

281 BILLION
metric tons of ice lost from the Greenland Ice Cap annually

406 PARTS PER MILLION
of carbon dioxide, the highest levels in 650,000 years

1928

2004

Upsala Glacier in Argentina's Glaciers National Park has been retreating for decades due to global climate change. Glacial runoff is an important water source for many regions. *SDG6: Clean Water and Sanitation* calls for better management of freshwater ecosystems. (Ch. 4)

Map labels indicate Earth's 12 world regions, as defined by *Globalization and Diversity*

Sources: World Food Programme, United Nations FAO, World Health Organization, NASA, NOAA

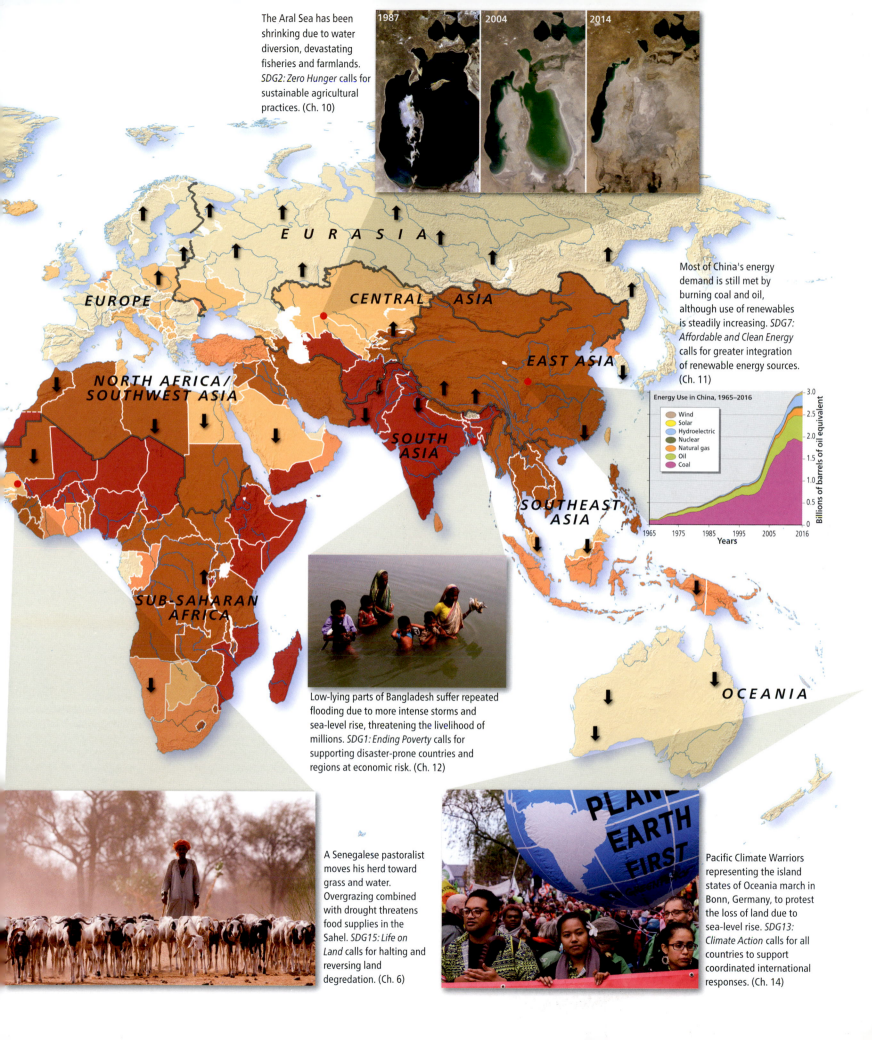

The Aral Sea has been shrinking due to water diversion, devastating fisheries and farmlands. *SDG2: Zero Hunger* calls for sustainable agricultural practices. (Ch. 10)

1987   2004   2014

EURASIA

EUROPE

CENTRAL ASIA

EAST ASIA

NORTH AFRICA/ SOUTHWEST ASIA

SOUTH ASIA

SUB-SAHARAN AFRICA

SOUTHEAST ASIA

OCEANIA

Most of China's energy demand is still met by burning coal and oil, although use of renewables is steadily increasing. *SDG7: Affordable and Clean Energy* calls for greater integration of renewable energy sources. (Ch. 11)

Energy Use in China, 1965–2016

- Wind
- Solar
- Hydroelectric
- Nuclear
- Natural gas
- Oil
- Coal

Billions of barrels of oil equivalent

3.0
2.5
2.0
1.5
1.0
0.5
0

1965 1975 1985 1995 2005 2016
Years

Low-lying parts of Bangladesh suffer repeated flooding due to more intense storms and sea-level rise, threatening the livelihood of millions. *SDG1: Ending Poverty* calls for supporting disaster-prone countries and regions at economic risk. (Ch. 12)

A Senegalese pastoralist moves his herd toward grass and water. Overgrazing combined with drought threatens food supplies in the Sahel. *SDG15: Life on Land* calls for halting and reversing land degradation. (Ch. 6)

Pacific Climate Warriors representing the island states of Oceania march in Bonn, Germany, to protest the loss of land due to sea-level rise. *SDG13: Climate Action* calls for all countries to support coordinated international responses. (Ch. 14)

Empowering Students to Address Global Issues

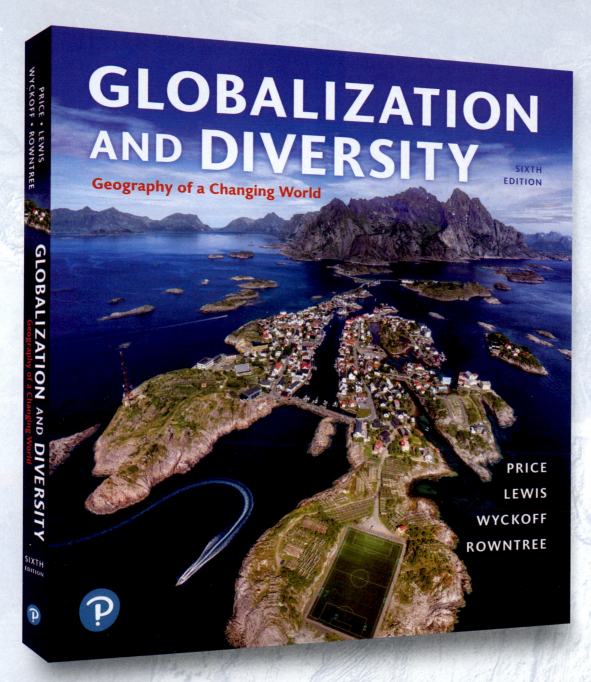

GLOBALIZATION AND DIVERSITY

Geography of a Changing World

SIXTH EDITION

PRICE
LEWIS
WYCKOFF
ROWNTREE

Globalization and Diversity, **Sixth Edition**

The sixth edition has a deep emphasis on humanitarian geography and sustainability, and encourages students to explore the sights, sounds, and tastes of world regions with embedded links to online digital resources.

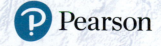

How Geographers Help Make the World a Better Place

HUMANITARIAN GEOGRAPHY
Unmasking the Tragedy in Flint, Michigan

▲ Figure 3.1.1 Rick Sadler

Rick Sadler, a geographer and GIS expert at Michigan State University, helped to uncover the roots of the water crisis that has recently plagued Flint, Michigan (Figure 3.1.1). In an effort to save money, the state directed the cash-strapped city to switch its water supply to the local Flint River in 2014. Tragically, the naturally more turbid water was not treated properly. Flint's new water supply corroded lead from delivery and home service lines, resulting in lead poisoning for thousands of residents, including many children. But initially, no one knew the source of the problem. A local pediatrician suspected that the elevated lead levels in Flint children were linked to Flint River water, and called on Sadler to examine the data.

Remapping the Data Sadler quickly realized that the data that state officials used were based on Flint ZIP codes, larger units that did not match city boundaries or represent the residents drinking the tainted water. "One-third of the addresses with a Flint ZIP code weren't in the city," Sadler recalls, and used a different water source. Remapping the data by the actual street addresses of children with elevated lead levels revealed a strong correlation with use of Flint River water. These older homes often house lead pipes that proved vulnerable to corrosion and leaching. Sadler's evidence confirmed the cause of the poisoning.

A Flint native, Sadler has been interested in mapping "as long as I can remember" and enrolled in GIS courses as an undergraduate. After returning to Flint, he became known as "the map guy" while working with different community groups. Sadler says his heart is in issues relevant to the city: "The more I learned about issues that drove Flint's decline…the more I felt compelled to not just understand them, but to uncover some of the spatial patterns—to use the tools that I had learned."

Based on Sadler's findings and intense public outcry surrounding the crisis, Flint returned to a safer water supply. Yet the crisis continues: less than 25 percent of the 20,000 affected lead-rich pipes had been upgraded by 2017, and contaminated water and its associated effects remain a daily challenge for thousands (Figure 3.1.2). Meanwhile, the spatial tools and geographic analysis that Sadler employed to confirm the source of the Flint tragedy have aided legal and criminal proceedings to address liability and seek environmental justice for affected residents.

Sadler notes that GIS is applicable to many public health issues, and that a geographer's multidisciplinary approach can be invaluable: "It's like being a goalie as opposed to being a forward. It's a special position that not everyone does, but it's absolutely essential."

1. Why do you think ZIP code zones are often used to map U.S. public health issues?
2. Find a map of local ZIP codes and argue why they may or may not be useful in studying environmental or social problems. What might be a better unit of analysis?

▲ Figure 3.1.2 Flint Residents Receive Bottled Water Volunteers from Full Gospel Churches in Michigan deliver bottled water to residents of Flint.

GOOGLE EARTH
Virtual Tour Video
https://goo.gl/cbEoGt

NEW! Humanitarian Geography features demonstrate how geographic tools and approaches improve the human condition when confronted with current challenges such as natural disasters, disease outbreaks, crisis and humanitarian mapping, and Sustainable Development Goals (SDGs).

UPDATED! Working Toward Sustainability features explore how the theme of sustainability plays out across world regions, looking at initiatives and positive outcomes of environmental and cultural sustainability. All features are integrated with Quick Response (QR) links to Google Earth® Virtual Tour Videos.

WORKING TOWARD SUSTAINABILITY
Saving the Great Barrier Reef

▲ Figure 14.1.1 Australia's Great Barrier Reef Off the eastern coast of Australia, the Great Barrier Reef stretches for more than 1400 miles (2300 km) through the Coral Sea.

Scientists call it the world's single largest expression of a living organism. Stretching through the azure- and turquoise-tinted waters of the Coral Sea for more than 1400 miles (2300 km) off the coast of Queensland, the Great Barrier Reef (GBR) includes more than 900 small islands and a myriad of underwater coral reefs (Figure 14.1.1). This remarkable ecosystem is home to 1500 species of fish, 400 species of coral, whales, dolphins, sea turtles, sea eagles, terns, and plant species found nowhere else on Earth (Figure 14.1.2). Taking more than 10,000 years to form, the reef has been a UN World Heritage Site since 1981, and much of the area is protected by Australia's Great Barrier Reef Marine Park.

Fighting for Survival Today, however, the GBR is in the fight of its life. Thanks to global climate change, the reef has lost more than half its coral cover since 1985, much of it damaged by warmer ocean temperatures that have accelerated rates of seawater acidification and coral bleaching. Bleaching occurs when the organism experiences increased stress from changing temperature, light, or nutrient conditions. The stressed corals expel helpful algae, causing them to turn white. Bleached corals may survive but are more susceptible to disease and death.

Coastal development also has added to its watery woes: More intensive agriculture in Queensland has produced ocean-bound sediment and increased runoff of toxic agricultural chemicals. Recent plans for expanding coal-loading depots at Abbot Point include potentially dumping waste rock onto the reef, a practice sure to disrupt the purity of local waters. On a hopeful note, the reef contributes to a huge tourist industry in Queensland, amounting to more than $4.6 billion annually. That translates into powerful economic interests that are actually committed to preserving the reef's environmental health.

Adapting to Change Recently, the Australian government pledged over $100 million toward improving reef water quality and toward innovations designed to slow future bleaching episodes. Farmers in nearby Queensland will be paid subsidies to limit runoff of harmful sediments and agricultural chemicals. Another area of investment is the development of a floating, sunscreen-like film that could protect especially vulnerable reefs by limiting the harmful rays of direct sunlight (which hasten the bleaching process). Additional money has been allocated toward targeting the crown of thorns starfish which has widely invaded delicate reef systems, making them even more vulnerable to bleaching and death. Scientists are also working to develop new genetic strains of coral that may be more resistant to global climate change and could be used to repopulate damaged settings.

Still, the long-term story for the GBR will probably hinge upon responsible shoreline development along the Queensland coast as well as the further global-scale impacts of ocean warming and acidification that will no doubt affect the region in coming decades. In addition, better aerial surveillance (such as more satellite reconnaissance and use of high-tech drones) of the Marine Park may monitor rogue fishing vessels more effectively in the future. For now, the reef's survival hangs in the balance, a giant poster child for a long list of damaging human impacts that threaten the environmental health of the entire South Pacific.

1. How might economic development in Queensland proceed while at the same time preserving the environmental health of the GBR?
2. Cite a fragile and protected environmental area in your region and briefly outline future prospects for its survival.

▲ Figure 14.1.2 Great Barrier Reef's Diverse Ecosystem This undersea view of the Great Barrier Reef features yellow sea-fan corals, acroporas (hard and white corals), and purple anthias (fish).

GOOGLE EARTH
Virtual Tour Video
http://goo.gl/zfZtkw

Connecting the Global to the Local

GLOBALIZATION IN OUR LIVES

Putin May Want to Be Your Friend

Russian operatives, Internet trolls, and hackers conducted a sustained campaign to influence the U.S. presidential election in 2016, according to the CIA, FBI, and the National Security Agency. These efforts included running anti-Clinton propaganda on Russian media outlets, hacking Democratic Party emails, and releasing these materials via WikiLeaks, and posting pro-Trump and politically divisive ads on a variety of social media that many of us use every day.

The abuse of social media platforms included Twitter, Facebook, Google, and Instagram, and was a reminder of the vulnerability of these virtual communities. More than 2700 fake Twitter accounts and 36,000 bots churned out pro-Trump tweets and political posts. Russian operatives purchased 80,000 Facebook ads that reached over 125 million users. The Russian-backed "Internet Research Agency" was the source of many of these bogus ads, but a lack of regulations and oversight at the time prevented many people from questioning their veracity or sources. Finally, Facebook CEO Mark Zuckerberg, aware that the Russians had outfoxed him, put in place more security measures designed to catch future abuses (Figure 9.4.1).

Russia's global reach into national elections is nothing new, but the pace of cyber-interference in the world of social media and computer hacking has accelerated since 2014. Russia has also used its superiority in the cyberworld to attempt to influence elections in the United Kingdom, Germany, France, and elsewhere. Has it worked? No one knows, but the next time you jump on your favorite social media site, you may be closer to your Russian comrades than you know.

▲ Figure 9.4.1 Mark Zuckerberg, Facebook CEO   Following the 2016 U.S. Presidential election, Zuckerberg and Facebook were criticized after it was revealed that Russian operatives had misused American social media during the campaign.

1. What geopolitical advantages might Russians hope to gain by interfering with elections in western Europe and the United States?

2. Are you vulnerable to being influenced by unethical ads or posts on your social media sites? Why or why not?

NEW! Globalization in Our Lives features explore common familiar commodities, cultural norms, activities, or popular culture that could be in a college student's experience or social network, showing how globalization connects their behavior or consumption across world regions.

UPDATED! Exploring Global Connections features describe unexpected and often surprising connections across world regions, leveraging recent events and coverage of cultural and environmental topics. All features are integrated with QR links to Google Earth Virtual Tour Videos.

EXPLORING GLOBAL CONNECTIONS

South America's Lithium Triangle

High in a remote corner of the Andes, where Bolivia, Argentina, and Chile meet, is the largest known reserve of lithium in the world. This soft, silver-white metal is an essential element in lightweight batteries, like those that power cell phones and laptops. It is also a key metal for electric vehicle batteries and photovoltaic cells. Companies such as Tesla, Samsung, and Apple are keenly aware of the cost and scarcity of lithium, which could greatly benefit these developing economies. Yet possessing more than half of the world's lithium is only step one—being able to extract it for global markets has been the challenge.

Lithium is found under salt flats in South America's Altiplano region, at elevations of up to 13,000 ft (Figure 4.4.1). Miners must extract the lithium-bearing brine from wells sunk deep below the salt crust and then deposit the liquid into evaporation ponds to let the sun do its work. Once sun-baked, the concentrate is taken for processing into lithium carbonate. South America's lithium boom thus far has been hindered by a lack of technology and capital, as well as national laws that designate lithium a strategic metal and therefore limit investment from foreign companies. Bolivia and Argentina have the largest known reserves, but Australia is the leading producer, followed by Chile. China rounds out the top five lithium source countries.

For decades, Chile has been the region's export leader, sending lithium carbonate primarily to manufacturers in South Korea, China, and Japan. The Atacama salt flats have the highest quality reserves, and ports such as Antofagasta are relatively close (Figure 4.4.2). Moreover, Chile's neoliberal policies have been more open to foreign investment in mining. Argentina is trying to catch up through increased foreign investment in lithium extraction around Jujuy province. In 2016, it produced about half as much lithium as Chile. Bolivia, which may have the largest reserves under the Salar de Uyuni salt flat, has yet to become a significant producer. This is partly due to the state's tight control of the resource and the wariness of foreign investors to engage in this country, which is noted for nationalizing key resources such as natural gas. As far as Bolivians are concerned, they need only look to the nearby mountain of Potosí, whose silver financed Spain's colonization of the Americas, to understand that owning a resource does not mean profiting from it.

▲ Figure 4.4.2 Lithium Processing in San Pedro de Atacama, Chile   In the high Atacama desert, lithium-laden brine is pumped out of the ground and into evaporation ponds. Once dried, the powdery substance is shipped and processed into lithium.

1. What are the factors that make Chile the leading South American exporter of lithium?

2. What are the products that you use that need lithium to function?

▲ Figure 4.4.1 Lithium Mining in South America   The largest lithium deposits in the world are found where Bolivia, Chile, and Argentina converge. Lithium is a critical metal for lightweight batteries used in cell phones and laptops.

LITHIUM MINING
◯ Salt flats

La Paz

BOLIVIA

Potosí

Uyuni

Lithium Triangle

ATLANTIC
OCEAN

CHILE

Olaroz

Atacama

Antofagasta

Jujuy

ARGENTINA

Pedernales

Maricunga

GOOGLE EARTH
Virtual Tour Video
https://goo.gl/5NLv7s

A Structured Learning Path to Support Today's Students

Physical Geography and Environmental Issues
China has long experienced severe deforestation and soil erosion, and its current economic boom is generating some of the world's worst pollution problems. Japan, South Korea, and Taiwan, however, have extensive forests and relatively clean environments.

Population and Settlement
Low birth rates and aging populations are found throughout East Asia. China is currently undergoing a major transformation as tens of millions of peasants move from impoverished villages in the interior to booming coastal cities.

Cultural Coherence and Diversity
Despite several unifying cultural features, East Asia in general and China in particular are divided along striking cultural lines. Historically, however, the entire region was linked by Mahayana Buddhism, Confucianism, and the Chinese writing system.

Geopolitical Framework
China's growing power is generating tension with other East Asian countries, while Korea remains a divided nation. As China's global influence grows, Japan, South Korea, and Taiwan are responding by strengthening ties with the United States.

Economic and Social Development
East Asia has been a core area of the world economy for several decades, with China undergoing one of the world's most rapid economic expansions. North Korea, however, remains desperately poor, plagued by widespread malnutrition.

UPDATED! Critical Themes of Geography
Following two unique introductory chapters, each regional chapter is organized into five thematic sections, making navigation and cross-regional comparisons easy for students and instructors. Themes include Physical Geography and Environmental Issues, Population and Settlement, Cultural Coherence and Diversity, Geopolitical Framework, and Economic and Social Development.

UPDATED! Region-specific Learning Outcomes in each
chapter's opening pages outline the knowledge and skills that students should gain from each chapter.

LEARNING OBJECTIVES

After reading this chapter you should be able to:

11.1 Contrast the physical geographies of the islands of East Asia (Japan and Taiwan) and the mainland.

11.2 Describe the main environmental problems China faces today and compare them with environmental challenges faced by Japan, South Korea, and Taiwan.

11.3 Explain why China's population is so unevenly distributed, with some areas densely settled and others almost uninhabited.

11.4 Summarize the distribution of major urban areas on the map of East Asia and explain why the region's largest cities are continuing to grow.

11.5 Describe how religion and other systems of belief both unify and divide East Asia.

11.6 Explain the distinction between, and geographical distribution of, the Han Chinese and the other ethnic groups of China, paying particular attention to language.

11.7 Outline the geopolitical division of East Asia during the Cold War era and explain how that division still influences East Asian geopolitics.

11.8 Identify factors behind East Asia's rapid economic growth in recent decades and discuss possible limitations to continued expansion at such a rate.

11.9 Summarize the geographical differences in economic and social development found in China and across East Asia as a whole.

UPDATED! Two Review Questions at the end of each section help students check their
comprehension as they read, and are followed by a listing of the key terms from each section, reinforcing the key concepts from each chapter section.

REVIEW

11.3 Why does East Asia import so much of its food and natural resources from other parts of the world?

11.4 Describe how the urban landscape of China is currently changing.

KEY TERMS anthropogenic landscape, hukou, urban primacy, megalopolis

REVIEW

11.9 How has the process of economic development been similar in Japan, South Korea, Taiwan, and China since the end of World War II, and how has it differed in each country?

11.10 Why do levels of social and economic development vary so extensively from the coastal region of China to its interior provinces?

KEY TERMS chaebol, laissez-faire, Special Economic Zone (SEZ), World Trade Organization (WTO), One Belt, One Road, social and regional differentiation, rust belt

Develop 21st Century Skills

NEW! 2-page Review, Reflect, & Apply Sections at the end of each chapter provide a robust interactive review experience, including a concise chapter summary, *Review Questions* that bridge multiple chapter themes, *Image Analysis* questions, new *Join the Debate* activities, new *Geospatial Data Analysis* activities, as well as QR links to *Geographers at Work* profiles.

NEW! Join the Debate presents two sides of a complex topic to engage students in active debate around the most critical topics of geography today. *Join the Debate* can be used for homework, group work, and discussions.

Geography of a Changing World 1

REVIEW, REFLECT, & APPLY

Summary

- Geography is the study of Earth's varied and changing landscapes and environments. This study can be done conceptually in many different ways, by physical or human geography and either topically or regionally—or by using a combination of all these approaches.
- Globalization affects all aspects of world geography with its economic, cultural, and political interconnectivity. However, despite fears that globalization will produce a homogeneous world, a great deal of diversity is still apparent. Geographers use various tools that draw on information gathered on the ground and by satellites to examine the world at different scales, from an inner-city block to the entire planet.
- Human populations around the world are growing either quickly or slowly depending on natural increase and widely different migration patterns. Urbanization is also a major factor in settlement patterns as people continue to move from rural to urban locales.
- Culture is learned behavior. It includes a range of tangible and intangible behaviors and objects, such as language and architecture. Globalization is changing the world's cultural geography, producing new cultural hybrids in many places. In other places, people resist change by protecting (or even resurrecting) traditional ways of life.
- Varying political systems provide the world with a dynamic geopolitical framework that is stable in some places and filled with tension and violence in others. As a result, the traditional concept of the nation-state is challenged by separatism, insurgency, and even terrorism.
- Proponents of globalization argue that all people in all places gain from expanded world commerce. But instead, there appear to be winners and losers, resulting in a geography of growing income inequality. Social development of health care and education is also highly uneven, but many key indicators are improving.

Review Questions

1. Define geography. Then define globalization and explain its relevance to understanding the world's changing geography.

2. What are the benefits of GIS, GPS, and satellite imagery in being able to monitor change and improve sustainability in a given place?

3. Summarize general migration trends around the world; and explain how these are influenced by and are impacting demographic, cultural, economic, and political change.

4. Explain the nation-state concept and provide examples. Is it still relevant in the age of globalization?

5. What is the difference between economic and social development? How might a rapidly developing country's population indicators from Table 1.2 change due to increasing well-being of its people?

Image Analysis

1. The flow of investment capital to remote parts of the planet is a feature of economic globalization. Which regions of the world receive relatively high foreign direct investment when compared with their gross domestic product? What do you think investors find attractive in these settings?

2. Imagine if you mapped which countries received the most FDI in absolute terms. What would that map look like, and why would it be different?

▶ Figure IA1 Foreign Direct Investment FDI is private foreign capital that enters a country for purposes of resource extraction, infrastructure development, and industrialization.

FOREIGN DIRECT INVESTMENT, 2016 (NET INFLOW AS % OF GDP)
- 6.0 or more
- 3.0–5.9
- 1.0–2.9
- Less than 1.0
- No data

Join the Debate

Globalization is most often associated with economic activity, but it impacts all aspects of the world's physical and human landscapes. Global linkages are complex and can result in a variety of outcomes—some unexpected. Is globalization generally good or bad for the social and economic development?

Globalization advances social and economic development!

- Technological advances level the global playing field and allow more people to engage in economic activity and trade.
- With open markets, there are fewer barriers, increasing the efficiency of goods production and reducing the price of goods.
- Open economies tend to be more democratic and more tolerant of diversity and have less gender inequality.

Globalization has negative consequences for development!

- As trade increases, wages decline and income inequality is exacerbated. Digital globalization increases efficiency but creates fewer high-skilled jobs because less labor is required.
- A growth-at-all-costs argument often accelerates depletion of natural resources and unsustainable development. Fluctuations in commodity prices can lead to economic instability.
- The speed at which capital is transferred can lead to instability in global financial markets.

Key Terms

autonomous area (p. 30)
choropleth map (p. 16)
colonialism (p. 31)
core–periphery model (p. 35)
cultural assimilation (p. 27)
cultural imperialism (p. 25)
cultural landscape (p. 6)
cultural syncretism (p. 27)
culture (p. 25)
decolonization (p. 32)
demographic transition model (p. 23)
diversity (p. 14)
economic migrant (p. 23)
ethnicity (p. 30)
formal region (p. 6)
functional region (p. 7)
gender (p. 28)
gender inequality (p. 39)
gender roles (p. 28)
geographic information systems (GIS) (p. 17)
geography (p. 5)
geopolitics (p. 29)
globalization (p. 8)
global positioning systems (GPS) (p. 15)
glocalization (p. 8)

gross domestic product (GDP) (p. 36)
gross national income (GNI) (p. 37)
gross national income (GNI) per capita (p. 37)
Human Development Index (HDI) (p. 37)
human geography (p. 5)
human trafficking (p. 12)
industrialization (p. 34)
informal economy (p. 35)
insurgency (p. 33)
language family (p. 27)
latitude (parallels) (p. 15)
less developed country (LDC) (p. 35)
lingua franca (p. 27)
longitude (meridians) (p. 15)
map projection (p. 15)
map scale (p. 15)
megacity (p. 25)
more developed country (p. 35)
nationalism (p. 26)
nation-state (p. 30)
neocolonialism (p. 32)
neoliberalism (p. 32)
net migration rate (p. 24)
physical geography (p. 5)
place (p. 6)
population density (p. 19)

population pyramid (p. 21)
purchasing power parity (PPP) (p. 37)
rate of natural increase (RNI) (p. 20)
refugee (p. 24)
region (p. 6)
regional geography (p. 5)
religion (p. 27)
remittances (p. 24)
remote sensing (p. 17)
replacement rate (p. 20)
secularism (p. 28)
sovereignty (p. 29)
space (p. 6)
sustainable development (p. 19)
Sustainable Development Goals (p. 19)
territory (p. 30)
terrorism (p. 33)
thematic geography (systematic geography) (p. 5)
total fertility rate (TFR) (p. 20)
urban primacy (primate city) (p. 25)
urbanization (p. 25)
World Bank (p. 13)
World Trade Organization (p. 12)

Mastering Geography

Looking for additional review and test prep materials? Visit the Study Area in **Mastering Geography** to enhance your geographic literacy, spatial reasoning skills, and understanding of this chapter's content by accessing a variety of resources, including **MapMaster** interactive maps, geoscience animations, videos, flashcards, web links, self-study quizzes, and an eText version of *Globalization and Diversity*.

▲ Figure D1 Global Consumers A busy shopping mall in Guangzhou, China.

GeoSpatial Data Analysis

Happiness as a Development Measure Many different indices try to measure relative development, levels of globalization, and overall well-being. In the last decade, there has been an increased effort to measure *happiness* and understand the causes for happiness and misery. Take a look at the World Happiness Report for 2018 (http://worldhappiness.report/ed/2018/). The report uses 2015–2017 data from the Gallop World Poll and focuses on happiness at the national level as well as analyzes the happiness of migrants who move within and between countries.

Click on the report's Chapter 2 link to review Figure 2.2, which ranks the happiness of 156 countries, and Figure 2.4, which ranks the happiness of foreign-born individuals within the surveyed countries. Then go back to the main page and click on the Online Data: Chapter 2 link to download a dataset for these measures.

World Happiness Report 2018
https://goo.gl/QRtTRR

Open MapMaster 2.0 in the Mastering Geography Study Area. Prepare and import the happiness data (click on the Table 2.1 tab at the bottom of the dataset) to map the world's 10 happiest countries, and answer these questions:

1. Where are these countries located? What variables make up this happiness measure? How can you explain the relative happiness of these countries?

2. Now consider the world's 10 largest countries discussed in the tables in our chapter. Where do these countries fall in the happiness ranking? Is there any relationship between GNI per capita as shown in Table 1.2 and relative happiness? Now map the happiness of foreign-born individuals within the surveyed countries (data from Table 2.4 in the Happiness Report; click the 2.4 tab in the dataset).

3. In which countries are the foreign-born most happy? Least happy? Are these immigrants happier than the domestic born or not? Can you explain what you observe?

40

41

NEW! GeoSpatial Data Analysis activities send students to online data sources to collect, prepare, and analyze spatial data using MapMaster 2.0.

The Sights, Sounds, & Tastes of World Regions

▲ Figure 1.3 The Cultural Landscape Despite globalization, the world's landscapes still have great diversity, as seen in Prague, Czechia (Czech Republic). Red tile roofs, three-to-four-story buildings, organic street patterns, and open squares distinguish this historic capital city. Geographers use the cultural landscape concept to better understand how people interact with their environment.

Explore the SIGHTS of Historic Prague
https://goo.gl/1J9dbJ

NEW! Sights of the Region features link photos and maps to dynamic online Google Maps that include community contributed photos, empowering students to explore the places and spaces that make up world regions.

▼ Figure 2.16 Glacier Calving in Antarctica As Earth's climate warms, Antarctica's massive ice sheet is losing volume. Here, a large piece of ice is calving—breaking off—a glacier into the neighboring Southern Ocean.

Explore the SOUNDS of a Calving Glacier
https://goo.gl/iddu.zGW

VIEDMA GLACIER
Massive Glacier Wall Collapse

NEW! Sounds of the Region features give students access to audio of regional music, language, and nature.

▲ Figure 9.29 Osh This savory Central Asian dish—a mix of rice, meat, vegetables, and spices—is a common sight in Moscow as immigrants from Uzbekistan, Tajikistan, and Kyrgyzstan bring their Central Asian food traditions to the Russian capital.

Explore the TASTES of Uzbek Food
http://goo.gl/3ItQhc

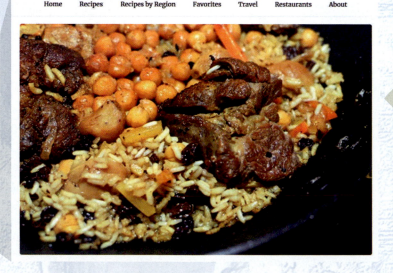

Home   Recipes   Recipes by Region   Favorites   Travel   Restaurants   About

NEW! Tastes of the Region features help students explore the geography and politics of food in each region, and include links to regional dishes and recipes.

MapMaster 2.0: Geospatial Tools in Your Hands

UPDATED! Mastering Geography is the teaching and learning platform that empowers you to reach *every* student. By combining trusted author content with digital tools developed to engage students and emulate the office-hour experience, Mastering personalizes learning and improves results for each student.

NEW! MapMaster 2.0 Interactive Map Activities Inspired by GIS, MapMaster 2.0 allows students to layer various thematic maps to analyze spatial patterns and data at regional and global scales. Now fully mobile, with enhanced analysis tools, MapMaster 2.0 allows students to upload their own data and geolocate themselves within the data. This tool includes zoom and annotation functionality, with hundreds of map layers leveraging recent data from sources such as the PRB, the World Bank, NOAA, NASA, USGS, United Nations, the CIA, and more. Available with assessment in Mastering Geography.

Transport Your Students to World Regions

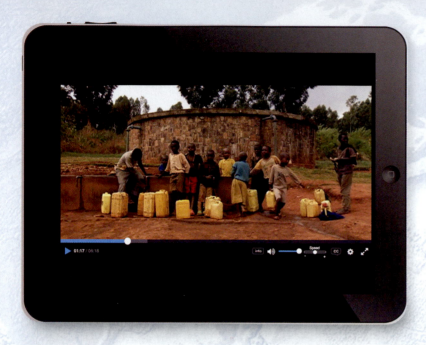

NEW! Video Activities from sources such as the BBC, Financial Times, and Television for the Environment's *Life* and *Earth Report* series are included in Mastering Geography. These videos provide students with applied real-world examples of geography in action, a sense of place, and allow students to explore a range of locations and topics.

NEW! Google Earth Virtual Tour videos give students brief narrated spatial explorations of places and people around the world, covering critical themes such as globalization and sustainability.

NEW! Mobile Field Trips videos for geography from Michael Collier give students another avenue for exploring U.S. landscapes and the major themes of physical geoscience concepts. These are embedded in the eText and available to assign in Mastering Geography.

Pearson eText:
An Integrated Learning Experience

NEW! Pearson eText is a simple-to-use, mobile-optimized, personalized reading experience available within Mastering. It allows students to easily highlight, take notes, and review key vocabulary all in one place—even when offline. Seamlessly integrated videos and other rich media engage students and give them access to the help they need, when they need it. Pearson eText is available within Mastering when packaged with a new book; students can also purchase Mastering with Pearson eText online.

UPDATED! Customizable for a Changing World
As an instructor you can add your own notes, embed videos and links, share highlights and notes with students, and rearrange chapters and sections to ensure the Pearson eText fits your unique course.

Resources for Instructors & Students

Instructor Resources are now found in Mastering Geography for your convenience. These resources provide everything you need to prep for your course and deliver a dynamic lecture, in one convenient place. Resources include:

Measuring Student Learning Outcomes

All of the Mastering Geography assignable content is tagged to Learning Outcomes from the book, the National Geography Standards, and Bloom's Taxonomy. You also have the ability to add your own learning outcomes, helping you track student performance against your course goals. You can view class performance against the specified learning outcomes and share those results quickly and easily by exporting to a spreadsheet.

PowerPoint & Lecture Assets for Each Chapter
- PowerPoint Lecture Outlines
- PowerPoint Clicker Questions
- All illustrations, tables, and photos from the text in PowerPoint and JPEG formats

TEST BANK
- The *Test Bank* in Microsoft Word formats
- TestGen Computerized *Test Bank*, which includes all the questions from the *Test Bank* in a format that allows you to easily and intuitively build exams and quizzes

ADDITIONAL RESOURCES
- *Instructor Resource Manual* in Microsoft Word and PDF formats
- Goode's *World Atlas*, 23rd Edition
- *Dire Predictions: Understanding Climate Change*, 2nd Edition by Mann and Kump

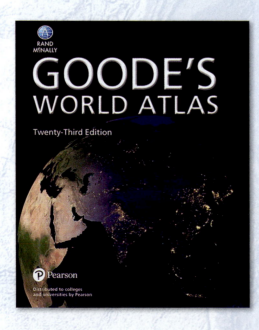

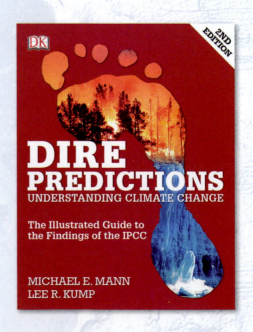

GLOBALIZATION AND DIVERSITY

MARIE PRICE
George Washington University

MARTIN LEWIS
Stanford University

WILLIAM WYCKOFF
Montana State University

LES ROWNTREE
University of California, Berkeley

with contributing author
WESLEY REISSER
George Washington University

Pearson

GLOBALIZATION AND DIVERSITY

GEOGRAPHY OF A CHANGING WORLD

SIXTH EDITION

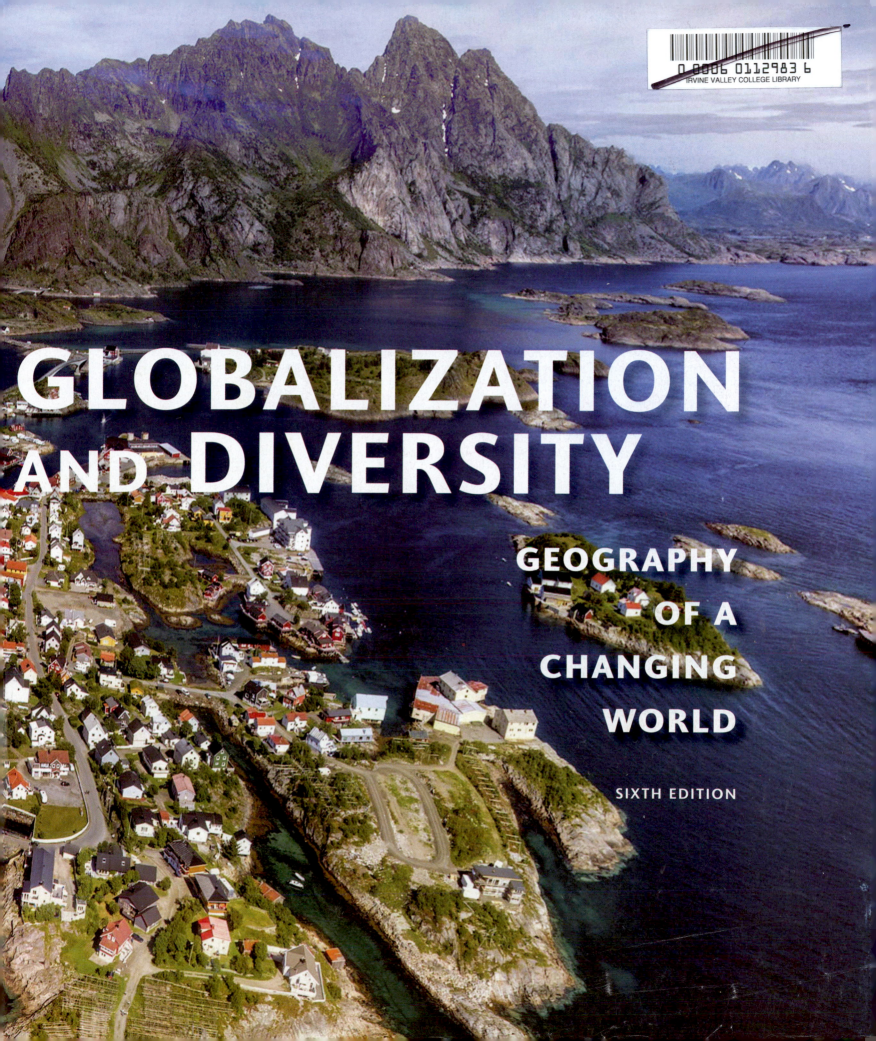

Executive Editor, Geosciences Courseware: Christian Botting
Director of Portfolio Management: Beth Wilbur
Content Producer: Brett Coker
Managing Producer: Michael Early
Courseware Director, Content Development: Ginnie Simione Jutson
Development Editor: Veronica Jurgena
Courseware Editorial Assistant: Sherry Wang
Rich Media Content Producer: Chloe Veylit
Full-Service Vendor: Pearson CSC
Full-Service Project Management: Pearson CSC, Erika Jordan
Cartography & Illustrations: International Mapping

Cartography & Art Coordinator: Kevin Lear, International Mapping
Design Manager: Mark Ong
Interior & Cover Designer: Jeff Puda
Rights & Permissions Project Manager: Pearson CSC, Matthew Perry
Rights & Permissions Management: Ben Ferrini
Photo Researcher: Kristin Piljay
Manufacturing Buyer: Stacey Weinberger, LSC Communications
Executive Field Marketing Manager: Mary Salzman
Product Marketing Manager: Alysun Estes
Cover Photo Credit: Ratnakorn Piyasirisorost / Getty Images

Library of Congress Cataloging-in-Publication Data

Names: Rowntree, Lester, 1938- author.
Title: Globalization and diversity : geography of a changing world / Les
   Rowntree, University of California, Berkeley, Martin Lewis, Stanford
   University, Marie Price, George Washington University, William Wyckoff,
   Montana State University.
Other titles: Globalisation and diversity
Description: Sixth Edition. | Hoboken, NJ : Pearson, [2019] | Revised edition of the authors' Globalization and diversity, [2017]
Identifiers: LCCN 2018053783 | ISBN 9780134898391 (pbk.)
Subjects: LCSH: Economic geography. | Globalization.
Classification: LCC HF1025 .G59 2019 | DDC 330.9--dc23
LC record available at https://lccn.loc.gov/2018053783

About Our Sustainability Initiatives

Pearson recognizes the environmental challenges facing this planet, as well as acknowledges our responsibility in making a difference. This book is carefully crafted to minimize environmental impact. The binding, cover, and paper come from facilities that minimize waste, energy consumption, and the use of harmful chemicals. Pearson closes the loop by recycling every out-of-date text returned to our warehouse.

Along with developing and exploring digital solutions to our market's needs, Pearson has a strong commitment to achieving carbon neutrality. As of 2009, Pearson became the first carbon- and climate-neutral publishing company, having reduced our absolute carbon footprint by 22% since then. Pearson has protected over 1,000 hectares of land in Columbia, Costa Rica, the United States, the United Kingdom and Canada.

In 2015, Pearson formally adopted *The Global Goals for Sustainable Development*, sponsoring an event at the United Nations General Assembly and other ongoing initiatives. Pearson sources 100% of the electricity we use from green power and invests in renewable energy resources in multiple cities where we have operations, helping make them more sustainable and limiting our environmental impact for local communities.

The future holds great promise for reducing our impact on Earth's environment, and Pearson is proud to be leading the way. We strive to publish the best books with the most up-to-date and accurate content, and to do so in ways that minimize our impact on Earth. To learn more about our initiatives, please visit

https://www.pearson.com/corporate/sustainability.html

ISBN 10: 0-134-89839-7; ISBN 13: 978-0-134-89839-1 (Student Edition)
ISBN 10: 0-135-20387-2; ISBN 13: 978-0-135-20387-3 (Loose-leaf Edition)
www.pearson.com

Brief Contents

Contents

Geography of a Changing World 1

Physical Geography and the Environment 2

The Caribbean   5

Sub-Saharan Africa   6

Europe 8

Southwest Asia and North Africa 7

Eurasia 9

Central Asia 10

East Asia   11

South Asia   12

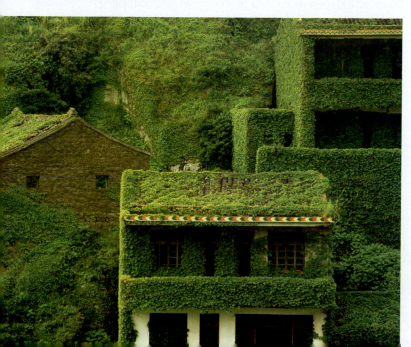

Oceania 14

Southeast Asia 13

Preface

Globalization and Diversity: Geography of a Changing World, sixth edition, is an issues-oriented textbook for college and university world regional geography classes that explicitly recognizes the vast geographic changes taking place because of globalization. With this focus we join the many scholars who consider globalization to be the most fundamental reorganization of the world's socioeconomic, cultural, and geopolitical structure since the Industrial Revolution. That provides the point of departure and thematic structure for our book.

As geographers, we think it essential for our readers to understand and critique two interactive themes: the consequences of converging environmental, cultural, political, and economic systems inherent to globalization, and the persistence—and even expansion—of geographic diversity and differences in the face of globalization. These two opposing forces, homogenization and diversification, are reflected in our book's title, *Globalization and Diversity*.

NEW & UPDATED IN THE SIXTH EDITION

- In this edition we welcome Dr. Wesley Reisser as a NEW contributing author. Dr. Reisser is a scholar and diplomat. He earned his PhD from the University of California, Los Angeles (UCLA) and teaches political geography and energy resources as an adjunct professor at the George Washington University. For over 15 years he has worked at the U.S. Department of State, most recently in the Bureau of Democracy, Human Rights, and Labor. Dr. Reisser brings to *Globalization and Diversity* a strong interest in political geography and human rights, extensive experience in Europe and Southwest Asia, and scholarly work on the social, political, and environmental implications of energy production and consumption.

- NEW *Humanitarian Geography* sidebars examine how geospatial tools and geographic analysis contribute to addressing humanitarian issues and natural disasters in each world region. Topics include finding and relocating refugee populations, teaching mapping skills to volunteers to respond to natural disasters or to build legal cases for human rights abuses, or uncovering and addressing environmental challenges. Some of the *Humanitarian Geography* examples feature individual geographers in the field, describing their work and their thoughts on geography's role in humanitarian efforts.

- NEW *Sights of the Region* features provide mobile-ready Quick Response (QR) links from photos to online Google Maps, enabling students to browse web maps and community-contributed photos of the diverse geographies featured in the print book. Students use mobile devices to scan Quick Response (QR) codes to get immediate online access and connect print images with dynamic online web maps and photos.

- NEW *Sounds of the Region* features provide QR links to sound clips that help give students a sense of culture and natural environments around the world, highlighting language, music, and the soundscapes of both natural and urban environments.

- NEW *Tastes of the Region* features in each regional chapter explore culinary traditions and innovations associated with different parts of the world. These QR links to websites or videos provide recipes and other pertinent information on food production and consumption, as well as material on cultural aspects of regional cuisines.

- The NEW end-of-the-chapter format—*Review, Reflect, & Apply*—asks students to answer broad-based questions spanning concepts and regions. Two of the three components of this feature, *Image Analysis* and the NEW *GeoSpatial Data Analysis*, provide concrete exercises based on the analysis of graphic images and demographic or socioeconomic data. *GeoSpatial Data Analysis* features invite students to prepare and visualize data using NEW MapMaster 2.0 in Mastering Geography. The NEW *Join the Debate* feature frames two opposing viewpoints on controversial issues and asks students to assess their claims and weigh in on their own. The end of chapter also features NEW QR links to online *Geographers at Work* profiles of geographers who specialize in that world region. Most describe their fieldwork as well as their insights on the discipline.

- NEW *Mobile Field Trip Videos* have students accompany renowned geoscience photographer Michael Collier in the air and on the ground to explore iconic landscapes that have shaped North America and beyond. Students scan QR codes in the print book to get instant access to these media, which are also available for assignment with quizzes in Mastering Geography.

- Many NEW *Key Concepts* and terms introduced in the first two chapters are revisited in at least two regional chapters. The overall number of terms have been reduced and major concepts that are critical to understanding globalization and diversity in a changing world are emphasized throughout the book.

- NEW chapter opener vignettes and photos highlight recent events and global linkages, with accompanying maps that pinpoint vignette locations. This edition also features more focused and consistent introductions in Chapters 3–14, placed under the heading "Defining the Region."

- UPDATED *Globalization in Our Lives* sidebars (previously known as *Everyday Globalization*) explore the daily items we use, from the cell phones that never seem to leave our sides, to the foods we eat, the clothing we wear, the music we listen to, and the technology that connects us to each other and the world.

- UPDATED & REVISED Tables in each chapter present *Population Indicators* and *Development Indicators* for the world's 10 most populous countries (Chapter 1) and for each country in the various world regions (Chapters 3–14). New indicators measuring development include Secondary School Enrollment Rates for males and females and an overall Freedom Rating developed by Freedom House, an independent watchdog organization.

New & Updated Features in Chapter 1: Geography of a Changing World

- This chapter has been retitled and the introduction is focused more sharply on geography in order to better integrate globalization processes and the discipline. In the section *Geography Matters*, the discussion of areal differentiation and connectivity has been revised. A *Mobile Field Trip* QR link, "Introduction to Geography," has been added.

- *Converging Currents of Globalization* section is retitled *Globalization and New Geographies* with new examples and figures, including a new International Migration diagram and new Global Arms Sales maps. The "Thinking Critically about Globalization" discussion has been revised and shortened.

- The thematic sections of this introductory chapter have also been updated, revised, and enhanced with new examples and photos. The *Geopolitical Framework* section introduces the Freedom Rating (included in all chapter data tables) with a new *Freedom in the World* map, while the *Economic and Social Development* section updates the discussion of poverty measures (with a new map of Morocco) and introduces secondary school enrollment (now included in all chapter data tables).

- Several key terms have been added, including *economic migrant, refugee*, and *sustainable development*. Existing sidebars have been revised, and a new *Humanitarian Geography* sidebar titled "Tools for Service" has been added.

New and Updated Features in Chapter 2: Physical Geography and the Environment

- The discussion of plate tectonics in the *Geology* section has been revised and shortened, with new examples and photos. NEW *Mobile Field Trip* QRs link to concise videos on climate change, volcanic activity, and cloud dynamics.

- The *Global Climates* section has been revised, with enhanced discussions of both climate change itself and international mitigation efforts. The 2015 Paris Agreement goals are introduced, as are various national climate plans.

- New key terms include *ecosystem, geothermal, Paris Agreement*, and *watershed*. New sidebars highlight climate change on the Greenland Ice Sheet; Saudi Arabian plans to acquire water rights abroad; developed and developing countries aiming for carbon neutrality, and the problem of plastics pollution.

ORGANIZATION

Globalization and Diversity: Geography of a Changing World is organized around the conventional world regions of Sub-Saharan Africa, Europe, Latin America, East Asia, South Asia, and so on. We have, however, added two distinctive regions that are often excluded from the standard world regional scheme: Central Asia and the Caribbean. Also in this edition Chapter 9 has been renamed Eurasia and Chapter 14, Oceania. Our 12 regional chapters further depart from the treatment found in traditional world regional textbooks by employing a thematic framework that avoids extensive descriptions of each individual country.

Globalization and Diversity opens with two substantive introductory chapters that provide the geographic fundamentals of both human and physical geography. *Chapter 1: Geography of a Changing World* begins by introducing the discipline of geography and its major concepts, followed by a section on the geographic dimensions of globalization, including discussion of the costs and benefits of globalization according to proponents and opponents. Next is a section called "The Geographer's Toolbox," where students are introduced to map-reading, cartography, aerial photos, remote sensing, and GIS. This initial chapter concludes with a discussion of the concepts and tabular data that are used throughout the regional chapters.

Chapter 2: Physical Geography and the Environment builds an understanding of physical geography and environmental issues with discussions of geology and environmental hazards; weather, climate, and global climate change; global bioregions and biodiversity; hydrology and water stress; and energy issues. Both introductory chapters introduce key concepts that are revisited in the regional chapters.

Each regional chapter is structured around five geographic themes:

- **Physical Geography and Environmental Issues**, in which we not only describe the physical geography of each region, but also environmental issues, including climate change and energy.

- **Population and Settlement**, where we examine the region's demography, migration patterns, land use, and settlement, including cities.

- **Cultural Coherence and Diversity** covers the traditional topics of language and religion, but also examines the ethnic and cultural tensions resulting from globalization. New to this edition is a focus on regional foodways. Gender issues and popular culture topics such as sports and music (with *Sounds of the Region* QR links) are also included in this section.

- **Geopolitical Framework** examines the political geography of the region, taking on such issues as postcolonial tensions, ethnic conflicts, separatism, micro-regionalism, and global terrorism.

- **Economic and Social Development** explores each region's economic framework at both local and global scales and examines such social issues as health, education, and gender inequalities.

CHAPTER FEATURES

- **Structured learning path.** Every chapter begins with an explicit set of *learning objectives* to provide students with the larger context of each chapter. *Review questions* after each section allow students to test their learning. Each chapter ends with an innovative *Review, Reflect, & Apply* section, where students are asked to apply what they have learned from the chapter using an active-learning framework: broad questions integrating material across sections and chapters, image analysis and debate, and mapping real-world data.

- **Comparable regional maps.** Of the many maps in each regional chapter, many are constructed on the same themes and with similar data so that readers can easily draw comparisons between regions. Most regional chapters have maps of physical geography, climate, environmental issues, population density, migration, language, religion, and geopolitical issues.

- **Other chapter maps pertinent to each region.** The regional chapters also contain many additional maps illustrating important geographic topics such as global economic issues, social development, and ethnic tensions.

- **Regional data sets integrated with MapMaster 2.0.** Two thematic tables in each regional chapter provide insights into the characteristics of each region and facilitate comparisons between regions. The first table provides population data on a number of issues, including fertility rates and proportions of the population under 15 and over 65 years of age, as well as net migration rates for each country within the region. The second table presents economic and social development data for each country, including gross national income per capita, gross domestic product growth, percentage of the population living on less than $3.10 per day, child mortality rates, secondary school enrollment, the international gender inequality index and the Freedom Rating. Each table now includes a QR link to MapMaster 2.0 in Mastering Geography, so that students gain experience mapping data and analyzing the map to answer questions about the region.

- **Sidebar features.** Each chapter has four sidebars that expand on geographic themes:

  - New ***Humanitarian Geography*** explores the geospatial tools and geographic analysis employed to address pressing issues such as responses to natural disasters, assistance to refugees, monitoring human rights abuses, and tracking environmental issues.

  - ***Globalization in Our Lives*** (previously known as *Everyday Globalization*) shows examples of how globalization influences our daily lives from the clothing we wear, the foods we consume, the technology we rely upon, and the activities we enjoy.

  - ***Working Toward Sustainability*** feature case studies of sustainability projects throughout the world, emphasizing positive environmental and social initiatives and their results. Each includes a QR link to an online Google Earth Virtual Tour Video.

  - ***Exploring Global Connections*** investigate the many ways in which activities in different parts of the world are linked so that students understand that in globalized world regions are neither isolated nor discrete. Each includes a QR link to an online Google Earth Virtual Tour Video.

ACKNOWLEDGEMENTS

We have many people to thank for the conceptualization, writing, rewriting, and production of *Globalization and Diversity*, sixth edition. First, we'd like to thank the thousands of students in our world regional geography classes who have inspired us with their energy, engagement, and curiosity; challenged us with their critical insights; and demanded a textbook that better meets their need to understand the contemporary geography of their dynamic and complex world.

This is also the first edition of *Globalization and Diversity* in which Dr. Les Rowntree has not contributed new materials and revisions. Dr. Rowntree led this textbook team since its inception in the mid-1990s. After a long and productive career as professor, scholar, and author, he has retired. Dr. Rowntree assembled the author team for this book, and collectively the authors have enjoyed over two decades of fruitful collaboration, scholarship, and friendship.

Finally, we are deeply indebted to many professional geographers and educators for their assistance, advice, inspiration, encouragement, and constructive criticism as we labored through the different stages of this book. Among the many who provided invaluable comments on various drafts and editions of *Globalization and Diversity* or who worked on supporting print or digital material are: Gilian Acheson (S. Illinois Univ., Edwardsville), Joy Adams (Humboldt State Univ.), Victoria Alapo (Metropolitan Comm. College), Dan Arreola (Arizona State Univ.), Bernard BakamaNume (Texas A&M Univ.), Brad Baltensperger (Michigan Tech. Univ.), Max Beavers (Samford Univ.), Laurence Becker (Oregon State Univ.), Dan Bedford (Weber State Univ.), James Bell (Univ. of Colorado), Katie Berchak (Univ. of Louisiana, Lafayette), William H. Berentsen (Univ. of Connecticut), Kevin Blake (Kansas State Univ.), Mikhail Blinnikov (St. Cloud State Univ.), Sarah Blue (Texas State Univ., San Marcos), Michelle Brym (Univ. of Central Oklahoma), Karl Byrand (Univ. of Wisconsin, Sheboygan County), Michelle Calvarese (California State Univ., Fresno), Craig Campbell (Youngstown State Univ.), G. Scott Campbell (College of DuPage), Elizabeth Chacko (George Washington Univ.), Philip Chaney (Auburn Univ.), Xuwei Chen (N. Illinois Univ.), David B. Cole (Univ. of Northern Colorado), Amanda Coleman (N. State Univ.), Malcolm Comeaux (Arizona State Univ.), Jonathan C. Comer (Oklahoma State Univ.), Deborah Corcoran (Missouri State Univ.), Jeremy Crampton (George Mason Univ.), Kevin Curtin (Univ. of Texas at Dallas), James Curtis (California State Univ., Long Beach), Dydia DeLyser (California State Univ., Fullerton), Francis H. Dillon (George Mason Univ.), Jason Dittmer (Georgia Southern Univ.), Jerome Dobson (Univ. of Kansas), Caroline Doherty (N. Arizona Univ.), Vernon Domingo (Bridgewater State College), Roy Doyon (Ball State Univ.), Dawn Drake (Missouri W. State Univ.), Jane Ehemann (Shippensburg Univ.), Steven Ericson (Univ. of Alabama, Tuscaloosa), Chuck Fahrer (Georgia College and State Univ.), Dean Fairbanks (California State Univ., Chico), Emily Fekete (Univ. of Kansas), Caitie Finlayson (Florida State Univ.), Colton Flynn (Univ. of Arkansas, Fort Smith), Doug Fuller (Univ. of Miami), Douglas Gamble (Univ. of North Carolina, Wilmington), Sherry Goddicksen (California State Univ., Fullerton), Sarah Goggin (Cypress College), Mark Guizlo (Lakeland Community College), Reuel Hanks (Oklahoma State Univ.), Megan Hoberg (Orange Coast College & Golden West College), Steven Hoelscher (Univ. of Texas, Austin), Erick Howenstine (N. Illinois University), Tyler Huffman (E. Kentucky Univ.), Peter J. Hugil (Texas A&M Univ.), Eva Humbeck (Arizona State Univ.), Shireen Hyrapiet (Oregon State Univ.), Drew Kapp (Univ. of Hawaii, Hilo), Ryan S. Kelly (Univ. of Kentucky), Richard H. Kesel (Louisiana State Univ.), Cadey Korson (Kent State Univ.), Rob Kremer (Front Range Community College), Robert C. Larson (Indiana S. Univ.), Chi Kin Leung (California State Univ., Fresno), Alan A. Lew (N. Arizona Univ.), Elizabeth Lobb (Mt. San Antonio College), Catherine Lockwood (Chadron State College), Max Lu (Kansas State Univ.), Luke Marzen (Auburn Univ.), Daniel McGowin (Auburn Univ.), James Miller (Clemson Univ.), Bob Mings (Arizona State Univ.), Wendy Mitteager (SUNY, Oneonta), Sherry D. Morea-Oakes (Univ. of Colorado, Denver), Anne E. Mosher (Syracuse Univ.), Julie Mura (Florida State Univ.), Tim Oakes (Univ. of Colorado), Nancy Obermeyer (Indiana State Univ.), Karl Offen (Univ. of Oklahoma), Daniel Olsen (Brigham Young Univ.), Thomas Orf (Las Positas College), Kefa Otiso (Bowling Green State Univ.), Joseph Palis (Univ. of North Carolina),

Jean Palmer-Moloney (Hartwick College), Bimal K. Paul (Kansas State Univ.), Michael P. Peterson (Univ. of Nebraska, Omaha), Richard Pillsbury (Georgia State Univ.), Brandon Plewe (Brigham Young Univ.), Jess Porter (Univ. of Arkansas at Little Rock), Patricia Price (Florida International Univ.), Erik Prout (Texas A&M Univ.), Claudia Radel (Utah State Univ.), David Rain (George Washington Univ.), Rhonda Reagan (Blinn College), Joshua Regan (Western Connecticut State Univ.), Kelly Ann Renwick (Appalachian State Univ.), Craig S. Revels (Portland State Univ.), Pamela Riddick (Univ. of Memphis), Scott M. Robeson (Indiana State Univ.), Paul A. Rollinson (S. Missouri State Univ.), Jessica Salo (Univ. of N. Colorado), Yda Schreuder (Univ. of Delaware), Kathy Schroeder (Appalachian State Univ.), Kay L. Scott (Univ. of Central Florida), Patrick Shabram (S. Plains College), Duncan Shaeffer (Arizona State Univ.), Dimitrii Sidorov (California State Univ., Long Beach), Susan C. Slowey (Blinn College), Andrew Sluyter (Louisiana State Univ.), Christa Smith (Clemson Univ.), Joseph Spinelli (Bowling Green State Univ.), William Strong (Univ. of N. Alabama), Philip W. Suckling (Univ. of N. Iowa), Curtis Thomson (Univ. of Idaho), Suzanne Traub-Metlay (Front Range Community College), James Tyner (Kent State Univ.), Nina Veregge (Univ. of Colorado), Jonathan Walker (James Madison Univ.), Fahui Wang (Louisiana State Univ.), Gerald R. Webster (Univ. of Alabama), Keith Yearman (College of DuPage), Emily Young (Univ. of Arizona), Bin Zhon (S. Illinois Univ., Edwardsville), Henry J. Zintambia (Illinois State Univ.), Sandra Zupan (Univ. of Chicago).

In addition, we wish to thank the many publishing professionals who have made this book possible. We start with Christian Botting, Pearson's Executive Editor for Geosciences, a consummate professional and good friend whose leadership, high standards, and enduring patience has been laudable, inspiring, and necessary. Next in line is Brett Coker, Content Producer, for his daily attention to production matters as well as his graceful and diplomatic interaction with four demanding and often cranky authors. Veronica Jurgena, our outstanding Development Editor, whose insights, guidance, and encouragement (much of it coming from her education as a geographer) have been absolutely crucial to this revision. Also to be thanked are a number of behind-the-curtain professionals: Kristin Piljay, photo researcher; Ben Ferrini, Manager, Rights and Permissions, Erika Jordan, SPi Project Manager, for somehow turning thousands of pages of manuscript into a finished product; and Kevin Lear, *International Mapping* Senior Project Manager, for his outstanding work creating and revising our maps. Thanks are due to Zachary McGinley for his timely production of all data tables.

Not to be overlooked are the many professional geographers and scientists who allowed us to pry into their personal and professional lives so we could provide profiles about them in the various sidebars, both in print and online: Fenda Akiwumi (Univ. of S. Florida), Holly Barcus (Macalester College), Sarah Blue (Texas State Univ.), Laura Brewington (East-West Center), Karen Culcasi (W. Virginia Univ.), Corrie Drummond Garcia (USAID), Carol Farbotko (Commonwealth Scientific and Industrial Research Org.), M Jackson (Univ. of Oregon), Cary Karacas (College of Staten Island), P.P. Karan (Univ. of Kentucky), Weronika Kusek (N. Michigan Univ.), Lucia Lo (York Univ.), Chandra Mitra (Auburn Univ.), Timothy Mousseau (Univ. of S. Carolina), Kelsey Nyland (Michigan State Univ.), Temidayo Isaiah Oniosun (Federal Univ. of Tech, Akure), Diane Papineau (Montana State Library), Rick Sadler (Michigan State Univ.), Rachel Silvey (Univ. of Toronto), Laurence Smith (Univ. of California, Los Angeles), Dmitry Streletskiy (The George Washington Univ.), Brian Tomaszewski (Rochester Institute of Tech.), Gregory Veeck (W. Michigan Univ.), Susan Wolfinbarger (American Association for the Advancement of Science).

Last, the authors want to thank that special group of friends and family who were there when we needed you most—early in the morning and late at night; in foreign countries and at home; when we were on the verge of tears and rants, but needed lightness and laughter; for your love, patience, companionship, inspiration, solace, enthusiasm, and understanding. Words cannot thank you enough: Rob, Joseph, and James Crandall, Marie Dowd, Evan and Eleanor Lewis, Karen Wigen, Linda, Tom, and Katie Wyckoff, Robert Geremia, and Kurt and Susan Reisser.

Marie Price

Martin Lewis

William Wyckoff

Wesley Reisser

About the Authors

Marie Price is a Professor of Geography and International Affairs at George Washington University. A Latin American specialist, Dr. Price has conducted research in Belize, Mexico, Venezuela, Panama, Cuba, and Bolivia. She has also traveled widely throughout Latin America and Sub-Saharan Africa. Her studies have explored human migration, natural resource use, environmental conservation, and sustainability. She is President of the American Geographical Society and a nonresident fellow of the Migration Policy Institute, a nonpartisan think tank that focuses on migration issues. Dr. Price brings to *Globalization and Diversity* a special interest in regions as dynamic spatial constructs that are shaped over time through both global and local forces. Her publications include the co-edited book *Migrants to the Metropolis: The Rise of Immigrant Gateway Cities* (2008) and numerous academic articles and book chapters.

Martin Lewis is a Senior Lecturer in History at Stanford University, where he teaches courses on global geography. He has conducted extensive research on environmental geography in the Philippines and on the intellectual history of world geography. His publications include *Wagering the Land: Ritual, Capital, and Environmental Degradation in the Cordillera of Northern Luzon, 1900–1986* (1992), and, with Karen Wigen, *The Myth of Continents: A Critique of Metageography* (1997). Dr. Lewis has traveled extensively in East, South, and Southeastern Asia. His most recent book, co-written with Asya Pereltsvaig, is *The Indo-European Controversy: Facts and Fallacies in Historical Linguistics* (2015). In April 2009, Dr. Lewis was recognized by *Time* magazine as one of America's most favorite lecturers.

William Wyckoff is a geographer in the Department of Earth Sciences at Montana State University specializing in the cultural and historical geography of North America. He has written and co-edited several books on North American settlement geography, including *The Developer's Frontier: The Making of the Western New York Landscape* (1988), *The Mountainous West: Explorations in Historical Geography* (1995) (with Lary M. Dilsaver), *Creating Colorado: The Making of a Western American Landscape 1860–1940* (1999), and *On the Road Again: Montana's Changing Landscape* (2006). His most recent book, entitled *How to Read the American West: A Field Guide*, appeared in the Weyerhaeuser Environmental Books series and was published in 2014 by the University of Washington Press. A World Regional Geography instructor for 26 years, Dr. Wyckoff emphasizes in the classroom the connections between the everyday lives of his students and the larger global geographies that surround them and increasingly shape their future.

Les Rowntree is currently a Research Associate at the University of California, Berkeley, where he writes about global and local environmental issues. This career change comes after 35 years teaching both Geography and Environmental Studies at San Jose State University. As an environmental geographer, Dr. Rowntree's interests focus on international environmental issues, biodiversity conservation, and climatic change. He sees world regional geography as way to engage and inform students by providing them with the conceptual tools to critically and constructively assess the contemporary world. His current writing projects include a natural history book and website about California's Coast Ranges, and several essays on different European environmental topics. Along with these writings he maintains an assortment of web-based blogs and websites.

Wesley Reisser is an adjunct professor of Geography at the George Washington University specializing in political geography and energy. Since 2003, Dr. Reisser has served at the U.S. Department of State in a variety of positions working on human rights, the United Nations, the Israeli-Palestinian conflict, and responding to crisis situations abroad. Dr. Reisser received the United Nations Association Tex Harris Award for Human Rights and Diplomacy in 2015. Dr. Reisser's first book, *The Black Book: Woodrow Wilson's Secret Plan for Peace* (2013), is the only comprehensive analysis of the maps and plans used by the United States at the end of World War I. His second book, written with his brother Colin, is *Energy Resources: From Science to Society* (2018), the first interdisciplinary textbook on global energy issues. Dr. Reisser is a Councilor of the American Geographical Society, the founding Artistic Director of Washington, DC's central and eastern European Carpathia Folk Dance Ensemble, and is the 2007 World Geography Bowl MVP.

Digital & Print Resources

This edition provides a complete world regional geography program for students and teachers.

FOR STUDENTS & TEACHERS

Mastering Geography with Pearson eText for *Globalization and Diversity*.

The Mastering platform is the most widely used and effective online homework, coaching, and assessment system for the sciences. It delivers self-paced coaching activities that provide individualized coaching, focus on course objectives, and are responsive to each student's progress. The Mastering system helps teachers maximize class time with customizable, easy-to-assign, and automatically graded assessments that motivate students to learn outside of class and arrive prepared for lecture. Mastering Geography offers:

- **Assignable activities** that include GIS-inspired MapMaster 2.0 interactive maps, *Encounter World Regional Geography* Google Earth Explorations, Videos, GeoTutors, Thinking Spatially & Data Analysis activities, end-of-chapter questions, reading quizzes, *Test Bank* questions, map labeling activities, and more.

- **Student study area** with GIS-inspired **MapMaster 2.0** interactive maps, videos, geoscience animations, web links, glossary flash cards, reference maps, an optional Pearson eText, and more.

  www.pearson.com/mastering/geography

FOR TEACHERS

Instructor Resource Manual (Download Only from Mastering Geography) (9780135225653).

Updated for the sixth edition, the *Instructor Resource Manual* is intended as a resource for both new and experienced instructors. It includes lecture outlines, teaching tips, additional references, advice about how to integrate Mastering Geography media and assessment, and various other ideas for the classroom.

www.pearson.com/mastering/geography

Test Bank (Download Only from Mastering Geography) (9780135224281).

This *Test Bank* includes over 1000 multiple choice and short answer/essay questions. Questions are correlated to learning outcomes from the book, as well as to the updated U.S. National Geography Standards and Bloom's Taxonomy, helping instructors better map the assessments against both broad and specific teaching and learning objectives. The *Test Bank* is available in the Mastering Geography item library, as well as in TestGen, a computerized test generator that lets instructors view and edit *Test Bank* questions, transfer questions to tests, and print the test in a variety of customized formats. The *Test Bank* is also available in Microsoft Word as well as a version that is importable into Blackboard and other LMS.

www.pearson.com/mastering/geography

Instructor Resource Materials (Download Only from Mastering Geography) (9780135224274).

The *Instructor Resource Materials* include high-quality electronic versions of photos, maps, illustrations, and tables from the book in JPEG and PowerPoint formats, as well as customizable PowerPoint lecture presentations, Classroom Response System questions in PowerPoint, and the *Instructor Resource Manual* and *Test Bank* in Word and TestGen formats. For easy reference and identification, all resources are organized by chapter.

www.pearson.com/mastering/geography

FOR STUDENTS

Goode's World Atlas, 23rd Edition (0133864642).

Goode's World Atlas has been the world's premier educational atlas since 1923, and for good reason. It features over 250 pages of maps, from definitive physical and political maps to important thematic maps that illustrate the spatial aspects of many important topics. The 23rd edition includes digitally-produced reference maps, as well as new thematic maps on demography, global climate change, sea level rise, CO_2 emissions, polar ice fluctuations, deforestation, extreme weather events, infectious diseases, water resources, and energy production. The atlas is also available in Pearson Collections and in various eText formats, including an upgrade option from Mastering Geography courses. www.pearsonhighered.com/collections

Encounter World Regional Geography Workbook & Website by Jess C. Porter (0321681754).

For classes that do not use Mastering Geography, *Encounter World Regional Geography* provides rich, interactive explorations of world regional geography concepts through Google Earth. Students explore the globe through themes such as population, sexuality and gender, political geography, ethnicity, urban geography, migration, human health, and language. All chapter explorations are available in print format as well as online quizzes, accommodating different classroom needs. All worksheets are accompanied with corresponding Google Earth KMZ media files, available for download for those who do not use Mastering Geography, from www.mygeoscienceplace.com.

Dire Predictions: Understanding Climate Change, 2nd edition, by **Michael Mann and Lee R. Kump** (0133909778).

Periodic reports from the Intergovernmental Panel on Climate Change (IPCC) evaluate the risk of climate change brought on by humans. But the sheer volume of scientific data remains inscrutable to the general public, particularly to those who may still question the validity of climate change. In just over 200 pages, this practical text presents and expands upon the essential findings of the IPCC's Fifth Assessment Report in a visually stunning and undeniably powerful way to the lay reader. Scientific findings that provide validity to the implications of climate change are presented in clear-cut graphic elements, striking images, and understandable analogies. The second edition covers the latest climate change data and scientific consensus from the IPCC Fifth Assessment Report and integrates links to online media. The text is also available in various eText formats, including an eText upgrade option from Mastering Geography courses.

Practicing Geography: Careers for Enhancing Society and the Environment (0321811151).

This book examines career opportunities for geographers and geospatial professionals in business, government, nonprofit, and educational sectors. A diverse group of academic and industry professionals share insights on career planning, networking, transitioning between employment sectors, and balancing work and home life. The book illustrates the value of geographic expertise and technologies through engaging profiles and case studies of geographers at work.

Television for the Environment *Earth Report* Geography Videos on DVD (0321662989).

This three-DVD set is designed to help students visualize how human decisions and behavior have affected the environment and how individuals are taking steps toward recovery. With topics ranging from the poor land management promoting the devastation of river systems in Central America to the struggles for electricity in China and Africa, these 13 videos from Television for the Environment's global *Earth Report* series recognize the efforts of individuals around the world to unite and protect the planet.

Geography of a Changing World

1

New York City, New York

People stroll along New York City's High Line, an elevated park of a former railway, taking in the view of this dynamic urban space. The High Line has become a symbol of repurposing old infrastructure for public uses and gentrification in New York City. It has spurred other cities to consider similar elevated parks. The quintessential global metropolis, New York City encapsulates the processes of globalization and diversity examined in this book.

Geography Matters

Globalization and New Geographies

The Geographer's Toolbox

Population and Settlement

Cultural Coherence and Diversity

Geopolitical Framework

Economic and Social Development

Manhattan, the hub of New York City, has experienced a demographic and economic rebirth in the past two decades. A prominent symbol of the creative redesign of the city is the High Line, an elevated linear park on a former abandoned railroad line, which first opened to the public in 2009. With spectacular views of the skyline, the High Line meanders through neighborhoods such as Chelsea and the Garment District, revitalizing interest in these parts of the city. The park has become a highly desirable place to walk, enjoy green space, observe art installations, or just sit and watch the world go by. Its success has inspired other cities to create "elevated" parks by repurposing outdated or abandoned infrastructure. The High Line has also driven a real estate gold rush, with developers competing to construct luxury properties in neighborhoods that used to be known for gritty tenement buildings.

New York City is the quintessential global city; a center of global finance, marketing, and entertainment, it is also a long-standing immigrant gateway with people from every country of the world. In the city that never sleeps, nearly 40 percent of the population is foreign-born, with the Dominican Republic, China, Mexico, Jamaica, Guyana, and Ecuador being the top sending countries. While its industrial output has declined, New York City has made up for it through financial services and creative industries such as fashion, design, art, advertising, and education. It's a city of 8.5 million, in a metropolitan area of 20 million that attracts 60 million tourists a year. Many businesses and institutions in New York City are leaders in the economic, cultural, and political aspects of globalization. At the same time, globalization has transformed New York City's economy, which has attracted a diverse range of people with varying skills who call this place home.

Through the lens of *geography*, a discipline that examines Earth's physical and human dimensions, *Globalization and Diversity* investigates these global interactions and patterns. The analysis is by world regions, which invites consideration of long-term cultural and environmental practices that characterize and shape these distinct areas. Yet we contend that *globalization*—the increasing interconnectedness of people and places through converging economic, political, and cultural activities—is one of the most important forces reshaping the world today. Pundits say globalization is like the weather: It's everywhere, all the time. It is a ubiquitous part of our lives and landscapes that is both beneficial and negative, depending on our needs and point of view.

While some people in some places embrace the changes brought by globalization, others resist and push back, seeking refuge in traditional habits and places. Thus, globalization's impact is highly uneven across space, which invites the need for a geographic (or spatial) understanding. As you will see in the pages that follow, geographers, who study places and phenomena around the globe and seek to explain the similarities and differences among places, are uniquely suited to analyze the impacts of globalization in different cities, countries, or world regions. In our opening New York City example, consider how millions of migrants over decades have transformed this city by bringing with them different languages, foods, musical traditions, and ways of organizing their environments. Collectively, this diverse yet highly unequal city has been a driving force for cultural, political, and economic change.

> Globalization's impact is highly uneven across space, which invites the need for a geographic understanding.

As a counterpoint to globalization, *diversity* refers to the state of having different forms, types, practices or ideas, as well as the inclusion of distinct peoples, in a particular society. We live on a diverse planet with a mix of languages, cultures, environments, political ideologies, and religions that influence how people in particular localities view the world. At the same time, the intensification of communication, trade, travel, and migration that result from global forces have created many more settings in which people from vastly different backgrounds live, work, and interact. For example, in metropolitan Toronto, Canada's largest city, over half of the area's 5.5 million residents were born in another country. Increasingly, modern diversifying societies must find ways to build social cohesion among distinct peoples. Confronting diversity can challenge a society's tolerance, trust, and sense of shared belonging. Yet, diverse societies also stimulate creative exchanges and new understandings that are beneficial, building greater inclusion. The regional chapters that follow provide examples of the challenges and opportunities that diverse societies in an interconnected world experience today. We begin by introducing the discipline of geography and then examine this ongoing diversity in the context of globalization from a geographer's perspective.

LEARNING OBJECTIVES

After reading this chapter you should be able to:

1.1 Describe the conceptual framework of world regional geography.

1.2 Identify the different components of globalization, including controversial aspects, and list several ways in which globalization is changing world geographies.

1.3 Summarize the major tools used by geographers to study Earth's surface.

1.4 Explain the concepts and metrics used to document changes in global population and settlement patterns.

1.5 Describe the themes and concepts used to study the interaction between globalization and the world's cultural geographies.

1.6 Explain how different aspects of globalization have interacted with global geopolitics from the colonial period to the present day.

1.7 Identify the concepts and data important to documenting changes in the economic and social development of more and less developed countries.

Geography Matters: Environments, Regions, Landscapes

Geography is a foundational discipline, inspired and informed by the long-standing human curiosity about our surroundings and how we are connected to the world. The term *geography* has its roots in the Greek word for "describing the Earth," and this discipline is central to all cultures and civilizations as humans explore their world, seeking natural resources, commercial trade, military advantage, and scientific knowledge about diverse environments. In some ways, geography can be compared to history: Historians describe and explain what has happened over time, whereas geographers describe and explain the world's spatial dimensions—how it differs from place to place.

Given the broad scope of geography, it is no surprise that geographers have different conceptual approaches to investigating the world. At the most basic level, geography can be broken into two complementary pursuits: *physical* and *human geography*. **Physical geography** examines climate, landforms, soils, vegetation, and hydrology. **Human geography** concentrates on the spatial analysis of economic, social, and cultural systems.

Mobile Field Trip:
Introduction to
Geography
https://goo.gl/VYnVqW

A physical geographer, for example, studying the Amazon Basin of Brazil, might be interested primarily in the ecological diversity of the tropical rainforest or the ways in which the destruction of that environment changes the local climate and hydrology. A human geographer, in contrast, would focus on the social and economic factors explaining the migration of settlers into the rainforest or the tensions and conflicts over resources between new migrants and indigenous peoples. Both human and physical geographers share an interest in human–environment dynamics, asking how humans transform the physical environment and how the physical environment influences human behaviors and practices. Thus, they learn that Amazon residents may depend on fish from the river and plants from the forest for food (Figure 1.1) but raise crops for export and grow products such as black pepper or soy, rather than wheat, because wheat does poorly in humid tropical lowlands.

Another basic division in geography is the focus on a specific topic or theme as opposed to analysis of a specific place or a region. The theme approach is termed

▲ **Figure 1.1 Rio Itaya Settlement in the Amazon Basin** A woman and child peer out the doorway of their newly built waterfront home near Iquitos, Peru. Settlers in the Amazon Basin have relied upon the vast forests and rivers of this region for their food, livelihood, and transport.

thematic or **systematic geography**, while the regional approach is called **regional geography**. These two perspectives are complementary and by no means mutually exclusive. This textbook, for example, utilizes a regional scheme for its overall organization, dividing Earth into 12 separate world regions. It then presents each chapter thematically, examining the topics of environment, population and settlement, cultural differentiation, geopolitics, and socioeconomic development in a systematic way. In doing so, each chapter combines four kinds of geography: physical, human, thematic, and regional geography.

Areal Differences and Connections

As a spatial science, geography is charged with the study of Earth's surface. A central theme of that responsibility is describing and explaining

▶ **Figure 1.2 Areal Differences** The oasis village of Tingher on the southern slope of Morocco's Atlas Mountains illustrates dramatic landscape change over short distances. Agricultural fields and date palms in the foreground are irrigated by a river that flows from the high mountains. Irrigated land in an arid environment is precious, so the village settlements are nearby in the dry areas. In the background, the desert and mountains loom.

what distinguishes one piece of the world from another. These differences can be about the physical Earth, or about cultural features such as building designs, transportation systems, or language groups. Why is one part of Earth humid and lush, while another, just a few hundred kilometers away, is arid (Figure 1.2)? Or, why are people in one setting more affluent, while those in an adjoining area are poor?

Geographers are not only interested in place differences, but also in how these distinct localities are interconnected within and among each other. This concern for understanding *integration* and *connectivity* is fundamental to geographic analysis. For example, a geographer might ask how and why the economies of Singapore and the United States are closely intertwined, even though the two countries are situated in entirely different physical, cultural, and political environments. Questions of linkages over space are becoming increasingly important because of the new global connections inherent in globalization.

Scale: Global to Local

All systematic inquiry considers *scale*, whatever the discipline. In biology, some scientists study the very small units such as cells, genes, or molecules, while others take a larger view, analyzing plants, animals, or whole ecosystems. Geographers also work at different scales. While one may concentrate on analyzing a local landscape—perhaps a single village in the Philippines—another might focus on the broader regional picture, examining patterns of trade throughout Southeast Asia. Other geographers do research on a still larger global scale, perhaps studying emerging trade networks between southern India's center of information technology in Bengaluru and North America's Silicon Valley, or investigating how the Indian monsoon might be connected to and affected by the Pacific Ocean's El Niño phenomenon. But even though geographers may work at different scales, they never lose sight of the interactivity and connectivity among local, regional, and global scales. They will note the ways that the village in the Philippines might be linked to world trade patterns, or how the late arrival of the monsoon could affect agriculture and food supplies in Bangladesh.

The Cultural Landscape: Space into Place

Humans transform space into distinct places that are unique and heavily loaded with significance and symbolism. **Place**, as a geographic concept, is not just the characteristics of a location; it also encompasses the meaning that people give to such areas, as in the sense of place. This diverse fabric of *placefulness* is of great interest to geographers because it tells us much about the human condition throughout the world. Places can tell us how humans interact with nature and how they interact among themselves; where there are tensions, and where there is peace.

A common tool for the analysis of place is the concept of the **cultural landscape**, which is the tangible, material expression of human settlement, past and present. Thus, the cultural landscape visually reflects the most basic human needs—shelter, food, and work. Additionally, the cultural landscape acts to bring people together (or keep them apart) because it is a marker of cultural values, attitudes, history, and symbols. As cultures vary greatly around the world, so do cultural landscapes (Figure 1.3).

Geographers are also interested in spatial analysis and the concept of space. **Space** represents a more abstract, quantitative, and model-driven

▲ **Figure 1.3 The Cultural Landscape** Despite globalization, the world's landscapes still have great diversity, as seen in Prague, Czechia (Czech Republic). Red tile roofs, three-to-four-story buildings, organic street patterns, and open squares distinguish this historic capital city. Geographers use the cultural landscape concept to better understand how people interact with their environment.

Explore the SIGHTS of Historic Prague
https://goo.gl/1J9dbJ

approach to understanding how objects and practices are connected to and impact each other. For example, a geographer interested in economic development may measure income inequality and examine how it differs from one location to another to better understand how poverty might be addressed. Similarly, a geographer interested in the impacts of climate change might model the effects of sea-level change on coastal settlements based on different warming scenarios. An appreciation for space and place is critical in understanding geographic change.

Regions: Formal and Functional

The human intellect seems driven to make sense of the universe by lumping phenomena together into categories that emphasize similarities. Biology has its taxa of living organisms, while history marks off eras and periods of time. Geography, too, organizes information about the world into units of spatial similarity called **regions**—each a contiguous bounded territory that shares one or many common characteristics.

Sometimes, the unifying threads of a region are physical, such as climate and vegetation, resulting in a regional designation like the *Sahara Desert* or *Siberia*. Other times, the threads are more complex, combining economic and social traits, as in the use of the term *Rust Belt* for parts of the northeastern United States that have lost industry and population. Think of a region as spatial shorthand that provides an area with some signature characteristic that sets it apart from surrounding areas. In addition to delimiting an area, generalizations about society or culture are often embedded in these regional labels.

Geographers designate two types of regions: formal and functional. A **formal region** is defined by some long-term aspect of physical form, such as a climate type or mountain range. The Rocky Mountains or the Amazon Basin are two examples of formal regions. Cultural features, such as the dominance of a particular language or religion, can also be used to define formal regions. Belgium can be divided into Flemish-speaking Flanders and French-speaking Wallonia. Many of the maps in this book denote formal regions. In contrast, a

▶ **Figure 1.4 U.S. Rust Belt** The Rust Belt is an example of a functional region. It is delimited to show an area that has lost manufacturing jobs and population over the last four decades. By constructing this region, a set of functional relationships is highlighted. **Q: In what formal and functional regions do you live?**

functional region is one where a certain activity (or cluster of activities) takes place. The earlier example of North America's Rust Belt is such a region because it encompasses a triangle from Milwaukee to Cincinnati to Syracuse, where manufacturing dominated through the 1960s and then experienced steady decline as factories shut down and people left (Figure 1.4). Geographers designate functional regions to show changing regional associations, such as the spatial extent of a sports team's fan base or the commuter shed of a major metropolitan area like Los Angeles. Delimiting such regions can be valuable for marketing, planning transportation, or thinking about the ways that people identify with an area.

Regions can be defined at various scales. In this book, we divide the world into 12 *world regions* based on formal characteristics such as physical features, language groups, and religious affiliations, but also relying on functional characteristics such as trade groups and regional associations (Figure 1.5). In Chapter 3, we will begin with a region familiar to most of our readers—North America—and then move on to Latin America, the Caribbean, Sub-Saharan Africa, North Africa and Southwest Asia, Europe, Eurasia, and the different regions of Asia, before concluding with Oceania. Each regional chapter employs the same five-part thematic structure—physical geography and environmental issues, population and settlement, cultural coherence and diversity, geopolitical framework, and economic and social development.

Some of these regional terms are in common use, such as Europe or East Asia. Understandings and characteristics of these regions have often evolved over centuries. Yet the boundaries of these regions do shift. For example, during the Cold War, it made sense to divide Europe into east and west, with eastern Europe closely linked to the former Soviet Union. With the 1991 collapse of the Soviet Union and the expansion of the European Union in the 2000s, that divide became less meaningful. In this edition, the regions of Europe (Chapter 8) and Eurasia (Chapter 9, which includes Russia) reflect

▼ **Figure 1.5 World Regions** The boundaries shown here are the basis for the 12 regional chapters in this book. Countries or areas within countries that are treated in more than one chapter are designated on the map with a striped pattern. For example, western China is discussed in both Chapter 10, on Central Asia, and Chapter 11, on East Asia. Also, three countries on the South American continent are discussed as part of the Caribbean region because of their close cultural similarities to the island region.

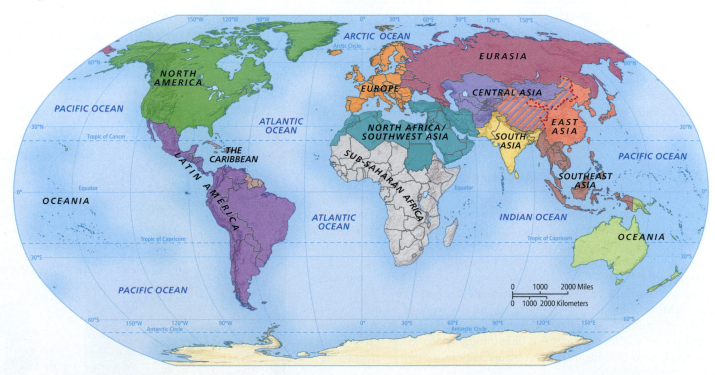

this long-standing west–east divide. Working at the world regional scale invariably creates regions that are not homogeneous, with some states fitting better into regional stereotypes than others. Yet understanding world regional formations is an important way to explore the impact of globalization on environments, cultures, political systems, and development.

REVIEW

1.1 Explain the difference between place and space in geographic understanding and analysis.

1.2 How is the concept of the cultural landscape related to place?

1.3 How do functional regions differ from formal regions?

KEY TERMS geography, physical geography, human geography, thematic geography (systematic geography), regional geography, place, cultural landscape, space, region, formal region, functional region

---

Globalization and New Geographies

One of the most important features of the 21st century is **globalization**—the increasing interconnectedness of people and places. Once-distant regions and cultures are now increasingly linked through commerce, communications, and travel. Although earlier forms of globalization existed, especially during Europe's colonial period, the current degree of planetary integration is stronger than ever. In fact, many observers argue that contemporary globalization is the most fundamental reorganization of the world's socioeconomic structure since the Industrial Revolution (see *Exploring Global Connections: A Closer Look at Globalization*).

Economic activities may be the major force behind globalization, but the consequences of globalization affect all aspects of land and life: Human settlement patterns, cultural attributes, political arrangements, and social development are all undergoing profound change. Because natural resources are viewed as global commodities, the planet's physical environment is also affected by globalization. Financial decisions made thousands of miles away now affect local ecosystems and habitats, often with far-reaching consequences for Earth's health and sustainability. For example, gold mining in the Peruvian Amazon is profitable for the corporations involved and even for individual miners, but it may ruin biologically diverse ecosystems and threaten indigenous communities.

The Environment and Globalization

The expansion of a globalized economy is creating and intensifying environmental problems throughout the world. Transnational firms conducting business through international subsidiaries disrupt ecosystems around the globe with their incessant search for natural resources and manufacturing sites. Landscapes and resources previously used by only small groups of local peoples are now considered global commodities to be exploited and traded in the world marketplace.

On a larger scale, globalization is aggravating worldwide environmental problems such as climate change, air pollution, water pollution, and deforestation. Consequently, it is only through global cooperation, such as the United Nations treaties on biodiversity protection or greenhouse gas reductions, that these problems can be addressed. Environmental degradation and efforts to address it are discussed further in Chapter 2.

Globalization and Changing Human Geographies

Globalization changes cultural practices. The spread of a global consumer culture, for example, often accompanies globalization and frequently hurts local economies. It sometimes creates deep and serious social tensions between traditional cultures and new, external global culture. Television shows and movies available via satellite, along with online videos and social media such as Facebook and Twitter, often implicitly promote Western values and culture that are then imitated by millions throughout the world (Figure 1.6).

Fast-food franchises are changing—some would say corrupting—traditional diets, with explosive growth in most of the world's cities. Although these foods may seem harmless to North Americans because of their familiarity, they are an expression of deep cultural changes for many societies and are also generally unhealthy and environmentally destructive. Yet some observers contend that even multinational corporations have learned to pay attention to local contexts. **Glocalization** (which combines globalization with locale) is the process of modifying an introduced product or service to accommodate local tastes or cultural practices. For example, a McDonald's in Japan may serve shrimp burgers along with Big Macs.

Although the media give much attention to the rapid spread of Western consumer culture, nonmaterial culture is also dispersed and homogenized through globalization. Language is an obvious example—American tourists in far-flung places are often startled to hear locals speaking an English made up primarily of movie or TV clichés. However, far more than speech is involved, as social values also are dispersed globally. Changing expectations about human rights, the role of women in society, and the intervention of nongovernmental organizations are also expressions of globalization that may have far-reaching effects on cultural change.

▼ **Figure 1.6 Global Communications** The effects of globalization are everywhere, even in remote villages in developing countries. This rural family in a small village in southwestern India earns a few dollars a week by renting out viewing time on its globally linked television set.

A Closer Look at Globalization

Globalization comes in many shapes and forms as it connects far-flung people and places. Many of these interactions are common knowledge, such as the global reach of multinational corporations like H&M and Zara transforming how young people dress. Others may be rather surprising. Who would expect to find Bosnians transforming a St. Louis neighborhood, or Filipino contract workers employed in nearly every world region? Would you predict that Saudi investors are leasing large tracts of land in the arid U.S. Southwest for hay exports to the Arabian Peninsula?

Indeed, global connections are ubiquitous and often complex—so much so that understanding the many different shapes, forms, and scales of these interactions is a key component of the study of world geography. To complement that study, each chapter of this book contains an *Exploring Global Connections* sidebar that presents a globalization case study.

The Chapter 4 sidebar, for example, explains how and why a remote area of the South American Andes has become a focus for foreign capital investment to extract lithium. This lightweight medal is essential for the small batteries that run laptops and cell phones, and this is the region of the world where the largest reserves of lithium are concentrated (Figure 1.1.1). Other examples include Dubai Airport as a global travel hub (Chapter 7); India's expanding video game industry (Chapter 12); and the spread and influence of the Armenian diaspora (Chapter 9). A Google Earth virtual tour video supplements each sidebar.

1. **Consider complex global connections based on your own experiences. For example, what food from another part of the world did you** buy today, and how did it get to your store?

2. **Now choose a foreign place in a completely different part of the world,** either a city or a rural village, then suggest how globalization affects the lives of people in that place.

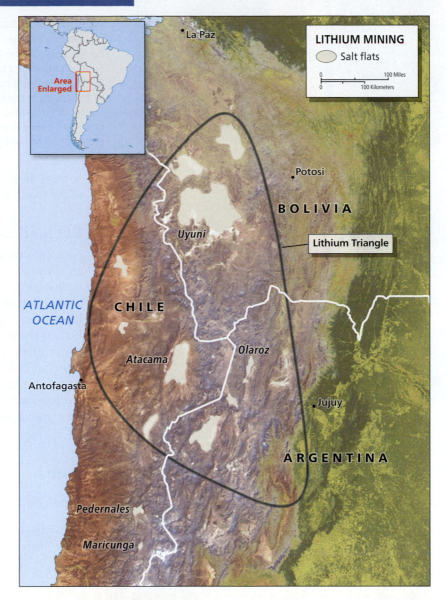

▲ **Figure 1.1.1 Lithium Triangle** The world's largest lithium deposits lie where the three countries of Bolivia, Chile, and Argentina converge. Lithium is a critical metal for lightweight batteries used in cell phones and laptops.

GOOGLE EARTH
Virtual Tour Video
http://goo.gl/Uorj2U

In return, cultural products and ideas from around the world greatly impact U.S. culture. The large and diverse immigrant population in the United States has contributed to heightened cultural diversity and exchange. The internationalization of American food and music, and the multiple languages spoken in American cities, are all expressions of globalization (Figure 1.7).

Globalization also clearly influences population movements. International migration is not new, but increasing numbers of people from all parts of the world are now crossing national boundaries, legally and illegally, temporarily and permanently. The United Nations (UN) estimates that there are over 250 million immigrants in the world (people who live in a country other than their country of birth). Figure 1.8 shows the major migration flows from regions of origin designated as Africa, Asia, Europe, Latin America and the Caribbean, North America, and Oceania. One of the most striking aspects of the figure is that many of the largest international flows are

◄ **Figure 1.7** Ethiopian Culture in Washington, DC   While many view globalization as the one-way spread of North American and European socioeconomic traits into the developing world, one needs only to look around his or her own neighborhood to find expressions of global culture within the United States. For example, the largest concentration of Ethiopians in the United States is in Metropolitan Washington, DC, where Ethiopian cuisine and music are a visible presence in the nation's capital.

in more detail later in the chapter) and demographic changes in parts of the world with aging populations and shortages of labor.

Geopolitics and Globalization

Globalization also has important geopolitical components. To many, an essential dimension of globalization is that it is not restricted by territorial or national boundaries. For example, the creation of the United Nations (UN) following World War II was a step toward creating an international governmental structure in which all nations could find representation (Figure 1.9). The simultaneous emergence of the Soviet Union as a military and political superpower led to a rigid division into Cold War blocs that slowed further geopolitical integration. However, with the peaceful end of the Cold War in the early 1990s, the former communist countries of eastern Europe and the Soviet Union were opened almost immediately to global trade and cultural exchange, changing those countries immensely.

intraregional (60 million within Asia, 40 million within Europe, and 18 million within Africa). Yet there are also substantial *interregional* flows, such as 26 million from Latin America and the Caribbean to North America, 20 million from Asia to Europe, and 17 million from Asia to North America. Attempts to control the movement of people are evident throughout the world—much more so, in fact, than control over the movement of goods or capital. Yet this growing flow of immigrants is propelled, in part, by the uneven economic development (discussed

Although the Cold War ended nearly 30 years ago, the two largest arms exporters in the world are the United States and Russia, with China and France rounding out the top four. Figure 1.10 reveals distinct geopolitical relationships surrounding large arms sales (over $100 million) between 2011 and 2015. The United States ($46 billion in arms exports) supplies Southwest Asia (especially Saudi Arabia,

► **Figure 1.8** International Migration   Globalization in its many forms is connected to the largest migration in human history, as people are drawn to centers of economic activity in hopes of a better life. This diagram shows that nearly half of the world's immigrants move within major world regions (such as Europe and Asia). But there are major interregional flows from Asia to Europe or from Latin America to North America. **Q: What international groups are found in your city?**

◄ **Figure 1.9 UN Peacekeepers in Africa** A convoy rolls past displaced people walking towards a UN camp outside of Malakal, South Sudan. Conflict in South Sudan has displaced tens of thousands. The town of Malakal was destroyed by fighting; its former residents sought shelter in a UN camp.

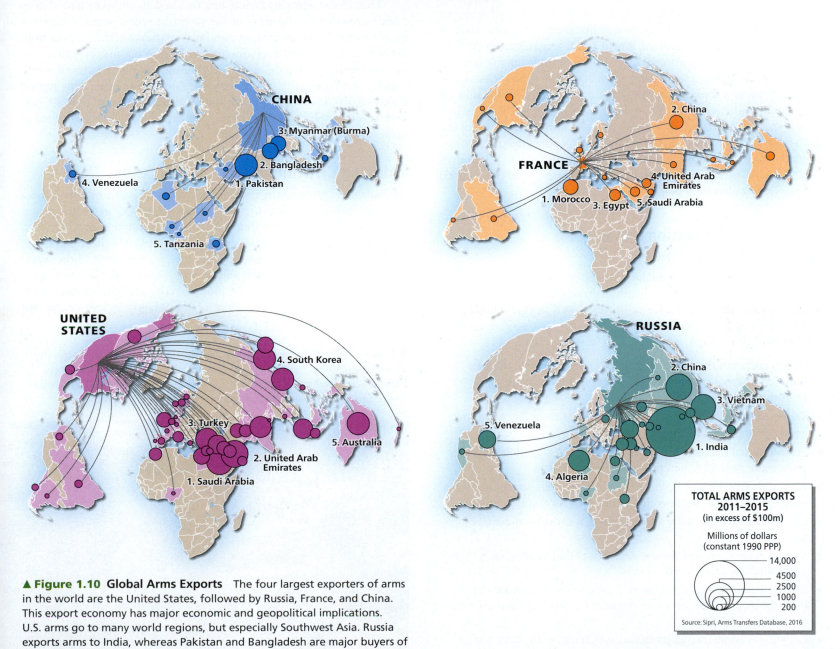

CHINA
3. Myanmar (Burma)
2. Bangladesh
1. Pakistan
4. Venezuela
5. Tanzania

FRANCE
2. China
4. United Arab Emirates
1. Morocco
3. Egypt
5. Saudi Arabia

UNITED STATES
4. South Korea
3. Turkey
5. Australia
2. United Arab Emirates
1. Saudi Arabia

RUSSIA
2. China
3. Vietnam
5. Venezuela
1. India
4. Algeria

TOTAL ARMS EXPORTS 2011–2015
(in excess of $100m)

Millions of dollars
(constant 1990 PPP)

14,000
4500
2500
1000
200

Source: Sipri, Arms Transfers Database, 2016

▲ **Figure 1.10 Global Arms Exports** The four largest exporters of arms in the world are the United States, followed by Russia, France, and China. This export economy has major economic and geopolitical implications. U.S. arms go to many world regions, but especially Southwest Asia. Russia exports arms to India, whereas Pakistan and Bangladesh are major buyers of arms from China.

the United Arab Emirates, and Turkey); Russia ($35 billion) exports to India, China, and Vietnam; France largely exports arms to Morocco and China; and China is the top supplier of armaments to Pakistan and Bangladesh in South Asia. These maps suggest evolving geopolitical relations that may differ from more formal geopolitical or economic ties.

At the same time, there are globalized criminal networks that trade in weapons, drugs, prostitution, pornography, wildlife, money laundering, and forced labor. These illegal networks can incorporate often impoverished and remote places such as Afghanistan, Myanmar (Burma), or Zimbabwe into thoroughly integrated circuits of global exchange. **Human trafficking** is the illegal trade of humans for the purpose of forced labor, sexual slavery, or commercial sexual exploitation that is often integrated into these illegal networks. The International Labor Organization (ILO) estimated that in 2014, trafficking of people was a $150 billion industry.

Ironically, many observers argue that globalization—almost by definition—has weakened the political power of individual states by strengthening regional economic and political organizations, such as the European Union (EU) and the **World Trade Organization** (WTO), an institution that deals with the global rules of trade among nations. In some world regions, a weakening of traditional state power has led to stronger local and separatist movements, as illustrated by the turmoil on Russia's southern border. Yet even established regional blocs such as the EU may be contested, as witnessed by the surprising result of the 2016 referendum in the United Kingdom to leave the European Union. Similarly, many view the election of U.S. President Donald J. Trump in 2016 as a vote against open trade and open borders, as he campaigned aggressively against such policies.

Economic Globalization and Uneven Development Outcomes

Most scholars agree that the major component of globalization is the economic reorganization of the world. Although different forms of a world economy have existed for centuries, a well-integrated and truly global economy is primarily the product of the past several decades. Attributes of this system, while familiar, bear repeating:

- Global communication systems and the digital flow of information that links all regions and most people instantaneously (Figure 1.11)
- Transportation systems that can quickly and inexpensively move goods by air, sea, and land
- Transnational business strategies that have created global corporations more powerful than many sovereign nations
- New and more flexible forms of capital accumulation and international financial institutions that make 24-hour trading possible
- Global and regional trade agreements that promote more free trade
- Market economies and private enterprises that have replaced state-controlled economies and services
- An emphasis on producing more goods, services, and data at lower costs to fulfill consumer demand for products and information (Figure 1.12)
- Growing income inequality between rich and poor, both within and between countries

◀ **Figure 1.11 Global Use of Cell Phones** Mobile technologies have revolutionized the way people communicate, acquire information, and interact in a globalized world. In Nairobi, Kenya, the majority of the city's adult population now uses M-Pesa, a cell-phone-based money transfer service, to pay for everything from street food to rides on the city's privately owned minibuses.

This global reorganization has resulted in unprecedented economic growth in some areas of the world in recent years; China is a good example, with an average annual growth rate of 8.2 percent from 2009 to 2015. But not everyone profits from economic globalization, as continuing wage gaps within China indicate, nor have all world regions shared equally in the benefits. During the same time period (2009–2015), Latin America and the Caribbean averaged only 2.7 percent annual growth, while the financial troubles in Greece resulted in an annual average of −4.5 percent.

Thinking Critically About Globalization

Globalization, particularly its economic aspects, is one of today's most contentious issues. Supporters believe that globalization creates greater economic efficiency that will eventually result in rising prosperity for the entire world. In contrast, critics claim that globalization largely benefits those who are already prosperous, leaving most of the world poorer than before as the rich and powerful exploit the less fortunate. Increasingly, scholars discuss the pros and cons of digital globalization, which is less about the movement of capital, goods, and people but

▼ **Figure 1.12 Chinese Factories** Workers sew denim jeans in the city of Shenzhen, China. Typically from rural parts of China where wages are lower, these workers live in factory-owned dorms and work six days a week. The products they sew are shipped around the world.

instead describes the accelerated movement of data to facilitate daily demands for information, searches, financial transactions, communication, and video.

Pro-globalization Arguments   Economic globalization is generally applauded by corporate leaders and economists, and underlies pro-market reforms and fiscal discipline as exemplified by policies put forward by the World Bank, International Monetary Fund (IMF), and the World Trade Organization. The primary function of the **World Bank** is to make loans to poor countries so that they can invest in infrastructure and build more modern economic foundations. The IMF makes short-term loans to countries in financial difficulty—those having trouble, for example, making interest payments on loans that they had previously taken. The WTO, a much smaller organization than the other two, works to reduce trade barriers between countries to enhance economic globalization. The WTO also tries to mediate between countries and trading blocs engaged in trade disputes.

Beyond North America, moderate and conservative politicians in most countries generally support free trade and other aspects of economic globalization. Advocates argue that globalization is a logical and inevitable expression of contemporary international capitalism that benefits all nations and all peoples. Economic globalization can work wonders, they contend, by enhancing competition, increasing the flow of capital and employment opportunities to poor areas, and encouraging the spread of beneficial new technologies and ideas. To support their claims, pro-globalizers argue that countries that have embraced the global economy generally enjoy more economic success than those that have sought economic self-sufficiency. The world's most isolated country, North Korea, is an economic disaster with little growth and rampant poverty, whereas those that have opened themselves to global forces during the same period, such as Vietnam and Thailand, have seen rapid growth and substantial reductions in poverty.

Critics of Globalization   Opponents of globalization, such as labor and environmental groups, as well as many social justice movements, often argue that globalization is not a "natural" process. Instead, it is the product of an explicit economic policy promoted by free-trade advocates, capitalist countries (mainly the United States, but also Japan and the countries of Europe), financial interests, international investors, and multinational firms that maximize profits by moving capital and seeking low-wage labor. Further, because the globalization of the world economy is creating greater inequity between rich and poor, the trickle-down model of developmental benefits for all people in all regions has yet to be validated. On a global scale, the richest 20 percent of the world's people consume 86 percent of the world's resources, whereas the poorest 80 percent use only 14 percent. Critics also worry that a globalized economic system—with its instantaneous transfers of vast sums of money over nearly the entire world on a daily basis—is inherently unstable. The worldwide recession of 2008–2010 demonstrated that global interconnectivity can also increase economic vulnerability, as illustrated by the collapse of financial institutions in Iceland or the decline of remittances from Mexicans working in the United States to their families in Mexico.

There are growing concerns that an emphasis on export-oriented economies at the expense of localized ones has led to overexploitation of resources. World forests, for example, are increasingly cut for export timber rather than serving local needs. As part of their economic structural adjustment package, the World Bank and the IMF often encourage developing countries to expand their resource exports to earn more hard currency to pay off their foreign debts. When commodity prices are high, this strategy can stimulate growth, but as commodity prices decline, as they did in 2014–2016, growth rates in developing countries slow. Moreover, the IMF often requires countries that receive loans to adopt programs of fiscal austerity that substantially reduce public spending for education, health, and food subsidies. Adopting such policies, critics warn, will further impoverish a country's people (Figure 1.13).

A Middle Position   Advocates of a middle-ground position argue that economic globalization is indeed unavoidable and that, despite its promises and pitfalls, globalization can be managed at both the national and the international levels to reduce economic inequalities and protect the natural environment. These experts stress the need for strong yet efficient national governments, supported by international institutions (such as the UN, World Bank, and IMF) and globalized networks of nongovernmental environmental, labor, and human rights groups. Moreover, the global movement of goods has flattened in the last few years, whereas the digital flow of information has soared, creating new opportunities and pitfalls that require further study.

Unquestionably, globalization is one of the most important issues of the day—and certainly one of the most complicated. While this book does not pretend to resolve the controversy, nor does it take a position, it does encourage readers to reflect on these critical points as they apply to each world region.

Diversity in a Globalizing World

As globalization progresses, many observers foresee a world far more uniform and homogeneous than today's. The optimists among them imagine a universal global culture uniting all humankind into a single community untroubled by war, ethnic strife, or resource shortages—a global utopia of sorts.

A more common view is that the world is becoming blandly homogeneous as different places, peoples, and environments lose their

▼ **Figure 1.13 Protesting Greeks**   Cuts to pensions and health care prompted Athens residents to march in Athens in 2018. Greece's financial bailout from the IMF and EU in 2010 led to mandatory austerity measures, and workers and pensioners have experienced dramatic declines in income and services.

distinctive character and become indistinguishable from their neighbors. Yet even as globalization generates a certain degree of homogenization, the world is still a highly diverse place (Figure 1.14). We can still find marked differences in culture (language, religion, architecture, foods, and other attributes of daily life), economies, and politics—as well as in the physical environment—from place to place. Such **diversity** is so vast that it cannot readily be extinguished, even by the most powerful forces of globalization. Diversity may be difficult for a society to live with, but it also may be dangerous to live without. Nationality, ethnicity, cultural distinctiveness—all are defining expressions of humanity that are nurtured in specific places.

In fact, globalization often provokes a strong reaction on the part of local people, making them all the more determined to maintain what is unique about their way of life. Thus, globalization is understandable only if we also examine the diversity that continues to characterize the world and, perhaps most important, the tension between these two forces: the homogenizing power of globalization and the reaction against it, often through demands for protecting cultural distinctiveness.

The politics of diversity demand increasing attention as we try to understand such concepts as global terrorism, ethnic identity, religious practices, and political independence. Groups of people throughout the world seek self-rule of territory they can call their own. Today most wars are fought *within* countries, not *between* them. As a result, our interest in geographic diversity takes many forms and goes far beyond simply celebrating traditional cultures and unique places. People have many ways of making a living throughout the world, and it is important to recognize this fact as the globalized economy becomes increasingly focused on mass-produced consumer goods. Furthermore, a stark reality of today's economic landscape is uneven outcomes: While some people and places prosper, others suffer from unrelenting poverty (Figure 1.15). To analyze these patterns of unevenness and change, the next section considers the tools used by geographers to better know the world.

▼ **Figure 1.14 Shopping in Isfahan**   Young women shop in the grand bazaar in Isfahan, Iran, in preparation for Eid al-Fitr, the celebration at the end of Ramadan. While few places are beyond the reach of globalization, it is also true that distinct cultures, traditions, and landscapes exist in the world's various regions.

▲ **Figure 1.15 Landscape of Economic Inequality**   The gap between rich and poor is painfully obvious in many of the world's large cities. In Rio de Janeiro, Brazil, the wealthy reside in highrises that offer security, modernity, and ocean views, whereas the city's poor live in *favelas*—sprawling self-built shantytowns where crime, violence, and poverty are all too common. Brazil, the world's fifth largest country, suffers from extreme economic inequality.

REVIEW

1.4   Provide examples of how globalization impacts the culture of a place or region.

1.5   Describe and explain five components of economic globalization.

1.6   Summarize three elements of the controversy about globalization.

KEY TERMS   globalization, glocalization, human trafficking, World Trade Organization (WTO), World Bank, diversity

The Geographer's Toolbox: Location, Maps, Remote Sensing, and GIS

Geographers use many different tools to represent the world in a convenient form for examination and analysis. Different kinds of images and data are needed to study vegetation change in Brazil or mining activity in Mongolia; population density in Tokyo or language regions in India; religions practiced in Southwest Asia or rainfall distribution in southwestern Australia. Knowing how to display and interpret information in map form is part of a geographer's skill set. In addition to traditional maps, today's modern satellite and communications systems provide an array of tools not imagined 50 years ago.

Latitude and Longitude

To navigate their way through daily tasks, people generally use a mental map of *relative locations* to locate specific places in terms of their relationship to other landscape features. The shopping mall is near the highway, for example, or the college campus is near the river. In contrast, map makers use *absolute location*, often called a mathematical location, which draws on a universally accepted coordinate system that gives every place on Earth a specific numerical address based on latitude and longitude. The absolute location for the Geography Department at the University of Oregon, for example, has the mathematical address of 44 degrees, 02 minutes, and 42.95 seconds north and 123 degrees, 04 minutes, and 41.29 seconds west. This is written 44° 02′ 42.95″ N and 123° 04′ 41.29″ W.

Lines of **latitude**, often called **parallels**, run east–west around the globe and are used to locate places north and south of the equator (0 degrees latitude). In contrast, lines of **longitude** (or **meridians**), run from the North Pole (90 degrees north latitude) to the South Pole (90 degrees south latitude). Longitude values locate places east or west of the *prime meridian*, located at 0 degrees longitude at the Royal Naval Observatory in Greenwich, England (just east of London) (Figure 1.16). The equator divides the globe into northern and southern hemispheres, whereas the prime meridian divides the world into eastern and western hemispheres; these latter two hemispheres meet at 180 degrees longitude in the western Pacific Ocean. The International Date Line, where each new solar day begins, lies along much of 180 degrees longitude, deviating where necessary to ensure that small Pacific island nations remain on the same calendar day.

Each degree of latitude measures 60 nautical miles or 69 land miles (111 km) and is made up of 60 minutes, each of which is 1 nautical mile (1.15 land miles). Each minute has 60 seconds of distance, each approximately 100 feet (30.5 meters).

From the equator, parallels of latitude are used to mathematically define the tropics: The Tropic of Cancer at 23.5 degrees north and the

▼ **Figure 1.16 Latitude and Longitude**   Latitude locates a point between the equator and the poles and is designated in degrees north or south. Longitude locates a point east or west of the prime meridian, located just east of London, England. **Q: What are the latitude and longitude of your school?**

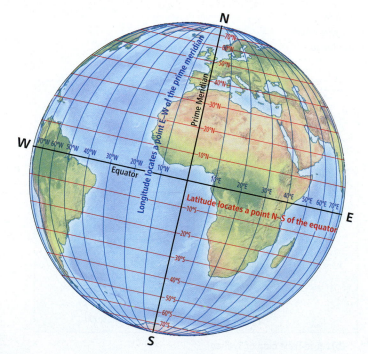

Tropic of Capricorn at 23.5 degrees south. These latitude lines denote where the Sun is directly overhead at noon on the solar solstices in June and December. The Arctic and Antarctic circles, at 66.5 degrees north and south latitude respectively, mathematically define the polar regions.

Global Positioning Systems (GPS)   Historically, precise measurements of latitude and longitude were determined by a complicated method of celestial navigation, based on one's location relative to the Sun, Moon, planets, and stars. Today absolute location on Earth (or in airplanes above Earth's surface) is determined through satellite-based **global positioning systems (GPS)**. These systems use time signals sent from your location to a satellite and back to your GPS receiver (which can be a smartphone) to calculate precise coordinates of latitude and longitude. GPS was first used by the U.S. military in the 1960s and then made available to the public in the later decades of the 20th century. Today GPS guides airplanes across the skies, ships across oceans, private autos on the roads, and hikers through wilderness areas. In the future, such systems will guide driverless cars. While most smartphones use locational systems based on triangulation from cell-phone towers, some smartphones are capable of true satellite-based GPS accurate to 3 feet (or 1 meter).

Map Projections

Because the world is a sphere, mapping the globe on a flat piece of paper creates inherent distortions in the latitudinal, or north–south, depiction of Earth's land and water areas. Cartographers (those who make maps) have tried to limit these distortions by using various **map projections**, defined as the different ways to project a spherical image onto a flat surface. Historically, the Mercator projection was the projection of choice for maps used for oceanic exploration. However, a glance at the inflated Greenlandic and Russian landmasses shows its weakness in accurately depicting high-latitude areas (Figure 1.17). Over time, cartographers have created literally hundreds of different map projections in their attempts to find the best ways to map the world while limiting distortions.

For the last several decades, cartographers have generally used the Robinson projection for maps and atlases. In fact, several professional cartographic societies tried unsuccessfully in 1989 to ban the Mercator projection for world maps because of its spatial distortions. Like many other professional publications, maps in this book utilize the Robinson projection.

Map Scale

All maps must reduce the area being mapped to a smaller piece of paper. This reduction involves the use of **map scale**, or the mathematical ratio between the map and the surface area being mapped. Many maps note their scale as a ratio or fraction between a unit on the map and the same unit in the area being mapped, such as 1:63,360 or 1/63,360. This means that 1 inch on the map represents 63,360 inches on the land surface; thus, the scale is 1 inch equals 1 mile. Although 1:63,360 is a convenient mapping scale to understand, the amount of surface area that can be mapped and fitted on a letter-sized sheet of paper is limited to about 20 square miles. At this scale, mapping 100 square miles would produce a bulky map 8 feet square. Therefore, the ratio must be changed to a larger number, such as 1:316,800. This means that 1 inch on the map now represents 5 miles (8 km) of distance on land.

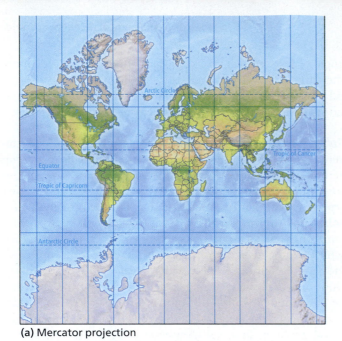

(a) Mercator projection

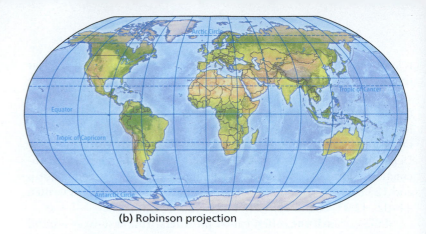

(b) Robinson projection

◀▲ **Figure 1.17** **Map Projections** Cartographers have long struggled with how best to accurately map the world given the distortions inherent in transferring features on a round globe to a flat piece of paper. Early map-makers commonly used the Mercator projection (a), which distorts features in the high latitudes but worked fairly well for seagoing explorers. (b) This map is the Robinson projection, developed in the 1960s and now the industry standard because it minimizes cartographic distortion.

Animation: Map Projections https://goo.gl/Y3z9rY

Based on the *representative fraction*, the ratio between the map and the area being mapped, maps are categorized as having either large or small scales (Figure 1.18). It may be easy to remember that large-scale maps make landscape features like rivers, roads, and cities *larger*, but because the features are larger, the maps must cover *smaller* areas. Conversely, small-scale maps cover *larger* areas, but must then make landscape features *smaller*.

Map scale is probably the easiest to interpret when it is a *graphic* or *linear scale*, which visually depicts distance units such as feet, meters, miles, or kilometers on a horizontal bar. Most of the maps in this book are small-scale maps of large areas; thus, the graphic scale is in miles and kilometers.

Map Patterns and Map Legends

Maps depict everything from the most basic representation of topo-graphic and landscape features to complicated patterns of population,

migration, economic conditions, and more. A map can be a simple *reference map* showing the location of certain features, or a *thematic map* displaying data such as religious affiliations or popular tourist attractions in a city. Most of the maps in this text are thematic maps illustrating complicated spatial phenomena. Every map has a *legend* that provides information on the categories used in the map, their values (when relevant), and other symbols that may need explanation.

One type of thematic map used often in this book is the **choropleth map** in which color shades represent different data values, with darker shades generally showing larger average values. Per capita income and population density are often represented by these maps, with data divided into categories and then mapped by spatial units such as countries, provinces, counties, or neighborhoods. The category breaks and

▼ **Figure 1.18** **Small- and Large-Scale Maps** A portion of Australia's east coast north of Sydney is mapped at two scales: (a) one at a small scale and (b) the other at a large scale. Note the differences in distance depicted on the linear scales of the two maps. There is more closeup detail in the large-scale map, but it covers only a small portion of the area mapped at a small scale.

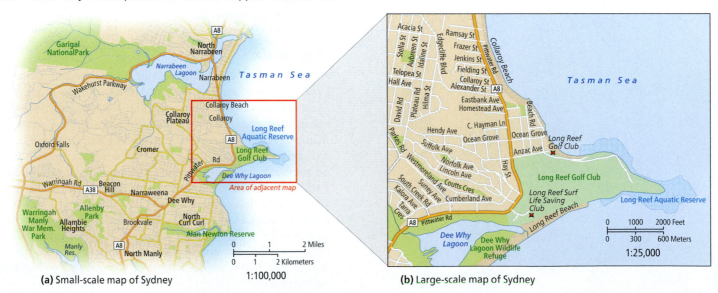

(a) Small-scale map of Sydney

1:100,000

(b) Large-scale map of Sydney

1:25,000

spatial units selected can have a dramatic impact on the patterns shown in a choropleth map (Figure 1.19).

Aerial Photos and Remote Sensing

Although maps are a primary tool of geography, much can be learned about Earth's surface by deciphering patterns on aerial photographs taken from airplanes, balloons, or satellites. Originally available only in black and white, today these images are digital and can exploit visible light (like a photograph) or other light wavelengths such as infrared that are not visible to the human eye.

Even more information about Earth comes from electromagnetic images taken from aircraft or satellites, termed **remote sensing** (Figure 1.20). Unlike aerial photography, remote sensing gathers electromagnetic data that must be processed and interpreted by computer software to produce images of Earth's surface. This technology has many applications, including monitoring the loss of rainforests, tracking the biological health of crops and woodlands, and even measuring

▼ **Figure 1.19 Choropleth Map** The population density of South Asia is mapped using different categories, from sparsely populated to very high densities, depicted with increasing intensity of colors so that you see immediately the gradients from low to high population density. This is an example of a choropleth map, which is commonly used to show variations across space of a particular phenomenon.

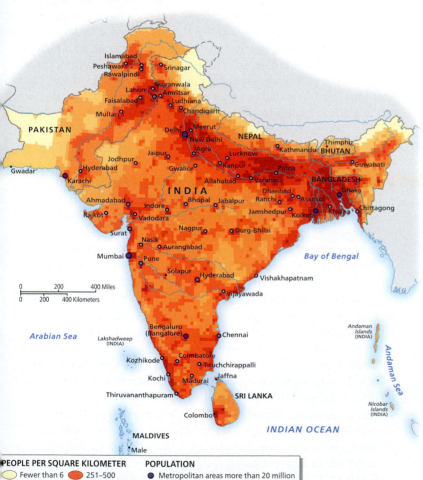

▲ **Figure 1.20 Remote Sensing of Dubai** This NASA satellite image of Dubai shows the extraordinary changes that have taken place along the arid gulf coast of the United Arab Emirates. Sprawling urbanization, construction of port facilities, new water features, and expensive island real estate (one shaped like a palm and the other shaped like the continents) are evident. Areas in red are irrigated green spaces for parks and golf courses.

Explore the SIGHTS of Palm Jumeirah
http://goo.gl/4akxhs

PEOPLE PER SQUARE KILOMETER		POPULATION	
Fewer than 6	251–500	●	Metropolitan areas more than 20 million
6–25	501–1000	●	Metropolitan areas 10–20 million
26–100	1001–12,800	●	Metropolitan areas 5–9.9 million
101–250	More than 12,800	●	Metropolitan areas 1–4.9 million
		•	Selected smaller metropolitan areas

the growth of cities. Remote sensing is also central to national defense, such as monitoring the movements of troops or the building of missile sites in hostile countries.

The Landsat satellite program launched by the United States in 1972 is a good example of both the technology and the uses of remote sensing. These satellites collect data simultaneously in four broad bands of electromagnetic energy, from visible through near-infrared wavelengths, that is reflected or emitted from Earth. Once these data are processed by computers, they display a range of images, as illustrated in Figure 1.20. The resolution on Earth's surface ranges from areas 260 feet (80 meters) square down to 98 feet (30 meters) square.

Commercial satellite companies such as DigitalGlobe now provide high-resolution satellite imagery down to 1.5 feet (or 0.5 meters) square. This means that a car, small structure, or group of people would be easily seen, but an individual person would not. Of course, cloud cover often compromises the continuous coverage of many parts of the world.

Geographic Information Systems (GIS)

Vast amounts of computerized data from sources such as maps, aerial photos, remote sensing, and census data are brought together in **geographic information systems (GIS)**. The resulting spatial databases are used to analyze a wide range of issues. Conceptually, GIS can be considered a computer system for producing a series of overlay maps showing spatial patterns and relationships (Figure 1.21). A GIS map, for example, might combine a conventional map with data on toxic waste sites, local geology, groundwater flow, and surface hydrology to determine the source of pollutants appearing in household water systems.

Although GIS dates back to the 1960s, it is only in the last several decades—with the advent of desktop computer systems and remote sensing data—that GIS has become absolutely central to geographic problem solving. It plays a central role in city planning, environmental science, public health, and real-estate development, to name a few of the many activities using these systems. GIS and other spatial tools and techniques are also critical in uncovering the patterns that allow geographers to address the themes discussed in the rest of this chapter and in the rest of the book.

Physical Geography and Environmental Issues: The Changing Global Environment

Chapter 2 provides background on world physical and environmental geography, outlining the global environmental elements fundamental to human settlement—landforms, climate, vegetation, hydrology, and energy. In the regional chapters, the physical geography sections explain the environmental issues relevant to each world region, covering such topics as climate change, sea-level rise, acid rain, energy and resource issues, deforestation, and wildlife conservation. Each regional chapter addresses specific environmental problems, but also discusses

▼ **Figure 1.21 GIS Layers** Geographic information systems (GIS) maps usually consist of many layers of information that can be viewed and analyzed separately or as a composite overlay. This is a typical environmental planning map where different physical features (such as wetlands and soils) are combined with zoning regulations.

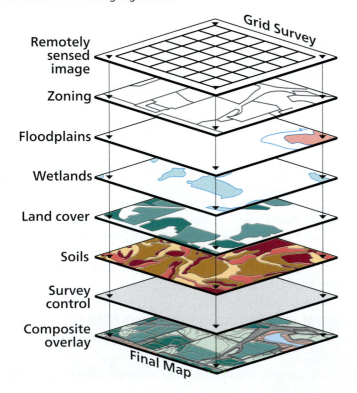

policies and plans to resolve those issues (see *Working Toward Sustainability: Meeting the Needs of Future Generations*).

REVIEW

1.7 Explain the difference between latitude and longitude, and describe how they are used to locate a place.

1.8 What does a map's scale tell us?

1.9 What is a choropleth map, and what might it depict?

1.10 What are geographic information systems (GIS), and how are they used today to address societal needs?

KEY TERMS latitude (parallels), longitude (meridians), global positioning systems (GPS), map projection, map scale, choropleth map, remote sensing, geographic information systems (GIS), sustainable development, Sustainable Development Goals (SDGs)

Population and Settlement: People on the Land

Currently, Earth has more than 7.5 billion people, with demographers (those who study human populations and population change) forecasting an increase to 9.8 billion by 2050 (Figure 1.22). Most of this increase will take place in Sub-Saharan Africa, North Africa and Southwest Asia, and South Asia. In contrast, the regions of Europe, Eurasia, and East Asia will likely experience no demographic growth between now and 2050. Population concerns vary, with some countries, such as Bangladesh trying to slow population growth, while others, like Ukraine, worry about population decline.

Population is a complex topic, but several points may help to focus the issues:

- The current rate of population growth is now half the peak rate experienced in the early 1960s, when the world population was around 3 billion. At that time, talk of a "population bomb" and "population explosion" was common, as scholars and activists voiced concern about what might happen if such high growth rates continued. Still, even with today's slower growth, demographers predict that over 2 billion more people will be added by 2050, with much of this growth taking place in the world's poorest countries.

- Population planning takes many forms, from the fairly rigid two-child policy of China to slow population growth to the family-friendly policies of no-growth countries, like South Korea, that would like to increase their natural birth rates. Over half of the world's married women use modern contraceptive methods, which has contributed to slower growth (Figure 1.23).

- Not all attention should be focused on natural growth because migration is increasingly a significant source of growth in some countries. International migration is often driven by a desire for a better life in a new country. Although much international migration is to developed countries in Europe, North America, and Oceania, there are comparable flows of migrants moving between developing countries, such as flows from South Asia to Southwest Asia or immigration within Latin America and Sub-Saharan

Meeting the Needs of Future Generations

The word *sustainable* seems to be everywhere, as we hear about sustainable cities, agriculture, forestry, business—even sustainable lifestyles. With so many different uses of the word, it is appropriate to revisit its original definition.

Sustainable has two meanings: The first is to endure or to maintain something at a certain level so that it lasts. The second means something that can be upheld or defended, such as a *sustainable idea* or *action*. Resource management has long used terms such as *sustained-yield forestry* to refer to timber practices in which tree harvesting is attuned to the natural rate of forest growth so that the resource is not exhausted, but can renew itself over time.

Moral and ethical dimensions were added to this traditional usage in 1987 when the UN World Commission on Environment and Development addressed the complicated relationship between economic development and environmental deterioration. The commission stated that **"sustainable development** is development that meets the needs of the present without compromising the ability of future generations to meet their own needs." This cautionary message expands the notion of sustainability from a narrow focus on resource management to include the whole range of human "needs," both now and in the future. In 2016, the United Nations Development Program launched the **Sustainable Development Goals** (SDGs), 17 universal goals aimed at ending poverty, protecting the planet, and promoting peace, in effect until 2030. These goals are priority areas in which governments should invest their resources, including poverty reduction, health, education, gender equality, clean water, clean energy, climate action, infrastructure, sustainable cities, responsible consumption/production, biodiversity, and strong institutions.

Sidebars in the following chapters explore the different ways that people around the world are working toward sustainability and sustainable development. Examples include offshore wind power in Denmark (Chapter 8); China's efforts at flood control (Chapter 11) (Figure 1.2.1); and preserving gibbon habitat in Cambodia (Chapter 13). Each sidebar links to a Google Earth virtual tour video.

1. **How might the concept of sustainability differ for a city or town in India or China compared to a U.S. city?**

2. **Does your college or community have a sustainability plan? If so, what are the key elements?**

▲ **Figure 1.2.1 Urban Wetlands**   This opera house near Harbin, China, was built on a large semi-urban wetland. China is designing model "sponge cities" to absorb and store runoff in wetland environments in order to reduce the risk of urban flooding.

GOOGLE EARTH
Virtual Tour Video
http://goo.gl/oGTPq9

Africa. In addition, the UN estimates that over 60 million people were displaced as a result of civil strife, political persecution, and environmental disasters in 2016, the largest number ever recorded. This includes both internally displaced people and refugees who have left their country of origin.

- The greatest migration in human history is going on now as millions of people move from rural to urban places. In 2009, a landmark was reached when demographers estimated that for the first time, more than half the world's population lived in towns and cities.

Population Growth and Change

Geographers make use of a variety of ways to define the population characteristics of a region. The most common measures and models are described here. Because of the central importance of demography in shaping localities, each regional chapter has a table of population indicators for the countries of that region. Table 1.1 includes key population indicators for the world's 10 largest countries by total population size in 2018. Keep in mind that one-third of the world's 7.5 billion people live in two countries—China and India. The next largest countries are the United States, followed by Indonesia and Brazil. Combined, the 10 largest countries account for over 60 percent of the world's population.

Population size alone tells only part of the story. **Population density**, for the purposes of this text, is the average number of people per square kilometer. Thus, China is the world's largest country demographically, but the population density of India, the second largest country, is three times that of China. Bangladesh's population density is far greater still at over 1200 people per square kilometer.

Population densities differ considerably across a large country and between rural and urban areas, making the gross national figure a bit misleading. Many of the world's largest cities, for example, have

▶ **Figure 1.22 World Population** This map emphasizes the world's different population densities. East and South Asia stand out as the most populated regions, with high densities in Japan, eastern China, northern India, and Bangladesh. In arid North Africa and Southwest Asia, settlements are often linked to the availability of water for irrigated agriculture, as is apparent with the population cluster along the Nile River. Higher population densities in Europe, North America, and other countries are usually associated with major metropolitan areas.

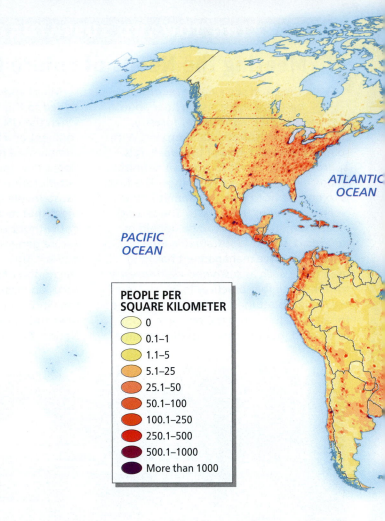

densities of more than 30,000 people per square mile (10,300 per square kilometer), with the central areas of São Paulo, Brazil, and Shanghai, China, easily twice as dense because of the prevalence of high-rise apartment buildings. In contrast, most North American cities have densities of fewer than 10,000 people per square mile (3800 per square kilometer), due largely to a cultural preference for single-family dwellings on individual urban lots.

The statistics in Table 1.1 might seem daunting, but this information is crucial to understanding general population trends, overall growth rates, age structure, patterns of settlement, and rates of migration among the countries that make up various world regions.

Natural Population Increase A common starting point for measuring demographic change is the **rate of natural increase (RNI)**, which provides the annual growth rate for a country or region as a percentage. This statistic is produced by subtracting the number of deaths from the number of births in a given year. Important to remember is that population gains or losses through migration are not considered in the RNI.

The RNI is a small number with major consequences. It can be positive as in the case of Nigeria, or stable as in the United States; some countries have negative rates, which means they are losing population. China's RNI is 0.5, whereas India's is 1.4. Yet if those rates are maintained, China's population will double in 140 years, whereas India's will double in 50 years. This is why demographers are confident that India will surpass China as the largest country in the next decade or so. The country with the highest RNI on Table 1.1 is Nigeria at 2.6. If Nigeria maintains that rate, it will double its size in 27 years. In fact, demographers forecast that in 2050 Nigeria will be the third largest country in the world, after India and China. Countries with a rate close to zero are demographically stable, but countries with persistent negative rates will experience slow declines in population unless immigration occurs.

▼ **Figure 1.23 Family Planning in Cambodia** Women in a reproductive health clinic in Kampong Cham, Cambodia learn about modern birth control options. The availability of modern contraception has brought total fertility rates down throughout the world.

For many years Japan was one of the demographic top 10, but due to negative RNI, the country has lost population and in 2015 was replaced by Mexico (Figure 1.24).

Total Fertility Rate Population change is impacted by the **total fertility rate (TFR)**, which is the average number of live births a woman has in her lifetime. The TFR is a good indicator of a country's potential for growth. A TFR of 2.1 is considered the **replacement rate** and suggests that it takes two children per woman, with a fraction more to compensate for infant and child mortality, to maintain a stable population. Where infant mortality is high, a country's actual replacement rate could be higher—say, 3.0. Clearly, women do not have 1.6 or 5.6 children; rather, women in some countries on average have 1 to 2 children versus 5 to 6 children, which means the potential for population growth is very different. In 1970, the global TFR was 4.7, but by 2017 that rate was nearly cut in half to 2.5. Around the world, fertility rates have been coming down for the last four decades as women move to cities, become better educated, work outside the home, control their fertility with modern contraception, and receive better medical care for themselves and their infants.

Four of the countries listed in Table 1.1 have a below-replacement TFR, meaning that over time their natural growth will slow as fewer children are born; and in some cases, population will decline if immigration does not occur. Even India's current TFR is 2.3, a dramatic change from 5.5 in 1970. India will still grow for many decades to come, but the potential for growth has been reduced as Indian women have smaller families. The countries with the highest total fertility rates are in Sub-Saharan Africa, where the average is about 5. Nigeria's TFR is slightly higher at 5.5.

PEOPLE PER SQUARE KILOMETER
- 0
- 0.1–1
- 1.1–5
- 5.1–25
- 25.1–50
- 50.1–100
- 100.1–250
- 250.1–500
- 500.1–1000
- More than 1000

PACIFIC OCEAN

ATLANTIC OCEAN

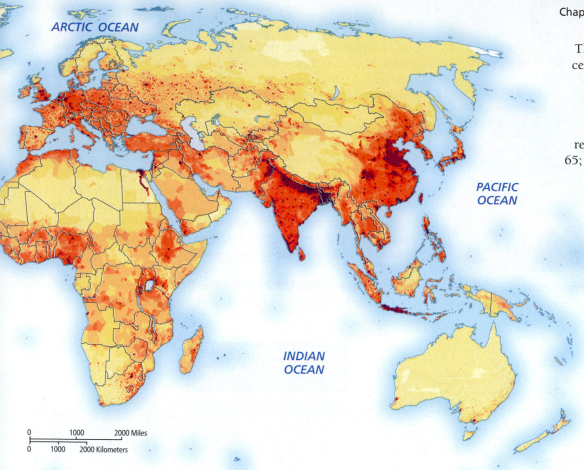

The other end of the age spectrum—the percentage of a population over age 65—is also important. Just 9 percent of the world's population is over age 65, yet the percentage is twice that in many developed countries. Japan distinguishes itself in this regard with 28 percent of its population over 65; in contrast, only 12 percent of its population is under 15, indicating that the average age of the population is increasing as well. An aging population is significant when calculating a country's need to provide social services for its senior citizens and pensioners. It also has implications for the size of the overall workforce that supports retired and elderly individuals.

Population Pyramids The best graphic indicator of a population's age and gender structure is the **population pyramid**, which depicts the percentage of a population (or, in some cases, the raw number) that is male or female in different age classes, from young to old (Figure 1.25). If a country has many more young people than old, the graph has a broad base and a narrow tip, thus taking on a pyramidal shape that commonly forecasts rapid population growth. In contrast, slow-growth or no-growth populations are top-heavy, with a larger number of seniors than younger age classes.

Not only are population pyramids useful for comparing different population structures around the world at a given point in time; they also capture the structural changes of a population as it transitions from fast to slow growth. In addition, population pyramids display gender differences within a population, showing whether or not there is a disparity in the numbers of males and females. In the mid-20th century,

Population Age Structure Another important indicator of a population's relative youthfulness, and its potential for growth, is the percentage of the population under age 15. Currently, 26 percent of the world's population is younger than age 15. However, in fast-growing Sub-Saharan Africa, that figure is 43 percent. This is another indicator of the population growth that will continue in this region for at least another generation. In contrast, only 16 percent of the populations of East Asia and Europe are under 15, suggesting slower growth and shrinking family sizes.

TABLE 1.1 Population Indicators				Life Expectancy					
Country	Population (millions) 2018	Population Density (per square kilometer)[1]	Rate of Natural Increase (RNI)	Total Fertility Rate	Male	Female	Percent Urban	Percent <15	Percent >65	Net Migration (rate per 1000)
China	1393.8	148	0.5	1.8	75	78	59	17	11	0
India	1371.3	450	1.4	2.3	67	70	34	28	6	0
United States	328.0	36	0.3	1.8	76	81	82	19	15	3
Indonesia	265.2	146	1.2	2.4	67	71	54	28	5	−1
Brazil	209.4	25	0.8	1.7	72	79	86	22	8	0
Pakistan	200.6	256	1.9	3.1	66	68	37	33	4	−4
Nigeria	195.9	210	2.6	5.5	53	54	50	44	3	0
Bangladesh	166.4	1265	1.4	2.1	70	73	37	29	5	−3
Russia	147.3	9	−0.1	1.6	68	78	74	18	14	2
Mexico	130.8	66	1.3	2.2	75	80	73	27	7	−1

Explore these data in MapMaster 2.0 https://goo.gl/Ab8mdY

Sources: Population Reference Bureau, *World Population Data Sheet, 2018.*
[1] World Bank Open Data 2018.

Log in to Mastering Geography & access MapMaster to explore these data!
1) Compare the maps for the overall Population and the RNI for each county in this table. How might the top 10 rankings change in the next 20 years?
2) What is the value of the replacement rate? Which countries are currently below the replacement rate of 2.1? What does that mean for their long-term growth?

▲ **Figure 1.24 Smaller Families, Declining Population** Japan has seen its family size shrink to one or two children. Consequently, the total population of the country has declined as well.

is nearly always higher than male by a few years. Some countries, such as Bangladesh, Iran, and Nepal, have seen average life expectancies increase by 20 years or more since 1970. Table 1.1 shows that women in China, the United States, Brazil, and Mexico all have life expectancies of 78–81 years; men's life expectancies are a few years less. Of the top 10 countries, Nigeria's life expectancy is the lowest. Russia has the widest gap between male and female life expectancy—10 years. The life expectancy of men in Russia is similar to that of men in India and Pakistan, although Russia has higher levels of economic development.

Because a large number of social factors—such as health services, nutrition, and sanitation—influence life expectancy, many researchers use life expectancy as a surrogate measure for development. When this figure is improving, it indicates that other aspects of development are occurring. Each regional chapter reports on male and female life expectancy.

The Demographic Transition The historical record suggests that population growth rates have slowed over time. More specifically, in for example, population pyramids for those countries that fought in World War II (such as the United States, Germany, the former Soviet Union, and Japan) showed a distinct deficit of males, indicating those lost to warfare. Similar patterns are found today in countries experiencing widespread conflict and civil unrest.

Cultural preferences for one sex or another, such as the preference for male infants in China and India, show up in population pyramids when there are more male children than female. Because of their utility in displaying population structures graphically, comparative population pyramids are found in many of the regional chapters of this book.

Life Expectancy A demographic indicator containing information about health and well-being in a society is *life expectancy*, which is the average number of years a typical male or female in a specific country can be expected to live. Life expectancy generally has been increasing around the world, indicating that conditions supporting life and longevity are improving. To illustrate, in 1970 the average life expectancy figure for the world was 58 years, whereas today it is 70 for men and 74 for women—female life expectancy

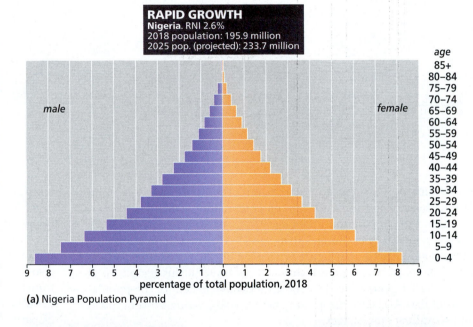

(a) Nigeria Population Pyramid

▶ **Figure 1.25 Population Pyramids** The term *population pyramid* comes from the shape of the graph representing a rapidly growing country such as (a) Nigeria, when data for age and sex are plotted as percentages of the total population. The broad base illustrates the high percentage of young people in the country's population, indicating that rapid growth will likely continue for at least another generation. Contrast the pyramidal shape with the narrow bases of slow- and negative-growth countries, such as (b) the United States and (c) Germany, which have fewer children and people in the childbearing years and a larger proportion of the population over age 65. Q: Think of two example countries that fit into each of these three categories: rapid growth, slow growth, and negative growth.

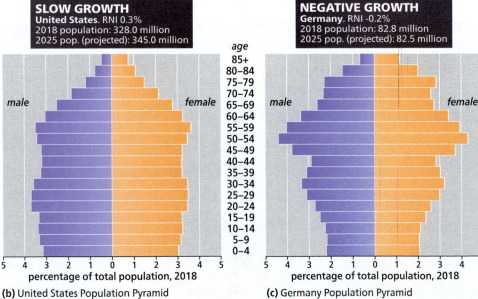

(b) United States Population Pyramid

(c) Germany Population Pyramid

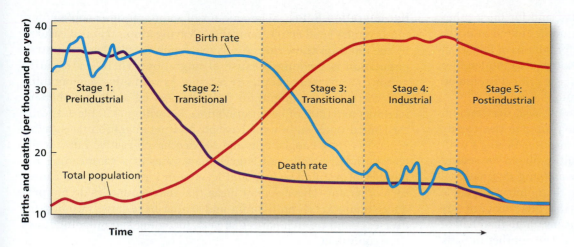

◀ Figure 1.26 Demographic Transition
As a country industrializes, its population moves through the five stages in this diagram, known as the *demographic transition model*. In Stage 1, population growth is low because high birth rates are offset by high death rates. Rapid growth takes place in Stage 2, as death rates decline. Stage 3 is characterized by a decline in birth rates. The transition was initially thought to end with low growth once again in Stage 4, resulting from a relative balance between low birth rates and low death rates. But with many developed countries now showing no natural growth, demographers have added a fifth stage to the traditional model to show no growth or even negative natural growth.

Europe and North America, population growth slowed as countries became increasingly industrialized and urbanized. From these historical data, demographers generated the **demographic transition model**, a conceptualization that tracks the changes in birth rates and death rates over time. Birth rates are the annual number of births per 1000 people in a country, and death rates are the annual number of deaths per 1000 people. When birth rates exceed death rates, natural increase occurs (Figure 1.26).

Stage 1 of the demographic transition model is characterized by both high birth rates and high death rates, resulting in a very low RNI. Historically, this stage is associated with Europe's preindustrial period, a time that predated common public health measures such as water and sewage treatment, an understanding of disease transmission, and the most fundamental aspects of modern medicine. Not surprisingly, death rates were high and life expectancy was short. Currently there are no Stage 1 countries in the world.

In Stage 2, death rates fall dramatically while birth rates remain high, thus producing a rapid rise in the RNI. In both historical and contemporary times, this decrease in death rates is commonly associated with the development of public health measures and modern medicine. Additionally, one of the assumptions of the demographic transition model is that these health services become increasingly available only after some degree of economic development and urbanization takes place.

However, even as death rates fall and populations increase, it takes time for people to respond with lower birth rates. This happens in Stage 3, the transitional stage in which people become aware of the advantages of smaller families in an urban and industrial setting, contrasted with the earlier need for large families in rural, agricultural settings or where children worked at industrial jobs (both legally and illegally).

Then in Stage 4, a low RNI results from a combination of low birth rates and low death rates. Until recently, this stage was assumed to be the static end point of change for developing and urbanizing populations. However, this does not seem to be the case. In many highly urbanized developed countries, such as those in Europe as well as Japan, the death rate now exceeds the birth rate, and the RNI falls below the replacement level, expressed as a negative number. This negative growth state can be considered a fifth stage of the traditional demographic transition model.

Remember, though, that the RNI is just that—the rate of *natural* increase—and does not capture a country's growth from immigration.

For example, even if RNI is negative, a country may demographically grow or stabilize due to immigration from other countries.

Global Migration and Settlement

Never before have so many people been on the move, either from rural areas to cities or across international borders. Today more than 250 million people live outside the country of their birth and thus are officially designated as immigrants by the UN and other international agencies. Much of this international migration is directly linked to the new globalized economy because the majority of these migrants live either in the developed world or in developing countries with vibrant industrial, mining, or petroleum extraction economies. In the oil-rich countries of the United Arab Emirates, Qatar, and Saudi Arabia, the labor force is composed primarily of foreign migrants, especially from South Asia (Figure 1.27), who are considered **economic migrants** (immigrants who arrive for employment opportunities). The top six destination countries, which account for 40 percent of the world's immigrants, are major industrial or mining economies: United States, Russia, Germany, Saudi Arabia, United Kingdom, and

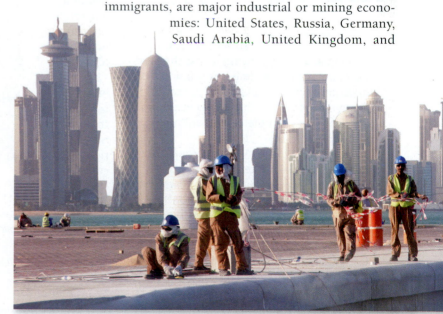

▲ Figure 1.27 Global Workforce South Asia laborers working on a construction site in Doha, Qatar. Many Persian Gulf countries rely on vast numbers of contract laborers, mostly from South Asia, to provide the labor necessary to build these modern cities and serve the populations living there.

United Arab Emirates. Within these countries, and for most other destinations, migrants are drawn to the opportunities found in major metropolitan areas. They are also responsible for billions of dollars in **remittances**, monies sent by individuals working abroad to families in the origin country. Many view remittances as a critical livelihood resource that is a catalyst for continued migration.

Not all migrants move for economic reasons. War, persecution, famine, and environmental destruction cause people to flee to safe havens elsewhere. Accurate data on **refugees** (migrants fleeing a well-founded fear of persecution) are often difficult to obtain for several reasons. Often these individuals are illegally crossing international boundaries, or countries deliberately obscure the number of people fleeing for political reasons. The UN estimates that there are currently 60 million refugees or internally displaced persons; more than half of these people are in Africa and Southwest Asia. The conflict in Syria has displaced over half of that country's population. Over half of the 11.6 million displaced people are scattered within Syria, but nearly 5 million live outside the territory—mostly in Turkey, Lebanon, Jordan, and Iraq (see *Humanitarian Geography: Tools for Service*).

Net Migration Rates The amount of immigration (people entering a country) and emigration (those leaving a country) is measured by the **net migration rate**. A positive figure means that a country's population is growing because of migration, whereas a negative number means more people are leaving. As with other demographic indicators, the net migration rate is expressed as the numbers of migrants per 1000 of base population. Returning to Table 1.1, only the United States and Russia have positive net migration rates. Indonesia, Pakistan, Bangladesh, and Mexico have negative rates, while China, India, Brazil, and Nigeria have rates at zero, meaning the number of people entering and leaving in a particular year cancel each other out. This does not mean that these countries do not produce immigrants—both India and China have large populations overseas—but for that particular year, incoming and outgoing flows were equivalent.

Countries with some of the highest net migration rates—such as Qatar, Bahrain, and the United Arab Emirates—depend heavily on migrants for their labor force. Countries with the highest negative migration rates include those in conflict—such as Syria and Somalia, and Pacific island nations with relatively small populations and weak economies, such as Samoa and Micronesia.

HUMANITARIAN GEOGRAPHY

Tools for Service

▲ **Figure 1.3.1**
Susan Wolfinbarger

Geographers regularly use satellite imagery, GIS, drones, and crowd-sourced data to address humanitarian crises and to seek justice. From mapping the extent of a landslide or flood, to identifying the source of a pollutant, to tracking a wildfire moving across the landscape, accurate real-time information on maps can save lives, plan a response, or document crimes. Each chapter in this book highlights an example of humanitarian geography.

For example, Susan Wolfinbarger, a geographer at the U.S. State Department, has utilized satellite imagery for years, documenting human rights abuses in conflict zones (Figure 1.3.1). When she worked at the American Association for the Advancement of Science, her team used high-resolution satellite imagery to analyze the increase in roadblocks in the Syrian city of Aleppo in 2013. Roadblocks demonstrate a decline in the circulation of people and goods in this densely settled city (Figure 1.3.2).

Examples of geospatial tools and analysis contributing to humanitarian efforts include mapping environmental hazards in Indonesia (Chapter 13); teaching map skills to defectors to pinpoint human rights crimes in North Korea (Chapter 11); monitoring the exclusion zone around the Chernobyl nuclear site (Chapter 9), and rescuing migrants in the Mediterranean Sea (Chapter 8).

1. **Explain how mapping skills would be useful during a natural disaster.**

DigitalGlobe/AAAS

▲ **Figure 1.3.2** **Monitoring Aleppo** This image shows the Syrian city in May 2013, where over 1000 roadblocks were detected. Roadblocks are an indicator of ongoing conflict and potential humanitarian concerns because they restrict the movement of people and goods throughout the city.

How might mapping be helpful in documenting crime?

2. **How have maps benefited you or your community?**

GOOGLE EARTH
Virtual Tour Video
https://goo.gl/SxgydT

Settlement in an Urbanizing World The focal points of today's globalizing world are cities—the fast-paced centers of deep and widespread economic, political, and cultural change. This vitality, and the options cities offer to impoverished and uprooted rural peoples, make them magnets for migration. In 2018, some 4.2 billion people lived in urban areas. The scale and rate of growth of some world cities are absolutely staggering. Between natural growth and in-migration, New Delhi, India's capital, grew from 14 million people in 2001 to 29 million people in 2018. Similarly, Shanghai expanded from 16 million people in 2000 to 26 million in 2018. Currently there are 33 **megacities** (a city with more than 10 million inhabitants), most of which are in the developing world, and geographers project that 10 more will be added by 2030.

Today 54 percent of the world's population lives in cities. **Urbanization** varies by region. Latin America and North America are highly urbanized regions, with 80 percent of the population in cities. Sub-Saharan Africa may be rapidly urbanizing, but 60 percent of the people are still in rural areas. And in India, just one-third of the population is urbanized; that means nearly 900 million people still live in rural areas.

Generally speaking, most countries with high rates of urbanization are also more developed and industrialized, because manufacturing tends to cluster around urban centers. Many countries are characterized by **urban primacy**, in which a **primate city** (often the capital) is three or four times larger than the country's next largest city. Seoul, Lagos, London, Manila, and Buenos Aires are all examples of primate cities.

REVIEW

1.11 What is the rate of natural increase (RNI), and how can it be a negative number?

1.12 Explain a high versus a low total fertility rate, and give examples.

1.13 Describe and explain the demographic transition model.

1.14 How is a population pyramid constructed, and what kind of information does it convey?

KEY TERMS population density, rate of natural increase (RNI), total fertility rate (TFR), replacement rate, population pyramid, demographic transition model, economic migrant, remittances, refugee, net migration rate, megacity, urbanization, urban primacy, primate city

Cultural Coherence and Diversity: The Geography of Change and Tradition

Culture binds people together in a diverse social fabric, but cultural differences can be a source of tension and conflict. As noted earlier, with the rise of digital communication, stereotypical Western culture has diffused at a rapid pace. Some cultures accept these new influences willingly, while others resist and push back against what some view as *cultural imperialism* through local protests, censorship, and even terrorism. Still others use digital technology to advance their own cultural or political agendas.

The geography of cultural cohesion and diversity studies tradition and change, new cultural forms produced by interactions between cultures, gender issues, and global languages and religions.

Culture in a Globalizing World

The dynamic changes connected with globalization have blurred traditional definitions of culture. A very basic definition provides a starting point. **Culture** is learned, not innate, behavior shared by a group of people, empowering them with what is commonly called a "way of life."

Culture has both abstract and material dimensions: speech, religion, ideology, livelihood, and value systems, but also technology, housing, foods, dress, and music. Even something like sports can have deep cultural meaning. Think of how 3 billion people watch the World Cup and support their "national" teams with near-religious devotion. These varied expressions of culture are relevant to the study of world regional geography because they tell us much about the way people interact with their environment, with one another, and with the larger world (Figure 1.28). Not to be overlooked is that culture is dynamic and ever changing, not static. Thus, culture is a process, not a condition—an abstract yet useful concept that is constantly adapting to new circumstances. This often results in tensions between the conservative, traditional elements of a culture and the newer forces promoting change.

When Cultures Collide Cultural change often takes place within the context of international tensions. Sometimes, one cultural system will replace another; at other times, resistance by one group to another's culture will stave off change. More commonly, however, a newer, hybrid form of culture results from an amalgamation of two cultural traditions (Figure 1.29). Historically, colonialism was the most important perpetuator of these cultural collisions; today globalization in its varied forms is a major vehicle for cultural exchange, tensions, and change (see *Globalization in Our Lives: Everyday Grains*).

The active promotion of one cultural system at the expense of another is called **cultural imperialism**. The most severe examples occurred during the colonial period, when European cultures spread

▼ **Figure 1.28 Culture as a Way of Life** A diverse crowd of young people enjoy the Art Murmur street festival in Oakland, California. Residents in major urban centers like Oakland influence popular culture through dress, design, music, and technology.

▲ **Figure 1.29 Culture Clash in Goa, India** Beach-loving western tourists get better acquainted with a sacred symbol of Hindu culture.

worldwide and overwhelmed, eroded, and even replaced indigenous cultures. During this period, Spanish and Portuguese cultures spread widely in Latin America, French culture diffused into parts of Africa and Southeast Asia, and British culture was imprinted on North America as well as much of South Asia and Sub-Saharan Africa. New languages

were mandated, new educational systems were implanted, and new administrative institutions replaced the old. Foreign dress styles, diets, gestures, and organizations were added to existing cultural systems. Many vestiges of colonial culture are still evident today.

Today's cultural diffusion is seldom linked to an explicit colonizing force; it more often comes as a byproduct of economic globalization. Though many expressions of cultural imperialism carry a Western (even U.S.) tone—such as McDonald's, MTV, KFC, Marlboro cigarettes, and the widespread use of English as the dominant language of the Internet—these facets result more from a search for new consumer markets than from deliberate efforts to spread modern U.S. culture throughout the world (Figure 1.30).

The reaction against cultural imperialism is heightened **nationalism**. This is the process of protecting and defending a cultural system against diluting or undesirable cultural expressions, while at the same time actively promoting national and local cultural values. Often nationalism takes the form of explicit legislation or official censorship that simply outlaws unwanted cultural traits. For example, France has long fought the Anglicization of its language by banning "Franglais" in official governmental French, thereby exorcising commonly used words such as *weekend, downtown, chat,* and *happy hour.* Many Muslim countries limit Western cultural influences by restricting or censoring international TV, an element they consider the source of many undesirable cultural influences. Most Asian countries are also increasingly protective of their cultural values, and many are

GLOBALIZATION IN OUR LIVES

Everyday Grains

Globalization is not just about multicultural corporations doing business all over the world; it is part of daily life, from your clothing to the smartphone in your hand and the coffee you drink. Whatever the product, it likely involves a complex, evolving world geography.

Food is a common cultural expression, yet even our daily grain consumption is shifting, in part due to global influences. From 1997 to 2015, U.S. wheat flour consumption per capita declined from 67 kilos to 60 kilos (148 to 132 pounds). Americans still consume plenty of flour but have replaced some of it with rice and alternative grains such as quinoa from South America. Conversely, rice is the staple grain throughout Asia; but wheat consumption is on the rise, especially in South Korea, Thailand, Vietnam, Nepal, Indonesia, and Bangladesh (Figure 1.4.1). And in West Africa, where sorghum and millet once dominated, rice has become the most consumed grain over the last decade, in part because it is faster to prepare and now more widely available.

A sidebar in each chapter illustrates how globalization influences our daily lives through the foods we eat, the music we listen to, the activities we enjoy, and the choices we make as consumers and citizens. Chapter 2 examines efforts to ban the ubiquitous plastic bag to reduce pollution, especially in our cities and oceans. Chapter 3 looks at the growing number of international students at American colleges and universities as an everyday expression of globalization, while Chapter 4 showcases Zumba—a group exercise with a Latin beat, which has spread globally.

1. How has globalization changed what your family eats?

▲ **Figure 1.4.1 Wheat Consumption in Asia** A street vendor in Bangkok, Thailand, hawks colorful sandwiches served on wheat rolls. Although rice is still the dominant daily grain in Asia, wheat consumption is on the rise.

2. Identify a commonplace item or activity in your life that has an interesting backstory involving globalization.

Explore the **TASTES** of Rice
https://goo.gl/MXDvQU

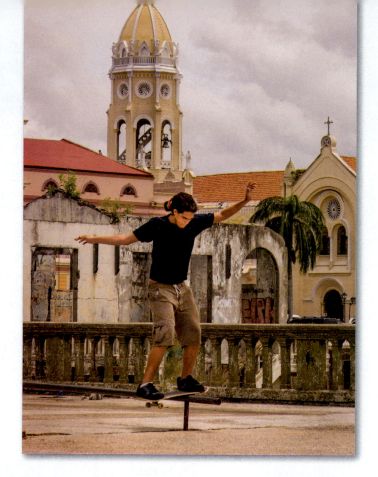

▲ **Figure 1.30 Grinding in Panama** A young man practices grinding his skateboard on a rail in Panama City. Skateboarding began in the United States but has made its way into popular youth culture throughout the world, especially in urban areas with paved surfaces. In the background looms a Catholic church, an indicator of the Spanish colonial influence in this region.

demanding changes to music videos to tone down the sexual content. In China, government censors block access to Facebook and Google.

Cultural Mixing and Assimilation The most common product of cultural exchange is the blending of forces to form a new, synergistic form of culture, a process called **cultural syncretism**. To characterize India's culture as British, for example, exaggerates England's colonial influence. Instead, Indians have adapted many British traits to their own circumstances, infusing them with their own meanings. India's use of English, for example, has produced a unique form of "Indlish" that often befuddles visitors to South Asia. Nor should we forget that India has added many words to our English vocabulary— *khaki, pajamas, veranda,* and *bungalow,* among others. Clearly, both the Anglo and the Indian cultures have been changed by the British colonial presence in South Asia. When immigrants move from one country to another, a process of **cultural assimilation** can occur, which is the adoption of the language, customs, or norms of the host society by the newcomer. This process takes time and is often completed over generations.

Language and Culture in Global Context

Language and culture are so intertwined that often language is the major characteristic that defines one cultural group and differentiates

it from another (Figure 1.31). Furthermore, as the primary means for communication, language folds together many other aspects of cultural identity, such as politics, religion, commerce, and customs. Language is fundamental to cultural cohesiveness and distinctiveness, for it not only brings people together, but it also sets them apart from nonspeakers of that language. Therefore, language is an important component of national or ethnic identity as well as a means for creating and maintaining boundaries for group and regional identity.

Because most languages have common historic (and even prehistoric) roots, linguists have grouped the thousands of languages found throughout the world into a handful of **language families**, based on common ancestral speech. For example, about half of the world's people speak Indo-European languages, a large family that includes not only European languages such as English and Spanish, but also Hindi and Bengali, the dominant languages of South Asia.

Within language families, smaller units also give clues to the common history and geography of peoples and cultures. Language branches and groups (also called *subfamilies*) are closely related subsets within a language family, usually sharing similar sounds, words, and grammar. Well known are the similarities between German and English and between French and Spanish. Additionally, individual languages often have distinctive *dialects* associated with specific regions and places. Think of the distinctions among British, Canadian, and Jamaican English, or the city-specific dialects that set apart New Yorkers from residents of Dallas, Berliners from inhabitants of Munich, or Parisians from villagers of rural France.

When people from different cultural groups cannot communicate directly in their native languages, they often employ a third language to serve as a common tongue, a **lingua franca**. Swahili has long served that purpose for speakers of the many tribal languages of eastern Africa, and French was historically the lingua franca of international politics and diplomacy. Today English is increasingly the common language of international communications, science, and air transportation.

Explore the SOUNDS of Swahili
https://goo.gl/CTDCrq

The Geography of World Religions

Another important defining trait of cultural groups is religion (Figure 1.32). Indeed, in this era of a comprehensive global culture, **religion** has become increasingly important in defining cultural identity. Recent ethnic violence based upon religious differences in far-flung places such as the Balkans, Iraq, Syria, and Myanmar illustrate the point.

Major religions—such as Christianity, Islam, and Buddhism—attempt to appeal to all peoples, regardless of location or culture, and are sometimes referred to as *universalizing religions*. These religions usually have a proselytizing or missionary program that actively seeks new converts throughout the world. In contrast, other religions are more closely identified with a specific ethnic, tribal, or national group. Judaism and Hinduism, for example, normally do not actively seek new converts; instead, people are born into these religious practices.

Christianity is the world's largest religion in both areal extent and number of adherents. Though broadly divided into Roman Catholic and Protestant Christianity and further fragmented into myriad branches and churches, Christianity as a whole has over 2.2 billion adherents. The largest numbers of Christians can be found in Europe, Africa, Latin America, and North America.

Islam, which has spread from its origins on the Arabian Peninsula east to Indonesia and the Philippines, has about 1.8 billion members. Although not as severely fragmented as Christianity, it is also split into separate groups. *Shi'a Islam* constitutes about 11 percent of the total Islamic population and represents a majority in Iran and southern Iraq, while the more dominant *Sunni Islam* is found from the Arab-speaking lands of North Africa to Indonesia. Probably in response to Western influences connected to globalization, both the Shi'a and the Sunni branches of Islam are currently experiencing fundamentalist revivals in which proponents seek to maintain purity of faith, separate from these Western influences.

Judaism, the parent religion of Christianity, is also closely related to Islam. Although tensions are often high between Jews and Muslims, these two religions, along with Christianity, actually share historical and theological roots in the Hebrew prophets and leaders. Judaism now numbers about 14 million adherents, having lost perhaps one-third of its total population to the systematic extermination of Jews during World War II.

Hinduism, which is closely linked to India, has about 1 billion adherents and is the world's third largest religion. Outsiders often regard Hinduism as polytheistic because Hindus worship many deities. Most Hindus argue, however, that all of their faith's gods are merely representations of different aspects of a single divine, cosmic unity. Historically, Hinduism is linked to the caste system, with its segregation of peoples based on ancestry and occupation. However, because India's democratic government is committed to reducing the social distinctions among castes, the connections between religion and caste are now much less explicit than in the past.

Buddhism, which originated as a reform movement within Hinduism 2500 years ago, is widespread in Asia, extending from Sri Lanka to Thailand and from Mongolia to Vietnam (Figure 1.33). There are two major branches of Buddhism: *Theravada*, followed throughout Southeast Asia and Sri Lanka, and *Mahayana*, found in Tibet and East Asia. In its spread, Buddhism came to coexist with other faiths in certain parts of Asia, making it difficult to accurately estimate the number of adherents. Estimates of the total Buddhist population range from 350 million to 900 million people.

Finally, in some parts of the world religious practice has declined, giving way to **secularism**, in which people consider themselves either nonreligious or outright atheistic. Though secularism is difficult to measure, social scientists estimate that about 1 billion people fit into this category worldwide. Perhaps the best example of secularism comes from the former communist lands of Russia and eastern Europe where, historically, there was overt hostility between government and church from the time of the 1917 Russian Revolution. Since the demise of Soviet control in the 1990s, however, many of these countries have experienced religious revivals.

Secularism has also grown more pronounced in western Europe. Although France was historically, and to some extent still is culturally, a Roman Catholic country, more people in France attend Muslim mosques on Fridays than attend Christian churches on Sundays. Japan and the other countries of East Asia are also noted for their high degree of secularization.

Culture, Gender, and Globalization

Culture includes not just the ways people speak or worship, but also embedded practices that influence behavior and values. **Gender** is a

LANGUAGE FAMILIES AND AREAL GROUPINGS

- Afro-Asiatic
- Altaic
- Amerindian
- Aboriginal Australian
- Austro-Asiatic
- Austronesian
- Caucasian
- Dravidian
- Eskimo-Aleut
- Indo-European
- Japanese
- Khoisan
- Korean
- Niger-Congo
- Nilo-Saharan
- Paleo-Siberian
- Papuan
- Sino-Tibetan
- Tai-Kadai
- Uralic
- Other

sociocultural construct, linked to the values and traditions of specific cultural groups that differentiate the characteristics of the two biological sexes, male and female. Central to this concept are **gender roles**, the cultural guidelines that define appropriate behavior within a specific context. In traditional tribal or ethnic groups, for example, gender roles might rigidly distinguish between women's work (often domestic tasks) and men's work (done mostly outside the home). Gender roles similarly guide many other social behaviors within a group, such as child rearing, education, marriage, and even recreational activities.

The explicit and often rigid gender roles of a traditional social unit contrast greatly with the less rigid, more implicit, and often flexible gender roles of a large, modern, urban industrial society. More to the point, globalization in its varied expressions is causing significant changes to traditional gender roles throughout the world. Nowhere is this more apparent than in the growing legal recognition of same-sex marriage worldwide (Figure 1.34). Since 2000, over 30 countries have recognized such unions, including the United States. Yet there is also a distinct geography of anti-gay legislation, especially in Africa, Southwest Asia, Russia, and South Asia. In extreme cases, gay expression can result in imprisonment and even death. Changes to the institution of marriage are part of a more globalized cultural discussion of what constitutes basic human rights. These shifting norms are embraced by some and rejected by others.

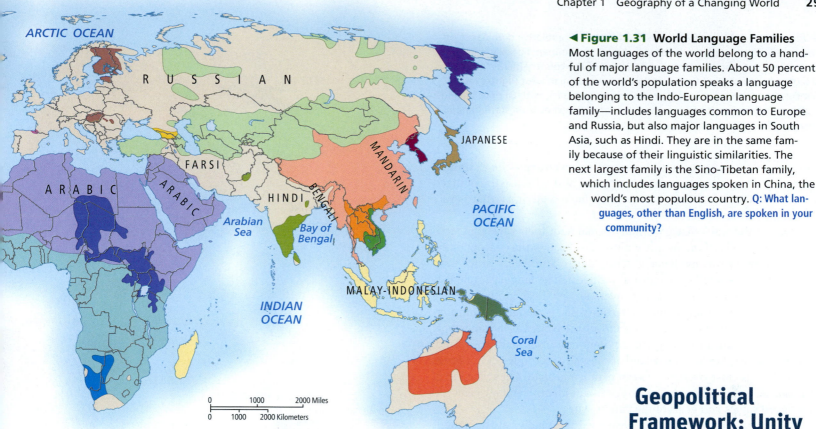

ARCTIC OCEAN

R U S S I A N

FARSI

A R A B I C

ARABIC

HINDI

BENGALI

MANDARIN

JAPANESE

Arabian
Sea

Bay of
Bengal

PACIFIC
OCEAN

INDIAN
OCEAN

MALAY-INDONESIAN

Coral
Sea

0   1000   2000 Miles
0 1000 2000 Kilometers

◀ **Figure 1.31 World Language Families**
Most languages of the world belong to a handful of major language families. About 50 percent of the world's population speaks a language belonging to the Indo-European language family—includes languages common to Europe and Russia, but also major languages in South Asia, such as Hindi. They are in the same family because of their linguistic similarities. The next largest family is the Sino-Tibetan family, which includes languages spoken in China, the world's most populous country. Q: What languages, other than English, are spoken in your community?

Globalization has also spread the notion of gender equality around the globe, calling into question and exposing those cultural groups and societies that blatantly discriminate against women. This topic is discussed later in the chapter as a measure of social development.

1.15 Define cultural imperialism and cultural assimilation. How are they similar and how are they different?

1.16 What is a lingua franca? Provide two examples.

1.17 Describe the geographies of Catholicism and Protestantism shown in Figure 1.32. What would explain these distinct realms of influence?

1.18 Discuss the patterns of acceptance and exclusion shown in Figure 1.33 with regard to gay rights.

KEY TERMS culture, cultural imperialism, nationalism, cultural syncretism, cultural assimilation, language family, lingua franca, religion, secularism, gender, gender roles

Geopolitical Framework: Unity and Fragmentation

The term **geopolitics** is used to describe the close link between geography and politics. More specifically, geopolitics focuses on the interactivity between political power and territory at all scales, from the local to the global. Unquestionably, the speed, scope, and nature of political change in various world regions over the last several decades are characteristics of globalization; thus, discussions of geopolitics are central to world regional geography.

The demise of the Soviet Union in 1991 brought opportunities for self-determination and independence in eastern Europe, Eurasia, and Central Asia, resulting in fundamental changes to economic, political, and even cultural alignments. Religious freedom helped drive national identities in some new Central Asian republics, whereas eastern Europe was primarily concerned with new economic and political links to western Europe. Russia itself still wavers perilously between different geopolitical pathways. Russia's justification for taking over parts of Ukraine in 2014 was based on the fact that ethnic Russians live there. Meanwhile, these acts have been condemned internationally as an affront to state sovereignty (Figure 1.35). All these topics are discussed further in Chapters 8, 9, and 10.

The Nation-State Revisited

A map of the world shows an array of nearly 200 countries ranging in size from microstates like Vatican City and Andorra to huge territorial and multiethnic states such as Russia, the United States, Canada, and China. All of these countries are regulated by governmental systems, ranging from democratic to autocratic. These different forms of government share a concern with **sovereignty**, which can be defined geopolitically as the ability (or inability) of a government to control activities within its borders. Integral to the

▶ **Figure 1.32 Major Religious Traditions**   The relative dominance of major religions in a particular area is shown on this map. For example, most Brazilians are Catholic, but the eastern half of the country (darker red) has a higher percentage of Catholics than the Amazonian west (lighter red). Similarly, Canada is a mix of Protestants and Catholics, with Québec having a strong Catholic majority. Yet there are large stretches in western Canada (gray) with no dominant religion.

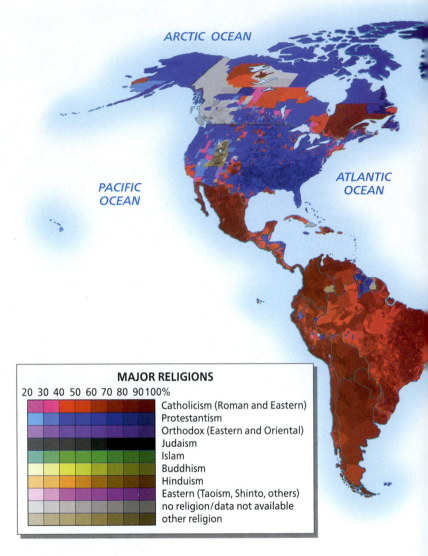

MAJOR RELIGIONS

20 30 40 50 60 70 80 90 100%

Catholicism (Roman and Eastern)
Protestantism
Orthodox (Eastern and Oriental)
Judaism
Islam
Buddhism
Hinduism
Eastern (Taoism, Shinto, others)
no religion/data not available
other religion

practice of sovereignty is the concept of **territory**—the delimited area over which a state exercises control and which is recognized by other states. A sovereign state must have a territory that is recognized by other states.

One of the ways that governments maintain their sovereign territory, and the unity of the people within it, is through the concept of the **nation-state**. In this hyphenated term, *nation* describes a large group of people with shared sociocultural traits, such as language, religion, and shared identity. The word *state* refers to a political entity that has a government and a clearly delimited territory that is maintained and controlled. Historically, France and England are often cited as the archetypal examples of a nation-state. Contemporary countries such as Iceland, Egypt, Czechia, Japan, and Poland are modern examples of countries where there is close overlap between nation and state. The related term *nationalism* is the sociopolitical expression of identity and allegiance to the shared values and goals of the nation-state.

Globalization, however, has weakened the vitality of the nation-state concept because today most countries are culturally diverse and their sense of nationhood comes from shared political values and common experiences, rather than the traditional definition of nation. International migration has led to many countries having large populations of ethnic minorities who may not share the national culture of the majority. In fact, **ethnicity**—a social group with a common or distinctive culture, religion, language, or history—can be a source of tension and strength among multiethnic states. In England, for example, large numbers of South Asians form their own communities, speak their own languages, practice their own religions, and dress to their own standards—practices that some Britons have criticized. At the same time, a majority of Londoners elected Sadiq Khan as their mayor in 2016; Khan is of Pakistani ancestry and is Muslim. In countries with large and diverse immigrant populations, the presence of ethnic groups over time changes the very nature of "national culture." The fact that countries such as the United States, Canada, the United Kingdom, and Germany officially embrace cultural diversity, with policies declaring themselves as multicultural states, shows how nationhood concepts can embrace diversity.

Decentralization and Devolution   Also residing within many nation-states are groups of people who seek autonomy from the central government and argue for the right to govern themselves by creating **autonomous areas**. This autonomy can range from the simple *decentralization* of power from a central government to smaller governmental units, such as the indigenous territory of Nunavut in Canada. At the far end of the spectrum is outright political separation and full governmental autonomy. This transfer of power is termed *devolution*, and the extent of autonomy can vary, as does support within a country for autonomy. For example, the citizens of Scotland held a referendum in 2014 to separate from the United Kingdom; although the referendum failed, 45 percent of voters chose independence. In 2017, residents of Catalonia voted to separate from Spain, but the Spanish state did not recognize the vote (Figure 1.36).

◀ **Figure 1.33 Ananda Temple, Myanmar**   This magnificent Buddhist temple was constructed over 900 years ago in Bagan, Myanmar. Thousands of temples were constructed in this region during the time of the Pagan Kingdom from the 9th to 13th centuries. Today the area attracts tourists.

Explore the SIGHTS of Ananda Temple
https://goo.gl/ZdnF8v

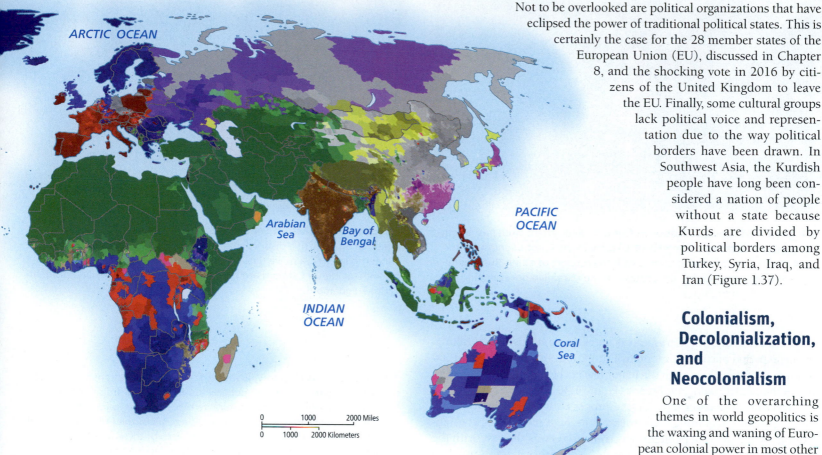

ARCTIC OCEAN

Arabian
Sea

Bay of
Bengal

PACIFIC
OCEAN

INDIAN
OCEAN

Coral
Sea

0   1000   2000 Miles

0   1000   2000 Kilometers

Not to be overlooked are political organizations that have eclipsed the power of traditional political states. This is certainly the case for the 28 member states of the European Union (EU), discussed in Chapter 8, and the shocking vote in 2016 by citizens of the United Kingdom to leave the EU. Finally, some cultural groups lack political voice and representation due to the way political borders have been drawn. In Southwest Asia, the Kurdish people have long been considered a nation of people without a state because Kurds are divided by political borders among Turkey, Syria, Iraq, and Iran (Figure 1.37).

Colonialism, Decolonialization, and Neocolonialism

One of the overarching themes in world geopolitics is the waxing and waning of European colonial power in most other regions of the world. **Colonialism** consists of the formal establishment of rule over a foreign population. A colony has no independent standing in the world community but instead is seen only as an appendage of the colonial power. The historic Spanish presence and rule over parts of the United States,

▼ **Figure 1.34 Mapping Gay Rights**   Since 2000, nearly 30 countries have recognized same-sex marriage. From Australia to Mexico and from South Africa to Ireland, a major cultural shift has occurred. At the same time, there are countries where gay expression is illegal and, in the most extreme cases, punishable by death.

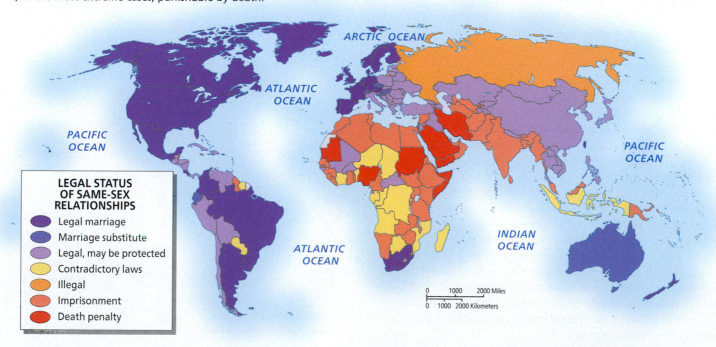

ARCTIC OCEAN

ATLANTIC
OCEAN

PACIFIC
OCEAN

PACIFIC
OCEAN

INDIAN
OCEAN

ATLANTIC
OCEAN

LEGAL STATUS OF SAME-SEX RELATIONSHIPS
- Legal marriage
- Marriage substitute
- Legal, may be protected
- Contradictory laws
- Illegal
- Imprisonment
- Death penalty

0   1000   2000 Miles

0   1000 2000 Kilometers

▲ **Figure 1.35 Russian Troops in Crimea** Pedestrians stroll past a combat vehicle on a street in Simferopol, Ukraine in the Crimea Peninsula. In 2014, Russia took control of Crimea with the support of ethnic Russians living in this territory, despite the protests of the Ukrainian government.

▲ **Figure 1.36 Catalan Separatists** Residents of Catalonia, a wealthy region of Spain with a distinctive culture, approved a referendum for independence in 2017, a move declared illegal by Spain's government. These pro-independence Catalans are listening to a speech by Catalan President Carles Puigdemont.

Latin America, and the Caribbean is an example. Generally speaking, the main period of colonialization by European countries was from 1500 through the mid-1900s, with the major players being England, Belgium, the Netherlands, Spain, and France (Figure 1.38).

Decolonialization refers to the process of a colony gaining (or, more correctly, regaining) control over its own territory and establishing a separate, independent government. As was the case with the Revolutionary War in the United States, this process often involves violent struggle. Similar wars of independence became increasingly common in the mid-20th century—particularly in South Asia, Southeast Asia, and Africa. Consequently, most European colonial powers recognized the inevitable and began working toward peaceful disengagement from their colonies. British rule ended in South Asia in 1947 and, in the late 1950s and early 1960s, Britain and France granted independence to their African colonies. This period of European colonialism symbolically closed in 1997 when England turned over Hong Kong to China.

However, decades and even centuries of colonial rule are not easily erased. The influences of colonialism are still found in the culture, government, educational systems, and economic life of the former colonies, evident in the many contemporary manifestations of British culture in India and the continuing Spanish influences in Latin America.

In the 1960s, the term **neocolonialism** came into popular usage to characterize the many ways that newly independent states, particularly those in Africa, felt the continuing control of Western powers, especially in economic and political matters. To receive financial aid from the World Bank, for example, former colonies were often required to revise their internal economic structures to become better integrated with the emerging global system. This economic restructuring is also called **neoliberalism**, as it emphasizes privatization, foreign investment, and free trade. Such policies may seem necessary from a global perspective, but the dislocations caused at national and local scales led critics of globalization to characterize these external influences as no better than the formal control by colonial powers.

▼ **Figure 1.37 A Nation Without a State** Not all nations or large cultural groups control their own political territories. The Kurdish people of Southwest Asia occupy a large cultural territory that lies in four different political states—Turkey, Iraq, Syria, and Iran. As a result of this political fragmentation, the Kurds are considered a minority in each of these four countries. **Q: Suggest issues that might result from the Kurds lacking a political state.**

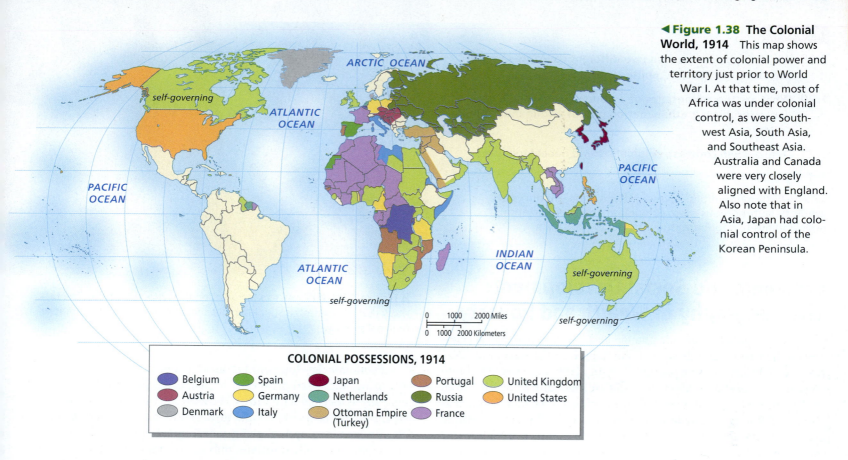

◀ **Figure 1.38 The Colonial World, 1914** This map shows the extent of colonial power and territory just prior to World War I. At that time, most of Africa was under colonial control, as were Southwest Asia, South Asia, and Southeast Asia. Australia and Canada were very closely aligned with England. Also note that in Asia, Japan had colonial control of the Korean Peninsula.

COLONIAL POSSESSIONS, 1914

- Belgium
- Austria
- Denmark
- Spain
- Germany
- Italy
- Japan
- Netherlands
- Ottoman Empire (Turkey)
- Portugal
- Russia
- France
- United Kingdom
- United States

Global Conflict and Political Freedom

As mentioned earlier, challenges to a centralized political state or authority have long been part of global geopolitics as rebellious and separatist groups seek independence, autonomy, and territorial control. These actions are termed **insurgency**. Armed conflict has also been part of this process; the American and Mexican revolutions were both successful wars for independence fought against European colonial powers. **Terrorism**, which can be defined as violence directed at nonmilitary targets, has also been common, albeit to a far lesser degree than today.

Until the September 2001 terrorist attacks on the United States by Al Qaeda, terrorism was usually directed at specific local targets and committed by insurgents with focused goals. The Irish Republican Army (IRA) bombings in Great Britain and Basque terrorism in Spain are illustrations. The attacks on the World Trade Center and the Pentagon (as well as the thwarted attack on the U.S. Capitol), however, went well beyond conventional geopolitics as a small group of religious extremists attacked the symbols of Western culture, finance, and power. Experts believe that Al Qaeda's goal was less about disrupting world commerce and politics and more about displaying the strength of their own convictions and power. Regardless of motives, those acts of terrorism underscore the need to expand our conceptualization of the linkages between globalization and geopolitics.

Many experts argue that global terrorism is both a product of and a reaction to globalization. Unlike earlier geopolitical conflicts, the geography of global terrorism is not defined by a war between well-established political states. Instead, the Al Qaeda terrorists appear to belong to a web of small, well-organized cells located in many different

countries. Boko Haram, a Muslim extremist group in Nigeria, has terrorized villages and kidnapped schoolchildren and is linked to Al Qaeda. Similarly Al-Shabaab, based in Somalia and responsible for attacks in Kenya, is affiliated with Al Qaeda.

One terrorist group, the Islamic State of Iraq and the Levant, also known as ISIS or IS), is unusual in that it aspired to form a new modern-day caliphate (a fundamentalist Islamic state) in Southwest Asia. ISIS's use of terror, kidnapping, public executions, extortion, and social media brought it international recognition and condemnation—as well as converts to its extremist cause (Figure 1.39). ISIS took advantage of the political vacuum left by years of conflict in Syria and Iraq that weakened those states and fostered the anti-Western, anti-secular, and anti-globalization sentiment that resonates in some parts of the world. In 2014, ISIS controlled a large territory, mostly within Syria and Iraq. Yet a coalition of states led by the United States over the past several years greatly reduced the power and territory under ISIS's control.

All regions undergo some conflict due to ethnic tensions, insurgencies, or authoritarian governments. Each year Freedom House, a nongovernmental organization, publishes an annual rating of every country's civil and political rights. This *freedom rating* takes into account election processes, political pluralism, government function, freedom of expression and belief, organizational rights, rule of law, and individual rights. States are rated from 1 to 7 (Figure 1.40) with 7 representing states in which citizens have the least amount of freedom. In 2018, China was rated 6.5 (not free) while India received a 2.5 and the United States a 1.5 (in the free category). In the next section, Table 1.2 presents development indicators for the world's 10 largest countries; the freedom rating is included for each entry to provide a snapshot of relative political freedom.

1.19 Why is it common to use two different concepts—nation and state—to describe political entities?

1.20 What are the advantages of autonomy versus independence for distinct ethnic/nation groups within multiethnic states?

1.21 Define, then contrast, colonialism and neocolonialism.

KEY TERMS geopolitics, sovereignty, territory, nation-state, ethnicity, autonomous area, colonialism, decolonialization, neocolonialism, neoliberalism, insurgency, terrorism

▲ **Figure 1.39 ISIS** Armed terrorists on the Syria-Iraq border. The territory that ISIS controls is shrinking, but these fighters are still a threat to regional security.

Economic and Social Development: The Geography of Wealth and Poverty

The pace of global economic change and development has accelerated in the past several decades, rising rapidly at the start of the 21st century and then slowing precipitously in 2008 as the world fell into an economic recession. Most countries are gradually recovering from the depths of the global recession, although major economies such as Brazil and Russia have faltered in the last couple of years. If nothing else, the 2008 recession and its unsteady recovery have highlighted the overarching question of whether the benefits of economic globalization outweigh the negative aspects. Responses vary considerably, depending on one's point of view, occupation, career aspirations, and socioeconomic status. Comprehending the contemporary world requires a basic understanding of global economic and social development. To that end, each regional chapter contains a substantive section on this topic, drawing on the concepts discussed in the following paragraphs.

Economic development is considered desirable because it generally brings increased prosperity to people, regions, and nations. It is typically associated with **industrialization** (the organization of society for the purposes of manufacturing), urbanization, and the growth of a service economy. Following conventional thinking, this economic development usually translates into social improvements such as better health care, improved educational systems, higher wages, and longer life expectancies. One of the most troubling expressions of global

▼ **Figure 1.40 Freedom Map** Each year Freedom House, a nongovernmental organization, assesses the civil and political rights of states on a scale of 1 to 7, with seven being the least free. One of their findings is a recent increase in the number of not free or partly free states.

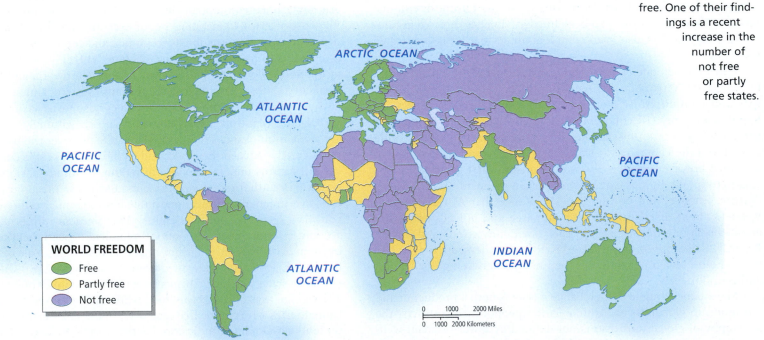

WORLD FREEDOM
- Free
- Partly free
- Not free

economic growth, however, has been the geographic unevenness of wealth and social improvement. That is, although some regions and places in the world prosper, others languish and even fall further behind the more developed countries. The gap between rich and poor regions has actually increased over the past several decades in many regions. This economic and social unevenness has, unfortunately, become one of the signatures of globalization. The numbers of people living in *extreme poverty*, defined as those living on less than $1.90 a day, have declined since the 1990s, largely due to economic growth in China. Yet if one considers those living on less than $3.10 a day, the poverty measure used by the World Bank and the UN, about 2.1 billion people (or three out of ten people in the world) still struggle for existence at this level. Many of these people live in Sub-Saharan Africa, South Asia, and Southeast Asia, and depend upon the **informal economy**, self-employed low-wage work that is usually unregulated, such as street vending and artisan manufacturing. Many also live in self-built housing, which is also part of the informal economy (Figure 1.41).

These inequities are problematic because they are intertwined with political, environmental, and social issues. For example, political instability and civil strife within a state are often driven by the economic disparity between a poor peripheral area and an affluent industrial core—between the haves and have-nots. Such instability, in turn, can strongly influence international economic interactions.

More and Less Developed Countries

Until the middle of the 20th century, economic development was centered in North America, Japan, Europe, and Australia, with most of the rest of the world gripped in poverty. This uneven distribution of economic power led scholars to devise a **core–periphery model** of the world. According to this model, these developed countries and regions constituted the global economic *core*, centered for the most part in the Northern Hemisphere, whereas most of the areas in the Southern Hemisphere made up a less developed *periphery*. Although oversimplified, this core–periphery dichotomy does contain some truth. All the G7 countries—the exclusive club of the world's major industrial nations, made up of the United States, Canada, France, England, Germany, Italy, and Japan—are located in the Northern Hemisphere. (Russia was excluded from the group in 2014 because of its invasion of Ukraine; China—unquestionably a Northern Hemisphere industrial power—is not a member.) Many critics contend that the developed countries achieved their wealth primarily by exploiting the poorer countries of the southern periphery, historically through colonial relationships and today through various forms of neocolonialism and neoliberal policies that focus on foreign direct investment.

Following this core–periphery model, much has been made of "north–south tensions," a phrase that distinguishes the rich and powerful countries of the Northern Hemisphere from the poor and less powerful countries primarily in the Southern Hemisphere. However, this model demands revision because over recent decades the global economy has become much more complicated. Several former colonies of the periphery or "south"—most notably, Singapore—have become very wealthy, while a few northern countries—notably, Russia—have experienced very uneven economic growth since 1989, with some parts of the country actually seeing economic declines. Additionally,

▲ **Figure 1.41 Extreme Poverty** A Bangladeshi woman and her child living in extreme poverty in a shanty town on the outskirts of Dhaka. Reducing extreme poverty by 2030 is a driving force behind the United Nations Sustainable Development Goals (SDGs).

the developed Southern Hemisphere countries of Australia and New Zealand never fit into the north–south framework. For these reasons, many global experts conclude that the designation *north–south* is outdated and should be avoided.

Third world is another term often erroneously used as a synonym for the developing world. Historically, the term was part of the Cold War vocabulary used to describe countries that were not part of either the capitalist Westernized first world or the communist second world dominated by the Soviet Union and China. Thus, in its original sense *third world* signified a political and economic orientation (capitalist vs. communist), not a level of economic development. With the Soviet Union's demise and China's considerably changed economic orientation, *third world* has lost its original political meaning. In this book, we use relational terms that capture a complex spectrum of economic and social development—**more developed country (MDC)** and **less developed country (LDC)**. This global pattern of MDCs and LDCs can be inferred from a map of gross national income per capita based on purchasing power parity (PPP) (Figure 1.42), a measure of economic health discussed in more detail later.

Indicators of Economic Development

The terms *development* and *growth* are often used interchangeably when referring to international economic activities. There is, however, value in keeping them separate. *Development* has both qualitative and quantitative dimensions. When we talk about economic development, the term usually implies structural changes, such as a shift from agricultural to manufacturing activity that also involves changes in the allocation of labor, capital, and technology. Along with these changes are assumed improvements in standard of living, education, and political organization. The structural changes experienced by Southeast Asian countries such as Thailand and Malaysia in the past several decades capture this process.

Growth, in contrast, is simply the increase in the size of a system. The agricultural or industrial output of a country may grow, as it has

▶ **Figure 1.42 More and Less Developed Countries**
Based on GNI per capita, purchasing power parity (PPP) adjusted, you can see the global pattern of more and less developed countries (MDCs and LDCs). Sub-Saharan Africa and South Asia stand out as the regions with the greatest number of low-income countries.

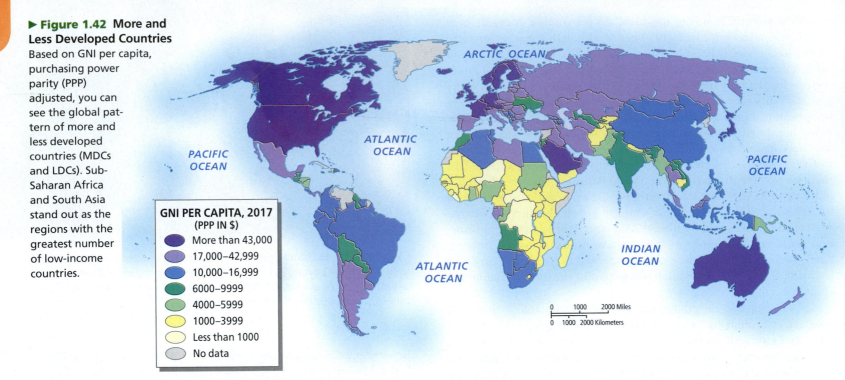

GNI PER CAPITA, 2017
(PPP IN $)

- More than 43,000
- 17,000–42,999
- 10,000–16,999
- 6000–9999
- 4000–5999
- 1000–3999
- Less than 1000
- No data

for India in the past decade, and this growth may—or may not—have positive implications for development. Many growing economies, in fact, have actually experienced increased poverty with economic expansion. When something grows, it gets bigger; when it develops, it improves. Critics of the world economy often say that we need less growth and more development.

Each of the 12 regional chapters has a table of development indicators. Table 1.2 highlights the development indicators used throughout the book for the 10 largest countries.

Gross Domestic Product and Income   A common measure of the size of a country's economy is the **gross domestic product (GDP)**,

the value of all final goods and services produced within its borders. Table 1.2 shows GDP average annual growth for 2009–2015. Most of the large countries saw growth in GDP during this period following the global recession. Compare the average annual growth of the United States with that of Indonesia. The U.S. growth rate is far lower than Indonesia's, but the U.S. economy is far larger and more diversified, and people in the high-income United States have far more resources than do residents of Indonesia, a lower-middle-income country. In general, the less developed countries shown in this table have higher growth rates than the more developed countries, with China having the highest annual growth rate at 8.2 percent.

TABLE 1.2   Development Indicators   Explore these data in MapMaster 2.0 https://goo.gl/PZFHk3

Country	GNI per Capita, PPP 2017[1]	GDP Average Annual Growth 2009-2015[2]	Human Development Index (2016)[3]	Percent Population Living Below $3.10 a Day[2]	Under Age 5 Mortality Rate (per 1000 live births), 1990[1]	Under Age 5 Mortality Rate (per 1000 live births), 2016[1]	Secondary School Enrollment Ratios[4] Male (2009–2016)	Secondary School Enrollment Ratios[4] Female (2009–2016)	Gender Inequality Index (2016)[3,6]	Freedom Rating (2018)[5]
China	16,807	8.2	0.738	19	54	10	93	96	0.164	6.5
India	7056	7.0	0.624	68	126	43	74	74	0.530	2.5
United States	59,532	2.1	0.920	–	11	7	97	98	0.203	1.5
Indonesia	12,284	5.7	0.689	39	84	26	86	86	0.467	3.0
Brazil	15,484	2.2	0.754	9	64	15	97	102	0.414	2.0
Pakistan	5527	3.7	0.550	44	139	79	49	39	0.546	4.5
Nigeria	5861	5.2	0.527	79	213	104	58	53	–	4.0
Bangladesh	3869	6.2	0.579	63	144	34	60	67	0.520	4.0
Russia	25,533	2.0	0.804	<2	22	8	106	103	0.271	6.5
Mexico	18,149	3.1	0.762	10	46	15	88	93	0.345	3.0

[1] World Bank Open Data, 2018.
[2] World Bank - *World Development Indicators*, 2017.
[3] United Nations, *Human Development Report*, 2016.
[4] Population Reference Bureau, *World Population Data Sheet, 2017*.
[5] Freedom House—Freedom in the World 2018. See pp. 33–34, for more info. on this scale (1–7, with 7 representing states with the least freedom).
[6] See p. 39, for more info. on this scale (0–1, with higher values representing less gender equality).

Log in to Mastering Geography & access MapMaster to explore these data!
1) Compare the countries in this table with Figure 1.45 showing levels of human development. Which countries would you classify as MDC or LDC, and why?
2) Which countries experienced the greatest improvement in Under Age 5 Mortality Rate?

When GDP is combined with net income from outside a country's borders through trade and other forms of investment, this constitutes a country's **gross national income (GNI)**. Although these terms are widely used, both GDP and GNI are incomplete and sometimes misleading economic indicators because they completely ignore nonmarket economic activity, such as bartering and household work, and do not take into account ecological degradation or depletion of natural resources. For example, if a country were to clear-cut its forests—an activity that would probably limit future economic growth—this resource usage would actually increase the GNI for that particular year, but then GNI would likely decline. Diverting educational funds to purchase military weapons might also increase a country's GNI in the short run, but its economy would likely suffer in the future because of its less-well-educated population. In other words, GDP and GNI are a snapshot of a country's economy at a specific moment in time, not a reliable indicator of continued vitality or social well-being.

Comparing Incomes and Purchasing Power

Because GNI data vary widely among countries, **gross national income (GNI) per capita** figures are used, allowing comparison of large and small economies regardless of population size (see Figure 1.42). An important qualification to these GNI per capita data is the concept of adjustment through **purchasing power parity (PPP)**, which accounts for the value of goods that can be purchased with the equivalent of one international dollar in a particular country. An international dollar has the same purchasing power parity over all GNI and is set at a designated U.S. dollar value. Thus, if a country's food costs are lower than the U.S. cost, the per capita purchasing power for that country increases. PPP was created to adjust comparisons between countries because (for example) an income of $5,000 in India can purchase more basic goods than the same amount of money in the United States (Figure 1.43).

In Table 1.2, the United States has the strongest GNI (PPP) at $59,532, followed by Russia at $25,533 and then Mexico at $18,149. The country with the lowest GNI (PPP) is Bangladesh at $3869. When contrasting the world's two largest countries, China and India, the GNI per capita of China is more than twice that of India.

Measuring Poverty

A principal goal of the UN and the World Bank is poverty reduction; this is the first of the Sustainable Development Goals that will influence development policy until 2030. As noted earlier, the UN definition of *poverty* is living on less than $3.10 per day, and *extreme poverty* is living on less than $1.90 per day. While the cost of living varies greatly around the world, the UN usually uses unadjusted per capita income figures when measuring poverty. Table 1.2 shows that 79 percent of Nigeria's population lives in poverty, followed by 63 percent of Bangladesh's population and 68 percent of India's. Note that this measure of poverty does not exist for the United States. While poverty data are usually presented at the country level, the World Bank and other agencies compile data at the sub-state level in order to better understand the poverty landscape within a country. The patterns of poverty in Morocco provide an instructive example, with areas around Rabat, Casablanca,

and Tangier having lower rates of poverty compared to higher rates in the inland communities around Fez and Marrakesh (Figure 1.44). Poverty mapping at this scale helps governments and development agencies decide which areas of a country require more aid and investment in order to reduce poverty.

Indicators of Social Development

Although economic growth is a major component of development, equally important are quality-of-life measures. As described earlier, the standard assumption is that economic development will spill over into the social infrastructure, leading to improvements in life expectancy, child mortality, gender equality, and education. Even some of the world's poorest countries have experienced significant improvements in all these measures. Much of the foreign development aid since 2000 goes to tracking and improving these development indicators.

The Human Development Index   For the past three decades, the UN has tracked social and economic development in the world's countries through the **Human Development Index (HDI)**, which combines data on life expectancy, literacy, educational attainment, gender inequality, and income (Figure 1.45). A 2015 analysis ranks the 188 countries that provided data to the UN from high to low, with Norway achieving the highest score, Australia in second place, and Canada and the United States tied for tenth place. Chad, Niger, and the Central African Republic are ranked the lowest. Countries with high or very high human development have HDI scores of .700 or better; Norway's score is .949. The low human development states score .540 or lower.

Although this measure is criticized for using national data that miss the diversity of development within a country, overall, the HDI conveys a reasonably accurate sense of a country's human and social

▼ **Figure 1.43 Purchasing Power**   Purchasing power parity (PPP) takes into consideration the strength or weakness of local currencies and relative value of what can be purchased with the equivalent of one international dollar in a particular country. The GNI per capita in India is about $1600 but the GNI PPP is over $6500.

▶ **Figure 1.44 Poverty Mapping** The World Bank regularly engages in poverty mapping to better understand where poverty is concentrated. This map of Morocco shows rates of poverty are higher around the inland cities of Marrakesh and Fez, and lower around the major coastal cities of Casablanca, Rabat, and Tangier. Mapping poverty at this scale is used to decide where scarce resources could be spent to improve overall development. **Q: Why are urban poverty rates lower in cities compared to rural areas?**

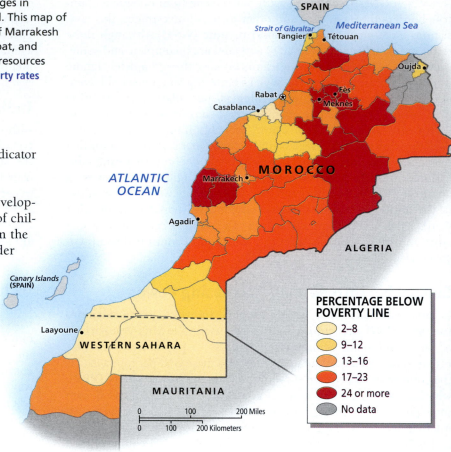

development. Thus, we include HDI data in our development indicator tables for each regional chapter.

Child Mortality Another widely used indicator of social development is data on *under age five mortality*, which is the number of children in that age bracket who die per 1000 children. Aside from the tragedy of infant death, child mortality also reflects the wider conditions of a society, such as the availability of food, health services, and public sanitation. If those factors are lacking, children under age five suffer most; therefore, their death rate is taken as an indicator of whether a country has the necessary social infrastructure to sustain life (Figure 1.46). In the social development table for each regional chapter, child mortality data are given for two points in time, 1990 and 2016, to indicate whether the social structure has improved in the intervening years. Every country has seen improvements over this period; especially significant in Table 1.2 are the improvements in China, India, Indonesia, and Bangladesh.

▶ **Figure 1.45**
Human Development Index This map depicts the most recent rankings assigned to four categories that make up parts of the Human Development index (HDI). In the numerical tabulation, Norway, Australia, Switzerland and Germany have the highest rankings, while several African countries are lowest on the scale. **Q: Compare this map with Figure 1.40 on political and civic freedom. Is there a relationship between human development and political freedom?**

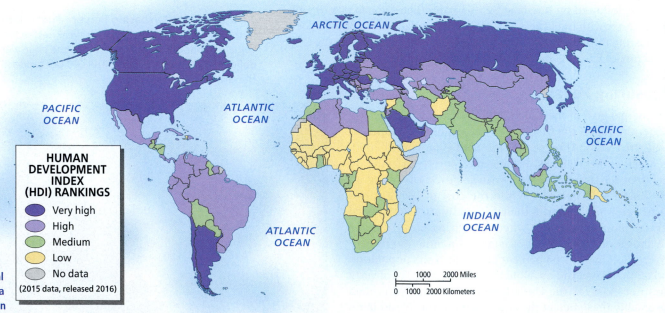

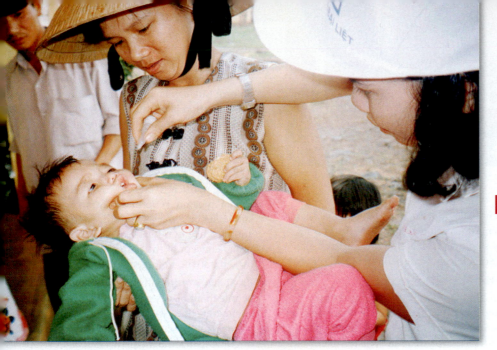

◀ **Figure 1.46** **Children's Health and Mortality**   The mortality rate of children under the age of 5 is an important indicator of social conditions such as food supply, public sanitation, and public health services. This child is receiving a polio vaccine in Vietnam.

REVIEW

1.22 What is the difference between development versus growth?

1.23 What is Purchasing Power Parity (PPP), and why is it useful?

1.24 How does the UN measure gender inequality? Explain why this is a useful metric for social development.

Education   Educating youth is critical in today's world, yet current data show about one-quarter of the world's population of young men and women (ages 12–17) are not enrolled in secondary school. The Sustainable Development Goals focus on youth education, especially secondary school enrollment. Table 1.2 contrasts secondary school enrollment of males and females. Pakistan has the lowest enrollment figures for boys and girls, while the highest figures come from Russia. (Enrollment figures can exceed 100 when current enrollment exceeds the population of the relevant age group in the country—this can happen due to foreign student enrollment and migration flows). Interestingly, in Brazil, Bangladesh, and Mexico, girls have higher enrollment rates than boys. This is because young men in these countries often seek employment within the country or abroad before finishing high school.

Gender Inequality   Discrimination against women takes many forms, from not allowing them to vote to discouraging school attendance (Figure 1.47). Given the importance of this topic, the United Nations calculates **gender inequality** among countries in order to measure the relative position of women to men in terms of employment, empowerment, and reproductive health (in terms of maternal mortality and adolescent fertility). The UN index ranges from 0 to 1; the higher the number, the greater the gender inequality. In Table 1.2, China has the lowest gender inequality at 0.164, followed by the United States and Russia. In contrast, the South Asian countries of India, Pakistan, and Bangladesh have much higher gender inequality. The UN gender inequality scores are reported in the development indicator table in each regional chapter.

Some countries may register reasonably high on the HDI (which is positive) yet also receive a relatively high gender inequality score (which is not so good). Qatar, for example, is a rich country that uses assets from its oil resources to provide many social benefits to its citizens, which explains its high HDI ranking; at the same time, its conservative Muslim culture produces a gender inequality score of .542 and a rank of 127 among countries for that indicator.

KEY TERMS   industrialization, informal economy, core–periphery model, more developed country (MDC), less developed country (LDC), gross domestic product (GDP), gross national income (GNI), gross national income (GNI) per capita, purchasing power parity (PPP), Human Development Index (HDI), gender inequality

▼ **Figure 1.47** **Women and Literacy**   Gender inequities in education lead to higher rates of illiteracy for women. However, when there is gender equity in education, female literacy has several positive outcomes. For example, educated women have a higher participation rate in family planning, which usually results in lower birth rates. This Iranian teacher is leading school children on a field trip to Isfahan. Iran has excellent youth literacy rates.

Explore the SIGHTS of Isfahan
https://goo.gl/xF5jWv

Geography of a Changing World

1

REVIEW, REFLECT, & APPLY

Summary

- Geography is the study of Earth's varied and changing landscapes and environments. This study can be done conceptually in many different ways, by physical or human geography and either topically or regionally—or by using a combination of all these approaches.

- Globalization affects all aspects of world geography with its economic, cultural, and political interconnectivity. However, despite fears that globalization will produce a homogeneous world, a great deal of diversity is still apparent. Geographers use various tools that draw on information gathered on the ground and by satellites to examine the world at different scales, from an inner-city block to the entire planet.

- Human populations around the world are growing either quickly or slowly depending on natural increase and widely different migration patterns. Urbanization is also a major factor in settlement patterns as people continue to move from rural to urban locales.

- Culture is learned behavior. It includes a range of tangible and intangible behaviors and objects, such as language and architecture. Globalization is changing the world's cultural geography, producing new cultural hybrids in many places. In other places, people resist change by protecting (or even resurrecting) traditional ways of life.

- Varying political systems provide the world with a dynamic geopolitical framework that is stable in some places and filled with tension and violence in others. As a result, the traditional concept of the nation-state is challenged by separatism, insurgency, and even terrorism.

- Proponents of globalization argue that all people in all places gain from expanded world commerce. But instead, there appear to be winners and losers, resulting in a geography of growing income inequality. Social development of health care and education is also highly uneven, but many key indicators are improving.

Review Questions

1. Define geography. Then define globalization and explain its relevance to understanding the world's changing geography.

2. What are the benefits of GIS, GPS, and satellite imagery in being able to monitor change and improve sustainability in a given place?

3. Summarize general migration trends around the world; and explain how these are influenced by and are impacting demographic, cultural, economic, and political change.

4. Explain the nation-state concept and provide examples. Is it still relevant in the age of globalization?

5. What is the difference between economic and social development? How might a rapidly developing country's population indicators from Table 1.2 change due to increasing well-being of its people?

Image Analysis

1. The flow of investment capital to remote parts of the planet is a feature of economic globalization. Which regions of the world receive relatively high foreign direct investment when compared with their gross domestic product? What do you think investors find attractive in these settings?

2. Imagine if you mapped which countries received the most FDI in absolute terms. What would that map look like, and why would it be different?

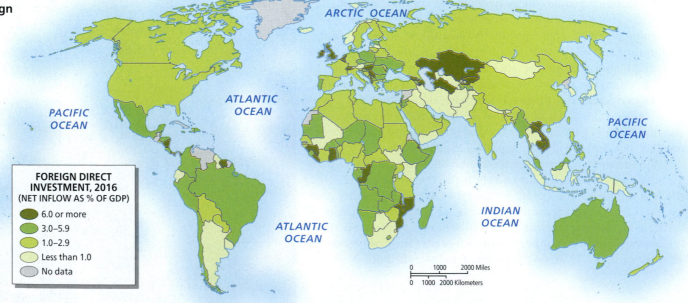

▶ Figure IA1 Foreign Direct Investment
FDI is private foreign capital that enters a country for purposes of resource extraction, infrastructure development, and industrialization.

FOREIGN DIRECT INVESTMENT, 2016
(NET INFLOW AS % OF GDP)
- 6.0 or more
- 3.0–5.9
- 1.0–2.9
- Less than 1.0
- No data

Join the Debate

Globalization is most often associated with economic activity, but it impacts all aspects of the world's physical and human landscapes. Global linkages are complex and can result in a variety of outcomes—some unexpected. Is globalization generally good or bad for the social and economic development?

Globalization advances social and economic development!

- Technological advances level the global playing field and allow more people to engage in economic activity and trade.

- With open markets, there are fewer barriers, increasing the efficiency of goods production and reducing the price of goods.

- Open economies tend to be more democratic and more tolerant of diversity and have less gender inequality.

Globalization has negative consequences for development!

- As trade increases, wages decline and income inequality is exacerbated. Digital globalization increases efficiency but creates fewer high-skilled jobs because less labor is required.

- A growth-at-all-costs argument often accelerates depletion of natural resources and unsustainable development. Fluctuations in commodity prices can lead to economic instability.

- The speed at which capital is transferred can lead to instability in global financial markets.

▲ **Figure D1 Global Consumers** A busy shopping mall in Guangzhou, China.

Key Terms

autonomous area (p. 30)
choropleth map (p. 16)
colonialism (p. 31)
core–periphery
  model (p. 35)
cultural assimilation (p. 27)
cultural imperialism (p. 25)
cultural landscape (p. 6)
cultural syncretism (p. 27)
culture (p. 25)
decolonialization (p. 32)
demographic transition
  model (p. 23)
diversity (p. 14)
economic migrant (p. 23)
ethnicity (p. 30)
formal region (p. 6)
functional region (p. 7)
gender (p. 28)
gender inequality (p. 39)
gender roles (p. 28)
geographic information
  systems (GIS) (p. 17)
geography (p. 5)
geopolitics (p. 29)
globalization (p. 8)
global positioning systems
  (GPS) (p. 15)
glocalization (p. 8)

gross domestic product
  (GDP) (p. 36)
gross national income
  (GNI) (p. 37)
gross national income (GNI)
  per capita (p. 37)
Human Development Index
  (HDI) (p. 37)
human geography (p. 5)
human trafficking (p. 12)
industrialization (p. 34)
informal economy (p. 35)
insurgency (p. 33)
language family (p. 27)
latitude (parallels) (p. 15)
less developed country
  (LDC) (p. 35)
lingua franca (p. 27)
longitude (meridians) (p. 15)
map projection (p. 15)
map scale (p. 15)
megacity (p. 25)
more developed country (p. 35)
nationalism (p. 26)
nation-state (p. 30)
neocolonialism (p. 32)
neoliberalism (p. 32)
net migration rate (p. 24)
physical geography (p. 5)
place (p. 6)
population density (p. 19)

population pyramid (p. 21)
purchasing power parity
  (PPP) (p. 37)
rate of natural increase
  (RNI) (p. 20)
refugee (p. 24)
region (p. 6)
regional geography (p. 5)
religion (p. 27)
remittances (p. 24)
remote sensing (p. 17)
replacement rate (p. 20)
secularism (p. 28)
sovereignty (p. 29)
space (p. 6)
sustainable development
  (p. 19)
Sustainable Development
  Goals (p. 19)
territory (p. 30)
terrorism (p. 33)
thematic geography (systematic
  geography) (p. 5)
total fertility rate (TFR)
  (p. 20)
urban primacy (primate
  city) (p. 25)
urbanization (p. 25)
World Bank (p. 13)
World Trade Organization
  (p. 12)

Mastering Geography

Looking for additional review and test prep materials? Visit the Study Area in **Mastering Geography** to enhance your geographic literacy, spatial reasoning skills, and understanding of this chapter's content by accessing a variety of resources, including **MapMaster** interactive maps, geoscience animations, videos, flashcards, web links, self-study quizzes, and an eText version of *Globalization and Diversity*.

GeoSpatial Data Analysis

Happiness as a Development Measure Many different indices try to measure relative development, levels of globalization, and overall well-being. In the last decade, there has been an increased effort to measure *happiness* and understand the causes for happiness and misery. Take a look at the World Happiness Report for 2018 (http://worldhappiness.report/ed/2018/). The report uses 2015–2017 data from the Gallop World Poll and focuses on happiness at the national level as well as analyzes the happiness of migrants who move within and between countries.

Click on the report's Chapter 2 link to review Figure 2.2, which ranks the happiness of 156 countries, and Figure 2.4, which ranks the happiness of foreign-born individuals within the surveyed countries. Then go back to the main page and click on the Online Data: Chapter 2 link to download a dataset for these measures.

World Happiness Report 2018
https://goo.gl/QRtTRR

Open MapMaster 2.0 in the Mastering Geography Study Area. Prepare and import the happiness data (click on the Table 2.1 tab at the bottom of the dataset) to map the world's 10 happiest countries, and answer these questions:

1. Where are these countries located? What variables make up this happiness measure? How can you explain the relative happiness of these countries?

2. Now consider the world's 10 largest countries discussed in the tables in our chapter. Where do these countries fall in the happiness ranking? Is there any relationship between GNI per capita as shown in Table 1.2 and relative happiness?

   Now map the happiness of foreign-born individuals within the surveyed countries (data from Table 2.4 in the Happiness Report; click the 2.4 tab in the dataset).

3. In which countries are the foreign-born most happy? Least happy? Are these immigrants happier than the domestic born or not? Can you explain what you observe?

Physical Geography and the Environment 2

Dead Sea

▲ The hypersaline Dead Sea sits in the Great Rift Valley between Jordan, Israel, and the West Bank. As it shrinks, the retreating saltwater reveals lakebeds covered in salt, leaving white crusts along the shores of this ever-smaller body of water.

Geology: A Restless Earth

Global Climates: Adapting to Change

Bioregions and Biodiversity: The Globalization of Nature

Water: A Scarce World Resource

Energy: The Essential Resource

The immense physical diversity of Earth, with its varied climates, deep oceans, towering mountain ranges, dry deserts, and rainy tropics, makes it unique in our solar system. Other planets are too warm (Venus) or too cold (Mars), but Earth contains the perfect mix of conditions for diverse forms of life to inhabit. In turn, life forms of all different sorts—plant, animal, and human—have interacted with the physical environment to produce the varied landscapes and habitats that make Earth our home. Despite its vastness, humankind's impact is found now on every part of the planet, from the middle of cities to the most remote parts of the high seas.

The Dead Sea, located at the nexus of Israel, Jordan, and the West Bank in Southwest Asia, demonstrates the changes that humans are making to Earth's unique ecosystems. The Dead Sea is the world's saltiest water body, so salty that you float in the water without even trying, and so salty that only specially adapted single-celled organisms can live in it. The shoreline of the Dead Sea is also the lowest point on dry land at 1412 feet (431 meters) below sea level (Earth's lowest point is deep below the Pacific Ocean in the Marianas Trench). The water level falls about 3 feet (1 meter) every year, due to a combination of climate change leading to less rainfall in the headwaters of the Jordan River, its sole source of water, and to increased extraction of the Jordan River's waters for agriculture and other human uses. The Dead Sea today is just one of many examples of a unique ecosystem under severe threat from human impacts.

Human geography is driven by the unique combination of physical attributes that make every place on planet Earth unique. Thus, a necessary starting point for the study of world regional geography is knowing more about Earth's physical environment—its geology, climate, diverse life forms, hydrology, and energy resources.

> **Human geography is driven by the unique combination of physical attributes that make every place on planet Earth unique.**

LEARNING OBJECTIVES

After reading this chapter you should be able to:

2.1 Describe those aspects of tectonic plate theory responsible for shaping Earth's surface.

2.2 Identify on a map those parts of the world where earthquakes and volcanoes are hazardous to human settlement.

2.3 List and explain the factors that control the world's weather and climate, and use these to describe the world's major climate regions.

2.4 Define the greenhouse effect, and explain how it is related to anthropogenic climate change.

2.5 Summarize the major issues underlying international efforts to address climate change.

2.6 Locate on a map and describe the characteristics of the world's major bioregions.

2.7 Name some threats to Earth's biodiversity.

2.8 Identify the causes of global water stress.

2.9 Describe the world geography of fossil fuel production and consumption.

2.10 List the advantages and disadvantages of the different kinds of renewable energy.

Geology: A Restless Earth

The world's continents, separated by vast oceans, are made up of an array of high mountains, deep valleys, rolling hills, and flat plains created over time by geologic processes originating deep within our planet and then sculpted on the surface by everyday processes such as wind, rain, and running water. Not only do these physical features give Earth its unique character, but the varied landscape also affects a wide range of human activities, creating resources in many places but posing daunting challenges in others, with destructive earthquakes and volcanic eruptions (Figure 2.1).

Plate Tectonics

The starting point for understanding geologic processes is the theory of **plate tectonics**, which states that Earth's outer layer, the lithosphere, consists of large plates that move very slowly across its surface. Driving the movement of these plates is a heat exchange deep within Earth; Figure 2.2 illustrates this complicated process.

On top of these plates sit continents and ocean basins; however, note in Figure 2.3 that the world's continents and oceans are not identical to the underlying plates, but rather have different margins and boundaries.

▼ **Figure 2.1 Kilauea Crater Erupting** Kilauea Volcano, located on the island of Hawaii, is the most active volcano on Earth. Despite being protected within the bounds of Hawaii Volcanoes National Park, areas beyond the park feel the destructive force of this dynamic natural feature, with towns on the island overrun by lava flows in 2018.

Mobile Field Trip: Kilauea Volcano https://goo.gl/IlcRez

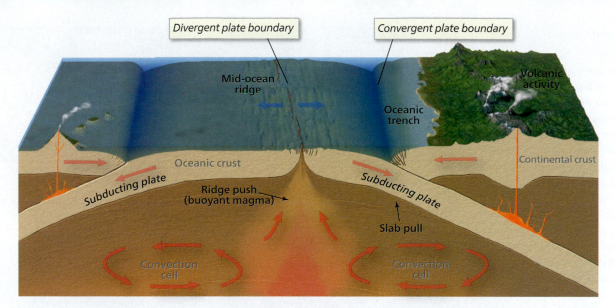

Divergent plate boundary

Convergent plate boundary

Mid-ocean ridge

Volcanic activity

Oceanic trench

Oceanic crust

Continental crust

Subducting plate

Ridge push (buoyant magma)

Subducting plate

Slab pull

Convection cell

Convection cell

▲ **Figure 2.2 Plate Tectonics** The driving force behind the theory of plate tectonics is the convection cells resulting from heat differences within Earth's mantle. These cells circulate slowly, producing surface movement in the crustal tectonic plates. New plate material reaches the surface in the mid-oceanic ridges and then moves away slowly from these divergent boundaries. As the plate material cools it tends to sink, creating subduction zones along convergent plate boundaries.

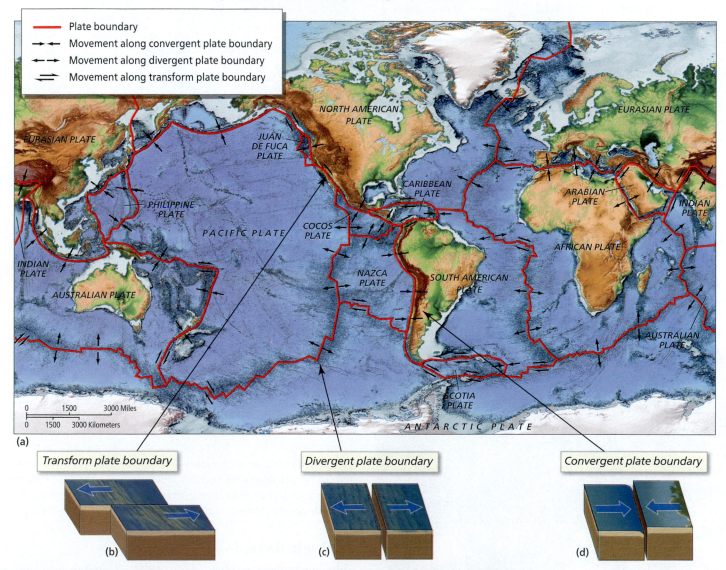

———	Plate boundary
→←—	Movement along convergent plate boundary
←—→	Movement along divergent plate boundary
⇄	Movement along transform plate boundary

NORTH AMERICAN PLATE

EURASIAN PLATE

EURASIAN PLATE

JUAN DE FUCA PLATE

CARIBBEAN PLATE

ARABIAN PLATE

INDIAN PLATE

PHILIPPINE PLATE

PACIFIC PLATE

COCOS PLATE

AFRICAN PLATE

INDIAN PLATE

NAZCA PLATE

SOUTH AMERICAN PLATE

AUSTRALIAN PLATE

SCOTIA PLATE

ANTARCTIC PLATE

AUSTRALIAN PLATE

0   1500   3000 Miles
0   1500   3000 Kilometers

(a)

Transform plate boundary

Divergent plate boundary

Convergent plate boundary

(b)

(c)

(d)

▲ **Figure 2.3 Tectonic Plate Boundaries** This world map shows the global distribution of the major tectonic plates, along with the general direction of plate movement. The different categories of plate boundaries are also shown. Note that continental boundaries do not always coincide with plate boundaries. Continents are not the same as tectonic plates but instead ride on top of the plates.

This is important because most earthquakes and volcanoes and their associated hazards are found along these plate boundaries. Some of these plates are converging upon each other, while others pull apart and still others grind past one another. The grinding of these plates along a **fault** triggers devastating earthquakes, like those along California's San Andreas Fault.

Along convergent boundaries, one plate sinks below another, creating a **subduction zone**. Deep trenches characterize these zones where the ocean floor has been pulled downward by sinking plates. Subduction zones exist off the west coast of South America, off the northwest coast of North America, offshore of eastern Japan, and near the Philippines, where the Marianas Trench is the world's lowest point at 35,000 feet (10,700 meters) below the ocean's surface. These subduction zones are also the locations of Earth's most powerful earthquakes, as evidenced by the magnitude 8.2 earthquake that killed hundreds around Mexico City in 2017 (Figure 2.4) and the magnitude 9.0 earthquake that devastated coastal Japan in 2011. Subduction zones are also home to many volcanoes, including those that surround the Pacific Ocean in the so-called "Ring of Fire" (Figure 2.5). On convergent boundaries where plates collide rather than subduct, towering mountains are formed. The best known of these are Asia's Himalayas, the world's highest mountains. Where plates diverge, magma

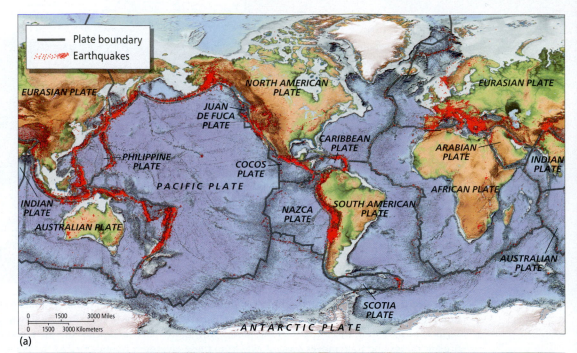

▼ Figure 2.4 Mexico City's 2017 Earthquake Mexico City, one of the world's largest metro areas, sits near the confluence of multiple plates, creating nearby volcanoes and subjecting the city to massive earthquakes. Here, rescuers search the rubble of a collapsed building looking for survivors following a 7.1 magnitude quake that shook central Mexico.

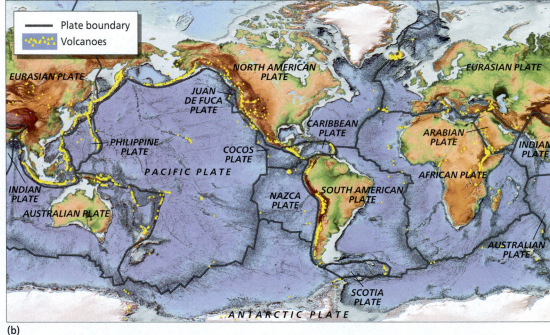

▲ Figure 2.5 The Geography of Earthquakes and Volcanoes (a) Most, but not all, earthquakes take place near plate boundaries. Further, most of the strongest and most devastating earthquakes are located near converging subduction zone boundaries. (b) While there is a strong correlation between the distribution of volcanoes, plate boundaries, and earthquakes, in many places in the world, volcanoes are far removed from plate boundaries. The island volcanoes of Hawaii, which lie on a hotspot in the Earth's crust, are an example.

from Earth's interior often flows to the surface to create mountain ranges and active volcanoes, as seen in Iceland, along the Mid-Atlantic Ridge (Figure 2.6). But at other divergent boundaries, deep depressions—called **rift valleys**—are formed, such as East Africa's Great Rift Valley.

Geologic Hazards

Although extreme weather events like floods and tropical storms typically take a higher toll of human life each year, earthquakes and volcanoes can significantly affect human settlement and activities. Nearly 20,000

▲ **Figure 2.6 Iceland's Divergent Plate Boundary** Volcanic activity is common in Iceland because of its location on the divergent plate boundary bisecting the Atlantic Ocean. This eruption took place on the Holuhraun Fissure near the Bardarbunga Volcano.

Explore the **SIGHTS** of Plate Divergence in Iceland
http://goo.gl/U7JIKZ
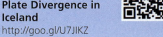

people died in March 2011 from the combination of an earthquake and massive tsunami in coastal Japan, and a year earlier (January 2010), over 230,000 people were killed in a magnitude 7.0 earthquake in Haiti. The vastly different effects of these two quakes underscore the fact that vulnerability to geologic hazards differs considerably around the world, depending on local building standards, population density, housing traditions, and the effectiveness of search, rescue, and relief organizations.

In addition to earthquakes, volcanic eruptions occur along certain plate boundaries and can also cause major destruction. But because volcanoes usually provide an array of warnings before they erupt, loss of life is generally far lower than that from earthquakes. In the 20th century, an estimated 75,000 people were killed by volcanic eruptions, whereas approximately 1.5 million died in earthquakes.

Volcanoes also provide some benefits. Volcanic soils are rich in nutrients, providing excellent sites for agriculture, and **geothermal** heat can be used for heating buildings and for generating electricity. Additionally, local economies benefit from tourists attracted to scenic volcanic landscapes in such places as Hawaii, Japan, and Yellowstone National Park (Figure 2.7).

REVIEW

2.1 Where are most of the world's earthquakes and volcanoes found? Why are they located where they are?

2.2 Despite their hazards, billions of people live in areas susceptible to earthquakes and volcanoes. Why might they live in such places?

KEY TERMS plate tectonics, fault, subduction zone, rift valley, geothermal

Global Climates: Adapting to Change

Many human activities are closely tied to weather and climate. Farming depends on certain conditions of sunlight, temperature, and precipitation to produce the world's food, while transportation systems are often disrupted by extreme weather events like snowstorms, typhoons, and even heat waves. Furthermore, a severe weather event in one location can affect far-flung places. Reduced harvests due to drought in Russia's grain belt, for example, ripple through global trade and food supply systems, with serious consequences worldwide.

Aggravating these interconnections is global climate change. Just what the future holds is not entirely clear, but even if the long-term forecast has some uncertainty, there is little question that all forms of life—including humans—must adapt to vastly different climatic conditions by the middle of the 21st century (see *Humanitarian Geography: Measuring Climate Change on the Greenland Ice Sheet*).

Climate Controls

The world's climates differ significantly from place to place and seasonally with greatly varying patterns of temperature and precipitation (rain and snow) that can be explained by physical processes termed climate controls.

Solar Energy The Sun's heating of Earth and its atmosphere is the most important factor affecting world climates. Not only does solar energy cause temperature differences between warmer and colder regions, but it also drives other important climate controls such as global pressure systems, winds, and ocean currents.

Incoming short-wave solar energy, called **insolation**, passes through the atmosphere and is absorbed by Earth's land and water surfaces. As these surfaces warm, they reradiate heat back into the lower atmosphere as infrared, long-wave energy. This reradiating energy, in turn, is absorbed by water vapor and other atmospheric gases such as carbon dioxide (CO_2), creating the envelope of warmth that makes life possible on our planet. Because there is some similarity between this heating process and the way a garden greenhouse traps warmth from the Sun, this natural process of atmospheric heating is called the

◄ **Figure 2.7 Geyser Erupting at Yellowstone** The world's first national park, Yellowstone lies on one of the world's biggest volcanoes. Tourists flocking to see the park's magnificent geothermal features seldom realize that beneath them lies a massive volcano. If Yellowstone were to erupt again, it would devastate the western half of the United States and Canada—but don't worry, it last erupted 70,000 years ago.

Measuring Climate Change on the Greenland Ice Sheet

An important challenge in modeling climate change is determining how quickly sea levels could rise. Nowhere is glacial melt accelerating more than on the Greenland Ice Sheet, home to enough water to raise global sea level over 20 feet (6 meters) if it all melts away. A team of geographers led by UCLA Professor Laurence C. Smith has spent several summers measuring melt across the Greenland Ice Sheet, and the results are startling.

Mobile Field Trip: Climate Change in the Arctic
https://goo.gl/4m3Py0

Measuring Melt Research on ice sheets tends to focus on their margins, where large blocks of ice have broken off, or calved, into the sea. However, much of Greenland's ice is melting on the ice sheet's surface into rivers that fall through giant ice holes (moulins) and then tunnel under the ice to flow all the way to the sea (Figure 2.1.1) Understanding this flow is essential to predicting future sea level rise that will impact coastal dwellers in far-flung places—from Miami to low-lying island countries like the Maldives and Kiribati. This rise will be most acutely felt by the poorest coastal communities, worlds away from this melt atop a sheet of Arctic ice.

Explore the SIGHTS of Greenland's Melting Ice Cap
http://goo.gl/ROOHtE

Dr. Smith's team measured the rate of water flowing in a supraglacial river off the melting ice sheet surface into a moulin, including groundbreaking work using Acoustic Doppler Current Profiler (ADCP) and global positioning satellite (GPS) measurements of flow velocity and discharge in the gushing river. The flows are over 400,000 gallons per minute. The findings will help researchers better model the ice sheet's hydrologic system as it undergoes intense evolution due to anthropogenic climate change.

Using these field methods combined with remote techniques such as satellite and drone measurements, geographers are on the frontline of understanding a critical impact of climate change. Smith is also deeply involved in the Surface Water and Ocean Topography satellite mission that will launch in 2021 or 2022 (https://swot.jpl.nasa.gov/). A joint project of NASA, the French Space Agency (CNES), and the Canadian Space Agency, this mission will greatly improve estimates of water volume and discharge in the world's water bodies.

A geologist by training, Dr. Smith was hired by UCLA's geography department and now considers himself a "geographer for life." He explains, "Geography's great strength is that it studies the natural world—with human beings in it. That is the obvious reality of the world today, and moving forward." Understanding and modeling surface water flow and storage and the impacts on sea-level rise will be essential to protecting coastal regions and the people who live there.

▲ **Figure 2.1.1 Laurence C. Smith** Dr. Smith doing field work in Greenland with former PhD student Mia Bennett, now an Assistant Professor at Hong Kong University.

1. **How does studying climate in Greenland apply to areas far away? How is this work linked to international efforts to address climate change?**

2. **Are there places you have lived or visited that will feel the effect of rising sea level? What places come to your mind as most impacted by this change?**

---

greenhouse effect (Figure 2.8). Without this process, Earth's climate would average about 60°F (33°C) colder, resulting in conditions much like those on Mars.

Latitude Because Earth is a tilted sphere with the North Pole facing away from the Sun for half of the year and toward the Sun for the other half, maximum solar radiation at a given location occurs seasonally (with summer marking the season of peak insolation for each hemisphere), and insolation strikes the surface at a true right angle only in the tropics. Therefore, solar energy is more intense and effective at heating land and water at low latitudes than at higher latitudes. Not only does this difference in solar intensity result in warmer tropical climates compared to the middle and high latitudes, but this heat also accumulates on a larger scale in these equatorial regions (Figure 2.9). This heat energy is then distributed away from the tropics through global pressure and wind systems, ocean currents, subtropical typhoons or hurricanes, and even midlatitude storms (Figure 2.10).

Interactions Between Land and Water Land and water areas differ in their ability to absorb and reradiate insolation; thus, the global arrangement of oceans and continents has a major influence on world climates. More solar energy is required to heat water than to heat land, so land areas heat and cool faster than do bodies of water (Figure 2.11). This explains why the temperature extremes of hot summers and cold winters are found in the continental interiors, such as the Great Plains of North America, while coastal areas experience more moderate winters

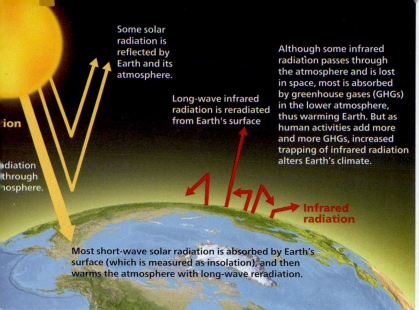

Some solar radiation is reflected by Earth and its atmosphere.

Although some infrared radiation passes through the atmosphere and is lost in space, most is absorbed by greenhouse gases (GHGs) in the lower atmosphere, thus warming Earth. But as human activities add more and more GHGs, increased trapping of infrared radiation alters Earth's climate.

Long-wave infrared radiation is reradiated from Earth's surface

...ion

...diation ...through ...nosphere.

Infrared radiation

Most short-wave solar radiation is absorbed by Earth's surface (which is measured as insolation), and then warms the atmosphere with long-wave reradiation.

▲ **Figure 2.8 Solar Energy and the Greenhouse Effect** Most incoming short-wave solar radiation is absorbed by land and water surfaces and then reradiated into the atmosphere as long-wave infrared radiation. It is this long-wave radiation absorbed by greenhouse gases—both natural and human-generated—that warms the lower atmosphere and affects Earth's weather and climate.

and cooler summers. These coastal–inland temperature differences also occur at smaller scales, often within just hundreds of miles of each other. In coastal San Francisco, for example, the average July maximum temperature is 60°F (15.6°C), whereas 80 miles (129 km) away in California's inland capital, Sacramento, the average maximum July temperature is 92.4°F (33.5°C).

The term **continental climate** describes inland climates with hot summers and cold winters, while locations where oceanic influences dominate are referred to as **maritime climates**. The island countries of Southeast Asia and the British Isles in Europe are good examples of

▼ **Figure 2.9 Solar Intensity and Latitude** Because of Earth's curvature, solar radiation is more intense and more effective at warming the surface in the tropics than at higher latitudes. The resulting heat buildup in the equatorial zone energizes global wind and pressure systems, ocean currents, and tropical storms.

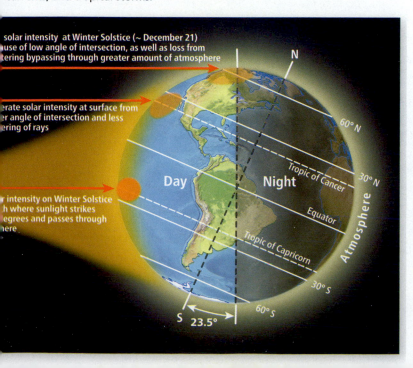

solar intensity at Winter Solstice (~ December 21) ...use of low angle of intersection, as well as loss from ...tering bypassing through greater amount of atmosphere

...erate solar intensity at surface from ...er angle of intersection and less ...ering of rays

... intensity on Winter Solstice ...h where sunlight strikes ...egrees and passes through ...ere

N

Day   Night

60° N

Tropic of Cancer   30° N

Equator

Atmosphere

Tropic of Capricorn

30° S

60° S

S   23.5°

▲ **Figure 2.10 Hurricane Maria** Atmospheric heat imbalances between the equatorial zone and the midlatitudes produce massive tropical cyclones called typhoons in the Pacific and hurricanes in the Atlantic. Both are capable of widespread damage through high winds, coastal flooding from waves, and heavy rainfall. This satellite image shows Hurricane Maria, which devastated the island of Puerto Rico in 2017, leaving this U.S. territory without electricity for months.

Explore the SOUNDS of a Hurricane
http://goo.gl/F86ijA

areas with maritime climates, while interior North America, Europe, and Asia have continental climates.

Global Pressure Systems The uneven heating of Earth due to latitudinal differences and the arrangement of oceans and continents produces a regular pattern of high- and low-pressure cells (Figure 2.12). These cells drive the movement of the world's wind and storm systems because air (in the form of wind) moves from high to low pressure. The interaction between high- and low-pressure systems over the North Pacific, for example, produces storms that are carried by winds onto the North American continent. Similar processes in the North Atlantic produce winter and summer weather for Europe. Farther south, over the subtropical zones, large cells of high pressure cause very different conditions. The subsidence (sinking) of warm air moving in from the equatorial regions causes the great desert areas at these latitudes. These high-pressure areas expand and shift during the warm summer months, producing the warm, rainless summers of Mediterranean climate areas in Europe, California, and parts of western South America. In the low latitudes, summer heating of the oceans also spawns the strong tropical storms known as typhoons or cyclones in Asia and as hurricanes in North America and the Caribbean.

Global Wind Patterns Several different wind patterns strongly influence Earth's weather and climate. At the global level are the polar and subtropical **jet streams**, powerful atmospheric rivers of eastward-moving air that drive storms and pressure systems in both the Northern and the Southern hemispheres (see Figure 2.12). These jets are products of Earth's rotation and global temperature differences. The two polar jets (north and south) are the strongest and the most variable, flowing 23,000–39,000 feet (7–12 km) above Earth's surface at speeds reaching 200 miles per hour (322 km/h).

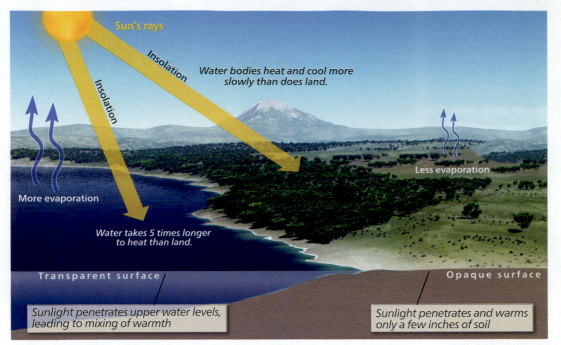

Sun's rays

Insolation

Insolation

Insolation

Water bodies heat and cool more slowly than does land.

More evaporation

Less evaporation

Water takes 5 times longer to heat than land.

Transparent surface

Opaque surface

Sunlight penetrates upper water levels, leading to mixing of warmth

Sunlight penetrates and warms only a few inches of soil

Nearer Earth's surface are continent-scale winds that, as mentioned earlier, flow from high- to low-pressure areas. Good examples are the **monsoon winds** of Asia and North America; summer monsoons bring welcome rainfall to the dry areas of interior South Asia and the Southwest United States (Figure 2.13).

▼ Figure 2.13 Monsoon Rains in India As the interior of South Asia warms during the northern hemisphere summer, heating creates thermal lows that drive moist air toward the Himalaya Mountains from the Indian Ocean. This leads to massive seasonal rainfall across India, Pakistan, Bangladesh, and Nepal that causes widespread flooding in built-up areas.

▲ Figure 2.11 Differential Heating of Land and Water Land heats and cools faster than does water through incoming and outgoing solar radiation. This is why inland temperatures are usually both warmer in the summer and colder in the winter than temperatures at coastal locations.

The subtropical jets are usually higher and somewhat weaker. While the northern jet stream has a major effect on the weather of North America, Eurasia and Europe, steering storms across the continents, the southern polar jet circles the globe in the sparsely populated areas near Antarctica.

▼ Figure 2.12 Global Pressure Systems and Winds Two jet streams, the Polar and Subtropical, are found in each hemisphere, northern and southern. In circling Earth, these jets often change position as they steer storms and air masses. The subtropical high-pressure cells are large areas of subsiding air that energize the midlatitude westerly winds and the tropical trade winds. The Inter-Tropical Convergence Zone (ITCZ) is a belt of low pressure that circles Earth and results from strong solar radiation in the equatorial zone.

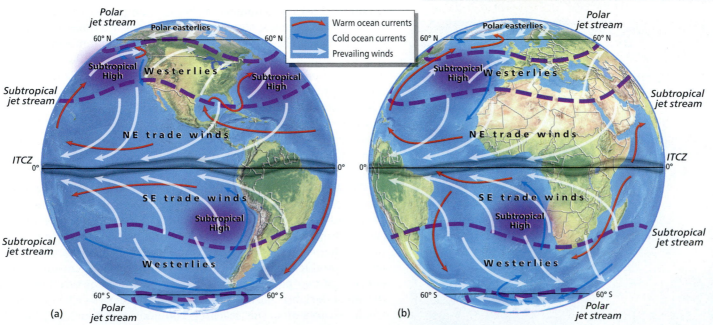

Polar jet stream

Polar easterlies

Warm ocean currents
Cold ocean currents
Prevailing winds

Polar easterlies

Polar jet stream

60° N

60° N

60° N

60° N

Subtropical High

Westerlies

Subtropical High

Subtropical High

Westerlies

Subtropical jet stream

Subtropical jet stream

NE trade winds

NE trade winds

ITCZ 0°

0°

0°

ITCZ 0°

SE trade winds

SE trade winds

Subtropical High

Subtropical High

Subtropical jet stream

Subtropical jet stream

Westerlies

Westerlies

60° S

60° S

60° S

60° S

(a)

Polar jet stream

(b)

Polar jet stream

Topography Weather and climate are affected by topography—an area's surface characteristics—in two ways: Cooler temperatures are found at higher elevations, and precipitation patterns are strongly influenced by topography.

Because the lower atmosphere is heated by solar energy reradiated from Earth's surface, air temperatures are warmer closer to the surface and become cooler with altitude. On average, the atmosphere cools by about 3.5°F for every 1000 feet gained in elevation (.65°C per 100 meters). This is called the **environmental lapse rate**. To illustrate, on a typical summer day in Phoenix, Arizona, at an elevation of 1100 feet (335 meters), the temperature often reaches 100°F (37.7°C). Just 140 miles away, in the mountains of northern Arizona at 7100 feet (2160 meters) in the small town of Flagstaff, the temperature is a pleasant 79°F (26°C). This difference of 21°F (11.7°C) results from 6000 feet (1825 meters) of elevation; this can be easily calculated by multiplying elevation in thousands (6) by the environmental lapse rate (3.5°F).

An area of rugged topography can "wring" moisture out of clouds when moist air masses cool as they are forced up and over mountain ranges in what is called the **orographic effect** (Figure 2.14). Cooler air cannot hold as much moisture as warm air, resulting in condensation and precipitation. An air mass moving up a mountain slope will cool at 5.5°F per 1000 feet (1°C per 100 meters) of elevation; this is called the **adiabatic lapse rate**. Note that the rising air mass cools (and warms upon descending) faster than the surrounding nonmoving air mass, for which the change in temperature with elevation is measured by the environmental lapse rate.

This process explains the common pattern of wet mountains and nearby dry lowlands. These arid areas are said to be in the **rain shadow** of the adjacent mountains. Rainfall lessens as downslope winds warm (the opposite of upslope winds, which cool), thus increasing an air mass's ability to retain moisture and deprive nearby lowlands of precipitation. Rain shadow areas are common in the mountainous regions of western North America, Andean South America, and many parts of South and Central Asia.

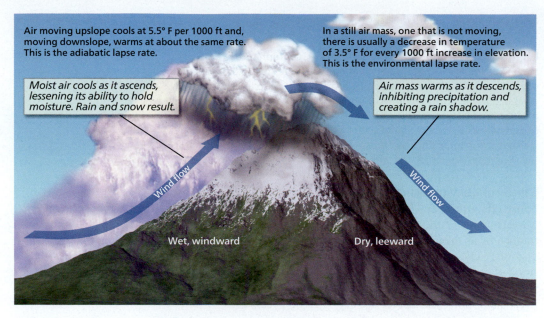

Air moving upslope cools at 5.5° F per 1000 ft and, moving downslope, warms at about the same rate. This is the adiabatic lapse rate.

In a still air mass, one that is not moving, there is usually a decrease in temperature of 3.5° F for every 1000 ft increase in elevation. This is the environmental lapse rate.

Moist air cools as it ascends, lessening its ability to hold moisture. Rain and snow result.

Air mass warms as it descends, inhibiting precipitation and creating a rain shadow.

Wind flow

Wind flow

Wet, windward

Dry, leeward

▲ **Figure 2.14 The Orographic Effect** Upland and mountainous areas are usually wetter than the adjacent lowland areas because of the orographic effect. This results from the cooling of rising air over higher topography, and as the air mass cools it loses its ability to hold moisture, resulting in rain and snowfall. In contrast, the leeward (downwind) side of the mountains is drier because downslope air masses warm, thus increasing their ability to retain moisture. These dry downwind areas are called rain shadows.

Q: Look at a map of the world and locate at least five different areas where the orographic effect would be found.

Mobile Field Trip: Clouds
https://goo.gl/2GoynZ

Climate Regions

Even though the world's weather and climate vary greatly from place to place, areas with similarities in temperature, precipitation, and seasonality can be mapped into global climate regions (Figure 2.15). Before going further, it is important to note the difference between these two terms. Weather is the daily expression of atmospheric processes; that is, weather can be rainy, cloudy, sunny, hot, windy, calm, or stormy, all within a short time period. As a result, weather is measured at regular intervals each day, usually hourly. These data are then compiled over a 30-year period to generate statistical averages that describe the typical meteorological conditions of a specific place, which is the climate. Simply stated, *weather* describes the short-term changes in atmospheric processes, and **climate** is the long-term average from daily weather measurements. Climate is what you expect, and weather is what you get.

We use a standard scheme of climate types throughout this text, and each regional chapter contains a map showing the different climates

of that region. In addition, these maps contain **climographs**, which are graphic representations of monthly average temperatures and precipitation. Two lines for temperature data are presented on each climograph: The upper line plots average high temperatures for each month, while the lower line shows average low temperatures. Besides these temperature lines, climographs contain bar graphs depicting average monthly precipitation. The total amount of rainfall and snowfall is important, as is the seasonality of precipitation. Figure 2.15 shows the climographs for Tokyo, Japan, and Cape Town, South Africa. When looking at these two climographs, remember that the seasons are reversed in the Southern Hemisphere.

Global Climate Change

Human activities connected to energy and land use changes have caused significant **climate change** worldwide, resulting in warmer temperatures, melting ice caps, desertification, rising sea levels, and more extreme weather events (Figure 2.16). Significant international action is necessary to limit atmospheric pollution; otherwise climatic changes will produce a challenging world environment by mid-century. Rainfall patterns may change so that agricultural production in traditional breadbasket areas such as the U.S. Midwest and the Eurasian plains may be threatened; low-lying coastal settlements in places like Florida and Bangladesh will be flooded as sea levels rise; increased heat waves will cause higher human death tolls in the world's cities; and freshwater will become increasingly scarce in many areas of the world. Ocean warming and acidification are also impacting global fisheries and coral reefs (see Chapter 14's *Working Toward Sustainability: Saving the Great Barrier Reef*).

▶ **Figure 2.15 Global Climate Regions** Geographers use a standard scheme called the Köppen system, named after the Austrian geographer who devised the plan in the early 20th century, to describe the world's diverse climates. Combinations of upper- and lowercase letters describe the general climate type, along with precipitation and temperature characteristics. Specifically, the *A* climates are tropical, the *B* climates are dry, the *C* climates are generally moderate and are found in the middle latitudes, and the *D* climates are associated with continental and high-latitude locations.

Causes of Climate Change As mentioned earlier, the natural greenhouse effect provides Earth with a warm atmospheric envelope; this warmth comes from incoming and outgoing solar radiation that is trapped by an array of such natural constituents as water vapor, carbon dioxide (CO_2), methane (CH_4), and ozone (O_3). Although the composition of these natural **greenhouse gases (GHGs)** in the atmosphere has varied somewhat over long periods of geologic time, it has been relatively stable since the last ice age ended 20,000 years ago.

However, the widespread consumption of fossil fuels associated with global industrialization has resulted in a huge increase in atmospheric carbon dioxide and methane. As a result, the natural greenhouse effect has been greatly magnified by **anthropogenic** or human-generated GHGs, which then trap increasing amounts of Earth's long-wave reradiation and thus warm the atmosphere and change our planet's climates. Figure 2.17 shows that in 1860, atmospheric CO_2 was measured at 280 parts per million (ppm); today it is 410 ppm. Even more troubling, these CO_2 emissions are forecast to reach 450 ppm by 2020, a level at which climate scientists predict irrevocable climate change.

Although the complexity of the global climate system leaves some uncertainty about exactly how the world's climates may change, climate scientists using high-powered computer models are reaching consensus on what can be expected.

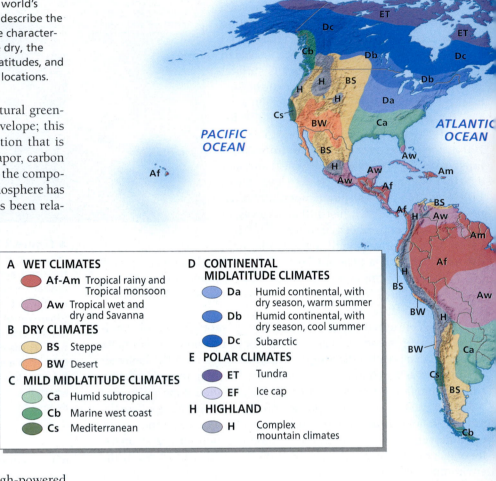

A WET CLIMATES	D CONTINENTAL MIDLATITUDE CLIMATES
Af-Am Tropical rainy and Tropical monsoon	**Da** Humid continental, with dry season, warm summer
Aw Tropical wet and dry and Savanna	**Db** Humid continental, with dry season, cool summer
B DRY CLIMATES	**Dc** Subarctic
BS Steppe	E POLAR CLIMATES
BW Desert	**ET** Tundra
C MILD MIDLATITUDE CLIMATES	**EF** Ice cap
Ca Humid subtropical	H HIGHLAND
Cb Marine west coast	**H** Complex mountain climates
Cs Mediterranean	

These models predict that average global temperatures will increase 3.6°F (2°C) by 2020, a temperature change of the same magnitude as the amount of cooling that caused ice-age glaciers to cover much of Europe and North America 30,000 years ago. Further, without international policies to limit emissions, this temperature increase is projected to double by 2100. The resulting melting of polar ice caps, ice sheets, and mountain glaciers will cause a sea-level rise currently estimated to be on the order of 4 feet (1.4 meters) by century's end (Figure 2.18). Climate change is also expected to increase both the frequency and intensity of extreme weather events, such as droughts, typhoons, or blizzards.

International Efforts to Limit Emissions Climate scientists have long expressed concern about fossil fuel emissions aggravating the natural greenhouse effect; as a result, in 1988 the United Nations (UN) began coordinating the study of global warming by creating the International Panel on Climate Change (IPCC). This group of international scientists is charged with providing the world with periodic Assessment Reports (ARs) of climate change science. The panel's first report, AR1, came out in 1990; AR5, published in late 2014, included a strongly worded statement that failure to reduce atmosphere emissions could threaten society with food shortages, refugee crises, the flooding of major cities and entire island chains, mass extinctions of plants and animals, and a climate so drastically altered it might become dangerous for people to work or play outside during the hottest times of the year. Further, the "continued emission of greenhouse gases will cause further warming and long-lasting changes in all components of the climate system, increasing the likelihood of severe, pervasive and irreversible impacts for people and ecosystems."

▼ **Figure 2.16 Glacier Calving in Antarctica** As Earth's climate warms, Antarctica's massive ice sheet is losing volume. Here, a large piece of ice is calving—breaking off—a glacier into the neighboring Southern Ocean.

Explore the **SOUNDS** of a Calving Glacier
https://goo.gl/dduzGW

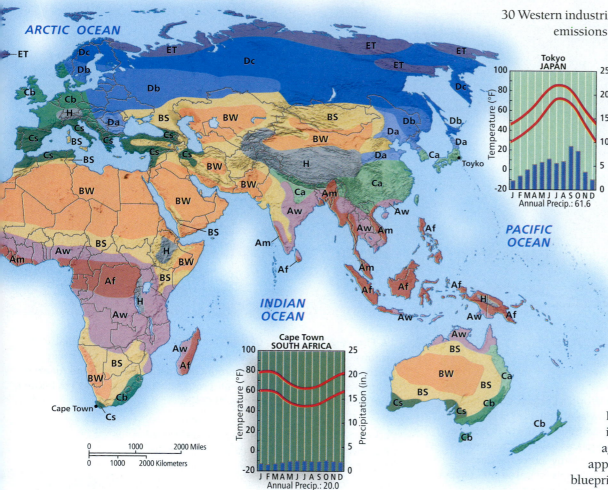

ARCTIC OCEAN

PACIFIC OCEAN

INDIAN OCEAN

Tokyo JAPAN
Annual Precip.: 61.6

Cape Town SOUTH AFRICA
Annual Precip.: 20.0

30 Western industrialized countries agreed to cut back their emissions to 1990 levels by 2012. The resulting **Kyoto Protocol** failed to live up to expectations, primarily because the United States, the world's largest polluter at that time, did not join. Additionally, the protocol did not place emission limits on booming developing economies like China. This had a major impact, as China overtook the United States as the world's largest GHG emitter in 2008 (Figure 2.19).

The Kyoto Protocol was extended to the end of 2015, due to slow negotiations on its successor agreement. In 2014, the UN asked all countries to submit their own strategies to limit future GHGs and address the challenges of climate change. These national plans, called "Nationally Determined Contributions" (NDCs), became the basis for negotiations at the IPCC Conference of Parties held in Paris, France, in December 2015. At that meeting, a new international greenhouse gas reduction agreement—the **Paris Agreement**—was approved and now serves as the world's blueprint for addressing the challenges of climate change. Unlike the Kyoto Protocol, the Paris Agreement calls for countries to submit new NDCs every five years, thus creating a lasting framework for international progress.

International efforts to reduce atmospheric emissions began in 1992, shortly after the IPCC's first report, when 167 countries meeting in Rio de Janeiro, Brazil, signed the Rio Convention to voluntarily limit their GHG emissions. However, because none of the Rio signatories reached its emission reduction targets, a more formal international agreement came from a 1997 meeting in Kyoto, Japan, where

▼ **Figure 2.18 Sea-Level Rise** One consequence of climate change will be a rise in the world's sea level due to a combination of polar ice cap melting and the thermal expansion of warmer ocean water. Rising sea levels are already affecting coastal cities like Miami Beach, Florida, shown here. At the current rate of climate warming, sea levels are forecast to rise about 4 feet (1.4 meters) by 2100. This rise will cause considerable flooding of low-lying coastal areas throughout the world.

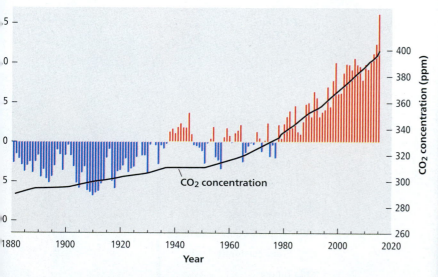

CO₂ concentration

▲ **Figure 2.17 Global Increase of CO₂ and Temperature** This graph shows the strong relationship between the recent increase in atmospheric CO₂ and the increase in the average annual temperature for the world. 2017 set the record as the hottest year in the United States.

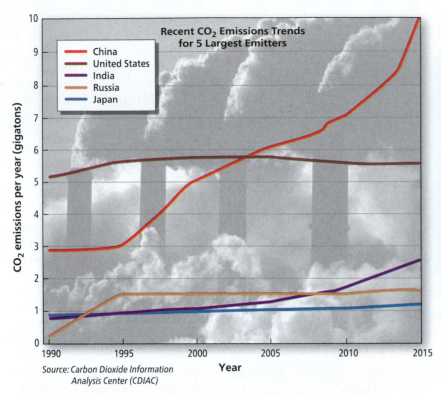

Source: Carbon Dioxide Information Analysis Center (CDIAC)

▲ **Figure 2.19** **Emission Trends for the World's Largest CO$_2$ Emitters** China's yearly CO$_2$ emissions continue to grow as hundreds of new coal-fired power plants come online. In contrast, emissions in the United States have stabilized recently because more power plants have switched from coal to natural gas as the price of that cleaner energy source has become increasingly competitive. The flat line of Russia's emissions after the collapse of the Soviet Union is somewhat of a mystery and may be a reporting problem. **Q: What are the similarities and differences between the CO$_2$ emission reduction plans of China, the United States, and India?**

The United States signed the Paris Agreement, but its greatest threat now is the 2017 U.S. withdrawal by the Trump administration. Every other UN member has signed the agreement, and its ultimate success or failure will have massive consequences on the future of humanity. Despite the U.S. withdrawal, many U.S. cities and states have pledged to follow the parameters of the agreement, and American philanthropists continue to support efforts to address climate change in developing countries.

Key components of the Paris Agreement are:

- It is an inclusive, international agreement, signed by 195 countries, covering the economic spectrum from most- to least-developed economies.

- Signatories are committed to reducing their emissions as presented in their 2015 NDCs. Additionally, each country must assess and revise their NDC every five years with the goal of further reducing GHG emissions.

- Countries commit to the goal of zero net emissions as soon as possible. This strategy combines emission reductions with carbon offsets, such as planting more trees to store carbon. An important part of this commitment is flexibility in achieving zero net emissions so that developing countries, such as India and Brazil, can move at their own pace toward this goal (Figure 2.20). One concern is the issue of transparency—how to monitor efforts and hold countries accountable for their commitments.

- Developed countries will contribute to a fund of $100 billion by 2020 to assist poorer countries in mitigating and adapting to

▲ **Figure 2.20** **Brazil's Carbon Footprint** An offshore oil platform is towed out to sea past Rio de Janeiro's iconic Sugarloaf Mountain. A combination of growing oil production, deforestation in the Amazon rainforest, and growing economic prosperity has caused Brazil to become the world's 11th largest emitter of GHGs.

climate change. Low-lying island nations endangered by sea-level rise are possible candidates for this aid (See Chapter 14's *Humanitarian Geography: Reimagining South Pacific Populations*).

PACIFIC OCEAN

ATLANTIC OCEAN

- Tropical rain forest
- Tropical seasonal forest
- Tropical savanna
- Desert and grassland
- Mediterranean shrubs and woodland
- Temperate deciduous forest
- Evergreen forest
- Tundra
- Ice

2.3 Define insolation and reradiation, and describe how these interact to warm Earth.

2.4 List the similarities and differences between maritime and continental climates. What causes the differences?

2.5 Explain how topography affects weather and climate.

2.6 What drives human-induced climate change? How are people addressing these challenges politically?

KEY TERMS insolation, greenhouse effect, continental climate, maritime climate, jet stream, monsoon wind, environmental lapse rate, orographic effect, adiabatic lapse rate, rain shadow, climate, climograph, climate change, greenhouse gases (GHGs), anthropogenic, Kyoto Protocol, Paris Agreement

▼ **Figure 2.21 Bioregions of the World** Although global vegetation has been greatly modified by clearing land for agriculture and settlements, as well as by cutting forests for lumber and paper pulp, there are still recognizable patterns to the world's bioregions, from tropical forests to arctic tundra. An important point is that each bioregion has its own unique array of ecosystems; also important is that these natural resources are used, abused, and conserved differently by humans.

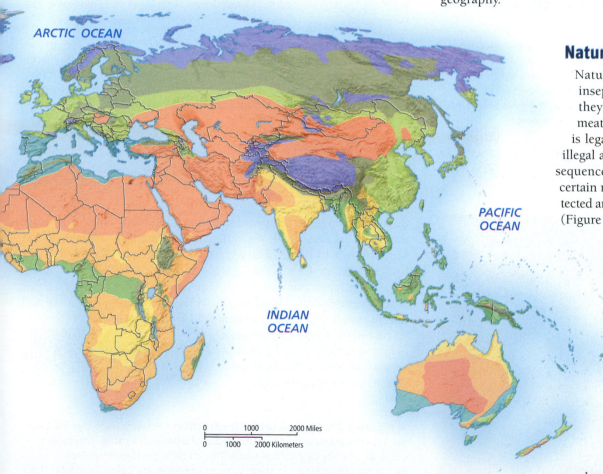

ARCTIC OCEAN

PACIFIC OCEAN

INDIAN OCEAN

0   1000   2000 Miles
0   1000   2000 Kilometers

Bioregions and Biodiversity: The Globalization of Nature

One aspect of Earth's uniqueness is the rich diversity of plants and animals covering its continents and oceans. This **biodiversity** can be thought of as the green glue that binds together geology, climate, hydrology, and life (Figures 2.21). Like climate regions, the world's biological resources can be broadly categorized into **bioregions**—areas defined by natural characteristics such as similar plant and animal life (see the bioregion photo essay, Figures 2.22 to 2.25).

Humans obviously are very much a part of this interaction. Not only are we evolutionary products of the African tropical savanna—a specific bioregion—but also our long human prehistory includes the domestication of plants and animals that led to modern agriculture and later our global food systems.

This human activity, however, combined with later urbanization and industrialization, has taken an immense toll on nature to the point where every bioregion carries strong markers of human impacts. Cultivated fields have replaced grasslands and woodlands; forests have been logged for wood products; wildlife has been hunted or subjected to habitat destruction; and far-flung environments are impacted by pollution and garbage (see *Globalization in Our Lives: Our Plastic Bag World*). As a result, our natural world has become a distinctly human, globalized one, dominated by various **ecosystems** that are completely new to Earth. Three pressing issues—the world economy, climate change, and the extinction crisis—make the study of biodiversity and bioregions an important part of world regional geography.

Nature and the World Economy

Natural plant and animal products are an inseparable part of the global economy, be they foodstuffs, wood products, or animal meat and fur. While most of this world trade is legal and appropriately regulated, much is illegal and has detrimental environmental consequences. Examples are the illegal logging of certain rainforest trees and the poaching of protected animal species, such as elephants for ivory (Figure 2.26).

Climate Change and Nature

The world map of natural bioregions (see Figure 2.21) shows a close relationship with the map of global climate regions (see Figure 2.15) because temperature and precipitation—the two major components of climate—are also the two most important influences on flora and fauna. Today, however, global climate change is causing possibly irreversible changes in the world's bioregions because plants and animals cannot adapt to

(a)

(a)

(b)

(b)

▲ **Figure 2.22 Tropical Rainforest and Savanna** (a) The tropical rainforest consists of a rich ecosystem of plants and animals adapted to differing levels of sunlight through multiple layers of vegetation. This ecosystem is adapted to heavy rainfall throughout the year. This tropi815cal rainforest is in Costa Rica. (b) A giraffe wanders through the Kenyan savanna. The tropical savanna ecosystem is less dense than the true tropical rainforest because it has adapted to less rain falling in only part of the year. As a result, the vegetation is often sparse, with widely spaced trees in a grassland.

Explore the **SOUNDS** of Rainforest Ambience
https://t2m.io/5gLBoG6f

▲ **Figure 2.24 Desert and Steppe** (a) Poleward of the tropics, in both the Northern and Southern Hemispheres, are true desert areas of sparse rainfall. In Figure 2.15, these are the areas of the BW climate found in Asia, Africa, North America, South America, and Australia, shown here. (b) Lush grasslands characterize the steppe bioregion, as depicted here in Mongolia. Steppes are found poleward of the true deserts where rainfall is commonly between 10 and 15 inches (254–381 mm) and more seasonal variability in temperatures.

Explore the **SIGHTS** of of the Australian Outback
https://goo.gl/QFEVLs

the rapid changes in temperature and precipitation. There is another important linkage as well: Vegetation and undisturbed soils take up and store carbon, which is released when plants die or soils are plowed. Long-lived forest trees, for example, are good storehouses of carbon, making responsible forest management an important part of atmospheric emission reduction plans. Therefore, the widespread practice of cutting and burning tropical rainforests to clear land for farming or cattle pastures is a major contributor of atmospheric CO_2 and a controversial component of a developing country's economic strategy.

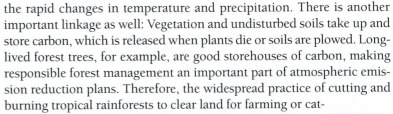
Explore the **TASTES** of the Amazon Rainforest
https://goo.gl/wWm3JR

◀ **Figure 2.23 Tropical Forest Destruction** Humans have cleared away much of the tropical rainforest to create grasslands for cattle, plantation crops, and family subsistence farms. This oil palm plantation utilizes destroyed rainforest on Borneo, where rainforests are being lost faster than any other place in the world.

Our Plastic Bag World

Looking for something to celebrate? How about International Plastic Bag–Free Day—every July 3, people around the world shop without plastic grocery bags, pick up trash from beaches and roadsides, even commit to banning plastic bags in their communities.

One ban-the-bag organization estimates that one million plastic bags are used each minute. Besides being a pollution problem, manufacturing new bags requires huge amounts of oil. In the United States alone, 12 million barrels of oil are used each year to make plastic bags.

True, some plastic bags can be recycled, but most are not, creating a global problem. Millions of plastic bags litter the countryside in China, where it's called "white pollution." Kenyans and South Africans refer to discarded white bags wafting about on the winds as their national flower. Floating plastic garbage has accumulated in oceans, creating the Great Pacific Garbage Patch (Figure 2.2.1). Even if these wayward bags were collected in landfills, they would take several hundred years to decompose.

Banning bags, or charging a fee, seems to reduce usage. Some 20 U.S. states and over 200 cities have plastic bag ordinances. China claims to have an outright ban on plastic bags, but enforcement is lax. Paris may be the biggest city with a full ban, and flood-prone Bangladesh may have the most compelling reason for a ban: Engineers say that loose plastic bags clog drains, worsening flooding.

1. **Does your community restrict or discourage the use of plastic shopping bags? If so, what do people use instead?**

2. **Does your community recycle plastics? Where do the plastics eventually end up?**

▲ **Figure 2.2.1 Pacific Garbage Patch** Massive amounts of floating plastic have become trapped by the ocean currents, with the largest patch of floating plastic located in the North Pacific Gyre (an area where currents circulate around). This Great Pacific Garbage Patch covers an area double the size of Texas.

The Current Extinction Crisis

Finally, as global climates change, environmental stresses will hasten the extinction of plants and animals. There have been five major extinction events over Earth's 4.5-billion-year history, all natural events that dramatically affected Earth's biological evolution. When climates changed naturally (such as at the end of the ice age 20,000 year ago), plants and animals migrated to find more favorable conditions. Today, however, humans are causing a sixth extinction event with more rapid, human-caused climate change; highways, cities, and farmland now form barriers to plant and animal migration, further aggravating habitat loss and extinction rates. Biologists estimate that habitat destruction and other changes are destroying several dozen species every day, resulting in an extinction rate that could see as much as 50 percent of Earth's species gone by 2050—reducing Earth's genetic resources, with potentially disastrous results.

(a)

(b)

▲ **Figure 2.25 Evergreen and Deciduous Forests** (a) Evergreen trees usually have needles rather than leaves and keep their foliage throughout the year, which is why they are called "evergreen." Most evergreens are cone-bearing, consist of softer interior wood, and thus are classified as "softwoods." Pictured here is an evergreen forest in Siberian Russia. (b) In contrast, deciduous trees drop their leaves during the winter season, often after providing a colorful display (like this forest in Vermont) that results from the tree's slowing physiology as it prepares to hibernate for the winter.

▲ **Figure 2.26 Elephant Tusks Being Burned** These ivory tusks were confiscated from poachers in Kenya, who illegally kill elephants solely for their tusks, to be sold in markets primarily in East Asia. The global ivory trade is highly regulated and mostly banned, but illegal flows of ivory threaten wild elephant populations with extinction. Governments destroy stocks of poached tusks to dissuade further trade in this natural commodity.

REVIEW

2.7 Use a map to locate and describe the bioregions found in the world's tropical climates.

2.8 What are some natural products you use regularly, and where did they come from?

2.9 Threatened species serve as symbols of climate change and other human impacts on our environment. Which animal symbolizes our changing biodiversity for you, and why?

KEY TERMS biodiversity, bioregions, ecosystems

Water: A Scarce World Resource

Water is central to all life, yet it is unevenly distributed around the world—plentiful in some areas, while distressingly scarce in others. As a result, around 1 billion people lack access to safe and reliable water sources. Water issues are not due simply to the distribution of diverse global climates that produce wet or dry conditions; they are also caused by a range of complex socioeconomic factors at all scales, local to global.

At first glance, Earth is indeed the water planet, with more than 70 percent of its surface area covered by oceans. But 97 percent of the total global water budget is saltwater, with only 3 percent freshwater. Of that small amount of freshwater, almost 70 percent is locked up in polar ice caps and mountain glaciers. Additionally, groundwater accounts for almost 30 percent of the world's freshwater. This leaves less than 1 percent of the world's water in more accessible surface rivers and lakes, with 20 percent of that water in Russia's massive Lake Baikal. Water moves within these river and lake systems, termed **watersheds**, with additional challenges encountered when these cross borders between countries.

Another way to conceptualize this limited amount of freshwater is to think of the total global water supply as 100 liters, or 26 gallons. Of that amount, only 3 liters (0.8 gallon) is freshwater; and of that small supply, a mere 0.003 liter, or only about half a teaspoon, is readily available to humans.

Water planners use the concept of **water stress** and scarcity to map where water problems exist and to predict where future problems will occur (Figure 2.27). These water stress data are generated by calculating the amount of freshwater available in relation to current and future population needs. Northern Africa stands out as the region of highest water stress; hydrologists predict that three-quarters of Africa's population will experience water shortages by 2025. Other problem areas are China, India, much of Southwest Asia, and even several countries in Europe (see *Exploring Global Connections: Acquiring Water Rights Abroad*). Although climate change will increase rainfall in some parts of the world as it shifts global rain patterns, some watersheds will benefit at the expense of others. Additionally, higher temperatures will further

▶ **Figure 2.27 Global Water Stress** This map shows where water planners are forecasting serious water problems in 2025. Although drought and highly variable rainfall regimes are major causes of water stress, socioeconomic factors can also limit supplies and access to water. That said, many water stress areas, such as western North America, the Mediterranean, the Sahel of Africa, and parts of South Asia, are linked to reduced or unreliable rainfall resulting from climate change.

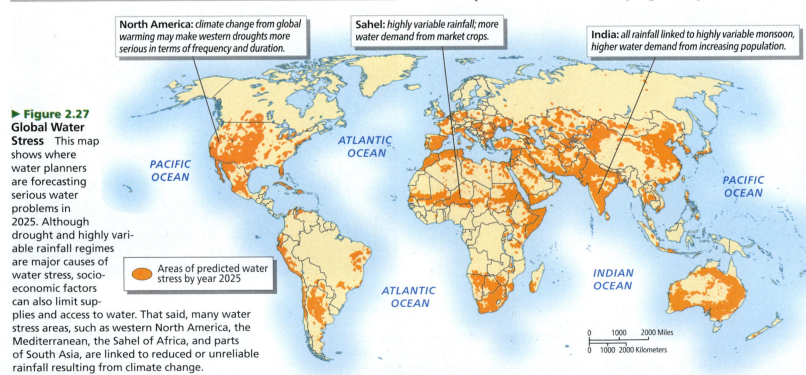

North America: *climate change from global warming may make western droughts more serious in terms of frequency and duration.*

Sahel: *highly variable rainfall; more water demand from market crops.*

India: *all rainfall linked to highly variable monsoon, higher water demand from increasing population.*

Areas of predicted water stress by year 2025

PACIFIC OCEAN

ATLANTIC OCEAN

ATLANTIC OCEAN

INDIAN OCEAN

PACIFIC OCEAN

0   1000   2000 Miles
0   1000   2000 Kilometers

Acquiring Water Rights Abroad

In a globalized world, water scarcity has led some countries with limited freshwater to outsource some of their water needs. In the past, this usually meant importing food products from water- and land-rich places, such as grain from the U.S. Midwest or Egypt's Nile Valley. Now water scarcity is leading to companies in arid regions buying up land abroad to produce products using the water from these foreign countries.

This process can create tension, especially when the lands being bought up are themselves experiencing water shortages. A recent controversy demonstrating our globalized water needs arose when Saudi Arabia's largest dairy company, Almarai (Figure 2.3.1), purchased farmland in Arizona and California.

Arid-Land Agriculture Saudi Arabia is one of the hottest and driest countries on Earth, with no year-round rivers or freshwater lakes in the entire country. Freshwater either comes from the desalinization of seawater, an energy-intensive process that is impractical for agriculture, or from "fossil" sources, meaning groundwater that was trapped in ancient times when the climate was wetter. Until recently, Saudi Arabia used this "fossil" water to grow crops like wheat and alfalfa, but the wells are running dry, leading the Saudi government to ban domestic production of these crops.

Arizona and southern California are also arid, but this region contains stretches of the Colorado River and its tributaries, allowing for irrigation-fed agriculture in the desert (Figure 2.3.2). Arizona also has groundwater that can be used for farming.

Water-Intensive Farming A dairy cow eats about 26 pounds of alfalfa hay per day. Growing alfalfa requires massive amounts of water, especially during the summer when desert temperatures are highest. Almarai plans to export this hay, essentially exporting U.S. water transformed into fodder, to feed over 170,000 cows back home to produce milk and yogurt for the Saudi and broader Southwest Asian market. Some irrigation wells will pump over 1.5 billion

▲ **Figure 2.3.1 Saudi Arabian Dairy**   Calves are fed at the world's largest dairy complex in Saudi Arabia. Much of the fodder needed for this massive operation must be imported, including hay imports from the Southwestern United States.

gallons (5.6 billion liters) of water into the alfalfa fields each year.

These land purchases have alarmed locals in Arizona and California, while also highlighting the relatively lax water regulations in parts of the American Southwest. Climate change is reducing the amount of water entering the Colorado River watershed, while at the same time local populations are increasing, attracted by the sunny, warm weather. This is straining a century-old water rights system that is ill-adapted to the globalized economy, as many local residents believe the water rights belong to Americans rather than people living halfway around the world.

Despite the controversy, wealthy dry places such as Saudi Arabia are likely to continue to secure their water needs abroad. Saudi Arabia and its neighbors are purchasing land not only in the United States, but also in such places as Australia, Sudan, Ukraine, and Brazil. As water becomes even scarcer, the global race to acquire water rights is likely to heat up further.

1. **List crops besides hay that can be outsourced from water-scarce places to overseas production. Does outsourcing of farming work for all countries facing future water scarcity, or only certain places? Why?**

2. **How is your own water use outsourced? What food products that you consume are produced in other countries?**

▲ **Figure 2.3.2 Farming in the Arizona Desert**   Agriculture is big business in the deserts of Arizona and Southern California. The arid conditions force farmers to rely mainly on water from the Colorado River and from groundwater sources that are quickly depleting.

GOOGLE EARTH
Virtual Tour Video
https://goo.gl/YP5bUx

aggravate global water problems because of higher evaporation rates and increased water usage.

Water Sanitation

Where clean water is not available, people use polluted water for their daily needs, resulting in a high rate of sickness and even death. More specifically, the UN reports that over half of the world's hospital beds are occupied by people suffering from illnesses linked to contaminated water. Further, more people die each year from polluted water than are killed in all forms of violence, including wars. This toll from polluted water is particularly high for infants and children, who have not yet developed resistance to or tolerance for contaminated water. The United Nations Children's Fund (UNICEF) reports that nearly 4000 children die each day from unsafe water and lack of basic sanitation facilities.

Water Access

The scarcity of freshwater in many locations creates adversity for the people living there, especially in less developed parts of the world. Women and children, for example, often bear the daily burden of providing water for family use, and this can mean walking long distances to pumps and wells and then waiting in long lines to draw water. Given the amount of human labor involved in providing water for crops, it is not surprising that some studies have shown that in certain areas, people expend as many calories of energy irrigating their crops as they gain from the food itself.

Ironically, some international efforts to increase people's access to clean water have in fact aggravated access problems instead. Historically, domestic water supplies have been public resources, organized and regulated—either informally by common consent or more formally as public utilities—resulting in free or low-cost water. In recent decades, however, the World Bank and the International Monetary Fund (IMF) have promoted privatizing water systems as a condition for providing loans and economic aid to developing countries. The agencies' goals have been laudable, trying to ensure that water is clean and healthful. However, the means have been controversial because the international engineering firms that typically upgrade rudimentary water systems by installing modern water treatment and delivery technology have increased the costs of water delivery to recoup their investment. Although locals may now have access to cleaner and more reliable water, in many cases the price is higher than they can afford, forcing them to either do without or go to other, unreliable and polluted sources.

In Cochabamba, Bolivia, for example, the privatization of the water system 20 years ago resulted in a 35 percent average increase in water costs. The people responded with demonstrations that tragically became violent. Eventually, the water system was returned to public control, but 40 percent of the city's population is still without a reliable water source today.

REVIEW

2.10 How much water is there on Earth, and how much is available for human usage? Answer using the analogy that Earth's water budget is just 100 liters.

2.11 Describe three major issues that cause water stress.

2.12 Where in the world are areas of the most severe water stress?

KEY TERMS watershed, water stress

Energy: The Essential Resource

The world runs on energy. While sunlight provides the natural world with its driving force, providing energy for the modern human world is much more complicated because of the uneven distribution of energy resources, the complex technologies required for extraction and, not least, the economic, environmental, and geopolitical dynamics involved in finding, exploiting, and transporting these essential resources (Figure 2.28).

Nonrenewable and Renewable Energy

Energy resources are commonly categorized as either nonrenewable or renewable. **Nonrenewable energy** sources do not regenerate within a short period of time, which is the case with the world's oil, coal, natural gas, and uranium/nuclear resources—resources that took millennia to create. In contrast, **renewable energy** depends on natural processes that are constantly renewed—namely, water (hydropower), wind, and solar energy. Currently, 90 percent of the world is powered by nonrenewable energy. While renewable energy powers the remaining 10 percent, this energy sector is increasing rapidly because of global concerns about reducing GHGs and the fast drop in the price of renewable technologies. Oil, coal, and natural gas—the fossil fuels—are considered "dirty" because of their carbon content, whereas renewable energy is generally "clean," with no GHG emissions resulting from its usage. Among the dirty fuels, coal adds the most harmful emissions, with natural gas emitting 60 percent less CO_2 than coal per unit of energy produced. Fossil fuels are also major sources of other air pollutants, causing acid rain and urban smog.

At a global level, oil and coal are the major fuels, currently making up 33 and 30 percent of all fuel usage respectively; natural gas trails at 24 percent, with nuclear far behind at 4 percent. Recent trends show an increase in natural gas at coal's expense because of environmental concerns and the fast dropping price of natural gas driven by the hydraulic fracturing boom.

Nonrenewable Energy Reserves, Production, and Consumption

As discussed at the start of this chapter, plate tectonics and other complex geologic forces have shaped ancient landscapes over millions of

▼ **Figure 2.28 Liquefied Natural Gas Tanker** Natural gas must be supercooled and compressed into a liquid for sea transport. Desire for cleaner burning fuels has led to a boom in the global trade of natural gas, which until recently was a commodity primarily sold only to locations that could be connected by pipelines.

years. As a result, fossil fuels are not evenly distributed around the world, but instead are clustered into specific geologic formations in specific locations, resulting in a complicated international pattern of supply and demand. Table 2.1 shows the varied world geography of energy reserves, production, and consumption—where the resources lie, who is mining and producing them, and which countries are primary consumers of coal, oil, and natural gas. More details on this global energy geography appear in each of the regional chapters of this text.

Because of the high degree of technological difficulty involved in extracting fossil fuels, the energy industry uses the concept of **proven reserves** of oil, coal, and gas to refer to deposits that can be extracted and distributed under current economic and technological conditions. An important component of this definition is "current," since economic, regulatory, and technological conditions change rapidly; as they do, this expands or contracts the amount of a resource deemed feasible to extract. If, for example, the price of oil on the global market is high, then oil reserves with relatively high drilling costs (such as in the Arctic) can be produced at a profit. Conversely, if the market price of oil is low, then only easily accessible reserves are economically feasible to collect.

Not only do the market prices for fossil fuels change over time, but also drilling and mining technologies become more efficient, changing the amount of proven reserves. A good example is the recent expansion of **hydraulic fracturing** or **fracking**, as it is called, a mining technique that releases oil and natural gas out of shale rock (Figure 2.29). Currently, fracking in North America has increased global oil and gas supplies to the point where the United States has overtaken Saudi Arabia and Russia as the world's biggest oil producer.

Additionally, nuclear power has not expanded in developed countries due to concerns over safety and high costs, but some developing countries, led by China, are expanding nuclear power as a means to generate electricity without the resulting air pollution caused by fossil fuels.

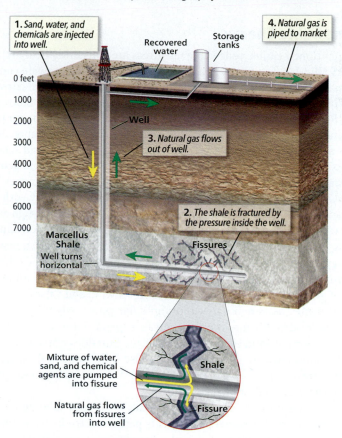

▲ **Figure 2.29 Hydraulic Fracturing**   This image illustrates the different components of hydraulic fracturing (fracking) for oil and gas. What it does not show are the environmental issues associated with fracking, which include the large amounts of freshwater required, problems in disposing of wastewater, and local noise and nuisance issues. Fracking has allowed for the development of massive natural gas deposits in shale rock, especially in the United States.

TABLE 2.1	2015 Geography of Fossil Fuels		Explore these data in MapMaster 2.0 https://goo.gl/Vck8nN			
Proven Reserves	World Share	Production	World Share	Consumption	World Share	
Oil		**Oil**		**Oil**		
Venezuela	17.7%	Saudi Arabia	13%	United States	19.7%	
Saudi Arabia	15.7%	United States	13%	China	12.9%	
Canada	10.1%	Russia	12.4%	India	4.5%	
Iran	9.3%	China	4.9%	Japan	4.4%	
Iraq	8.4%	Iraq	4.5%	Brazil	3.2%	
Coal		**Coal**		**Coal**		
United States	26.6%	China	47.7%	China	50%	
Russia	17.6%	United States	11.9%	India	10.6%	
China	12.8%	Australia	7.2%	United States	10.3%	
Australia	8.6%	Indonesia	6.3%	Japan	3.1%	
India	6.8%	India	7.4%	South Africa	2.2%	
Natural Gas		**Natural Gas**		**Natural Gas**		
Iran	18.2%	United States	22%	United States	22.8%	
Russia	17.3%	Russia	16.1%	Russia	11.2%	
Qatar	13.1%	Iran	5.4%	China	5.7%	
Turkmenistan	9.4%	Qatar	5.1%	Iran	5.5%	
United States	5.6%	Canada	4.6%	Japan	3.3%	
Log in to Mastering Geography & access MapMaster to explore these data!				**Nuclear Energy**		
1) If a country produces large quantities of the world's supply of oil, coal, or natural gas but is not among the world's largest consumers of that energy source, then it may be a major energy exporter. Which countries fall into this category?				United States	32.6%	
				France	17%	
2) What fuel is used to generate most of your community's power, and where does it come from?				South Korea	6.6%	
				China	4.4%	
				Canada	4%	

Source: BP Statistics, 2015 (June 2016)

Renewable Energy

As mentioned earlier, renewable resources are constantly being replenished. Wind and solar power are prime examples, but renewable energy sources also include hydroelectricity, geothermal power, tidal currents, and biofuels, which use the carbon in plants as their power source.

Renewable energy provides only 10 percent of the world's power but is expanding rapidly due to concerns over climate change and pollution. Some countries make considerable use of renewables in their power grid. Iceland, for example, is blessed with bountiful supplies of both water and geothermal resources and generates all its electricity from renewable sources.

Of the large industrial economies, Germany leads the way with over 20 percent of its power coming from renewables, primarily through extensive wind and solar power stations. China, the world's largest consumer of energy, reportedly drives a quarter of its massive economy with renewable energy, producing more than any other country (Figure 2.30). Unlike Germany, most of China's renewable energy comes from hydropower, although the country is rapidly expanding its wind and solar facilities.

Despite the benefits of wind and solar power, significant issues must be resolved before their usage can expand. Both wind and solar are intermittent sources of power, since the Sun does not always shine or the wind blow. To compensate for these lulls, large-scale wind and solar facilities need backups, such as fossil fuels, to generate power when the renewable sources cannot.

The intermittent nature of large-scale wind and solar generation also requires that national power grids perform a tricky balancing act between energy supply and demand. Power surges generated by sunny and windy periods must somehow be assimilated into a power grid built long ago under the assumption of steady and continual inputs of power from gas- or coal-fired power plants. Only when grid-level storage, such as through massive batteries, becomes possible will these resources be able to take over the majority of electricity production.

Renewable power can be more costly to develop and implement than fossil fuels because renewables lack the same degree of economic subsidies and tax incentives that are enjoyed by the oil, coal, gas, and nuclear power industries. At a global level, it is estimated that traditional fossil fuels receive six times the financial support through governmental incentives

▲ **Figure 2.31 Small Solar in Africa** Many villages in Africa are moving to small-scale solar systems to generate power that was formerly produced by gas or diesel generators. Not only is solar less expensive than fossil fuels, but it's also more adaptable to specific needs, such as the cell phone charging by these Malawian farmers.

that renewables do. When new technologies serve the public interest, they usually receive considerable economic support from national governments, and while this has happened in several notable cases (China and Germany), these subsidies must become more widespread before renewable energy can compete on a level playing field with fossil fuels (see *Working Toward Sustainability: Countries Aiming for Carbon Neutrality*).

Energy Futures

Global energy demand is forecast to increase 40 percent by 2030 as the large developing economies of China, India, and Brazil industrialize further. Not to be overlooked is the challenge of providing power to the 25 percent of the world's population that currently lacks access to an energy grid (Figure 2.31). In contrast, energy demand in the developed world (the United States, Europe, Japan) is expected to increase at a much slower rate, and perhaps even decrease, because of technological advancements in energy efficiency and deindustrialization.

Forecasts assume that fossil fuel reserves are adequate to meet future energy demands, but the expansion and falling costs of renewable technologies may actually decrease the demand for fossil fuels. Also important is whether unconventional drilling techniques, primarily hydraulic fracturing, will continue to expand.

Fossil fuel use will not come to an end anytime soon. While renewable energy can substitute for fossil fuels in electricity production, fossil fuels will continue to power much of world's transportation systems for many decades to come. The aviation sector may be the last one to wean itself off oil, as batteries are too heavy to power aircraft in the same way they fuel some new cars and trucks. The enormous costs of replacing fossil fuel–powered systems will also delay the transition from nonrenewable to renewable energy.

▼ **Figure 2.30 China's Three Gorges Dam** The Three Gorges Dam, along China's Chiang Jiang (Yangtze) River, is the world's largest hydroelectric power station. China now produces more renewable energy than any other country, although it also leads the world in coal consumption and greenhouse gas emissions.

Explore the SIGHTS of the Three Gorges Dam
https://goo.gl/s1ju2S

REVIEW

2.13 What is meant by fossil fuel proven reserves?

2.14 What are some of the problems associated with renewable energy?

2.15 Why will humans continue to use fossil fuels for many decades to come?

KEY TERMS nonrenewable energy, renewable energy, proven reserves, hydraulic fracturing (fracking)

Countries Aiming for Carbon Neutrality

In 2007, the small developing Central American nation of Costa Rica announced that it would become the world's first carbon-neutral country by 2021. Since then, other countries and subregions have set similar goals, including Iceland, Bhutan, Sweden, and the Canadian province of British Columbia.

Carbon neutrality doesn't mean zero GHG emissions; rather, it refers to balancing emissions through carbon offsets. A carbon offset is utilizing something else to absorb an amount of CO_2 from the atmosphere equal to that released from other practices. Planting additional trees and preserving forests, which take in carbon from the air, can be used to offset GHG emissions.

It is not surprising that Costa Rica and Bhutan were the first developing economies to aim for carbon neutrality. Costa Rica has long been a leader in progressive and environmental causes; it was the first country to abolish its military and is a world leader in protecting tropical habitats and biodiversity (see Chapter 4's *Working Toward Sustainability*). Bhutan has spurned traditional industrialization as a means to develop, focusing instead on preserving its traditional isolated way of life and privileging what it calls Gross National Happiness over economic changes that would increase carbon emissions and other pollution (Figure 2.4.1).

Developed locations like Iceland, Sweden, and British Columbia can aim for carbon neutrality because their geographies are ideal for renewable energy development. These locations, along with Costa Rica, are ideal for hydropower, as they enjoy wet climates and have significant elevation changes that create faster flowing rivers, key to efficient hydropower development. British Columbia and Costa Rica also sit along the Pacific Ring of Fire, which supplies geothermal energy. Iceland leads the world in the development of geothermal power, using it to generate electricity and to directly heat homes, melt snow, and heat swimming pools (Figure 2.4.2). Wind and solar power will also be used to reach 100 percent electricity production from renewables, a milestone already reached by Iceland.

The real challenge to reaching carbon neutrality is in the transportation sector. Oil remains the primary fuel for transportation purposes. Only in Bhutan is this less of an issue, as only the country's elite travel by car or plane. Electric vehicles with batteries charged by renewable power may someday move the transportation sector off fossil fuels, but these vehicles are expensive and have a short range, making them impractical for most transportation needs. Biofuels can be used, but biofuel production requires growing massive amounts of crops such as oil palm, sugarcane, or corn—and scaling up production of these crops would lead to deforestation, thereby releasing significant quantities of carbon into the atmosphere and threatening biodiversity.

As long as fossil fuels remain necessary for transportation, reaching carbon neutrality will be difficult. Carbon offsets, through the planting of trees and protecting additional lands from development, can make up for a small part of this fossil fuel consumption, but not enough. While the goal of carbon neutrality is laudable, places like Iceland, Sweden, British Columbia, and Costa Rica are likely decades away from a truly carbon-neutral economy.

1. **Why is the transportation sector the most reliant on fossil fuels? Are there particular challenges that developing economies will face in limiting the role of oil in their transportation sector?**

▲ **Figure 2.4.1 Bhutan, the Only Carbon-Negative Country** The tiny Himalayan Kingdom of Bhutan has chosen to forgo traditional development in favor of Gross National Happiness, which includes maintaining the traditional way of life. Bhutan's forests absorb more GHGs each year than the country produces.

▲ **Figure 2.4.2 Iceland's Blue Lagoon** Iceland's top tourist destination, the Blue Lagoon, is located adjacent to a geothermal power plant (in the background). Clean power from geothermal and hydroelectric sources make Iceland one of the cleanest developed countries in terms of its energy production.

2. **What fuels are used to generate electricity where you live? Do you live in a place where the geography is favorable to renewable electricity production?**

GOOGLE EARTH Virtual Tour Video
https://goo.gl/jMSsjk

Physical Geography and the Environment 2

REVIEW, REFLECT, & APPLY

Summary

- Earth is the water planet because of the large expanses of ocean covering the globe. These oceans, a critical part of Earth's physical geography, affect other physical phenomena as well as many aspects of human life.

- The arrangement of tectonic plates on Earth is responsible for diverse global landscapes, and the motion of these plates also causes earthquakes and volcanic hazards that threaten the safety of billions of people.

- Earth's climate is changing because of the addition of greenhouse gases (GHGs) from human activities. As a result of both historic and current emissions, global temperatures have warmed several degrees and, despite international efforts, will likely warm several more degrees by the century's end.

- Plants and animals throughout the world face an extinction crisis because of habitat destruction due to human activities. Tropical forests are a focus of these problems because of their biodiversity and capacity to both store and emit large amounts of carbon dioxide.

- Water is an increasingly scarce resource in parts of the world and will cause serious water stress problems in Sub-Saharan Africa, Southwest Asia, and western North America.

- Fossil fuels dominate the world's energy picture, with renewable energy currently providing only a fraction of the world's energy needs. While energy demand has leveled off (and even decreased) in developed countries, it continues to rise in China, India, and other developing economies.

Review Questions

1. Think about the different kinds of housing people build in different climates. Then, after looking at the map of geologic hazards, discuss whether or not certain climate regions of the world are more or less susceptible to earthquake damage.

2. Generally speaking, bioregions correspond closely to climate regions, but in some parts of the world, that's not the case. Where are those places, and what explains these exceptions?

3. Are water stress issues found only in dry climates? Give examples of where they are and where they are not, and defend your answers.

4. Describe some of the environmental and economic benefits to a country that is increasing renewable energy consumption. In what climate regions is the potential for renewable energy useage greatest?

5. Several rich countries such as Japan and Saudi Arabia have been buying up irrigated land along the Colorado River in North America's dry Southwest. Why?

Image Analysis

1. This image was taken from a plane flying over a clearcut forest, consisting primarily of Douglas firs, in British Columbia. How can you differentiate between older and newer clearcuts?

2. What other geographic patterns do you see? Are they close to or far away from settlements and major roads? What about the relationship between clearcuts and watersheds?

▶ **Figure IA2** Aerial View of Clear Cut Forest

Representatives of 195 countries participating in the 2015 United Nations Climate Change Conference voted to adopt the Paris Agreement to mitigate climate change. However, the agreement has been criticized for not going far enough to reduce greenhouse gas emissions, and for not being realistic about the financial support needed to reach this goal. Does the Paris Agreement go far enough to limit climate change?

The Paris Agreement goes far enough to mitigate climate change.

- Unlike the Kyoto Protocol, it involves many more countries, including the world's largest atmospheric polluter, China, and emphasizes limiting emissions from industrializing countries like India and Brazil, along with developed countries in Europe and North America.

- The Paris Agreement is flexible as to how countries achieve carbon reduction by emphasizing carbon management and sequestration.

- Global carbon management can be achieved by recognizing that rich countries can help poor countries through carbon offsets.

- The Paris Agreement will be revised and updated every 5 years, with reduction goals being increased regularly.

The Paris Agreement does not go far enough to avert disaster.

- Are you kidding me? This is simply wishful thinking about good intentions and lacks any sort of formal legal status.

- The Agreement is too flexible, and it has no standardized method for evaluating whether or not countries are achieving their carbon management goals.

- There is no mechanism other than "naming and shaming" countries that don't achieve their emission reduction goals.

- The Agreement doesn't even take effect until 2020, and by that time, the world will have emitted so much more atmospheric pollution that irrevocable climate change and catastrophic sea-level rise are inevitable.

▲ **Figure D2 Sea Level Rise** Sea-level rise from global climate change will be a serious problem in many of the world's coastal areas by 2030.

GeoSpatial Data Analysis

Global Oil Exports and Imports Oil remains the single largest global energy source, with exporting countries gaining great wealth from the sale of this commodity, while importers rely on access to this essential resource. This activity looks at the global pattern of oil exports and imports to identify the flows of this energy resource.

> **CIA World Factbook's Guide to Country Comparisons**
> https://goo.gl/VHHzA1

Open MapMaster 2.0 in the Mastering Geography Study Area. Now go to the *CIA World Factbook* at https://www.cia.gov/library/publications/the-world-factbook/ and select "Guide to Country Comparisons." Scroll down and click on "Energy" and select the page for crude oil exports. Download, import, and prepare this dataset to answer the following questions:

1. Which countries and regions account for the vast majority of global oil exports?

2. Are there any surprises for you among the major global oil exporters? Which countries did you expect to dominate this category?

Now go back to the *CIA World Factbook* and import and prepare the dataset for crude oil imports to create another map.

3. Which countries and regions account for the vast majority of global oil imports?

4. Are there any major oil importers that you did not expect?

5. Are there countries that play a major role both in oil exports and imports? Why do these make both lists?

Key Terms

adiabatic lapse rate (p. 51)
anthropogenic (p. 52)
biodiversity (p. 55)
bioregions (p. 55)
climate (p. 51)
climate change (p. 51)
climograph (p. 51)
continental climate (p. 49)
ecosystem (p. 55)
environmental lapse rate (p. 51)

fault (p. 46)
geothermal (p. 47)
greenhouse effect (p. 48)
greenhouse gases (GHGs) (p. 52)
hydraulic fracturing (fracking) (p. 61)
insolation (p. 47)
jet stream (p. 49)
Kyoto Protocol (p. 53)
maritime climates (p. 49)
monsoon winds (p. 50)

nonrenewable energy (p. 60)
orographic effect (p. 51)
Paris Agreement (p. 53)
plate tectonics (p. 44)
proven reserves (p. 61)
rain shadow (p. 51)
renewable energy (p. 60)
rift valley (p. 46)
subduction zone (p. 46)
watershed (p. 58)
water stress (p. 58)

Mastering Geography

Looking for additional review and test prep materials? Visit the Study Area in Mastering Geography to enhance your geographic literacy, spatial reasoning skills, and understanding of this chapter's content by accessing a variety of resources, including MapMaster interactive maps, videos, flashcards, web links, self-study quizzes, and an eText version of *Globalization and Diversity*.

> **Scan to read about Geographer at Work**
> M Jackson and her work studying glaciers in Iceland.
> https://goo.gl/Mfkhc3

North America

Physical Geography and Environmental Issues

Stretching from Texas to the Yukon, the North American region is home to an enormously varied natural setting and to an environment that has been extensively modified by human settlement and economic development.

Population and Settlement

Settlement patterns in North American cities reflect the diverse needs of an affluent, highly mobile population. Sprawling suburbs are designed around automobile travel and mass consumption, while many traditional city centers struggle to redefine their role within the decentralized metropolis.

Cultural Coherence and Diversity

Cultural pluralism remains strong in North America. More than 50 million immigrants live in the region, more than double the total in 1990. The tremendous growth in Hispanic and Asian immigrants since 1970 has fundamentally reshaped the region's cultural geography.

Geopolitical Framework

Cultural pluralism continues to shape political geographies in the region. Immigration policy remains hotly contested in the United States, and Canadians confront persistent regional and native peoples' rights issues.

Economic and Social Development

North America's economy has enjoyed moderate growth since 2010 following a harsh recession. Still, persisting poverty and social issues related to gender equity, aging, drug abuse, and health care challenge the region today.

▶ Sunset Arch is one of the many scenic highlights in southern Utah's Grand Staircase-Escalante National Monument.

Grand Staircase-Escalante National Monument

NORTH AMERICA

Southern Utah's dramatic red-rock landscapes have been the setting for a growing controversy over western land management. In both Canada and the United States, federally managed public lands have played key roles in shaping the region's landscape. Nowhere is this more obvious than in the American West, where national parks, monuments, forests, and Bureau of Land Management acreage dominate many states. But that could be changing. The Trump administration has spearheaded an effort to cut the size of several western national monuments. In an unprecedented move to open existing monument lands to more private energy development and motorized vehicle use, in 2017, the president cut the size of Utah's Bears Ears National Monument by 85 percent and Grand Staircase-Escalante National Monument by 46 percent.

Many residents applauded the move as an opportunity to promote more economic development and local control of these lands, but environmental activists, wilderness advocates, and tribal nation representatives protested the decision through a variety of legal moves designed to retain the monuments at their earlier size. Whatever the final outcome of the legal deliberations, the policy shift was a reminder that these spectacular western settings are valued in multiple, often contested ways.

In addition to traditions of public lands management, globalization has also transformed the North American landscape in powerful, enduring ways. Large foreign-born populations are found in many North American settings. Tourism brings in millions of additional foreign visitors and billions of dollars, which are spent everywhere from Las Vegas to Disney World. North Americans engage globally in more subtle ways: eating ethnic foods, enjoying the sounds of salsa and Senegalese music, and surfing the Internet from one continent to the next. Globalization is also a two-way street, and North American capital, popular culture, and power are ubiquitous. By any measure of multinational corporate investment and global trade, the region plays a role that far outweighs its population of 365 million residents.

Defining North America

North America is a culturally diverse and resource-rich region that has seen tremendous, sometimes destructive, modification of its landscape and extraordinary economic development over the past two centuries

(Figure 3.1). As a result, North America remains one of the world's wealthiest regions, with two highly urbanized, mobile populations that help drive the processes of globalization and have the highest rates of resource consumption on Earth. Indeed, the region exemplifies a **postindustrial economy** shaped by modern technology, innovative information services, and a popular culture that dominates both North America and the world beyond (Figure 3.2).

Politically, North America is home to the United States, the last remaining global superpower. In addition, North America's largest metropolitan area, New York City (20 million people), is home to the United Nations and other global political and financial institutions. North of the United States, Canada is the region's other political unit. Although slightly larger in area than the United States (3.83 million square miles [9.97 million square kilometers] versus 3.68 million square miles [9.36 million square kilometers]), Canada's population is only about 11 percent that of the United States.

The United States and Canada are commonly referred to as "North America," but that regional terminology can be confusing. As a physical feature, the North American continent commonly includes Mexico, Central America, and often the Caribbean. Culturally, however, the U.S.–Mexico border seems a better dividing line, although the large Hispanic (Latino/a) presence in the southwestern United States, as well as economic links across the border, makes even that regional division problematic. In addition, while Hawaii is a part of the United States (and included in this chapter), it is also considered a part of Oceania (and discussed in Chapter 14). Finally, Greenland (population 56,000), which often appears on the North American map, is actually an autonomous territory within the Kingdom of Denmark and is mainly known for its valuable, but diminishing, ice cap.

> North America is a culturally diverse and resource-rich region that has seen tremendous, sometimes destructive, modification of its landscape.

LEARNING OBJECTIVES

After reading this chapter you should be able to:

3.1 Describe North America's major landform and climate regions.

3.2 Identify key environmental issues facing North Americans and connect these to the region's resource base and economic development.

3.3 Analyze map data to identify and trace major migration flows in North American history.

3.4 Explain the processes that shape contemporary urban and rural settlement patterns.

3.5 List the five phases of immigration shaping North America, and describe the recent importance of Hispanic and Asian immigration.

3.6 Identify major cultural homelands (rural) and ethnic neighborhoods (urban) within North America.

3.7 Describe how the United States and Canada developed distinctive federal political systems, and identify each nation's current political challenges.

3.8 Discuss key location factors that explain why economic activities are located where they are in the region.

3.9 Summarize contemporary social issues that challenge North Americans in the 21st century.

ELEVATION IN METERS

- 4000+
- 2000–4000
- 500–1999
- 200–499
- 0–199
- Below sea level

Sea Level

ARCTIC OCEAN

PACIFIC OCEAN

ATLANTIC OCEAN

Gulf of Mexico

MEXICO

CANADA

UNITED STATES

EURASIAN PLATE

NORTH AMERICAN PLATE

PACIFIC PLATE

JUAN DE FUCA PLATE

Greenland (DEN.)

Baffin Bay

Baffin Island

Hudson Bay

Bering Strait

Bering Sea

Aleutian Islands

Anchorage

Brooks Range

Prudhoe Bay

Alaska Range

Gulf of Alaska

YUKON

Whitehorse

NORTHWEST TERRITORIES

NUNAVUT

Great Bear Lake

Great Slave Lake

Mackenzie R.

Peace R.

Athabasca R.

BRITISH COLUMBIA

ALBERTA

Edmonton

Calgary

SASKATCHEWAN

Saskatoon

Regina

MANITOBA

Lake Winnipeg

Winnipeg

ONTARIO

QUÉBEC

NEWFOUNDLAND AND LABRADOR

Newfoundland

Prince Edward Island

P.E.I.

NEW BRUNSWICK

NOVA SCOTIA

Saint John

Halifax

Québec

Montréal

Ottawa

St. Lawrence R.

L. Superior

L. Huron

L. Michigan

L. Erie

L. Ontario

Toronto

Hamilton

Buffalo

Cleveland

Detroit

Milwaukee

Chicago

Minneapolis-St. Paul

Missouri R.

Vancouver Island

Vancouver

Puget Sound

Seattle

Portland

Boise

Coast Mountains

Coast Ranges

Cascade Range

Sierra Nevada

Central Valley

ROCKY MOUNTAINS

GREAT PLAINS

Teton Range

Great Basin

Great Salt Lake

Salt Lake City

Sacramento

San Francisco

San Jose

Las Vegas

Los Angeles

Riverside

San Diego

Salton Sea

Phoenix

Tucson

Grand Canyon

Colorado Plateau

Colorado R.

Denver

Albuquerque

El Paso

Rio Grande

Austin

San Antonio

Houston

Dallas-Fort Worth

Oklahoma City

Kansas City

Platte R.

Arkansas R.

Red R.

St. Louis

Indianapolis

Cincinnati

Columbus

Pittsburgh

Louisville

Ohio R.

Mississippi R.

Ozark Plateau

Ouachita Mts.

Memphis

Nashville

Atlanta

New Orleans

Jacksonville

Orlando

Tampa-St. Petersburg

Miami

Florida Keys

Charlotte

Raleigh

Richmond

Virginia Beach

Washington, DC

Baltimore

Philadelphia

New York City

Bridgeport

Providence

Boston

Adirondack Mts.

APPALACHIAN HIGHLANDS

Piedmont

Chesapeake Bay

Arctic Circle

NORTH AMERICA
Political & Physical Map

- ⊛● Metropolitan areas more than 20 million
- ⊛● Metropolitan areas 10–20 million
- ⊛• Metropolitan areas 5–9.9 million
- ⊛• Metropolitan areas 1–4.9 million
- ⊛○ Selected smaller metropolitan areas
- ⌐ Plate boundaries

Anchorage

Honolulu

HAWAII

Kauai

Oahu

Maui

Hawaii

PACIFIC OCEAN

▲ **Figure 3.1 North America** With 365 million people and extensive economic development, North America plays a pivotal role in globalization. The region also contains one of the world's most highly urbanized and culturally diverse populations, and is also one of the largest consumers of natural resources on the planet.

▲ **Figure 3.2 Googleplex Campus, Silicon Valley** Google, a multinational technology company, is headquartered in California, and its sprawling Googleplex Campus is a major Silicon Valley employer.

Physical Geography and Environmental Issues: A Vulnerable Land of Plenty

North America's physical and human geographies are enormously diverse. In the past decade, the region has also witnessed a dizzying array of natural disasters and environmental hazards that suggest close connections between the region's complex physical setting, its human population, and the impacts of global climate change. Record-setting droughts struck California in multiple years between 2013 and 2016, and much of the Southwest experienced sustained below-average precipitation as well (Figure 3.3). In 2017, Hurricanes Harvey and Irma also devastated much of the Texas Gulf Coast and southwest Florida, respectively, with flooding rains and destructive winds. Harvey was especially remarkable in that the slow-moving storm dumped an amazing 19 trillion gallons of water on the region (including a dousing of more than 50 inches at localities near Houston) (Figure 3.4). Many climate experts agreed that the hurricane's sustained intensity may have been linked to changing climate conditions. In both 2017 and 2018, wide-ranging wildfires scorched vast portions of California, including destructive blazes in the densely settled Napa-Sonoma and Ventura-Santa Barbara county areas. These events remind us how the costs and impacts of both "natural" and "human" environmental disasters are intertwined with a region's broader cultural, social, and economic characteristics.

A Diverse Physical Setting

The North American landscape is dominated by interior lowlands bordered by more mountainous topography in the western part of the region (see Figure 3.1). In the eastern United States, extensive coastal plains stretch from southern New York to Texas and include a sizable portion of the lower Mississippi Valley. The Atlantic coastline is complex, made up of drowned river valleys, bays, swamps, and low barrier islands. The nearby Piedmont region, a transition zone, consists of rolling hills and low mountains that are older and less easily eroded than the lowlands. West and north of the Piedmont are the Appalachian Highlands, an internally complex zone reaching altitudes of 3000–6000 feet (900–1830 meters). To the southwest, Missouri's Ozark Mountains and the Ouachita Plateau of northern Arkansas resemble portions of the southern Appalachians.

Much of the North American interior is a vast lowland, extending east–west from the Ohio River valley to the Great Plains and north–south from west-central Canada to the lower Mississippi near the Gulf of Mexico (Figure 3.5). Glacial forces north of the Ohio and Missouri rivers once actively carved and reshaped this lowland, including the environmentally complex Great Lakes Basin.

In the West, mountain-building (including large earthquakes and volcanic eruptions), alpine glaciation, and erosion produce a regional topography quite unlike that of eastern North America. The Rocky Mountains reach more than 10,000 feet (3050 meters) in height and stretch from Alaska's Brooks Range to northern New Mexico's Sangre de Cristo Mountains (Figure 3.6). West of the Rockies, the Colorado Plateau is characterized by colorful sedimentary rock eroded into spectacular buttes and mesas. Nevada's sparsely settled basin and range country features north–south mountain ranges alternat-

Mobile Field Trip: Yosemite
https://goo.gl/jE56g2

ing with structural basins with no outlet to the sea. North America's western border is marked by the mountainous, rain-drenched coasts of southeast Alaska and British Columbia; the Coast Ranges of Washington, Oregon, and California; the lowlands of Puget Sound (Washington), Willamette Valley (Oregon), and Central Valley (California); and the complex uplifts of the Cascade Range and Sierra Nevada.

▼ **Figure 3.3 Shasta Lake, California**
California's recent droughts sent Shasta Lake, a reservoir that feeds into the Sacramento River, to record low levels.

Explore the **SIGHTS** of **Shasta Lake**
http://goo.gl/rGulSl

▲ Figure 3.4 Hurricane Harvey Inundates Houston Many of Houston's major highways were underwater when the city was drenched by Hurricane Harvey in 2017. Hurricane Florence similarly devastated much of North Carolina in 2018.

▲ Figure 3.6 Rocky Mountains This kayaker enjoys the spectacular glacial scenery from Two Medicine Lake, located in Montana's Glacier National Park.

Patterns of Climate and Vegetation

North America's climates and vegetation are diverse, mainly due to the region's size, latitudinal range, and varied terrain (Figure 3.7). As the climographs for Dallas, Texas, and Miami, Florida, suggest, much of the area south of the Great Lakes and east of the Rockies is characterized by a long growing season, 30–60 inches (75–150 cm) of precipitation annually, and a deciduous broadleaf forest (later cut down and replaced by crops). From the Great Lakes north, the coniferous evergreen forest, or **boreal forest**, dominates the continental interior.

▼ Figure 3.5 Satellite View of the Lower Mississippi Valley
This view of the Mississippi Delta shows how sediments from the interior lowlands have accumulated in the area. Note the sediment plume (right) extending into the Gulf of Mexico.

Mobile Field Trip: Mississippi River Delta https://goo.gl/B0kvSe

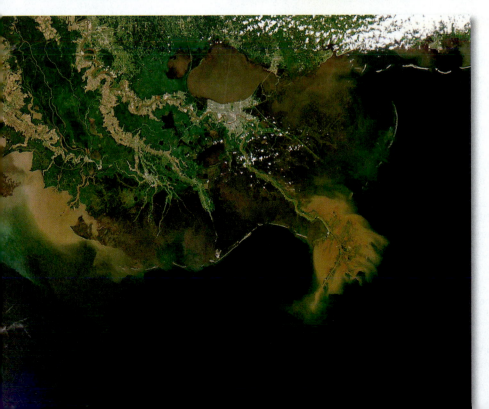

Near Hudson Bay and across harsher northern tracts, trees give way to **tundra**, a mixture of low shrubs, grasses, and flowering herbs that briefly flourish in the short growing season of the high latitudes. Drier climates from west Texas to Alberta feature large seasonal ranges in temperature and unpredictable precipitation, averaging 10–30 inches (25–75 cm) annually. The soils of much of this region are fertile and originally supported vast **prairies**, dominated by tall grasslands in the East and by short grasses and scrub vegetation in the West.

Western North American climates and vegetation are greatly complicated by the mountain ranges. The Rockies and the intermontane interior exhibit typical midlatitude seasonal variations, but topography greatly modifies climate and vegetation patterns. Interior arid settings lie in the dry rain shadow of the Cascade Range and Sierra Nevada. Farther west, marine west coast climates dominate from San Francisco to the Aleutian Islands, while a dry-summer Mediterranean climate occurs across central and southern California.

The Costs of Human Modification

North Americans have modified their physical setting in many ways. Processes of globalization and accelerated urban and economic growth have transformed the region's landforms, soils, vegetation, and climate. Indeed, problems such as acid rain, nuclear waste storage, groundwater depletion, and toxic chemical spills are manifestations of a way of life unimaginable only a century ago (Figure 3.8).

Transforming Soils and Vegetation While indigenous peoples have occupied North America for thousands of years, the arrival of Europeans dramatically affected the region's flora and fauna as countless new species were introduced, including wheat, cattle, and horses. As the number of settlers increased, forest cover was removed from millions of acres. Grasslands were plowed under and replaced with nonnative grain and forage crops. Unsustainable cropping and ranching practices increased soil erosion, and many areas of the Great Plains and South suffered lasting damage.

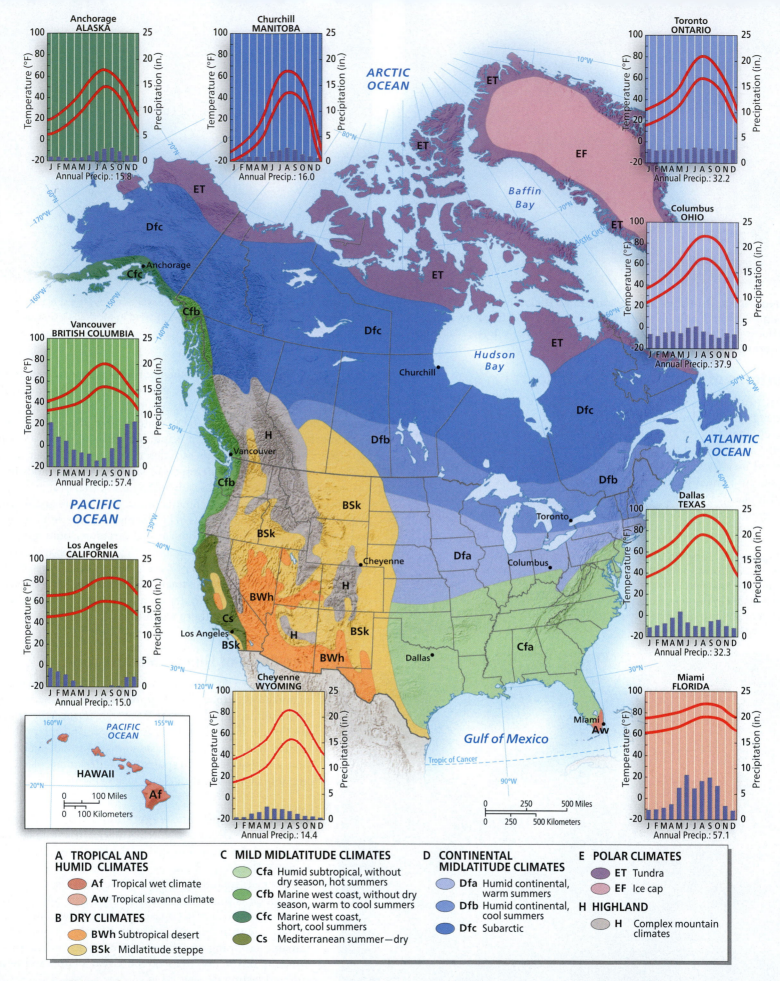

▲ Figure 3.7 Climate of North America The region's climates include everything from tropical savanna (Aw) to tundra (ET) environments. Most of the region's best farmland and densest settlements lie in the mild (C) or continental (D) midlatitude climate zones.

A TROPICAL AND HUMID CLIMATES

- **Af** Tropical wet climate
- **Aw** Tropical savanna climate

B DRY CLIMATES

- **BWh** Subtropical desert
- **BSk** Midlatitude steppe

C MILD MIDLATITUDE CLIMATES

- **Cfa** Humid subtropical, without dry season, hot summers
- **Cfb** Marine west coast, without dry season, warm to cool summers
- **Cfc** Marine west coast, short, cool summers
- **Cs** Mediterranean summer—dry

D CONTINENTAL MIDLATITUDE CLIMATES

- **Dfa** Humid continental, warm summers
- **Dfb** Humid continental, cool summers
- **Dfc** Subarctic

E POLAR CLIMATES

- **ET** Tundra
- **EF** Ice cap

H HIGHLAND

- **H** Complex mountain climates

Melting Sea Ice. *Recently, because of global climate change, the Arctic Ocean has seen dramatically reduced levels of sea ice in summer.*

Athabasca Oil Sands. *Gigantic deposits of oil sands in Alberta have generated controversial plans for long-distance pipelines, passing near some environmentally sensitive areas.*

Acid Precipitation. *Acid precipitation has devastated hundreds of sensitive lake environments across eastern Canada.*

Threatened Coastlines. *Increased shoreline development and recent coastal storms such as Hurricanes Harvey, Irma, and Sandy have combined to threaten many low-lying portions of the East Coast as well as the Gulf of Mexico.*

California Wildfires and Floods. *Since 2016, destructive floods and damaging wildfires have struck many portions of both northern and southern California.*

Legend:
- Areas affected by acid precipitation
- Desert
- Areas of groundwater depletion
- Vulnerable to sea-level rise
- Coastal pollution
- Endangered and polluted rivers
- Proposed pipeline routes
- Major hazardous waste sites
- Selected mining areas

▲ **Figure 3.8 Selected Environmental Issues in North America** Acid rain damage is widespread in regions downwind from industrial source areas. Elsewhere, widespread water pollution, cities with high levels of air pollution, and zones of accelerating groundwater depletion pose health dangers and economic costs to residents of the region. Since 1970, however, both Americans and Canadians have become increasingly responsive to the dangers posed by these environmental challenges.

Managing Water   North Americans consume huge amounts of water. While conservation efforts and technology have slightly reduced per capita rates of water use over the past 25 years, city dwellers still use an average of more than 175 gallons daily. Metropolitan areas such as New York City struggle with outdated municipal water supply systems. Flint, Michigan's municipal water disaster, first identified in 2014, continues to unfold today (see *Humanitarian Geography: Unmasking the Tragedy in Flint, Michigan*).

Farther west, beneath the Great Plains, the waters of the Ogallala Aquifer are being depleted; center-pivot irrigation systems have steadily lowered water tables by as much as 100 feet (30 meters) in the past 50 years. The fluctuating flows of the Colorado River are also chronically frustrating to residents in the rapidly growing southwestern United States, and

Mobile Field Trip: Moving Water Across California
https://goo.gl/rxtERk

HUMANITARIAN GEOGRAPHY

Unmasking the Tragedy in Flint, Michigan

▲ Figure 3.1.1 Rick Sadler

Rick Sadler, a geographer and GIS expert at Michigan State University, helped to uncover the roots of the water crisis that has recently plagued Flint, Michigan (Figure 3.1.1). In an effort to save money, the state directed the cash-strapped city to switch its water supply to the local Flint River in 2014. Tragically, the naturally more turbid water was not treated properly. Flint's new water supply corroded lead from delivery and home service lines, resulting in lead poisoning for thousands of residents, including many children. But initially, no one knew the source of the problem. A local pediatrician suspected that the elevated lead levels in Flint children were linked to Flint River water, and called on Sadler to examine the data.

Remapping the Data Sadler quickly realized that the data that state officials used were based on Flint ZIP codes, larger units that did not match city boundaries or represent the residents drinking the tainted water. "One-third of the addresses with a

Flint ZIP code weren't in the city," Sadler recalls, and used a different water source. Remapping the data by the actual street addresses of children with elevated lead levels revealed a strong correlation with use of Flint River water. These older homes often house lead pipes that proved vulnerable to corrosion and leaching. Sadler's evidence confirmed the cause of the poisoning.

A Flint native, Sadler has been interested in mapping "as long as I can remember" and enrolled in GIS courses as an undergraduate. After returning to Flint, he became known as "the map guy" while working with different community groups. Sadler says his heart is in issues relevant to the city: "The more I learned about issues that drove Flint's decline…the more I felt compelled to not

▲ Figure 3.1.2 **Flint Residents Receive Bottled Water**
Volunteers from Full Gospel Churches in Michigan deliver bottled water to residents of Flint.

just understand them, but to uncover some of the spatial patterns—to use the tools that I had learned."

Based on Sadler's findings and intense public outcry surrounding the crisis, Flint returned to a safer water supply. Yet the crisis continues: less than 25 percent of the 20,000 affected lead-rich pipes had been upgraded by 2017, and contaminated water and its associated effects remain a daily challenge for thousands (Figure 3.1.2). Meanwhile, the spatial tools and geographic analysis that Sadler employed to confirm the source of the Flint tragedy have aided legal and criminal proceedings to address liability and seek environmental justice for affected residents.

Sadler notes that GIS is applicable to many public health issues, and that a geographer's multidisciplinary approach can be invaluable: "It's like being a goalie as opposed to being a forward. It's a special position that not everyone does, but it's absolutely essential."

1. **Why do you think ZIP code zones are often used to map U.S. public health issues?**

2. **Find a map of local ZIP codes and argue why they may or may not be useful in studying environmental or social problems. What might be a better unit of analysis?**

GOOGLE EARTH
Virtual Tour Video
https://goo.gl/cbEoGt

---

climate-change scientists indicate that this arid region is particularly vulnerable to drier times ahead (see Figure 3.8). In California, many Central Valley farmers have dramatically increased their consumption of groundwater as winter snowpacks in the nearby Sierra Nevada have waned (farmers use 80 percent of the state's groundwater).

Altering the Atmosphere North Americans modify the very air they breathe, changing local and regional climates as well as the composition of the atmosphere. For example, built-up metropolitan areas create an **urban heat island** effect, in which infrastructure associated with cities can raise nighttime temperatures some 9–22°F (5–12°C) higher than nearby rural areas. At the local level, industries, utilities, and automobiles contribute carbon monoxide, sulfur, nitrogen

oxides, hydrocarbons, and particulates to the urban atmosphere. Overall air quality in urban North America has improved since the 1970s, but a 2014 American Lung Association report estimated that about 47 percent of U.S. residents still live in places with unsafe levels of air pollution, including high levels of particulates and ozone.

North America remains plagued by **acid rain**—industrially produced sulfur dioxide and nitrogen oxides in the atmosphere that combine with precipitation to damage forests, poison lakes, and kill fish. Many atmospheric pollution producers—factories, power plants, and motor vehicles—are located in the Midwest and southern Ontario, and prevailing winds transport pollutants and deposit damaging acid rain and snow across the Ohio Valley, the Northeast, and eastern Canada (see Figure 3.8).

Growing Environmental Initiatives

Many U.S. and Canadian environmental initiatives have addressed local and regional problems. For example, the improved water quality of the Great Lakes over the past 30 years is an achievement to which both nations contributed. Tougher air-quality standards have also reduced certain types of emissions in many North American cities. Similarly, the U.S. Superfund program (begun in 1980) and Canada's Environmental Protection Act (CEPA, begun in 1988) have significantly cleaned up hundreds of toxic waste sites.

Perhaps most important, North America is increasingly supporting green industries and technologies. An example is the growing popularity of **sustainable agriculture**, which combines organic farming principles, limited use of chemicals, and integrated crop and livestock management to offer both producers and consumers environmentally friendly alternatives.

The Shifting Energy Equation

Energy consumption in the region remains extremely high (the United States is still the source of almost 20 percent of Earth's greenhouse gas emissions), but growing incentives for energy efficiency may reduce per capita consumption in the future. The technological and economic appeal of **renewable energy sources**, such as hydroelectric, solar, wind, and geothermal, are likely to transform North America's economic geography in coming years as policymakers, industrial innovators, and consumers embrace their enduring availability and potentially lower environmental costs (Figure 3.9).

New fossil fuel discoveries and drilling technologies have fundamentally shifted the continent's energy equation. The Bakken Formation in North Dakota and Montana, thanks to new oil extraction methods, may one day produce more than 15–20 billion barrels of oil, making it one of the planet's great energy

Mobile Field Trip: Oil Sands https://goo.gl/48czBB

reserves (on par with Alaska's North Slope field). Canada alone has the world's third largest proven oil reserves: about 170 billion barrels of oil can be recovered just from its rich oil sands. Even with its high rates of consumption, North America may be a net energy export region by the 2020s. However, fluctuating global energy prices, such as the dramatic drop in crude oil prices in 2015 and 2016, makes such long-term projections complex, especially if a slowdown in the energy economy discourages investment in both alternative sources (such as wind and solar) and in shale-based fossil fuels.

Many issues complicate the clean development of North America's untapped fossil fuels. Moving fossil fuels involves huge investments and risks. Plans for transporting western coal (especially from Wyoming and Montana) to the Pacific Coast (for export to Asia) have stirred protests and, despite friendlier federal regulations for U.S. coal, its domestic use for electricity production continues to decline. Controversial pipelines such as the Keystone XL project (Alberta to the Gulf of Mexico) and the Northern Gateway Pipeline (Alberta to the Pacific Coast), designed to tap into Canada's rich Athabasca oil sands, also illustrate the tensions between increasing production (and job creation) and the potentially dangerous environmental consequences of moving fossil fuels long distances (see Figure 3.8). In the United States, despite lower energy prices, a Republican administration has backed the Keystone XL and other pipeline projects, but growing protests by western Canada's indigenous peoples have hampered construction of the Northern Gateway Pipeline.

In addition, the growing use of hydraulic fracturing, or **fracking** (a drilling technology in which a mix of water, sand, and chemicals is injected underground to release natural gas) in settings from North Dakota to Pennsylvania has been challenged by critics claiming that the practice leads to polluted groundwater and hazardous environmental conditions for nearby residents.

Climate Change and North America

North America continues to experience the accelerating impacts of climate change in myriad ways, and predicted changes suggest varying regional patterns. In the United States, southern California, the Southwest, and parts of Texas are likely to be hotter and drier, while Great Lakes states and the Northeast may see average precipitation, but heavier and more damaging precipitation events. Atlantic and Gulf of Mexico coastlines will be especially vulnerable to rising sea levels and more intense storms, including hurricanes.

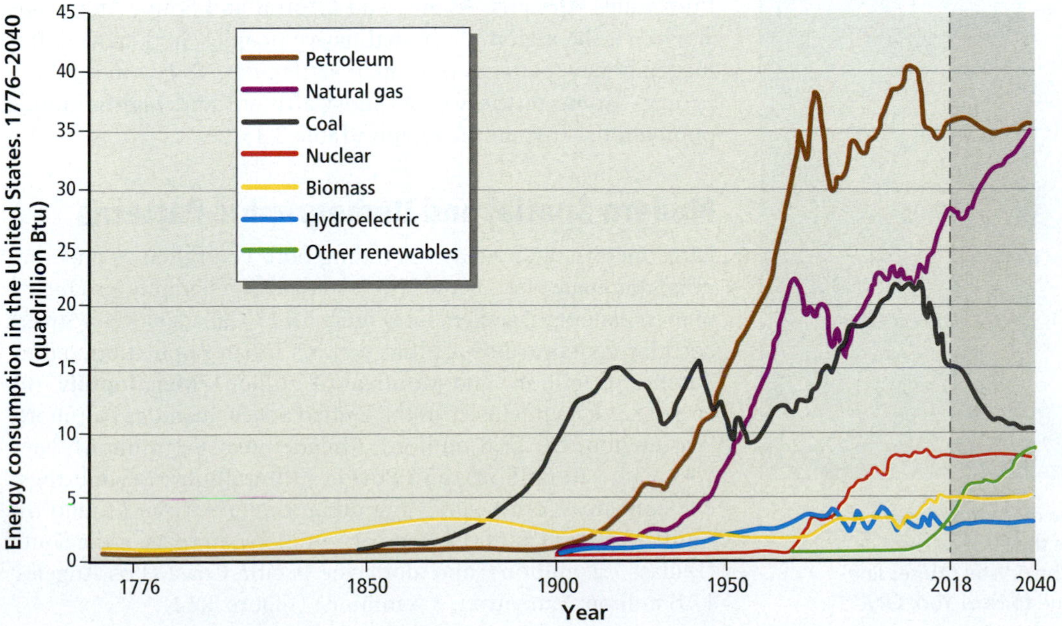

◀ **Figure 3.9 U.S. Energy Consumption, 1776–2040 (projected)** The growing popularity of fossil fuels is evident in U.S. energy consumption during the late 19th century as coal, oil, and then natural gas supplanted wood consumption. Projected trends suggest lower coal consumption and higher consumption of renewables and natural gas. Q: Approximately what percentage of U.S. energy consumption is currently accounted for by coal and renewable energy sources? What is projected for 2040?

The western mountains are particularly sensitive to climate change. Expanding bark beetle populations can survive milder winters and are infesting pine forests. Many of the region's spectacular alpine glaciers are rapidly disappearing. Earlier spring melting of reduced mountain snowpacks also impacts downstream fisheries, farms, and metropolitan areas that depend on these seasonal water resources. Seasonal wildfires are also growing in size and intensity across the region.

**Mobile Field Trip:
Forest Fires in the West**
https://goo.gl/v5RVD1

Polar Transformations   In North America's high latitudes, changes in temperatures, sea ice, permafrost conditions, and sea levels have increased coastal erosion, affected whale and polar bear populations, and made the Arctic a very different setting than it was just a decade ago. A recent government report concluded that North America's far north is warming twice as fast as the rest of the planet and that Arctic sea ice is very thin and melting at its fastest pace in the last 1500 years.

A more ice-free Arctic Ocean is also opening up potential for commercial shipping and resource development. Thanks to global climate change, since 2016, large luxury cruise ships having been crossing the Arctic, exploring a newly thawed and quite lucrative global connection between Anchorage and New York City. Following the so-called Northwest Passage around the northern perimeter of the continent, high-latitude tourists—for a mere $20,000 per person—can enjoy polar pleasures, including arctic wildlife, native villages, and iceberg-studded ocean vistas (Figure 3.10).

**Mobile Field Trip:
Climate Change
in the Arctic**
https://goo.gl/sh7QgZ

Shifting Policy Responses   Thanks to recent elections, Canada and the United States have adopted strikingly different strategies toward global climate change. With considerable controversy in energy-rich Canada, Prime Minister Justin Trudeau embraces global efforts to limit carbon emissions and promote renewable alternatives. South of the border, President Donald Trump—a persistent climate change skeptic—announced his intention to withdraw from the Paris Climate Accord and promote more fossil fuel–friendly energy policies. Many U.S. governors and big-city mayors reacted sharply to Trump's dramatic shift and urged their respective states and municipalities to move forward with more stringent guidelines on future carbon emissions. As a result, many long-term plans addressing climate change continue moving forward at state and local levels, even as the federal government turns a cold shoulder to such initiatives.

REVIEW

3.1 Describe North America's major landform regions and climates, and suggest how the region's physical setting has shaped patterns of human settlement.

3.2 Identify the key ways in which humans have transformed the North American environment since 1600.

3.3 Summarize four environmental problems that North Americans face in the early 21st century.

KEY TERMS   postindustrial economy, boreal forest, tundra, prairie, urban heat island, acid rain, sustainable agriculture, renewable energy sources, fracking

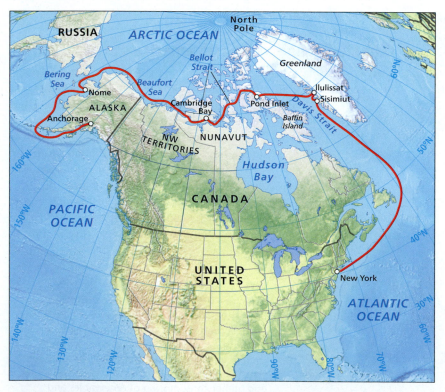
▲ Figure 3.10 The Fabled Northwest Passage   The map shows how modern cruise ships may increasingly ply arctic waters as they follow the fabled Northwest Passage through ice-free polar seas. The Crystal Cruises line now features a high-end, 32-day journey from Anchorage to New York City.

Population and Settlement: Reshaping a Continental Landscape

The North American landscape is the product of human settlement extending back at least 12,000–25,000 years. The pace of change for much of that period was modest and localized, but the last 400 years have witnessed an extraordinary transformation as Europeans, Africans, Asians, and Central and South Americans arrived in the region, disrupted native peoples, and created dramatically new patterns of human settlement. Today 365 million people—some of the world's most affluent and highly mobile populations—live in the region (Table 3.1).

Modern Spatial and Demographic Patterns

Large metropolitan areas (including both central cities and suburbs) dominate North America's population geography, producing uneven patterns of settlement (Figure 3.11). Canada's "Main Street" corridor contains most of that nation's urban population, led by Toronto (6 million) and Montreal (4 million). **Megalopolis**, the largest settlement cluster in the United States, includes Baltimore/Washington, DC (8.8 million), Philadelphia (6.1 million), New York City (20 million), and Boston (4.8 million). Beyond these two core areas, other sprawling urban centers cluster around the southern Great Lakes (Chicago, 9.5 million), in parts of the South (Dallas, 7.2 million), and along the Pacific Coast (Los Angeles, 17.8 million; Vancouver, 2.5 million) (Figure 3.12).

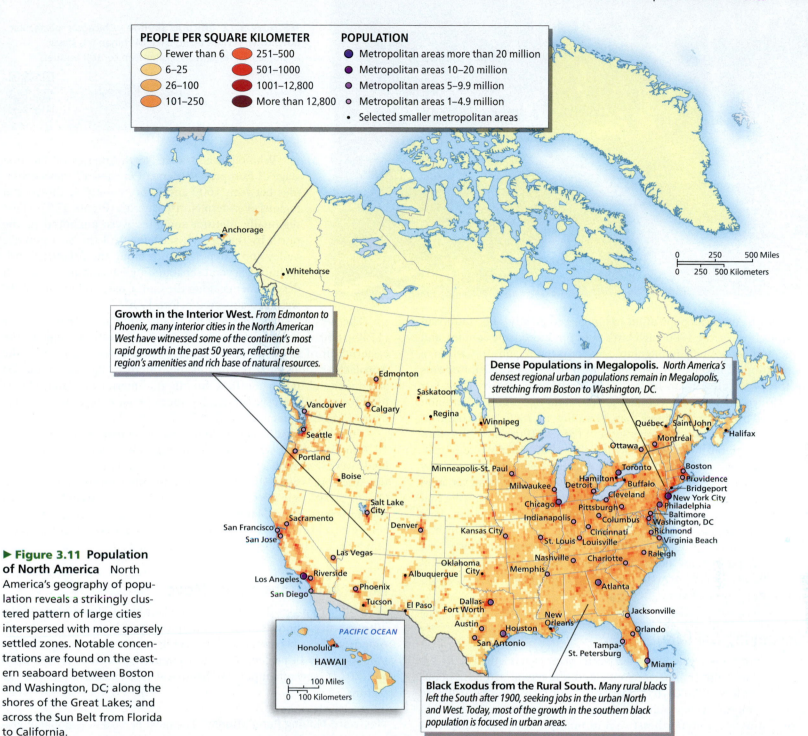

PEOPLE PER SQUARE KILOMETER

- Fewer than 6
- 6–25
- 26–100
- 101–250
- 251–500
- 501–1000
- 1001–12,800
- More than 12,800

POPULATION

- Metropolitan areas more than 20 million
- Metropolitan areas 10–20 million
- Metropolitan areas 5–9.9 million
- Metropolitan areas 1–4.9 million
- Selected smaller metropolitan areas

Growth in the Interior West. *From Edmonton to Phoenix, many interior cities in the North American West have witnessed some of the continent's most rapid growth in the past 50 years, reflecting the region's amenities and rich base of natural resources.*

Dense Populations in Megalopolis. *North America's densest regional urban populations remain in Megalopolis, stretching from Boston to Washington, DC.*

Black Exodus from the Rural South. *Many rural blacks left the South after 1900, seeking jobs in the urban North and West. Today, most of the growth in the southern black population is focused in urban areas.*

▶ **Figure 3.11 Population of North America** North America's geography of population reveals a strikingly clustered pattern of large cities interspersed with more sparsely settled zones. Notable concentrations are found on the eastern seaboard between Boston and Washington, DC; along the shores of the Great Lakes; and across the Sun Belt from Florida to California.

					Life Expectancy					
Country	Population (millions) 2018	Population Density (per square kilometer)[1]	Rate of Natural Increase (RNI)	Total Fertility Rate	Male	Female	Percent Urban	Percent <15	Percent >65	Net Migration (rate per 1000)
Canada	37.2	4	0.3	1.5	80	84	81	16	17	11
United States	328.0	36	0.3	1.8	76	81	82	19	15	3

TABLE 3.1 Population Indicators

Explore these data in MapMaster 2.0 https://goo.gl/PoSqWz

Source: Population Reference Bureau, *World Population Data Sheet*, 2018.
[1] World Bank Open Data 2018.

Log in to Mastering Geography & access MapMaster to explore these data!

1) Compare the populations and the population densities for Canada and the United States. What do these data suggest about the relative sizes of these two countries?
2) Using the demographic and urban data in the table, can you argue that there are differences in levels of development between these two nations?

▲ **Figure 3.12 Chicago Skyline** Chicago's spectacular downtown skyline along Lake Michigan is a classic example of a large North American central business district (CBD).

Explore the SIGHTS of Chicago's "Loop"
http://goo.gl/m47nZL

Who replaced Native North Americans? The first stage of European settlement created a series of colonies between 1600 and 1750, mostly in the coastal regions of eastern North America (Figure 3.13). These regionally distinct societies were anchored in the north by the French settlement of the St. Lawrence Valley and extended south along the Atlantic Coast, including several separate English colonies. Scattered developments along the Gulf Coast and in the Southwest also appeared before 1750.

North America's population has increased greatly since European colonization. Before 1900, high birth rates produced large families, and large numbers of new immigrants arrived in the region. In Canada, a population of fewer than 300,000 Native Americans and Europeans in the 1760s grew to an impressive 3.2 million a century later. For the United States, a late colonial (1770) total of around 2.5 million increased over tenfold to more than 30 million by 1860. Both countries saw even higher rates of immigration in the late 19th and 20th centuries, although birth rates gradually fell after 1900. After World War II, birth rates rose once again in both countries, resulting in the "baby boom" generation born between 1946 and 1965. Today, however, rates of natural increase in North America are below 1 percent annually, and the overall population is growing older. Still, the region attracts many immigrants. These growing numbers, along with higher birth rates among immigrant groups, recently led experts to increase long-term population projections. Indeed, UN predictions for a 2070 population of 467 million (419 million in the United States and 48 million in Canada) may prove conservative.

The second stage in the Europeanization of North America (1750–1850) featured settlement of better agricultural lands in the eastern half of the continent. Pioneers surged westward across the Appalachians following the end of the American Revolution (1783) and a series of Indian conflicts, finding the Interior Lowlands almost ideal for agricultural settlement. Southern Ontario, or Upper Canada, opened to development after 1791.

The third stage in North America's settlement accelerated after 1850 through 1910, when most of the region's remaining agricultural lands were settled by a mix of native-born and immigrant farmers. In the American West, settlers were drawn to opportunities in California, Oregon, Utah, and the Great Plains. In Canada, thousands occupied southern portions of Manitoba, Saskatchewan, and Alberta. The discovery of gold and silver led to development in areas such as Colorado, Montana, and British Columbia's Fraser Valley.

Occupying the Land

When Europeans began occupying North America more than 400 years ago, they were not settling an empty land; the region had been populated for at least 12,000–25,000 years by peoples as culturally diverse as the Europeans who conquered them. Native Americans migrated to North America from northeast Asia in multiple waves and dispersed across the region, adapting in diverse ways to its many natural environments. Cultural geographers estimate native populations in 1500 CE at 3.2 million for the continental United States and 1.2 million for Canada, Alaska, Hawaii, and Greenland.

These native peoples met many different fates, seeing their numbers reduced by more than 90 percent following European settlement. Some groups were exterminated by disease and war; others were expelled from their homelands and relocated on reservations, both in Canada and in the United States. Some also mixed with Europeans, losing parts of their cultural identity in the process. Today the majority of the region's Native Americans actually live in cities, often far removed from the land of their ancestors.

North Americans on the Move

From the legendary days of Davy Crockett and Calamity Jane to the 20th-century sojourns of John Steinbeck and Jack Kerouac, North Americans have been on the move. While recent trends suggest lower mobility (only 11 percent of the U.S. population moved between 2016 and 2017), several important historical movements dominate the picture.

Westward-Moving Populations The most persistent regional migration trend has been westward movement. By 1990, more than half of the U.S. population lived west of the Mississippi River, a dramatic shift from colonial times. Since 1990, some of the fastest-growing areas have been in the American West (including the states of Arizona and Nevada) and in the western Canadian provinces of Alberta and British Columbia. This trend was fueled by new job creation in high-technology, energy, and service industries as well as by the region's scenic, recreational, and retirement attractions.

Black Exodus from the South African Americans have generated distinctive patterns of interregional migration. Most blacks remained economically tied to the rural South after the Civil War, but by the

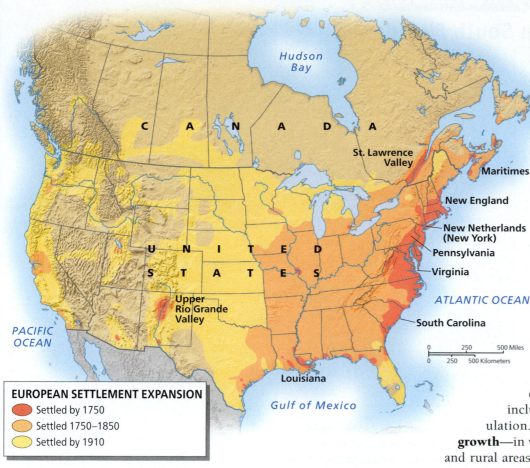

EUROPEAN SETTLEMENT EXPANSION
- Settled by 1750
- Settled 1750–1850
- Settled by 1910

◀ **Figure 3.13 European Settlement Expansion** Sizable portions of North America's East Coast and the St. Lawrence Valley were occupied by Europeans before 1750. The most remarkable surge of settlement occurred during the next century, as Europeans opened vast areas of land and dramatically disrupted Native American populations.

to most of the nation's fastest-growing counties. Factors contributing to the South's growth are its buoyant economy, growing global exports, modest living costs, adoption of air conditioning, attractive recreational opportunities, and appeal to snow-weary retirees (especially moves to Florida) (see *Working Toward Sustainability: Going Totally Solar in South Florida*).

Nonmetropolitan Growth   During the 1970s, some areas in North America beyond its large cities witnessed significant population gains, including rural settings that had previously lost population. Selectively, this pattern of **nonmetropolitan growth**—in which people leave large cities for smaller towns and rural areas—continues today. The growing retiree population in both Canada and the United States is part of this trend, but a substantial number are younger *lifestyle migrants*. Our electronically connected world allows them to work in affordable, amenity-rich smaller cities and rural settings, removed from perceived urban problems.

early 20th century, many African Americans migrated because of declining demands for labor in the agricultural South and growing industrial job opportunities in cities in the North and West. Boston, New York, Philadelphia, Detroit, Chicago, Los Angeles, and Oakland became major destinations for southern black migrants. Since 1970, however, more blacks have moved from North to South. Sun Belt jobs and federal civil rights guarantees now attract many northern urban blacks to growing southern cities. The net result is still a major change from 1900, when more than 90 percent of African Americans lived in the South; today just over half of the 47 million black Americans live in the region.

Rural-to-Urban Migration   Another continuing trend in North American migration has taken people from the country to the city. Two centuries ago only 5 percent of North Americans lived in urban areas (cities of more than 2500 people), whereas today more than 80 percent of the region's population is urban. Shifting economic opportunities account for much of the transformation: As farm mechanization reduced the demand for labor, many young people left for jobs in the city.

Sun Belt Growth   Late-20th-century moves to the American South are clearly related to other dominant trends in North American migration, yet the pattern deserves closer inspection. Particularly after 1970, states from the Carolinas to Texas grew much more rapidly than states in the Northeast and Midwest, and since 2010, the South has been home

Settlement Geographies: The Decentralized Metropolis

North American cities are characterized by **urban decentralization**, in which metropolitan areas sprawl in all directions and suburbs take on many of the characteristics of traditional downtowns. Although both Canadian and U.S. cities have experienced decentralization, the impact has been particularly profound in the United States, where inner-city problems, poor public transportation, widespread automobile ownership, and fewer regional-scale planning initiatives encourage middle-class urban residents to move beyond the central city.

Historical Evolution of U.S. Cities   Changing transportation technologies decisively shaped U.S. urban development (Figure 3.14). The pedestrian/horsecar city (pre-1888) was compact, essentially limiting growth to a 3- or 4-mile-diameter ring around downtown. The invention of the electric trolley in 1888 expanded the urbanized landscape into new "streetcar suburbs" that extended outward along streetcar lines, often for 5–10 miles from the city center. The biggest technological revolution came after 1920, with the mass production of cars. The automobile city (1920–1945) continued the

Going Totally Solar in South Florida

The first solar-powered city in the United States became a reality in early 2018 when South Florida's Babcock Ranch opened its doors to new residents (Figure 3.2.1). By move-in day, the 400-acre solar field—on land donated by the developers, but now owned by Florida Power and Light (FPL)—was also up and running (Figure 3.2.2). More than 300,000 individual panels (with an eventual capacity of 74.5 megawatts of electricity) have been installed in the solar field located on the northern edge of the 18,000-acre residential development.

Babcock Ranch was the brainchild of real estate developer and sustainable-living enthusiast Syd Kitson. The retired pro-football player purchased a large ranch east of Fort Myers in the early 2000s, donated 70,000 acres to a state-managed preserve (the Babcock Ranch Preserve), and then developed the remaining acres as an example of an eco-friendly development. With thousands of people continuing to move to Florida every year, Kitson wanted to offer a different model of suburban living that offered a balance between preservation and development. He hopes eventually to attract 50,000 people to the community.

The solar-energy arithmetic is impressive: the panels generate direct current (DC) electricity, which is then converted to alternating current (AC) power that can be used in the nearby development. On sunny days, FPL will use the electricity generated to power the community and export the remainder to nearby towns. At night, FPL can draw on other clean, natural gas–burning power

▲ **Figure 3.2.1 Babcock Ranch, Florida** Located just northeast of Fort Myers, the Babcock Ranch real estate development features an all-solar solution to its energy needs.

plants nearby. But overall, Babcock Ranch will produce more energy than it consumes.

Green Living What else can residents expect? The town has a well-defined commercial core and an already completed "green" school building accessible by walking and bike paths. Need a ride? Self-driving, electric-powered shuttle buses are already running. Other sustainable initiatives include locally grown food via a community garden project, plenty of residential options for additional rooftop solar panels to save even more money, loads of green space within the development, and proximity to the sprawling nature preserve, where alligators and native vegetation still offer some of the region's traditional environmental amenities.

Will Babcock Ranch emerge as economically viable model for future Sun Belt development? No one knows at this point, but the initiative may prove popular with a generation of environmentally sensitive baby boomers as well as younger families who see the bright side of living the solar-friendly version of the American dream.

1. Find a map of solar-energy potential in North America. Based on the map, what six states might be the best candidates for real estate developments like Babcock Ranch?

2. Can you cite any local real estate ventures that advertise "sustainable" or "eco-friendly" amenities? What might those amenities be?

▲ **Figure 3.2.2 A Solar-Powered Community** The impressive solar field at Babcock Ranch covers 400 acres of South Florida east of Fort Myers, Florida.

GOOGLE EARTH
Virtual Tour Video
https://goo.gl/aest8g

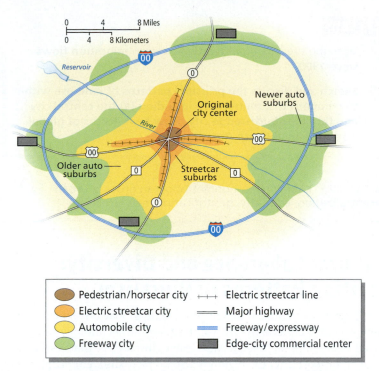

▲ **Figure 3.15 Costa Mesa, California** North America's edge-city landscape is nicely illustrated by Costa Mesa. Far from downtown Los Angeles, this sprawling complex of suburban offices and commercial activities reveals how and where many North Americans will live their lives in the 21st century.

Legend:
- Pedestrian/horsecar city
- Electric streetcar city
- Automobile city
- Freeway city
- ┼┼┼ Electric streetcar line
- ══ Major highway
- Freeway/expressway
- ▬ Edge-city commercial center

▲ **Figure 3.14 Growth of the American City** Many U.S. cities became increasingly decentralized as they moved through eras dominated by the pedestrian/horsecar, electric streetcar, automobile, and freeway. Each era left a distinctive mark on metropolitan America, including the recent growth of edge cities on the urban periphery.

expansion of middle-class suburbs. Post–World War II growth in the outer city (1945 to the present) promoted more decentralization along commuter routes as built-up areas appeared 40 to 60 miles from downtown.

Today's suburbs feature a mix of peripheral retailing (commercial strips, shopping malls, and big-box stores), industrial parks, office complexes, and entertainment facilities. This larger, peripheral node of activity, called an **edge city**, has fewer functional connections with the central city than it has with other suburban centers. Southern California's Costa Mesa office and retailing district, located south of Los Angeles, exemplifies an edge-city landscape on the expanding periphery of a North American metropolis (Figure 3.15).

The Consequences of Sprawl As suburbanization increased in the 1960s and 1970s, many inner cities, especially in the Northeast and Midwest, suffered absolute losses in population and a shrinking tax base. Unemployment rates remain above the national average. Many U.S. central cities also remain places of racial tension, the product of decades of discrimination, segregation, and poverty.

Amid these challenges, select inner-city landscapes are also experiencing a renaissance. Termed **gentrification**, the process involves higher-income newcomers displacing lower-income, central-city

residents, rehabilitating deteriorated inner-city landscapes, and constructing new shopping complexes, entertainment attractions, or downtown convention centers. Cosmopolitan, upscale residents are drawn to the more architecturally diverse housing and cultural amenities of the central city. Seattle's Pioneer Square, Baltimore's Harborplace, and Toronto's Yorkville district illustrate how such public and private investments shape the central city (Figure 3.16).

▼ **Figure 3.16 Yorkville Neighborhood, Toronto** Toronto's fashionable Yorkville neighborhood is home to gentrified housing, upscale shops, and well-manicured public spaces.

Explore the **SIGHTS** of Yorkville
https://goo.gl/KEA6WU

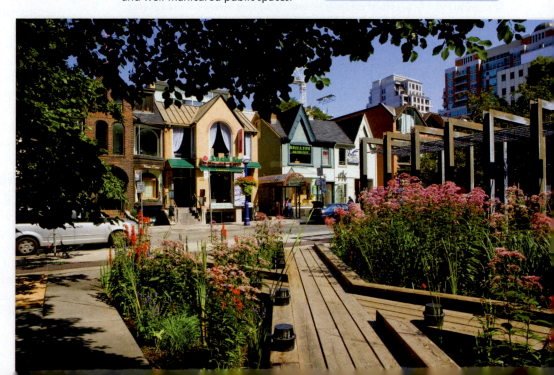

Many city planners and developers involved in such efforts also advocate **new urbanism**, an urban design movement stressing higher-density, mixed-use, pedestrian-scaled neighborhoods where residents can walk to work, school, and entertainment.

Settlement Geographies: Rural North America

The region's rural cultural landscapes trace their origins to early European settlement. These immigrants from Europe preferred dispersed settlement patterns as they created new farms. In portions of the United States settled after 1785, the federal government surveyed and sold much of the land. Organized around the rectangular pattern of the federal government's township-and-range survey system (Canada's system is similar), the surveys offered a convenient method of dividing and selling the public domain in 6-mile-square townships (Figure 3.17).

Commercial farming and technological changes further transformed the rural landscape. Railroads opened corridors of development, provided access to markets for commercial crops, and helped to establish towns. By 1900, several transcontinental lines spanned North America, radically transforming the farm economy and the pace of rural life. After 1920, however, even greater change accompanied the arrival of the automobile, farm mechanization, and better rural road networks. The need for farm labor declined with mechanization, and many smaller market centers withered as farmers drove automobiles and trucks farther and faster to larger towns. Typically, fewer but larger farms dot the modern rural scene, and many young people leave the land for urban employment.

Some rural settings show signs of growth, experiencing the effects of expanding edge cities. Others lie beyond direct metropolitan influence but attract new residents seeking amenities removed from city pressures. These trends are shaping the settlement landscape from British Columbia's Vancouver Island to Michigan's Upper Peninsula.

▼ **Figure 3.17 Minnesota Settlement Patterns**   The regular rectangles of this midwestern landscape are common features across the region. In the United States, the township-and-range survey system stamped such predictable patterns across vast portions of the North American interior. **Q: Can you identify any visual evidence of underlying land survey patterns in your local community?**

REVIEW

3.4   Describe the dominant North American migration flows since 1900.

3.5   Sketch and discuss the principal patterns of land use within the modern U.S. metropolis, including (a) the central city and (b) the suburbs/edge city. How have forces of globalization shaped North American cities?

KEY TERMS   Megalopolis, nonmetropolitan growth, urban decentralization, edge city, gentrification, new urbanism

Cultural Coherence and Diversity: Shifting Patterns of Pluralism

North America's cultural geography exerts global influence. At the same time, it is internally diverse. History and technology have produced a contemporary North American cultural force that is second to none in the world. Yet the region is also home to different peoples who retain part of their traditional cultural identities, celebrate their varied roots, and acknowledge the region's multicultural character.

The Roots of a Cultural Identity

Powerful historical forces formed a common dominant culture within North America. Although both the United States (1776) and Canada (1867) became independent from Great Britain, they remained closely tied to their Anglo roots. Key Anglo legal and social institutions solidified core values that North Americans shared with the British and, eventually, with one another. Traditional Anglo beliefs emphasized representative government, separation of church and state, liberal individualism, privacy, pragmatism, and social mobility. From those foundations, particularly within the United States, consumer culture blossomed after 1920, producing a shared set of experiences oriented around convenience, consumption, and the mass media.

But this cultural unity coexists with *pluralism*—the persistence and assertion of distinctive cultural identities. Closely related is the concept of **ethnicity**, in which people with a common background and history identify with one another, often as a minority group within a larger society. For Canada, the French colonization of Québec and the enduring power of its native peoples complicate its modern cultural geography. The greater diversity of ethnic groups within the United States produced different cultural geographies on both local and regional scales.

Peopling North America

North America became a region of immigrants. Decisively displacing Native Americans, immigrants created new cultural geographies of ethnic groups, languages, and religions. Though small in number, early migrants had considerable cultural influence. Over time, immigrant groups and their changing destinations produced a varied cultural geography across North America. Also varying among groups were the pace and degree of **cultural assimilation**, the process by which immigrants are absorbed by the larger host society.

Migration to the United States Variations in the number and source regions of migrants produced five distinctive chapters in U.S. history (Figure 3.18). In Phase 1 (prior to 1820), English and African influences dominated. Slaves, mostly from West Africa, added cultural influences in the South. Northwest Europe served as the main source region of immigrants between 1820 and 1870 (Phase 2). During this phase, Irish and Germans dominated the flow and provided more cultural variety.

As Figure 3.18 shows, immigration reached a much higher peak around 1900, when almost 1 million foreigners entered the United States *annually*. During Phase 3 (1870–1920), the majority of immigrants were southern and eastern Europeans escaping political strife and poor economies for available land and expanding industrialization in the United States. By 1910, almost 14 percent of the nation was foreign-born. Very few of these immigrants, however, settled in the job-poor U.S. South, creating a cultural divergence that still exists.

Between 1920 and 1970 (Phase 4), overall totals fell sharply, a function of more restrictive federal immigration policies (the Quota Act of 1921 and the National Origins Act of 1924), the Great Depression, and the disruption of World War II. Immigration sharply increased after 1970 (Phase 5), and now annual arrivals rival those of a century ago (see Figure 3.18). Most legal migrants since 1970 originated in Latin America or Asia. In 2000, about 60 percent of immigrants were Hispanic and only 20 percent were Asian, but by 2010, the balance shifted, with 36 percent from diverse Asian countries and only about 30 percent from Latin America. In percentage terms, black immigrant populations make up the fastest growing immigrant group (from widely varying settings) to the United States (Figure 3.19).

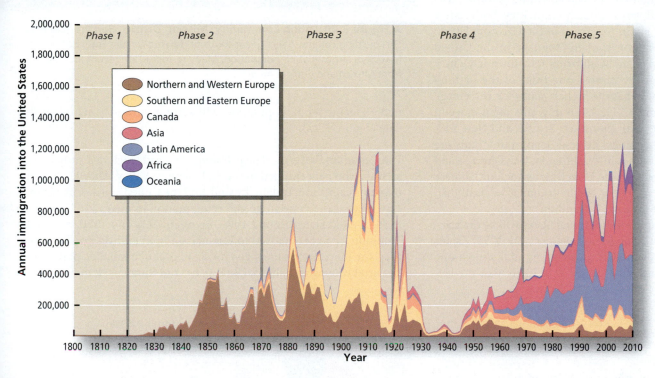

◄ **Figure 3.18 U.S. Immigration, by Year and Group** Annual immigration rates peaked around 1900, declined in the early 20th century, and then surged again, particularly since 1970. The source areas of these migrants have also shifted. Note the decreased role of Europeans versus the growing importance of Asians and Latin Americans.

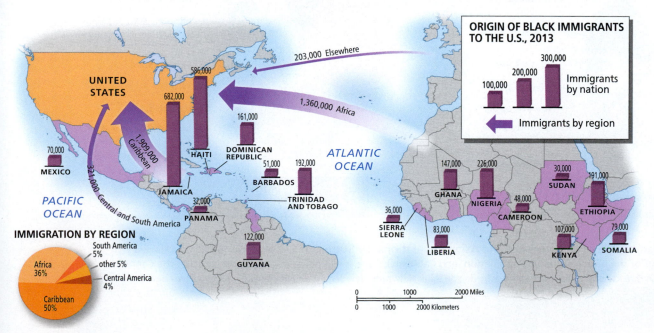

◄ **Figure 3.19 Black Immigrant Populations in the United States, 2013** The map shows the population and percentage of U.S. foreign-born blacks by birth region and birth countries that contributed at least 30,000 black immigrants.

The post-1970 surge of immigrants was due to economic and political instability abroad, a growing postwar American economy, and loosening immigration laws. Undocumented immigration, particularly from Mexico, rose after 1970, but since 2008 the pace has slowed appreciably, mostly because of fewer U.S. job opportunities. Today the United States is home to about 11–12 million undocumented immigrants.

The U.S. Hispanic population continues to grow, fundamentally reshaping North America's cultural and economic geography (Figure 3.20). In the next 25 years, most of the increase is projected to be fueled by births within the country rather than by new immigrants. Over 40 percent of the nation's 59 million Hispanics live in California or Texas, but they are increasingly moving to other states, such as Illinois, New Jersey, and Colorado (Figure 3.21).

In percentage terms, migrants from Asia constitute another fast-growing immigrant group, and various Asian ethnicities, both native and foreign-born, account for about 5 percent of the U.S. population. Chinese is the third most common spoken language in the United States (behind English and Spanish). California remains a key entry point for migrants and is home to almost one-third of the nation's Asian population, whereas Hawaii has the highest statewide percentage of Asian immigrants (see Figure 3.21). Asian migrants often move to large cities such as Los Angeles, San Francisco, Seattle, and New York City. Beyond these key gateway cities, diverse Asian immigrants are also moving to growing communities in Washington, DC, Chicago, and Houston. The largest Asian groups in the United States include Chinese (4.0 million), Filipino (3.4 million), Asian Indian (3.2 million), Vietnamese (1.7 million), and Korean (1.7 million).

The future cultural geography of the United States will be dramatically redefined by these recent immigration patterns. By 2050, Asians may total almost 10 percent of the U.S. population, and almost one in three Americans will be Hispanic. Indeed, it is likely that the U.S.

▲ **Figure 3.20 Latino Community, Mission District, San Francisco** This bustling commercial corridor south of downtown San Francisco serves that city's largest Latino population. Many residents are recent immigrants from Mexico and Central America.

non-Hispanic white population will achieve minority status by that date (Figure 3.22).

The Canadian Pattern The peopling of Canada included early French arrivals who concentrated in the St. Lawrence Valley. After 1765, many migrants came from Britain, Ireland, and the United States. Canada then experienced the same surge and reorientation in migration flows seen in the United States around 1900. Between 1900 and 1920, more than 3 million foreigners moved to Canada, an immigration rate far higher than for the United States given Canada's much smaller population. Eastern Europeans, Italians, Ukrainians, and Russians dominated these

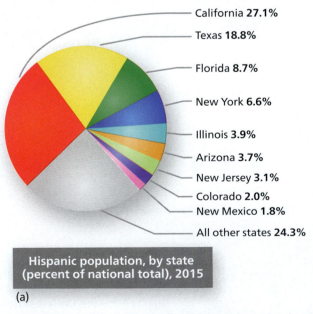

California **27.1%**
Texas **18.8%**
Florida **8.7%**
New York **6.6%**
Illinois **3.9%**
Arizona **3.7%**
New Jersey **3.1%**
Colorado **2.0%**
New Mexico **1.8%**
All other states **24.3%**

Hispanic population, by state (percent of national total), 2015

(a)

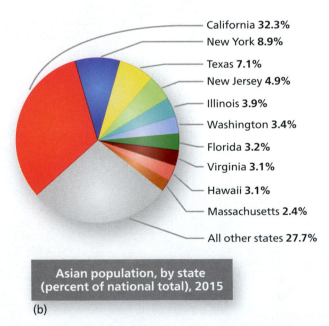

California **32.3%**
New York **8.9%**
Texas **7.1%**
New Jersey **4.9%**
Illinois **3.9%**
Washington **3.4%**
Florida **3.2%**
Virginia **3.1%**
Hawaii **3.1%**
Massachusetts **2.4%**
All other states **27.7%**

Asian population, by state (percent of national total), 2015

(b)

▲ **Figure 3.21 Distribution of U.S. Hispanic and Asian Populations, by State, 2015** California, Texas, and Florida claim more than half of the nation's Hispanic population, and California alone is still home to almost one-third of the country's Asian population. **Q: Outside of the West and South, why do you think New York, New Jersey, and Illinois are also important destinations for these immigrants?**

► **Figure 3.22 Projected U.S. Ethnic Composition to 2050** By the middle of the 21st century, almost one in three Americans will be Hispanic, and non-Hispanic whites will achieve minority status amid an increasingly diverse U.S. population.

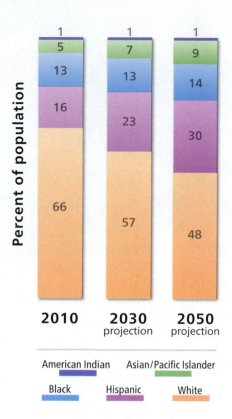

Percent of population

	2010	2030 projection	2050 projection
	1	1	1
	5	7	9
	13	13	14
	16	23	30
	66	57	48

American Indian   Asian/Pacific Islander

Black   Hispanic   White

glue that holds the homeland together. Policies adopted after 1976 strengthened the French language within the province by requiring French instruction in schools and national bilingual programming by the Canadian Broadcasting Corporation. Many Québécois feel that the greatest cultural threat comes not from Anglo-Canadians, but rather from recent immigrants to the province. Southern Europeans and Asians in Montreal, for example, show little desire to learn French, preferring instead to put their children in English-speaking private schools.

Another well-defined cultural homeland is the Hispanic Borderlands (see Figure 3.24), similar in size to French-Canadian Québec but significantly larger in total population and more diffuse in its cultural and political expression. Historical roots of the homeland are deep, extending back to the 17th century, when Spaniards opened the region to the European world. Spanish place names, earth-toned Catholic churches, and traditional Hispanic settlements dot the rolling highlands of northern New Mexico and southern Colorado. From California to Texas, other historical sites and place names also reflect this rich Hispanic legacy.

Unlike Québec, however, large 20th-century migrations from Latin America brought an entirely new wave of Hispanic settlement to the Southwest. About 59 million Hispanics live in the United States, with more than half in California, Texas, and Florida combined. Indeed, Hispanics outnumber non-Hispanic whites in California. New York City, Chicago, and Cuban South Florida serve as key points of Hispanic influence beyond this cultural homeland.

African Americans also retain a cultural homeland in the South, but it has become less important because of outmigration (see Figure 3.24). Dozens of rural counties in the Black Belt still have large African-American majorities, and the South remains home to many black folk traditions, including black spirituals and the blues, musical forms now popular far beyond their rural origins (Figure 3.25). Outside the South, African Americans have also created large, vibrant communities primarily in cities of the Northeast, Midwest, and West.

later movements. Today, about 60 percent of Canada's recent immigrants are Asians, and its 21 percent foreign-born population is among the highest in the developed world. In Toronto, the city's 45 percent foreign-born population reveals a slight bias toward European backgrounds, although the city's Chinese Canadian population remains a vibrant part of the community. On Canada's west coast, Vancouver (45 percent foreign-born) has been a key destination for Asian immigrants, particularly people from China, India, and the Philippines (Figure 3.23).

Culture and Place in North America

Cultural and ethnic identity is often strongly tied to place. North America's cultural diversity is expressed geographically in two ways. First, people with similar backgrounds congregate near one another and derive meaning from the territories they occupy together. Second, these distinctive cultures leave their mark—artifacts, habits, language, and values—on the everyday landscape. Boston's Italian North End simply looks and feels different from nearby Chinatown, and rural French Québec is a world away from a Hopi village in Arizona.

Cultural Homelands French-Canadian Québec is an excellent example of a **cultural homeland**: It is a culturally distinctive settlement in a well-defined geographic area, and its ethnicity has survived over time, stamping the landscape with an enduring personality (Figure 3.24). About 80 percent of Québec's population speaks French, and language remains the cultural

▼ **Figure 3.23 Vancouver's Chinatown** This traditional Chinatown is located just east of downtown Vancouver, but the city's ethnic Chinese population resides throughout the metropolitan area.

▶ **Figure 3.24 Selected Cultural Regions of North America** From northern Canada's Nunavut to the Southwest's Hispanic Borderlands, different North American cultural groups strongly identify with traditional local and regional homelands. Shaded portions of the map display a sampling of these regions across North America. Dotted areas suggest general locations for surviving ethnic islands of rural European settlement. A mother tongue other than English characterizes some cultural regions, both urban and rural.

▼ **Figure 3.25 Delta Blues Club, Mississippi** Red's Lounge Blues Club in Clarksdale, Mississippi regularly features live performers who entertain audiences with Delta Blues jazz music.

Explore the SOUNDS of the Delta Blues
http://goo.gl/ucQKOW

Another rural homeland, Acadiana, is a zone of persisting Cajun culture in southwestern Louisiana (see Figure 3.24). This homeland was founded in the 18th century, when French settlers expelled from eastern Canada (an area known as Acadia) relocated to Louisiana. Known today through food and music, Cajun culture is strongly linked to Louisiana's bayous and swamps.

Native American Signatures Native peoples are also strongly tied to their homelands. Indeed, many maintain intimate relationships with their surroundings, weaving elements of the natural environment into their material and spiritual lives. Over 5 million Native Americans, Inuits, and Aleuts live in North America, claiming allegiance to more than 1100 tribal bands. Place names, landscape features, and family ties cement this connection between people and place.

Particularly in the American West and the Canadian and Alaskan North, native peoples also control sizable reservations, although less than 25 percent of native populations reside on these lands. The largest block of native-controlled land in the lower 48 states is the

▲ **Figure 3.26 Native American Poverty, South Dakota** South Dakota's Pine Ridge Indian Reservation is home to the Oglala Sioux nation. The reservation faces a severe housing shortage, made worse by recent floods and persisting poverty.

Explore the **SIGHTS** of **Pine Ridge Reservation** http://goo.gl/De7ZSW

Navajo Reservation in the Southwest. About 300,000 people claim allegiance to the Navajo Nation. To the north, Canada's self-governing Nunavut Territory (population about 35,000) is another reminder of the enduring presence of native cultural influence and emergent political power within the region (see Figure 3.24). Although these homelands preserve traditional ties to the land, they are also settings for pervasive poverty, health problems, and increasing cultural tensions (Figure 3.26).

Explore the **SOUNDS** of **the Navajo Language** http://goo.gl/pwXM9D

A Mosaic of Ethnic Neighborhoods North America's cultural mosaic is characterized by smaller-scale ethnic signatures that shape both rural and urban landscapes (see Figure 3.24). During settlement of the agricultural interior, immigrants often established close-knit communities. Among others, German, Scandinavian, Slavic, Dutch, and Finnish settlements took shape, held together by common origins, languages, and religions. Rural landscapes in Wisconsin, Minnesota, the Dakotas, and the Canadian prairies still display these cultural imprints in the form of folk architecture, distinctive settlement patterns, ethnic place names, and rural churches.

Ethnic neighborhoods are also a part of the urban landscape and reflect both global-scale and internal North American migration patterns. The ethnic geography of Los Angeles is an example of both economic and cultural forces at work (Figure 3.27a). Because most of its economic expansion took place during the 20th century, the city's ethnic patterns reflect the movements of more recent migrants. African-American communities on the city's south side (Compton and Inglewood) are a legacy of black migration out of the South. Hispanic (East Los Angeles) and Asian (Alhambra and Monterey Park) neighborhoods are a reminder that about 40 percent of the city's population is foreign-born (Figure 3.27b).

Patterns of North American Religion

Distinctive religious traditions also shape North America's cultural geography. Reflecting its colonial roots, Protestantism dominates within the United States, accounting for about 60 percent of the population (Figure 3.28). In some settings, hybrid American religions sprang from broadly Protestant roots. By far the most successful is the Church of Jesus Christ of Latter-Day Saints (Mormons), which claims more than 6 million North American members, concentrated in Utah and Idaho. Although many traditional Catholic neighborhoods have lost population in the urban Northeast, Catholic numbers are growing in the West and South, reflecting both domestic migration patterns and larger Hispanic populations. Almost 40 percent of Canadians are Protestant, with the United Church of Canada claiming large numbers of followers (see Figure 3.28). French-Canadian Québec is a bastion of Catholic tradition and makes Canada's population (39 percent) distinctly more Catholic than that of the United States (22 percent).

Millions of other North Americans practice religions outside of the Protestant and Catholic traditions or are unaffiliated with Christianity. Orthodox Christians congregate in the urban Northeast, where many Greek, Russian, and Serbian Orthodox communities were established between 1890 and 1920. Ukrainian Orthodox churches also dot the Canadian prairies of Alberta, Saskatchewan, and Manitoba. More than 5 million Jews live in North America, concentrated in East and West Coast cities. Many Muslims (3.6 million), Buddhists (3.9 million), and Hindus (2.2 million) also live in the United States.

◄ **Figure 3.27 Racial Diversity in Los Angeles** (a) In many portions of Los Angeles, different racial and ethnic signatures overlap. (b) Southern California's Little India community features South–Asian owned businesses along busy Pioneer Boulevard.

No majority group
Hispanic
White
African American
Asian

0 3 6 Miles
0 3 6 Kilometers

San Fernando
Burbank
Pasadena
Alhambra
Covina
Santa Monica
Monterey Park
East Los Angeles
Culver City
Los Angeles
Florence
Marina Del Rey
Hawthorne
LOS ANGELES CO.
ORANGE CO.
Manhattan Beach
Compton
Long Beach
Rancho Palos Verdes
PACIFIC OCEAN

N

(a)

(b)

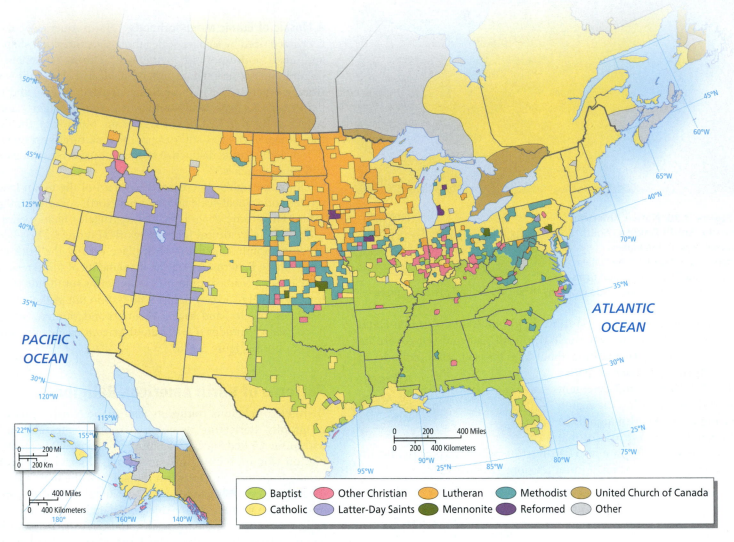

Baptist · Catholic · Other Christian · Latter-Day Saints · Lutheran · Mennonite · Methodist · Reformed · United Church of Canada · Other

▲ **Figure 3.28 Christian Denominations of North America**   Although many portions of North America feature great religious diversity, Roman Catholicism or various Protestant denominations dominate selected regions. Portions of rural Utah and Idaho dominated by the Mormon faith display some of the West's highest concentrations of any single religion.

The Globalization of American Culture

Simply put, North America's cultural geography is becoming more global at the same time that global cultures are becoming more North American (influenced particularly by the United States). But cultural globalization processes are complex; rather than simple flows of foreign influences into North America or of U.S. cultural dominance invading every traditional corner of the globe, the story of 21st-century cultural globalization increasingly mixes influences that flow in many directions at once, resulting in new hybrid cultural creations.

North Americans: Living Globally   More than ever, North Americans in their everyday lives are exposed to people from beyond the region. With more than 50 million foreign-born migrants living across the region, diverse global influences mingle in new ways. Millions of international visitors come to the region annually, both for business and for pleasure. In U.S. colleges and universities, although their numbers have recently dipped, more than one million international students add

global flavor to the classroom (see *Globalization in Our Lives: International Students and the American College Scene*).

Globalization presents challenges for North Americans. In the United States, one key issue revolves around the English language, which some have described as the "social glue" holding the nation together. Increasing use of **Spanglish**, a hybrid combination of English and Spanish spoken by Hispanic Americans, illustrates the complexities of North American globalization. Spanglish, an example of "code switching," where a speaker alternates between two or more languages, includes interesting hybrids such as *chatear*, which means "to have an online conversation."

North Americans are going global in other ways. In 2019, the vast majority of Americans and Canadians have Internet access, launching far-reaching journeys in cyberspace. For many, social media such as Facebook, Twitter, and Instagram have redefined the communities and networks that shape their daily lives. The popularity of ethnic cuisine has peppered the region with a bewildering variety of Cuban, Ethiopian, Basque, and South Asian eateries. Americans also consume

GLOBALIZATION IN OUR LIVES

International Students and the American College Scene

Over one million international students attend American colleges and universities, an increase of 85 percent since 2007. The surge has brought globalization into the everyday lives of millions of college students. Having Chinese, Nigerian, and Brazilian students in a geography classroom adds a whole new dimension to learning about these parts of the world. International student food fairs and cultural celebrations also provide campus diversity. Most importantly, friendships made during college years often endure. The overall impacts on the U.S. economy are massive: these students bring in almost $40 billion annually.

Where do international students study? Not surprisingly, large, university-rich states such as California (150,000+ students), New York (120,000), Texas (85,000), and Massachusetts (60,000) attract the world's brightest young minds. New York University, University of Southern California, Columbia, Northeastern, and Arizona State all have more than 13,000 international students on campus.

Major source regions are nations undergoing rapid economic growth and expanded demand for skilled, technically trained professionals: China (32.5 percent of the total) and India (17 percent) dominate, but a constellation of other, mostly Asian, countries are also key contributors (Figure 3.3.1). Most Americans studying abroad (325,000 in 2016) still target European universities, although that pattern is likely to change.

Newly arriving international student numbers declined in 2017, however. College officials fear the uncertain U.S. political and social climate may be a factor, as well as selective foreign cutbacks (in Brazil and Saudi Arabia, for example) in government scholarship programs. But in the long run, it seems likely that American campuses will continue to benefit from these students who bring enriching diversity to the nation's classrooms and dorms.

1. **How might global patterns of international student visitation change by 2050? Why?**

2. **What are important international student populations on your campus? What academic programs attract them and why?**

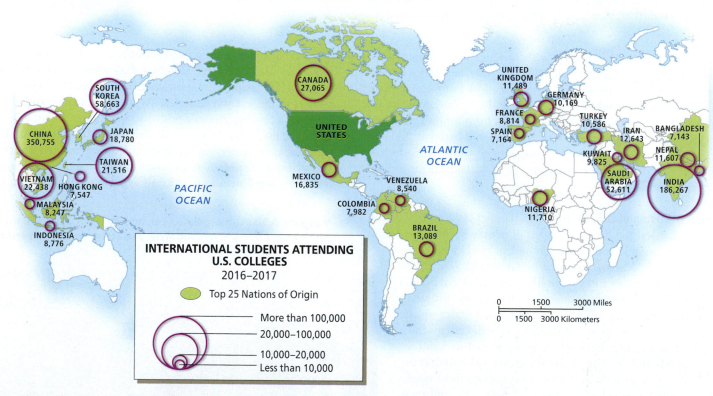

▲ **Figure 3.3.1 Top 25 Places of Origin of U.S. International Students, 2017**   China and India combine to make up half the international student population in the United States, but a global potpourri of nations contributes to the American campus scene.

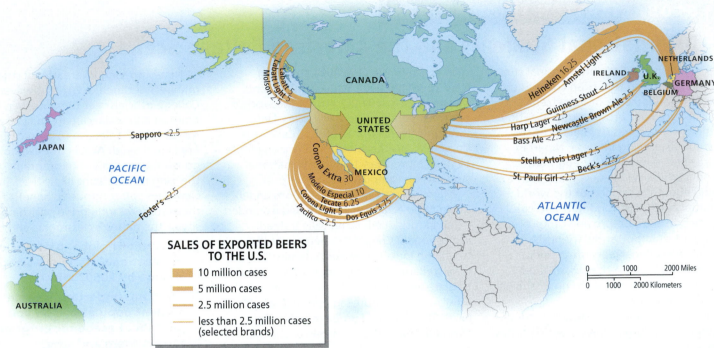

SALES OF EXPORTED BEERS TO THE U.S.

- 10 million cases
- 5 million cases
- 2.5 million cases
- less than 2.5 million cases (selected brands)

▲ **Figure 3.29** **Annual Sales (in millions of dollars) of Imported Beers to the United States, 2015** North Americans are increasingly eating and drinking globally. Rising beer imports, including many more expensive foreign brands, exemplify the pattern. The nation's beer drinkers know no bounds to their thirsts. Mexico dominates, along with varied European, Asian, and Australian producers.

Explore the **TASTES** of Cerveza
http://goo.gl/RSdXvv

imported beer in record quantities, with varieties from Mexico proving especially popular (Figure 3.29). *Gucci*, *Prada*, and *Dolce & Gabbanna* are household words for millions who follow European fashion, while German techno bands, K-pop tunes, and Latin rhythms are the soundtrack of daily life. Professional athletes from around the globe migrate to North America to compete in basketball, baseball, and soccer. Indeed, from acupuncture and massage therapy to soccer and New Age religions, North Americans tirelessly borrow, adapt, and absorb the larger world around them.

The Global Diffusion of U.S. Culture   In parallel fashion, U.S. culture has forever changed the lives of billions of people beyond the region. Although the economic and military power of the United States was notable by 1900, it was not until after World War II that American popular culture fundamentally reshaped global human geographies. The Marshall Plan and Peace Corps initiatives exemplified the growing presence of the United States on the world stage, even as European colonialism waned. Perhaps most critical was the marriage between growing global demand for consumer goods and the rise of the multinational corporation, which was superbly structured to meet and cultivate those needs. Global corporate advertising, distribution networks, and mass consumption bring Cokes and Big Macs to Moscow and Mumbai, golf courses to Thai jungles, and Mickey and Minnie Mouse to Tokyo and Paris. In one example, a wealthy enclave of Yichang, located in central China near the Yangtze River, features a complete community with North-American style homes and condominiums. Such developments reflect China's hunger, especially among its more affluent citizens, for housing styles and conveniences that resemble those in the United States (Figure 3.30).

But challenges to American cultural control illustrate the varied consequences of globalization. Hollywood's dominance within the global film industry has declined dramatically as filmmakers build their own movie studios in India, Latin America, West Africa, China, and elsewhere. As Internet access has grown worldwide, the online dominance of English-speaking users has dramatically declined, while Hindi and Mandarin users increase. Active resistance to U.S. cultural influence is also notable. For example, Canadian and French media interests chastise their radio, television, and film industries for allowing too much American influence.

▼ **Figure 3.30** **Yichang Gated Community** This upscale Chinese development features housing styles and conveniences that resemble those in the United States.

3.6 What are the distinctive eras of immigration in U.S. history, and how do they compare with those of Canada?

3.7 Identify four enduring North American cultural regions, and describe their key characteristics.

KEY TERMS ethnicity, cultural assimilation, cultural homeland, Spanglish

Geopolitical Framework: Patterns of Dominance and Division

North America is home to two of the world's largest states. Their creation, however, was neither simple nor preordained, but rather the result of historical processes that might have created quite a different North American map. Once established, these two states have coexisted in a close relationship of mutual economic and political interdependence.

Creating Political Space

The United States and Canada have very different political roots. The United States broke cleanly and violently from Great Britain. Canada, in contrast, was a country of convenience, born from a peaceful separation from Britain and then assembled as a collection of distinctive regional societies that only gradually acknowledged their common political destiny.

Europe imposed its own political boundaries on a future United States. The 13 English colonies, sensing their common destiny after 1750, united two decades later in the Revolutionary War. The Louisiana Purchase (1803) nearly doubled the national domain, and by the 1850s the remainder of the West had been added. The acquisition of Alaska (1867) and Hawaii (1898) rounded out what became the 50 states.

Canada was created under quite different circumstances. After the American Revolution, England's remaining territories in the region were controlled by administrators in British North America, and in 1867 the provinces of Ontario, Québec, Nova Scotia, and New Brunswick were united in an independent Canadian Confederation. Within a decade, the Northwest Territories, Manitoba, British Columbia, and Prince Edward Island joined this confederation, and the continental dimensions of the country took shape. Later infilling added Alberta, Saskatchewan, and Newfoundland. The creation of Nunavut Territory (1999) represents the latest change in Canada's political geography.

Continental Neighbors

Geopolitical relationships between Canada and the United States have always been close: Their common 5525-mile (8900-km) boundary requires both nations to pay attention to one another. In 1909, the Boundary Waters Treaty created the International Joint Commission, an early step in the common regulation of cross-boundary issues involving water resources, transportation, and environmental quality. The St. Lawrence Seaway (1959) opened the Great Lakes region to better global trade connections (Figure 3.31). The signing of the Great Lakes Water Quality Agreement (1972) and the U.S.–Canada Air Quality Agreement (1991), launched joint efforts to clean up Great Lakes pollution and reduce acid rain in eastern North America.

▲ **Figure 3.31** St. Lawrence Seaway This aerial view of the St. Lawrence Seaway shows a portion of the Welland Canal in Ontario that links Lakes Erie and Ontario.

Explore the SIGHTS of the St. Lawrence Seaway http://goo.gl/u0HP22

Close political ties also have strengthened trade. The United States receives about three-quarters of Canada's exports and supplies almost two-thirds of its imports. Conversely, Canada accounts for roughly 20 percent of U.S. exports and 15 percent of its imports. A bilateral Free Trade Agreement, signed in 1989, paved the way for the larger **North American Free Trade Agreement (NAFTA)** five years later, which extended the alliance to Mexico. Paralleling the success of the European Union (EU), NAFTA has forged the world's largest trading bloc, including almost 500 million consumers and a huge free-trade zone stretching from beyond the Arctic Circle to Latin America.

Political conflicts occasionally still divide North Americans (Figure 3.32). NAFTA itself was renegotiated by the Trump administration in 2018 and the resulting United States-Mexico-Canada Agreement (USMCA) still leaves many cross-border trade issues unresolved for both agricultural and industrial goods. Trans-boundary water issues also persist: Canada has protested North Dakota's plans to control the north-flowing Red River (which leads into Manitoba), while Montana residents are nervous that Canadian logging and mining interests in British Columbia will increase pollution on the south-flowing North Flathead River. Long-standing agreements on dams within the shared Columbia Basin are also being renegotiated amid new demands by indigenous groups on both sides of the border for expanded salmon habitat.

Agricultural and natural resource issues occasionally cause tensions. Canadian wheat and potato growers are periodically accused of dumping their products into U.S. markets, thus depressing prices and profits for U.S. farmers. Wisconsin and New York milk producers are

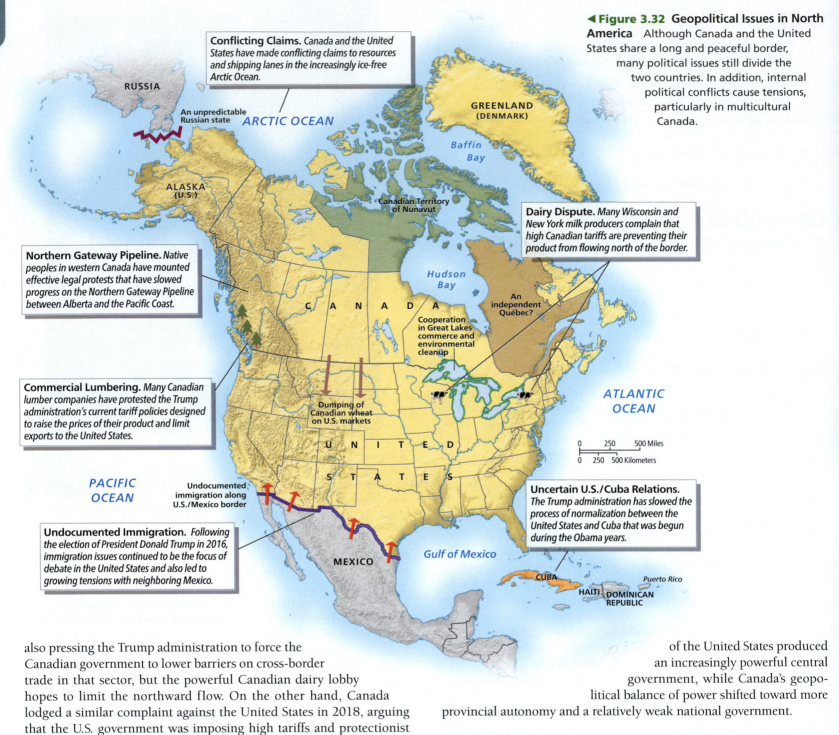

◀ **Figure 3.32** **Geopolitical Issues in North America** Although Canada and the United States share a long and peaceful border, many political issues still divide the two countries. In addition, internal political conflicts cause tensions, particularly in multicultural Canada.

Conflicting Claims. *Canada and the United States have made conflicting claims to resources and shipping lanes in the increasingly ice-free Arctic Ocean.*

An unpredictable Russian state

Northern Gateway Pipeline. *Native peoples in western Canada have mounted effective legal protests that have slowed progress on the Northern Gateway Pipeline between Alberta and the Pacific Coast.*

Commercial Lumbering. *Many Canadian lumber companies have protested the Trump administration's current tariff policies designed to raise the prices of their product and limit exports to the United States.*

Dairy Dispute. *Many Wisconsin and New York milk producers complain that high Canadian tariffs are preventing their product from flowing north of the border.*

Cooperation in Great Lakes commerce and environmental cleanup

An independent Québec?

Dumping of Canadian wheat on U.S. markets

Undocumented immigration along U.S./Mexico border

Undocumented Immigration. *Following the election of President Donald Trump in 2016, immigration issues continued to be the focus of debate in the United States and also led to growing tensions with neighboring Mexico.*

Uncertain U.S./Cuba Relations. *The Trump administration has slowed the process of normalization between the United States and Cuba that was begun during the Obama years.*

RUSSIA

ARCTIC OCEAN

GREENLAND (DENMARK)

Baffin Bay

ALASKA (U.S.)

Canadian Territory of Nunavut

C A N A D A

Hudson Bay

ATLANTIC OCEAN

U N I T E D

S T A T E S

PACIFIC OCEAN

0   250   500 Miles

0   250   500 Kilometers

MEXICO

Gulf of Mexico

CUBA

HAITI DOMINICAN REPUBLIC

Puerto Rico

---

also pressing the Trump administration to force the Canadian government to lower barriers on cross-border trade in that sector, but the powerful Canadian dairy lobby hopes to limit the northward flow. On the other hand, Canada lodged a similar complaint against the United States in 2018, arguing that the U.S. government was imposing high tariffs and protectionist barriers on Canadian lumber.

The Legacy of Federalism

The United States and Canada are **federal states** in that both nations allocate considerable political power to subnational units of government. Other nations, such as France, have traditionally been **unitary states**, in which power is centralized at the national level. Federalism leaves many political decisions to local and regional governments and often allows distinctive cultural and political groups to be recognized within a country. The U.S. Constitution (1787) limited centralized authority, giving all unspecified powers to the states or the people. In contrast, the Canadian Constitution (1867) created a federal state under a parliamentary system, giving most powers to central authorities. Ironically, the evolution

of the United States produced an increasingly powerful central government, while Canada's geopolitical balance of power shifted toward more provincial autonomy and a relatively weak national government.

Québec's Challenge   The political status of Québec remains a major issue in Canada (see Figure 3.32). Economic disparities between the Anglo and French populations have reinforced cultural differences between the two groups, with French Canadians often suffering when compared with their wealthier neighbors in Ontario. Beginning in the 1960s, a separatist political party (the Parti Québécois) increasingly voiced French-Canadian concerns. When the party won provincial elections in 1976, it declared French the official language of Québec. Formal provincial votes over the question of Québec's independence were held in 1980 and 1995. Both measures failed. Since then, support for separation has ebbed in favor of a more modest strategy of increased "autonomy" within Canada.

Native Peoples and National Politics   Another challenge to federal political power comes from North American Indian and Inuit populations. Within the United States, Native Americans asserted their political power in the 1960s, marking a decisive turn away from assimilation policies. The Indian Self-Determination and Education Assistance Act, passed in 1975, increased Native Americans' control of their economic and political destiny. In Alaska, native peoples acquired title to 44 million acres (18 million hectares) of land in 1971 under the Alaska Native Claims Settlement Act. The Indian Gaming Regulatory Act (1988) offered potential economic independence for many tribes. In 2016, Indian gaming operations (primarily gambling casinos) nationally netted tribes about $32 billion. In the western American interior, where Native Americans control roughly 20 percent of the land, tribes are also solidifying their hold on resources, reacquiring former reservation acreage, and participating in political interest groups, such as the Native American Fish and Wildlife Society and the Council of Energy Resource Tribes. However, tribes suffered a major defeat in 2017 when the Trump administration rejected protests and legal efforts to block the Dakota Access Pipeline, an underground conduit that now carries crude oil across native land from North Dakota's Bakken region to the Midwest.

In Canada, challenges by native peoples have yielded dramatic results. Agreements with native peoples in Québec, Yukon, and British Columbia returned millions of acres of land to aboriginal control and increased native participation in managing remaining public lands. By far the most ambitious agreement created Nunavut out of the eastern portion of the Northwest Territories in 1999 (see Figure 3.32), representing a new level of native self-government in North America. Nunavut is home to 35,000 people (85 percent Inuit) and is the largest territorial/provincial unit in Canada. Agreements between the Canadian Parliament and British Columbia tribes (the Nisga'a) have resulted in similar moves toward more native self-government (see Figure 3.24).

A growing number of indigenous groups are also using this power to control the pace of natural resource development in Canada. In 2016, 26 First Nations communities (Canadian indigenous groups other than Inuit or Métis), working with environmental groups, protected a Pacific Coast area from commercial logging. Other coastal and inland First Nations groups, citing potential harm to salmon fisheries, collectively voiced strong opposition to the $11 billion Pacific Northwest Liquefied Natural Gas (LNG) project planned for Canada's west coast.

Shifting Immigration and Refugee Policies

Recent elections in both Canada (2015) and the United States (2016) brought immigration and refugee policies into the spotlight. The two countries seem to be headed in quite different directions. In Canada, the Trudeau government has emphasized the positive economic and humanitarian benefits of a liberal immigration policy. With immigrants making up 22 percent of the population, the country is an inviting destination for many (especially from Asia and the Middle East). Since 2015, Canadian policies toward political refugees have also emphasized an open door, including a pledge to resettle 40,000 displaced Syrians. In addition, U.S. immigrants, some with uncertain and temporary visas, are flocking north to the Canadian border to seek asylum in friendlier territory. Thousands of Haitian refugees in the United States, fearing repatriation to their Caribbean homeland by the Trump administration, have already fled to Québec and Ontario.

Citing jobs and national security, the United States has made a pronounced shift toward more stringent border controls and the return of undocumented immigrants, especially convicted criminals. This shift began during the Obama administration with large increases in the number of border patrol agents and in the volume of deportations (2.7 million

deportees between 2009 and 2016). The Trump administration has adopted an even more aggressive border policy, with calls to "build a wall" along the nation's southern border and to severely restrict unauthorized entry, an argument that has created sharp debate and contributed to deteriorating relations with neighboring Mexico (Figure 3.33). The U.S. government has also adopted a more stringent and controversial immigration and travel policy that curtails travel from many other global regions (mostly Muslim-dominated countries). New refugee restrictions designed to carefully limit, vet, and control the inflow of future asylum seekers contrast with more open policies that allowed almost 900,000 refugees to settle in the country between 2001 and 2016 (see *Exploring Global Connections: Bosnian Refugees Reshape a St. Louis Neighborhood*). One especially contested issue involves the fate of the children of undocumented migrants (termed "Dreamers"). Under the Obama administration, this group was protected by the DACA (Deferred Action for Childhood Arrivals) program, but the Trump administration has argued for ending the program or severely restricting it.

A Global Reach

The geopolitical reach of the United States, in particular, extends far beyond the region's borders. World War II and its aftermath forever redefined the U.S. role in world affairs. The United States emerged from the conflict as the world's dominant political power. It also developed multinational political and military agreements, such as the North Atlantic Treaty Organization (NATO) and the Organization of American States (OAS). Conflicts in Korea (1950–1953) and Vietnam (1961–1975) pitted U.S. political interests against communist attempts to extend control beyond the Soviet Union and China. Even as the Cold War faded during the late 1980s, the geopolitical reach of the United States expanded. Direct involvement in conflicts within Central America, the Middle East, Serbia, and Kosovo exemplified the country's global agenda. Recent controversial wars in Iraq (2003–2011) and Afghanistan (2001–present) offer further evidence of America's global political presence.

In the United States, a more combative Trump administration has challenged the status quo with NATO and the United Nations.

▼ **Figure 3.33 International Border**   North America's southwestern landscape is boldly divided by an increasingly hardened international border that separates the United States and Mexico. This view is south of San Diego, California. **Q: In your opinion, what are the key (a) advantages and (b) disadvantages to building a much more visible and difficult-to-cross southern border wall?**

EXPLORING GLOBAL CONNECTIONS

Bosnian Refugees Reshape a St. Louis Neighborhood

How did the largest collection of Bosnian Americans in the United States end up in a rundown working-class neighborhood on the south side of St. Louis? The journey was long and tortuous, beginning with a dreadful war in southeastern Europe that consumed the former Yugoslavia for much of the 1990s. Bosnia-Herzegovina in particular witnessed massive genocide, and the "ethnic cleansing" practiced by multiple groups produced disruptive waves of desperate refugees. The first Bosnians relocated to St. Louis in 1993 under a federal refugee resettlement program. As the neighborhood grew and became known as "Little Bosnia," many more immigrants came from elsewhere in Europe and other U.S. cities. By 2015, the Bevo Mill community (named after an ornamental mill on a major commercial street) contained more than 70,000 Bosnian Americans (Figure 3.4.1).

A Revitalized Community The immigrants transformed the low-rent, crime-ridden district of abandoned buildings, older factories, and rundown housing into a vibrant and healthy ethnic enclave. Most Bosnians learned English, many created small businesses, and many more gradually invested their sweat equity to fix up their modest early-20th-century brick homes. Most of the existing residents welcomed the new arrivals, although there were a few issues with Bosnian families roasting whole lambs in backyard smokehouses. Overall, the Bosnians have thrived: their annual incomes of more than $80,000 are 25 percent higher than those of average St. Louis residents. Today, the neighborhood also has lower crime and unemployment rates than the city overall. While many immigrants are

homesick, they welcome the opportunity to begin new lives in a setting where they can still enjoy elements of the old country.

Landscape Signatures Perhaps the most visible sign of this largely Muslim population is an old branch bank building that was converted into an Islamic Community Center with a 107-foot-tall Turkish-style minaret added in 2007. Aside from the backyard smokehouses, other changes are fairly subtle. The main commercial avenue supports businesses catering to the immigrant community. Grocery stores, restaurants, bars, and community centers often advertise with bilingual signs, and the colorful blue and gold colors of the flag of Bosnia and Herzegovina adorn many storefront windows. Many of these businesses, especially the local cafes serving Bosnian specialties such as stuffed cabbage, janjetina (lamb), and sausage, also cater to visitors from beyond the neighborhood.

No surprise that the St. Louis mayor has welcomed these arrivals from southeastern Europe. In fact, he has been an outspoken advocate of resettling a newer generation of refugees from war-torn Syria in the city, no doubt figuring they might repeat the Bosnian miracle that has so impressively reinvigorated the Bevo Mill community.

▲ **Figure 3.4.1 St. Louis Bosnian Community** This modern minaret stands outside an Islamic Community Center in the Bevo Mill community. The Ottoman-style tower was added in 2007, six years after the Community Center was established on the site of a bank branch building.

1. **What might be the three greatest challenges you would be confronted with as a new Bosnian immigrant to St. Louis?**

2. **In your community, can you identify landscape signatures of immigrant populations, either from the recent past or earlier?**

GOOGLE EARTH
Virtual Tour Video
http://goo.gl/zSs92n

Tensions have also increased with Russia (including debates about that country's interference with the 2016 election), China (its own geopolitical designs in the western Pacific have raised American concerns), and Mexico (due to its negative response to more stringent U.S. border policies). In the Middle East, U.S. foreign policy has tilted sharply toward Israel, including controversial plans to move the American embassy to Jerusalem. Defense expenditures of more than $800 billion in 2018 (nearly as much as the rest of the world combined) suggest the United States will continue to play a highly visible role in global affairs.

REVIEW

3.8 How do the political origins of the United States and Canada differ, and what issues divide these nations today?

3.9 What are the key issues surrounding U.S. immigration and refugee policies, and how do these contrast with Canada?

KEY TERMS North American Free Trade Agreement (NAFTA), federal state, unitary state

Economic and Social Development: Geographies of Abundance and Affluence

North America possesses the world's most powerful economy and its wealthiest population. Its 365 million people consume huge quantities of global resources but also produce some of the world's most sought-after manufactured goods and services. The region's *human capital*—the skills and diversity of its population—has enabled North Americans to achieve high levels of economic development (Table 3.2).

An Abundant Resource Base

North America is blessed with numerous natural resources that provide diverse raw materials for development. Indeed, the direct extraction of natural resources still makes up 3 percent of the U.S. economy and more than 6 percent of Canada's economy. Some of these resources are exported to global markets, while other raw materials are imported to the region.

Agriculture remains a dominant land use across much of the region (Figure 3.34), but in a highly commercialized, mechanized, and specialized form that emphasizes efficient transportation, global markets, and large capital investments in farm machinery. Agriculture employs only a small percentage of the labor force in both the United States (1 percent) and Canada (2 percent), and the number of farms has sharply dropped, while average farm size has steadily risen.

The geography of North American farming represents the combined impacts of (1) diverse environments; (2) varied continental and global markets; (3) historical patterns of settlement and agricultural evolution; and (4) the role of **agribusiness**, or corporate farming. Agribusiness refers to large-scale enterprises that control closely integrated segments of food production, from farm to grocery store. In the Northeast, dairy operations and truck farms take advantage of proximity to major cities in Megalopolis and southern Canada. Corn, soybeans, and livestock production dominate the Midwest and western Ontario. To the south, the old Cotton Belt has been largely replaced by subtropical specialty crops; poultry, catfish, and livestock production; and commercial logging. Extensive grain-growing operations stretch from Kansas to Saskatchewan and Alberta, while irrigation allows agricultural production in the far West, depending on surface and groundwater resources. Indeed, California, nourished by large agribusiness operations in the irrigated Central Valley, accounts for more than 10 percent of the U.S. farm economy.

North Americans produce and consume huge quantities of other natural resources. The region consumes 40 percent more oil than all of the European Union, and oil and gas production is on the rise, especially on the Gulf Coast, the Central Interior (North Dakota became a major producing state after 2000), Alaska's North Slope, and central Canada (Alberta's oil sands) (Figure 3.35). The most abundant fossil fuel in the United States is coal (27 percent of the world's total), but its relative importance in the overall energy economy declined in the 21st century as industrial technologies changed and environmental concerns grew.

North America also remains a major producer of metals, although global competition, rising extraction costs, and environmental concerns pose challenges for this sector of the economy.

Creating a Continental Economy

The timing of European settlement was critical in North America's rapid economic transformation. The region's abundant resources came under the control of Europeans possessing new technologies that reshaped the landscape and reorganized its economy. By the 19th century, North Americans actively contributed to those technological changes. New resources were developed in the interior, and new immigrants arrived in large numbers. In the 20th century, although natural resources remained important, industrial innovations and more service-sector jobs added to the economic base and extended the region's global reach.

Connectivity and Economic Growth   North America's economic success was a function of its **connectivity**, or how well its different locations became linked with one another through vastly improved transportation and communications networks. Those links greatly facilitated the interaction between locations and dramatically reduced the cost of moving people, products, and information, thereby laying the foundation for urbanization, industrialization, and the commercialization of agriculture.

Technological breakthroughs revolutionized North America's economic geography between 1830 and 1920. By 1860, more than 30,000 miles (48,000 km) of railroad track had been laid in the United States, and the network grew to more than 250,000 miles (400,000 km) by 1910. The telegraph brought similar changes to information: Long-distance messages flowed across eastern North America by the late 1840s, and 20 years later, undersea cables linked the region to Europe, another milestone in the process of globalization.

Transportation and communications systems were modernized further after 1920. Automobiles, mechanized farm equipment, paved highways, commercial air links, national radio broadcasts, and dependable transcontinental telephone service reduced the cost of distance across

TABLE 3.2	**Development Indicators**						Explore these data in MapMaster 2.0	https://goo.gl/XfYK97		
Country	GNI per Capita, PPP 2017[1]	GDP Average Annual Growth 2009–2015[2]	Human Development Index (2016)[3]	Percent Population Living Below $3.10 a Day[2]	Under Age 5 Mortality Rate (per 1000 live births), 1990[1]	Under Age 5 Mortality Rate (per 1000 live births), 2016[1]	Secondary School Enrollment Ratios[4] Male (2009–2016)	Female (2009–2016)	Gender Inequality Index (2016)[3,6]	Freedom Rating (2018)[5]
Canada	46,378	2.4	0.920	–	8	5	110	110	0.098	1
United States	59,532	2.1	0.920	–	11	7	97	98	0.203	1.5

[1] World Bank Open Data, 2018.
[2] World Bank—*World Development Indicators*, 2017.
[3] United Nations, *Human Development Report*, 2016.
[4] Population Reference Bureau, *World Population Data Sheet*, 2017.
[5] Freedom House, Freedom in the World 2018. See Ch.1, pp. 33–34, for more info. on this scale (1–7, with 7 representing states with the least freedom).
[6] See Ch. 1, p. 39, for more info. on this scale (0–1, with higher values representing less gender equality).

Log in to Mastering Geography & access MapMaster to explore these data!

1) Compare GNI per Capita and Life Expectancy for these two countries. What might explain why Canadians earn less income but live longer?
2) While Gender Inequality Index values for both the United States and Canada are low by global standards, why might values for Canada suggest substantially less gender inequality in that country versus the United States?

▶ **Figure 3.34** **Major Agricultural and Manufacturing Activities of North America** Varied environmental settings, settlement histories, and economic conditions have produced the modern map of North American agriculture and manufacturing. A growing number of major metropolitan areas play increasingly visible roles in the global economy.

Legend:
- Specialty crop or livestock farming
- Mixed farming
- Commercial wheat and other small grain farming
- Dairy farming
- General farming
- Livestock ranching
- Irrigated agriculture
- Mediterranean agriculture (with irrigation)
- Nonfarming
- Major manufacturing region
- City with key global links
- Other center of manufacturing

▼ **Figure 3.35** **An Energy Landscape in the Permian Basin, West Texas** In addition to the visible signs of center-pivot irrigation evident in this West Texas landscape, note the numerous roads and development pads associated with oil drilling in this energy-rich portion of the Permian Basin.

North America. Perhaps most important, the region has taken the lead in the global information age, integrating computer, satellite, telecommunications, and Internet technologies in a web of connections that facilitates the flow of knowledge both within the region and beyond.

Sectoral Transformation Changes in employment structure signaled North America's economic modernization just as surely as its increasingly interconnected society. **Sectoral transformation** refers to the evolution of a nation's labor force from one dependent on the *primary* sector (natural resource extraction) to one with more employment in the *secondary* (manufacturing or industrial), *tertiary* (services), and *quaternary* (information processing) sectors. For example, mechanized agriculture reduced demand for primary-sector workers but opened new opportunities in the growing industrial sector. In the 20th century, new services (trade, retailing) and information-based activities (education, data processing, research) created other employment opportunities. Today the tertiary and quaternary sectors employ more than 70 percent of U.S. and Canadian workers.

Regional Economic Patterns North America's industries show important regional patterns, influenced by various **location factors** that explain *why* an economic activity is located where it is and *how* patterns of economic activity are shaped. Patterns of industrial location illustrate the concept (see Figure 3.34). The historical manufacturing core includes the cities of Megalopolis and southern Ontario, and the industrial Midwest. This core's proximity to *natural resources* (farmland, coal, and iron ore); its increasing *connectivity* (canals and railroad networks, highways, air traffic hubs, and telecommunications centers); its ready supply of *productive labor*; and a growing national, then global, *market demand* for its industrial goods encouraged continued *capital investment* and allowed this core to dominate steel, automobile, and equipment production as well as financial and insurance services.

In the last half of the 20th century, industrial- and service-sector growth shifted to the South and West. Cities of the South's Piedmont manufacturing belt (Greensboro to Birmingham) grew after 1960 as lower labor costs and Sun Belt amenities attracted new investment. North Carolina's "research triangle" area (Raleigh, Durham, and Chapel Hill) is the nation's third-largest biotech cluster, behind California and Massachusetts. The Gulf Coast industrial region is strongly tied to nearby fossil fuels that provide raw materials for the energy-refining and petrochemical industries.

The varied West Coast industrial region stretches from Vancouver, British Columbia, to San Diego, California (and beyond into northern Mexico), demonstrating the increasing importance of Pacific Basin trade. Large western aerospace firms also reflect the role of *government spending* as a location factor. Silicon Valley is now a leading region of manufacturing exports, and its proximity to Stanford, Berkeley, and other universities demonstrates the importance of *access to innovation and research* for many fast-changing high-technology industries (see Figure 3.2). Silicon Valley's location also shows the advantages of *agglomeration economies,* in which companies with similar, often integrated manufacturing operations locate near one another. Smaller places such as Provo, Utah, and Austin, Texas, specialize in high-technology industries and demonstrate the growing role of *lifestyle amenities* in shaping industrial location decisions, both for entrepreneurs and for skilled workers attracted to such amenities (Figure 3.36).

North America and the Global Economy

Together with Europe and East Asia, North America is a key player in the global economy and is home to a growing number of truly "global cities" that serve as connecting points and decision-making centers in the world economy (see Figure 3.34).

Creating the Modern Global Economy
The United States, with Canada's firm support, played a formative role in creating much of the new global economy and in shaping its key institutions. In 1944, allied nations met at Bretton Woods, New Hampshire, to discuss economic affairs. Under U.S. leadership, the group set up the International Monetary Fund (IMF) and the World Bank and gave these global organizations the responsibility for defending the world's monetary system. The United States was also the driving force for the 1948 creation of the General Agreement on Tariffs and Trade (GATT), renamed the **World Trade Organization (WTO)** in 1995. Its 164 member states are dedicated to reducing global trade barriers. The United States and Canada also participate in the **Group of Seven (G7)**, a collection of economically powerful countries (Japan, Germany, Great Britain, France, and Italy) that regularly meets to discuss key global economic and political issues.

▲ **Figure 3.36 Provo, Utah** Located about 40 miles south of Salt Lake City, Provo, Utah, has become a major center of technology-oriented businesses in the West. Its setting, nestled beneath the Wasatch Mountains, has attracted a large number of amenity migrants.

Explore the SIGHTS of Provo, Utah http://goo.gl/DtRfcQ

Patterns of Global Investment Patterns of capital investment and corporate power place North America at the center of global money flows and economic influence. The region's relative stability attracts huge inflows of foreign capital, both as investments in North American stocks and bonds and as foreign direct investment (FDI) by international companies.

Another form of foreign investment in the region is immigrant entrepreneurs who play an outsized role in producing economic growth, particularly in large cities (Figure 3.37). Whether it is the Chinese in Vancouver or the Cubans in Miami, immigrants in many of North America's largest, most global cities have made huge capital

▲ **Figure 3.37 Immigrant Entrepreneurs** Minority and immigrant entrepreneurs have transformed the economic geographies of North American cities. This Miami woman owns Margarita Flowers, a thriving small business in this culturally diverse south Florida city.

and human investments in their adopted communities. The economic consequences of these skilled people are enormous. One study of the 50 largest U.S. metropolitan areas found that immigrants owned 58 percent of dry cleaning and laundry businesses, 40 percent of motels, and 43 percent of liquor stores. A Center for an Urban Future report argues that "immigrants have been the entrepreneurial spark plugs of cities from New York to Los Angeles."

The impact of U.S. investments in foreign stock markets also suggests how outbound capital flows transform the way business is done throughout the world. Aging baby boomers have poured billions of pension fund and investment dollars into global stock and bond markets. In addition, multinational corporations based in the United States make investments around the world.

However, the geography of 21st-century multinational corporations is changing, illustrating three recent shifts in broader patterns of globalization. These shifts have important consequences for North Americans. First, U.S.-based multinational corporations are adopting a new, more globally integrated model. For example, IBM now has more than 150,000 employees in India. Second, multinational corporations based elsewhere in the world—especially in places such as China, India, Russia, and Latin America—are buying up companies and assets once controlled by North American or European capital. Third, many of these same multinational companies are making huge investments of their own in the less developed world, from Africa to Southeast Asia, bypassing North American control altogether. Today, more than one-third of FDI in emerging market nations comes from these newer multinationals. Simply put, the late-20th-century, top-down model of multinational corporate control and investment, traditionally based in North America, Europe, and Japan, is being replaced by a more globally distributed model of corporate control. This new model has many origins, many destinations, and new patterns of labor, capital, production, and consumption.

In addition, **outsourcing**, a business practice that transfers portions of a company's production and service activities to lower-cost, often overseas settings, has shifted millions of jobs in manufacturing, textiles, semiconductors, and electronics to countries such as China, India, and Mexico, which offer low-cost, less regulated settings for both local and foreign firms. The results are complex: North American consumers benefit from cheap imports but may find their own jobs threatened by the corporate restructurings that make such bargains possible.

Enduring Social Issues

Profound socioeconomic problems shape North America's human geography. Despite the region's continental wealth, great differences persist between rich and poor. Race, particularly within the United States, continues to be an issue of overwhelming importance. Both nations also face challenges related to gender inequity and aging populations.

Wealth and Poverty   Elite northeastern suburbs, gated California neighborhoods, upscale shopping malls, and posh alpine ski resorts are expressions of private, exclusive landscapes characterizing wealthy North America. In contrast, substandard housing, abandoned property, aging infrastructure, and unemployed workers illustrate the gap between rich and poor. Rural poverty remains a major social issue

in the Canadian Maritimes, Appalachia, the Deep South, the Southwest, and agricultural California. While many of the poorest Americans live in central cities, poverty rates are actually growing much more rapidly in suburbs than in either inner cities or rural areas. About 13–15 percent of North Americans live in poverty, and in the United States, about 26 percent of the country's African Americans and 24 percent of Hispanics live below the poverty line.

Food Deserts in a Land of Plenty   One revealing measure of poverty involves **food deserts**, where people do not have ready access to supermarkets and fresh, healthy, and affordable food. Many North American cities have three times as many supermarkets in wealthier versus poor neighborhoods. At the same time, unhealthy fast food restaurants often concentrate in low-income districts. Thus urban poverty is frequently associated with poor nutrition and relatively higher food expenditures. The U.S. Agriculture Department considers an urban area a food desert if at least 20 percent of residents live in poverty and at least one-third of residents live more than a mile from a supermarket. A rural food desert is defined as a rural setting where residents must drive more than 10 miles to the nearest supermarket. More than 2.3 million people (often elderly and less mobile) are impacted in such areas.

Health Care and Aging   Aging and other health care issues are key concerns for a region of graying baby boomers. A recent report on aging predicted that 20 percent of Americans will be older than 65 by 2050. Poverty rates are also higher for seniors. Whole sections of the United States—from Florida to southern Arizona—are increasingly oriented around retirement, with communities catering to seniors with special assisted-living arrangements, health care facilities, and recreation (Figure 3.38).

The health care systems in both countries are costly by global standards: Canadians spend about 12 percent of their gross domestic product (GDP) on health care, and costs are even higher in the United States (over 15 percent of GDP). For several decades, Canada has offered an enviable system of government-subsidized universal health care to its residents (who pay higher taxes to fund it). In the United States,

▼ **Figure 3.38 The Villages, near Orlando, Florida**   This planned retirement community near Orlando has witnessed some of Florida's most rapid population growth in the 21st century.

the Patient Protection and Affordable Care Act was signed into law in 2010, but many of its provisions were challenged or weakened during the Trump administration, leaving many health care policy issues unanswered.

Hectic lives, often oriented around fast food, have contributed to rapidly growing rates of obesity in North America since 1975. Almost two-thirds of adult Americans are overweight (over one-third are considered obese), contributing to higher rates of heart disease and diabetes. Combined with more sedentary lifestyles, the convenience-oriented diets that shape the everyday routine of millions of North Americans no doubt add to the long-term cost of health care. Still, there are bright spots, including declining rates of sugary soda consumption and a trend among more affluent and well-educated North Americans to eat fresh and locally produced foods.

Chronic alcoholism and substance abuse are also widespread. For example, about half of U.S. college students who drink engage in harmful binge drinking, and more than 150,000 students develop alcohol-related health problems annually. Similarly, an epidemic of drug-related overdoses and deaths (many from legal prescriptions of opioid painkillers) has hit many communities, including small towns and rural areas. Heroin addiction, once linked to poor inner-city populations, has spread to the wealthier suburbs and to the Farm Belt in record numbers.

Another critical health care issue has been the care and treatment of the region's 1.2 million HIV/AIDS victims. The price of the disease will be broadly borne in the 21st century, but particularly among poorer black (44 percent of AIDS cases in the United States) and Hispanic (19 percent) populations.

Gender, Culture, and Politics Since World War II, both the United States and Canada have seen great improvements in the role that women play in society. However, the **gender gap** is yet to be closed when it comes to differences in salary, working conditions, and political power. Women make up more than half of the North American workforce and are often more educated than men but still earn only about 80 cents for every dollar that men earn (Figure 3.39).

Although women have played critical roles in deciding recent national elections, political power remains largely in male hands. Canadian women have voted since 1918 and U.S. women since 1920, but females in the early 21st century remain clear minorities in both the Canadian Parliament (27 percent in 2018) and the U.S. Congress (20 percent in 2017).

New attention was focused on the larger issue of sexual harassment in the American workplace and beyond in 2017 and 2018 as dozens of high-profile allegations of abuse surfaced. The visible and dynamic #MeToo movement reflected growing resistance to unacceptable workplace behavior and also helped to highlight the broader economic disparities that remain within the region.

REVIEW

3.10 Define sectoral transformation. How does it help explain economic change in North America?

3.11 Cite five types of location factors, and illustrate each with examples from your local economy.

3.12 What common social issues do both Canadians and Americans face? How do they differ?

KEY TERMS agribusiness, connectivity, sectoral transformation, location factor, World Trade Organization (WTO), Group of Seven (G7), outsourcing, food desert, gender gap

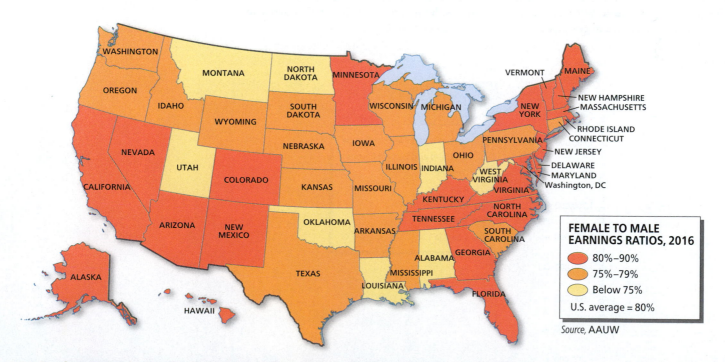

▲ **Figure 3.39 Earnings Ratios, by Gender, United States, 2016** This map shows the relative median annual earnings for women versus men, by state. Note the relatively higher earnings for women in the Northeast and in California versus selected southern and midwestern settings. **Q: What variables might help explain how your own state fits into the larger national pattern?**

North America

3

REVIEW, REFLECT, & APPLY

Summary

- North America's affluence comes with a considerable price tag. Today the region's environmental challenges include combating air and water pollution, improving the efficiency of its energy economy, and adjusting to the realities of global climate change.

- In a remarkably short time, a unique and changing mix of peoples from around the world radically disrupted indigenous populations and settled a huge, resource-rich continent that is now one of the world's most urbanized regions.

- North America is home to one of the world's most culturally diverse societies, and the region's contemporary popular culture has had an extraordinary impact on almost every corner of the globe.

- The region's two societies are closely intertwined, yet they face distinctive political and cultural issues. Canada's multicultural identity remains problematic, and it must deal with both the costs and benefits of living next door to its powerful continental neighbor.

- For the United States, social and political challenges linked to its ethnic pluralism, immigration issues, and enduring poverty and racial discrimination remain central concerns, particularly in its largest cities.

Review Questions

1. Explain how "natural hazards" can be shaped by human history and settlement. In other words, what role do humans play in shaping the distribution of hazards?

2. How have the major North American migration flows since 1900 influenced contemporary settlement patterns, cultural geographies, and political issues within the region?

3. Summarize and map the ethnic background and migration history of your own family. How do these patterns parallel or depart from larger North American trends?

4. Describe the strengths and weaknesses of federalism, and cite examples from both the United States and Canada.

5. The environmental price for North America's economic development has been steep. Suggest why it may or may not have been worth the price, and defend your answer.

6. Who will be North America's leading trade partner in 2050? Explain the reasons for your answer.

Image Analysis

1. This chart shows annual immigration to the United States by region of origin. Note the sharp peaks clustered in Phase 3 and Phase 5. Which immigrant groups dominated immigration during these peak years? What common economic or cultural factors might explain both surges?

2. Which of the immigrant groups shown in the graphic make up important portions of your local community, and why did they settle there? Did they settle in your community as foreign-born immigrants or did they arrive in later generations?

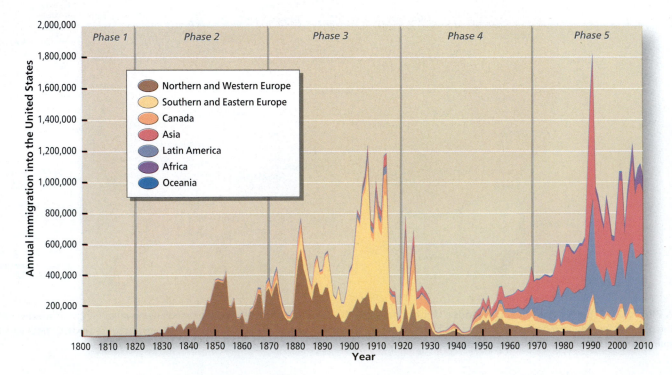

► Figure IA3 U.S. Immigration, by Year and Group

Legend:
- Northern and Western Europe
- Southern and Eastern Europe
- Canada
- Asia
- Latin America
- Africa
- Oceania

Y-axis: Annual immigration into the United States
X-axis: Year (1800–2010)
Phases: Phase 1, Phase 2, Phase 3, Phase 4, Phase 5

Join the Debate

Fracking, a relatively new addition to the American energy economy, is a highly controversial technology in which millions of gallons of water, sand, and other chemicals are injected deep into the earth. Are the economic benefits worth the potential costs?

Fracking has brought an amazing set of benefits to North America!

- Many shale-rich regions of the United States, including Texas, Oklahoma, Ohio, Pennsylvania, and North Dakota, enjoy economic growth as energy-related jobs are created and as landowners benefit from the leasing of their acreage to energy companies.

- Fracking has made the United States less vulnerable to foreign energy producers, and someday the country may even be a major energy exporter.

- Fracking has dramatically lowered the cost of clean-burning natural gas, a boon to consumers and to the entire economy as it outcompetes older, dirtier coal-fired power plants.

The costs of fracking far outweigh the benefits and should be stopped until we know consequences!

- Fracking wells have a notoriously short life, and the drilling process at the site seriously impacts the environment.

- Much of the water injected into the shale formations remains there, forever removed from other uses. Furthermore, contaminated water from fracking waste pits has leached into groundwater and increased contaminants such as methane gas and benzene, causing serious health problems for nearby residents.

- In places such as Ohio and Oklahoma, fracking has been strongly connected to increased earthquake activity. We don't know the long-term geological consequences of this technology or who will pay for damages. We need strong, uniform federal standards.

▲ **Figure D3 Fracking Rig** This drilling platform is located in North Dakota's Bakken Formation, near the Little Missouri River.

GeoSpatial Data Analysis

Homeownership Rates in the United States

Homeownership rates are defined as the percent of "owner-occupied housing units" divided by the "total occupied housing units" in a given area. This figure is about 64 percent for the entire United States but varies from place to place. Many Americans aspire to owning a home, and the homeownership rate is often seen as a measure of success.

Open MapMaster 2.0 in the Mastering Geography Study Area and add the "Persons Living in Poverty" data layer for U.S. counties. Next go to the U.S. Census Bureau's website (https://www.census.gov/) and click on **Browse by Topic,>Housing>Housing Vacancies.** Click on Annual Statistics, then scroll down to the table for Homeownership Rates by State. Click on the table, and import and prepare the data for the most current year into MapMaster. Compare the Poverty and Homeownership maps in split-screen mode and answer the following:

1. Identify the top five and top bottom states for homeownership.
2. Where do you see the largest clusters of poverty in the United States?
3. Do you see any relationship between these two patterns? What might complicate the relationship between these two variables?
4. Examine poverty and homeownership rates in your home state. Explain what you see.

Key Terms

acid rain (p. 74)
agribusiness (p. 95)
boreal forest (p. 71)
connectivity (p. 95)
cultural assimilation (p. 82)
cultural homeland (p. 85)
edge city (p. 81)
ethnicity (p. 82)
federal states (p. 92)
food deserts (p. 98)
fracking (p. 75)
gender gap (p. 99)
gentrification (p. 81)
Group of Seven (G7) (p. 97)
location factors (p. 96)
Megalopolis (p. 76)
new urbanism (p. 82)
nonmetropolitan growth (p. 79)
North American Free Trade Agreement (NAFTA) (p. 91)
outsourcing (p. 98)
postindustrial economy (p. 68)
prairie (p. 71)
renewable energy sources (p. 75)
sectoral transformation (p. 96)
Spanglish (p. 88)
sustainable agriculture (p. 75)
tundra (p. 71)
unitary states (p. 92)
urban decentralization (p. 79)
urban heat island (p. 74)
World Trade Organization (WTO) (p. 97)

Mastering Geography

Looking for additional review and test prep materials? Visit the Study Area in **Mastering Geography** to enhance your geographic literacy, spatial reasoning skills, and understanding of this chapter's content by accessing a variety of resources, including **MapMaster** interactive maps, videos, flashcards, web links, self-study quizzes, and an eText version of *Globalization and Diversity*.

Scan to read about Geographer at Work
Lucia Lo and her work studying Canada's Chinese immigrant businesses.
https://goo.gl/P7Lo9W

Latin America

4

Physical Geography and Environmental Issues

Tropical ecosystems in Latin America are one of planet's greatest reserves of biological diversity. A critical question is how to manage this diversity while extracting mineral wealth, building roads, and converting forests to farm or pasture.

Population and Settlement

Latin America is the developing world's most urbanized region, with 80 percent of the population in cities. Urban primacy is common, and five megacities (>10 million people) are found here. Emigration to North America, long a top destination for the region's migrants, are down.

Cultural Coherence and Diversity

Amerindian activism is on the rise. Indigenous peoples from Central America to the Andes and the Amazon demand territorial and cultural recognition. Catholicism remains the dominant faith, and Pope Francis is an Argentinian; even so, other Christian sects, especially evangelical congregations, are gaining popularity.

Geopolitical Framework

After 200 years of independence, most Latin American countries are fully democratic. Yet democratic institutions are weak, and heightened violence, especially in Venezuela and northern Central America, drive new migration flows. The Pacific Alliance is the newest regional agreement shaping trade in Latin America.

Economic and Social Development

The region falls in the middle-income category, but economic downturns have slowed growth and serious income inequality persists. Conditional cash transfer programs such as Brazil's Bolsa Familia have reduced extreme poverty and improved social development.

▶ Protestors holding Venezuelan flags stand before a wall of national police in Caracas in 2017. As access to food, electricity, and cash worsened in 2017 and 2018, protests became a regular occurrence in the cities of Venezuela. Frustration with a failing economy and political repression has led tens of thousands to flee the country.

Venezuela

LATIN AMERICA

Most Latin American states fall in the middle-income category, their economies are growing, and extreme poverty is declining. For decades, Venezuela was the region's economic star, but it is now mired in a crisis that has left two-thirds of its people poor, hungry, and fearful. Violence is common; hyperinflation undermines markets; and basic foods, medicine, and even toilet paper are in short supply. How can a country of 32 million people, with the world's largest known oil reserves and a founder of OPEC, be a failing state?

Several factors combined over the last few years to create this crisis. Populist leaders (Presidents Chavez and Maduro) promoted a socialist vision that sought to redistribute wealth in a country where income inequality was extreme. That worked when oil prices were high in the late 2000s. But when oil prices fell, debts mounted, nationalized industries faltered, and fiscal policies that limited access to capital meant there was no money to buy replacement parts, medicines, or food. Politically, the leadership surrounded itself with loyalists, undermined democratic institutions, and imprisoned opposition leaders or banned them from running for office. Professionals began to leave the country over a decade ago. By 2017, record numbers of Venezuelans from all classes were crossing into Colombia and Brazil in search of food, medicine, and survival. In 2018 President Maduro was re-elected for another term, despite widespread fraud and low voter turnout. As the situation worsens, many fear that Venezuela could be the most serious humanitarian crisis in the Americas.

Yet unlike war or natural disaster, most observers see this as a self-inflicted tragedy, where a ill-conceived fiscal policies, political corruption, weak institutions, and a decline in oil prices combined to fail Venezuelans. How the region and the world will respond to this crisis remains to be seen, but it is clear that the economic development that this country once boasted has been lost for a generation.

Defining Latin America

The concept of Latin America as a distinct region has been accepted for nearly a century. The boundaries of this region are straightforward, beginning at the Rio Grande (called the *Rio Bravo* in Mexico) and ending at Tierra del Fuego (Figure 4.1). French geographers are credited with coining the term *Latin America* in the 19th century to distinguish the Spanish- and Portuguese-speaking republics of the Americas plus Haiti from English-speaking territories. There is nothing particularly "Latin" about the area, other than the predominance of Romance languages. The term is vague enough to be inclusive of different colonial histories while also offering a clear cultural boundary from Anglo-America, the region called North America in this book.

This chapter describes the Spanish- and Portuguese-speaking countries of Central and South America, including Mexico. This division emphasizes Indian and Iberian influences on mainland Latin America while separating it from the unique colonial and demographic history of the Caribbean and the Guianas, discussed in Chapter 5.

Roughly equal in area to North America, Latin America has a much larger and faster-growing population of 600 million. Its most populous state, Brazil, has 209 million people, making it the world's fifth largest country by population; the next largest state, Mexico, has a population of nearly 131 million, making it the tenth largest.

Through colonialism, immigration, and trade, the forces of globalization are embedded in the Latin American landscape. The Spanish Empire focused on extracting precious metals, sending galleons laden with silver and gold across the Atlantic. The Portuguese became prominent producers of dyewoods, sugar, and gold. By the late 19th and early 20th centuries, exports to North America and Europe fueled the region's economy. Most countries specialized in one or two products: bananas and coffee, meats and wool, wheat and corn, petroleum and copper. Latin American states have since industrialized and diversified production but continue to be major producers of primary goods for North America, Europe, and East Asia.

Today, neoliberal policies that encourage foreign investment, export production, and privatization have been adopted by many states. Extractive industries continue to prevail, in part due to the area's impressive natural resources. Latin America is home to Earth's largest rainforest, the greatest river by volume, and massive reserves of natural gas, oil, gold, and copper. It is also a major exporter of grains, especially soy. With its vast territory, tropical location, and relatively low population density—half the population of India in nearly seven times the area—Latin America is also one of the world's great reserves of biological diversity. How this diversity will be managed in the face of global demand for natural resources is an important question for the countries of this region.

Forces of globalization are embedded in the Latin American landscape

LEARNING OBJECTIVES

After reading this chapter you should be able to:

4.1 Explain the relationships among elevation, climate, and agricultural production, especially in tropical highland areas.

4.2 Identify the major environmental issues of Latin America and how countries are addressing them.

4.3 Summarize the demographic issues impacting this region, such as rural-to-urban migration, urbanization, smaller families, and emigration.

4.4 Describe the cultural mixing of European and Amerindian groups in this region and indicate where Amerindian cultures thrive today.

4.5 Explain the global reach of Latino culture through immigration, sport, music, and television.

4.6 Describe the Iberian colonization of the region and how it affected the formation of today's modern states.

4.7 Identify the major trade blocs in Latin America and how they are influencing development.

4.8 Summarize the significance of primary exports from Latin America, especially agricultural commodities, minerals, wood products, and fossil fuels.

4.9 Describe the neoliberal economic reforms taken by Latin American countries and how they have influenced the region's development.

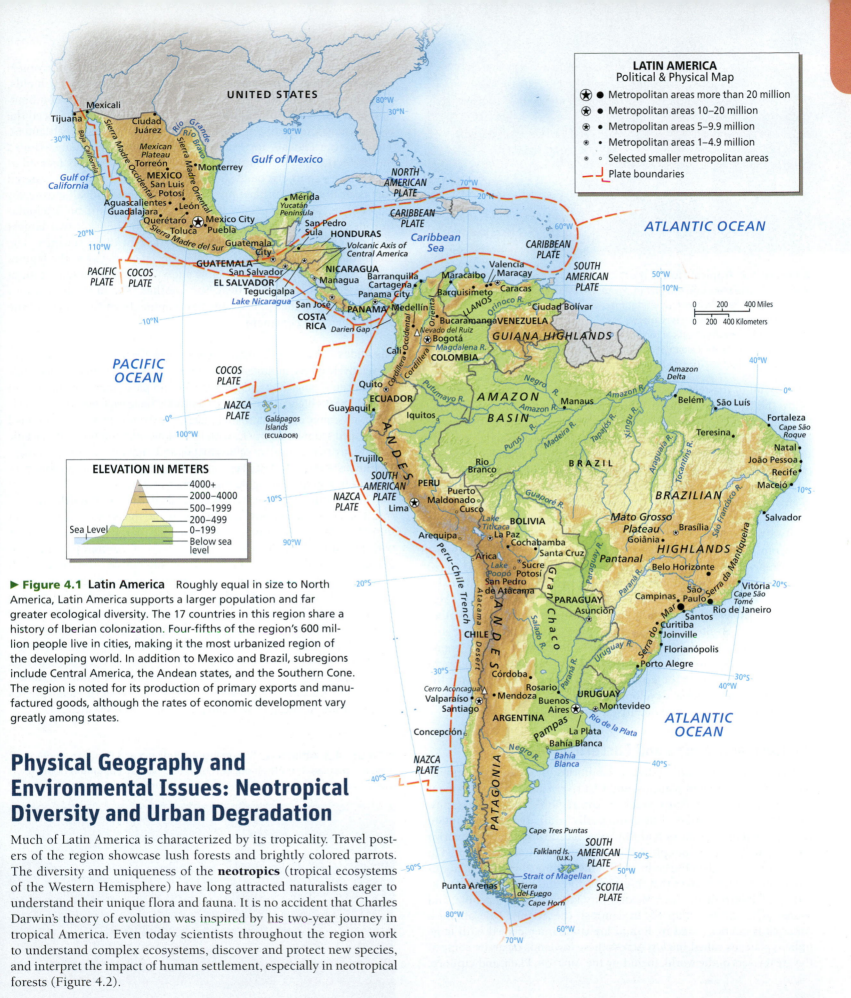

► Figure 4.1 Latin America Roughly equal in size to North America, Latin America supports a larger population and far greater ecological diversity. The 17 countries in this region share a history of Iberian colonization. Four-fifths of the region's 600 million people live in cities, making it the most urbanized region of the developing world. In addition to Mexico and Brazil, subregions include Central America, the Andean states, and the Southern Cone. The region is noted for its production of primary exports and manufactured goods, although the rates of economic development vary greatly among states.

Physical Geography and Environmental Issues: Neotropical Diversity and Urban Degradation

Much of Latin America is characterized by its tropicality. Travel posters of the region showcase lush forests and brightly colored parrots. The diversity and uniqueness of the **neotropics** (tropical ecosystems of the Western Hemisphere) have long attracted naturalists eager to understand their unique flora and fauna. It is no accident that Charles Darwin's theory of evolution was inspired by his two-year journey in tropical America. Even today scientists throughout the region work to understand complex ecosystems, discover and protect new species, and interpret the impact of human settlement, especially in neotropical forests (Figure 4.2).

▲ Figure 4.2 Tropical Flora The Osa Peninsula in Costa Rica is one of the most biodiverse places on the planet. Protected rainforest, notably in Corcovado National Park, is home to scarlet macaws, jaguars, tapir, and squirrel monkeys in addition to a vast variety of flora. With lowland tropical forest reaching to the Pacific Ocean, it is also a popular destination for tourists.

Mobile Field Trip:
Cloud Forest
https://goo.gl/wxHN1z

Not all of the region is tropical. Important population centers lie below the Tropic of Capricorn—most notably Buenos Aires, Argentina, and Santiago, Chile. Much of northern Mexico, including the city of Monterrey, is north of the Tropic of Cancer. Highlands and deserts exist throughout the region. Yet Latin America's tropical climate and vegetation define the region's image. Given its large size and relatively low population density, Latin America has not experienced the same levels of environmental degradation witnessed in East Asia and Europe. Huge areas remain relatively untouched, supporting an incredible diversity of plant and animal life. Throughout the region, national parks offer some protection to unique plant and animal communities. A growing environmental movement in countries such as Costa Rica and Brazil has yielded both popular and political support for "green" initiatives. In short, Latin Americans have entered the 21st century with a real opportunity to avoid many of the environmental mistakes seen in other world regions. At the same time, global market forces are driving governments to exploit minerals, fossil fuels, forests, shorelines, transportation routes, and soils. The region's biggest resource management challenge is to balance the economic benefits of extraction with the principles of **sustainable development**. Another major challenge is to improve the environmental quality of Latin American cities.

Western Mountains and Eastern Lowlands

Latin America is a region of diverse landforms, including high mountains, extensive upland plateaus, and vast river basins. The movement of tectonic plates explains much of the region's basic topography, including the formation of its geologically young western mountain ranges, such as the Andes and the Volcanic Axis of Central America (see Figure 4.1). For example, Chile's Villarica and Calbuco erupted in 2015 and Guatemala's Fuego in 2018. This tectonically active region is also prone to earthquakes that threaten people and damage property. In 2017, a 7.1 earthquake near Mexico City killed over 300 people, and scores of buildings collapsed. In contrast, the Atlantic side of South America is characterized by humid lowlands interspersed with large upland plateaus called *shields*. Across these lowlands meander some of the great rivers of the world, including the Amazon, Plata, and Orinoco.

Historically, the most important areas of settlement in tropical Latin America were not along the major rivers but across its shields, plateaus, and fertile mountain valleys. In these places, the combination of arable land, mild climate, and sufficient rainfall produced the region's most productive agricultural areas and its densest settlement. The Mexican Plateau, for example, is a massive upland area ringed by the Sierra Madre mountains. The Valley of Mexico is located at the plateau's southern end. Similarly, the elevated and well-watered basins of Brazil's southern mountains provide an ideal setting for agriculture. These especially fertile areas can support high population densities, so it is not surprising that the region's two largest cities, Mexico City and São Paulo, emerged in these settings. The Latin American highlands also lend a special character to the region. Lush tropical valleys nestled below snow-covered mountains hint at diverse ecosystems found near one another. The most dramatic of these highlands, the Andes, runs like a spine down the length of the South American continent.

The Andes From northwestern Venezuela to Tierra del Fuego, the Andes are relatively young mountains that extend nearly 5000 miles (8000 km). They are an ecologically and geologically complex mountain chain, with some 30 peaks higher than 20,000 feet (6000 meters). Created by the collision of oceanic and continental plates, the Andes are a series of folded and faulted sedimentary rocks with intrusions of crystalline and volcanic rock. Many rich veins of precious metals and minerals are found in these mountains. In fact, the initial economic wealth of many Andean countries came from mining silver, gold, tin, copper, and iron.

The lengthy Andean chain is typically divided into northern, central, and southern components. In Colombia, the northern Andes split into three distinct mountain ranges before merging near the border with Ecuador. High-altitude plateaus and snow-covered peaks distinguish the central Andes of Ecuador, Peru, and Bolivia. The Andes reach their greatest width here. Of special interest is the treeless high plain of Peru and Bolivia, the **Altiplano**. The floor of this elevated plateau ranges from 11,800 feet (3600 meters) to 13,000 feet (4000 meters) in altitude, limiting its usefulness for grazing. Two high-altitude lakes—Titicaca on the Peruvian and Bolivian border and the smaller Poopó in Bolivia—are located in the Altiplano, as are many mining sites (Figure 4.3). The highest peaks are found in the southern Andes, shared by Chile and Argentina, including the Western Hemisphere's highest peak, Aconcagua, at almost 23,000 feet (7000 meters).

▼ Figure 4.3 Altiplano Straddling the Bolivian and Peruvian Andes is an elevated plateau, the Altiplano. This high and windswept land is home to many Amerindian peoples and native species such as the llama and alpaca.

The Uplands of Mexico and Central America   The Mexican Plateau and the Volcanic Axis of Central America are the most important Latin American uplands in terms of settlement, as many major cities are found here. The Mexican Plateau is a large, tilted block with its highest elevations, about 8000 feet (2500 meters), in the south around Mexico City and its lowest, just 4000 feet (1200 meters), at Ciudad Juárez. The southern end of the plateau, the Mesa Central, contains several flat-bottomed basins interspersed with volcanic peaks that have long been significant areas for agricultural production, such as corn or agave (Figure 4.4). It also contains Mexico's megalopolis—a concentration of the largest population centers, such as Mexico City, Puebla, and Guadalajara.

Along Central America's Pacific coast lies the Volcanic Axis, a chain of volcanoes stretching from Guatemala to Costa Rica. It is a handsome landscape of rolling green hills, elevated basins with sparkling lakes, and volcanic peaks. More than 40 volcanoes, many still active, have produced a rich volcanic soil that yields a wide variety of domestic and export crops. Most of Central America's population is also concentrated in this zone, in the capital cities or surrounding rural villages. The bulk of the agricultural land is tied up in large holdings that produce beef, cotton, and coffee for export. However, in terms of numbers, most of the farms are small subsistence properties that produce corn, beans, squash, and assorted fruits.

The Shields   South America has three major **shields**—large upland areas of exposed crystalline rock similar to upland plateaus found in Africa and Australia. The Brazilian and Patagonian shields (the Guiana Shield is discussed in Chapter 5) vary in elevation from 600 to 5000 feet (200 to 1500 meters). The Brazilian Shield is larger and more important in terms of natural resources and settlement. Far from a uniform land surface, this shield covers much of Brazil from the Amazon Basin in the north to the Plata Basin in the south. In the southeast corner of the plateau is São Paulo, the largest urban conglomeration in South America. The other major population centers are on the plateau's coastal edge, where large protected bays made the sites of Rio de Janeiro and Salvador attractive to Portuguese colonists. Finally, the Paraná basalt plateau on the southern end of the Brazilian Shield is famous for its fertile red soils (*terra roxa*), which yield coffee, oranges, and soybeans. So fertile is this area that the economic rise of São Paulo is attributed to the expansion of commercial agriculture, especially coffee, into this area.

The Patagonian Shield lies in the southern tip of South America. Beginning south of Bahia Blanca and extending to Tierra del Fuego, the region to this day is sparsely settled and hauntingly beautiful. It is treeless, covered by scrubby steppe vegetation, and home to wildlife such as the guanaco and condor. Sheep were introduced to Patagonia in the late 19th century, spurring a wool boom. More recently, offshore oil production has renewed the economic importance of Patagonia and increased tourism to national parks such as Torres del Paine (Figure 4.5).

River Basins   Three great river basins drain the Atlantic lowlands of South America: the Amazon, Plata, and Orinoco. The Amazon drains an area of roughly 2.3 million square miles (5.9 million square kilometers), making it the largest river system in the world by volume and area and the second largest by length. The Amazon Basin is home to the world's largest rainforest; annual rainfall is more than 60 inches (150 centimeters) everywhere in the basin and close to 100 inches (250 centimeters) in the basin's largest city, Belem. The mighty Amazon drains eight countries, but two-thirds of the watershed is within Brazil. Active settlement of the Brazilian portion of the Amazon since the 1960s has boosted the population, and today some 35 million people live in this Basin, equal to 8 percent of the total population in South America. The basin's development—most notably through towns, roads, dams, farms, and mines—is forever changing what was viewed as a vast tropical wilderness just a half-century ago. The Brazilian government has plans to build up to 30 new dams in its portion of the Amazon to meet the country's growing energy demand for resource extraction. Perhaps the most contested dam is Belo Monte on the Xingu River, a tributary of the Amazon (Figure 4.6). When completed, it will be the world's third largest hydroelectric dam, generating more than 11,000 megawatts of electricity.

▼ **Figure 4.4 Mexico's Mesa Central** Mexico's elevated central plateau has long been the demographic and agricultural core of the country. Here a farmer in Jalisco harvests the core of the blue agave plant, used for tequila production. Tequila, a traditional drink in Mexico, has a growing export market. This traditional landscape is recognized as a United Nations World Heritage Site.

▼ **Figure 4.5 Torres del Paine National Park** Situated in the Patagonia portion of southern Chile, this park includes mountains, glaciers, lakes and rivers that attract hikers from around the world. The three granite peaks in the upper right are over 8000 ft (2500 m) and are a unique feature of the park.

◄ **Figure 4.6** **Belo Monte Dam** Currently the largest infrastructure project in Brazil, when completed, it will be the world's fourth-largest hydroelectric dam. The construction firm building the dam has experienced ongoing court battles from indigenous communities being displaced or threatened by the project.

Explore the SIGHTS of Xingu River
http://goo.gl/VJA9gi

The region's second largest watershed, the Plata Basin, begins in the tropics and discharges into the Atlantic in the midlatitudes near Buenos Aires. Several major rivers make up this system: the Paraná, the Paraguay, and the Uruguay. Unlike the Amazon Basin, much of the Plata Basin is now economically productive through large-scale mechanized agriculture, especially soybean production. The basin contains several major dams, including the region's largest hydroelectric plant, the Itaipú on the Paraná, which generates electricity for all of Paraguay and much of southern Brazil. As agricultural output in the watershed grows, sections of the Paraná have been canalized and dredged to enhance the river's capacity for barge and boat traffic.

The third largest basin by area is the Orinoco in northern South America. Although its watershed is only one-seventh the size of the Amazon watershed, the Orinoco's discharge is roughly equal to that of the Mississippi River. The Orinoco River meanders through much of southern Venezuela and part of eastern Colombia, giving character to the sparsely settled tropical grasslands called the *Llanos*. Since the colonial era, these grasslands have supported large cattle ranches. Although cattle are still important, the Llanos are also a dynamic area of petroleum production for both Colombia and Venezuela.

Latin American Climates

In tropical Latin America, average monthly temperatures in settings such as Managua (Nicaragua), Quito (Ecuador), and Manaus (Brazil) show little variation (see the climographs in Figure 4.7). Precipitation

patterns, however, are variable and create distinct wet and dry seasons. In Managua, for example, January is typically a dry month, and June is a wet one. The tropical lowlands of Latin America, especially east of the Andes, are usually classified as tropical humid climates that support forest or savanna, depending on rainfall totals. The region's desert climates are found along the Pacific coasts of Peru and Chile and in Patagonia, northern Mexico, and Bahia of Brazil. Because of the extreme aridity of the Peruvian coast, a city such as Lima, which is clearly in the tropics, averages only 1.5 inches (4 cm) of rainfall. Some sections of the Atacama Desert of Chile get no measurable rainfall (Figure 4.8). Yet the discovery of resources such as nitrates in the 19th century and copper in the 20th century made this hyper-arid region a source of conflict among Chile, Bolivia, and Peru.

Midlatitude climates, with hot summers and cold winters, prevail in Argentina, Uruguay, and parts of Paraguay and Chile (see the climographs for Buenos Aires and Punta Arenas in Figure 4.7). Recall that the midlatitude temperature shifts in the Southern Hemisphere are the opposite of those in the Northern Hemisphere (cold Julys and warm Januarys). In the mountain ranges, complex climate patterns result from changes in elevation. To appreciate how humans adapt to tropical mountain ecosystems, the concept of **altitudinal zonation**, which is the relationship between cooler temperatures at higher elevations and changes in vegetation, is important.

Altitudinal Zonation First described in the scientific literature by Alexander von Humboldt in the early 1800s, altitudinal zonation has practical applications that are intimately understood by the region's native inhabitants. Humboldt systematically recorded declines in temperature as he ascended to higher elevations, a phenomenon known as the **environmental lapse rate**. According to Humboldt, temperature declines approximately 3.5°F for every 1000 feet in higher elevation, or 6.5°C for every 1000 meters. Humboldt also noted changes in vegetation by elevation, demonstrating that plant

▼ **Figure 4.8** **Atacama Desert** This is one of the driest places on earth, with almost no vegetation; many visitors liken it to a moonscape. Yet the soils of the Atacama contain a wealth of copper and nitrates. Here is the Valley of the Moon in northern Chile.

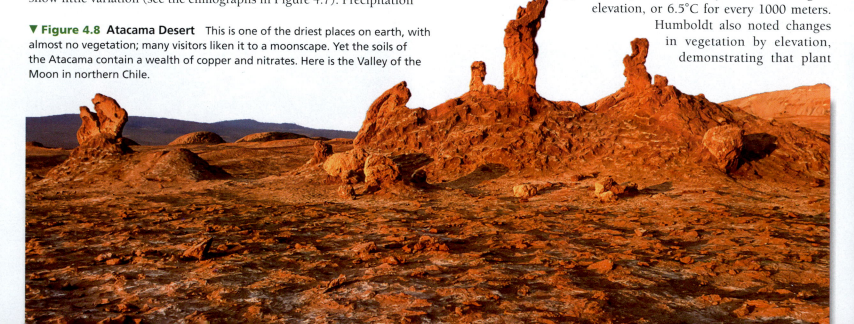

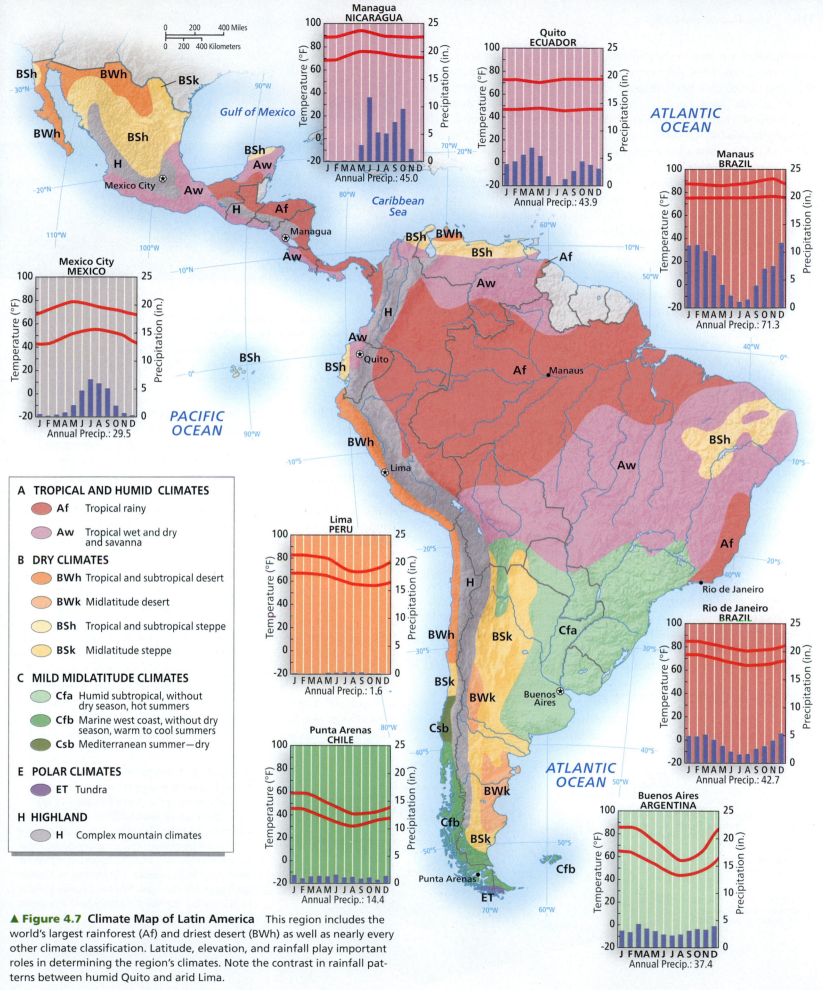

▲ Figure 4.7 Climate Map of Latin America This region includes the world's largest rainforest (Af) and driest desert (BWh) as well as nearly every other climate classification. Latitude, elevation, and rainfall play important roles in determining the region's climates. Note the contrast in rainfall patterns between humid Quito and arid Lima.

Legend:

A TROPICAL AND HUMID CLIMATES
- **Af** Tropical rainy
- **Aw** Tropical wet and dry and savanna

B DRY CLIMATES
- **BWh** Tropical and subtropical desert
- **BWk** Midlatitude desert
- **BSh** Tropical and subtropical steppe
- **BSk** Midlatitude steppe

C MILD MIDLATITUDE CLIMATES
- **Cfa** Humid subtropical, without dry season, hot summers
- **Cfb** Marine west coast, without dry season, warm to cool summers
- **Csb** Mediterranean summer—dry

E POLAR CLIMATES
- **ET** Tundra

H HIGHLAND
- **H** Complex mountain climates

communities common to the midlatitudes could thrive in the tropics at higher elevations. These different altitudinal zones are commonly termed the *tierra caliente* (hot land), from sea level to 3000 feet (900 meters); the *tierra templada* (temperate land), at 3000–6000 feet (900–1800 meters); the *tierra fría* (cold land), at 6000–12,000 feet (1800–3600 meters); and the *tierra helada* (frozen land), above 12,000 feet (3600 meters). Exploitation of these zones allows agriculturists, especially in the uplands, access to a great diversity of domesticated and wild plants (Figure 4.9).

The concept of altitudinal zonation is most relevant for the Andes, the highlands of Central America, and the Mexican Plateau. For example, traditional Andean farmers might use the high pastures of the Altiplano for grazing llamas and alpacas, the *tierra fría* for growing potatoes and quinoa, and the lower temperate zone for corn production. All the great pre-colonial civilizations, especially the Incas and the Aztecs, systematically extracted resources from these zones, thus ensuring a diverse and abundant resource base. Yet these complex ecosystems are extremely fragile and have become important areas of research on the effects of climate change in the tropics.

El Niño   One of the most studied weather phenomena in Latin America, **El Niño** (referring to the Christ child), occurs when a warm Pacific current arrives along the normally cold coastal waters of Ecuador and Peru in December, around Christmastime. This change in ocean temperature, which happens every few years, produces torrential rains, signaling the arrival of an El Niño year. The 1997–1998 El Niño was especially strong, but the 2015–2016 El Niño also proved to be a major event for Latin America. In the Plata Basin, heavy rains in December and January caused the worst flooding in 50 years and displaced more than 200,000 people. Heavy rains in April flooded thousands of acres of cropland, which drastically reduced harvests of corn and soy.

The less-talked-about result of El Niño is drought. While parts of South America experienced record rainfall in 2015–2016, northern Brazil and Central America were in the grip of drought. Guatemala, Honduras, Nicaragua, and El Salvador are affected by one of the worst droughts in decades. Brazil, one of the world's leading soy producers, saw exports plummet due to drought. In addition to crop and livestock losses, estimated to be in the billions of dollars, some 3.5 million rural people in Central America face food shortages. Lack of rain also led to hundreds of bush and forest fires leaving their mark on the landscape.

Impacts of Climate Change for Latin America

Global climate change has both immediate and long-term implications for Latin America. Of greatest immediate concern is how climate

(a)

(b)

▲ **Figure 4.10 Andean Glacial Retreat**   The Upsala Glacier in southern Argentina's Glaciers National Park has been in retreat for decades. The tongue of Upsala Glacier is seen in the bottom photo, lower right. It drains into Lake Argentina. (a) Upsala Glacier in 1928. (b) The same site in January 2004 (the southern hemisphere summer).

change will influence agricultural productivity, water availability, changes in the composition and productivity of ecosystems, and incidence of vector-borne diseases such as malaria, dengue fever, and now Zika. Changes attributable to climate change are already apparent in higher-elevation regions, making these concerns more pressing. For example, coffee growers in the Colombian Andes have seen a decline in productivity over the past five years, which they attribute to higher temperatures and longer dry spells. The long-term effects of global climate change on lowland tropical forest systems is less clear; for example, some areas may experience more rainfall, others less.

Climate change research indicates that highland areas are particularly vulnerable to warmer temperatures. Tropical mountain systems are projected to experience average annual temperature increases of 2–6°F (1–3°C) as well as lower rainfall. This will raise the altitudinal limits of various ecosystems, impacting the range of crops and arable land available to farmers and pastoralists. Research has documented the dramatic retreat of Andean glaciers—some no longer exist, such as Chacaltaya in Bolivia, and others have been drastically reduced, such as Argentina's Upsala glacier (Figure 4.10). This visible indicator of climate change also has pressing human repercussions, as many Andean villages and cities get their water from glacial runoff.

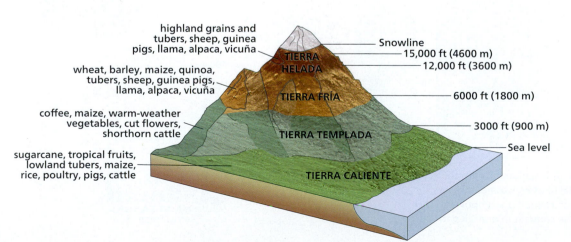

highland grains and
tubers, sheep, guinea
pigs, llama, alpaca, vicuña

Snowline
15,000 ft (4600 m)
12,000 ft (3600 m)

TIERRA HELADA

wheat, barley, maize, quinoa,
tubers, sheep, guinea pigs,
llama, alpaca, vicuña

TIERRA FRÍA

6000 ft (1800 m)

coffee, maize, warm-weather
vegetables, cut flowers,
shorthorn cattle

3000 ft (900 m)

TIERRA TEMPLADA

sugarcane, tropical fruits,
lowland tubers, maize,
rice, poultry, pigs, cattle

Sea level

TIERRA CALIENTE

◄ **Figure 4.9 Altitudinal Zonation**   Tropical highland areas support a complex array of ecosystems. In the *tierra fría* zone (6000 to 12,000 feet, or 1800 to 3700 meters), for example, midlatitude crops such as wheat and barley can be grown. The diagram depicts the range of crops and animals found at different elevations in the Andes.   **Q: Quinoa, an Andean grain grown in the *tierra fría*, has become globally popular in the last two decades. Where else might quinoa be grown?**

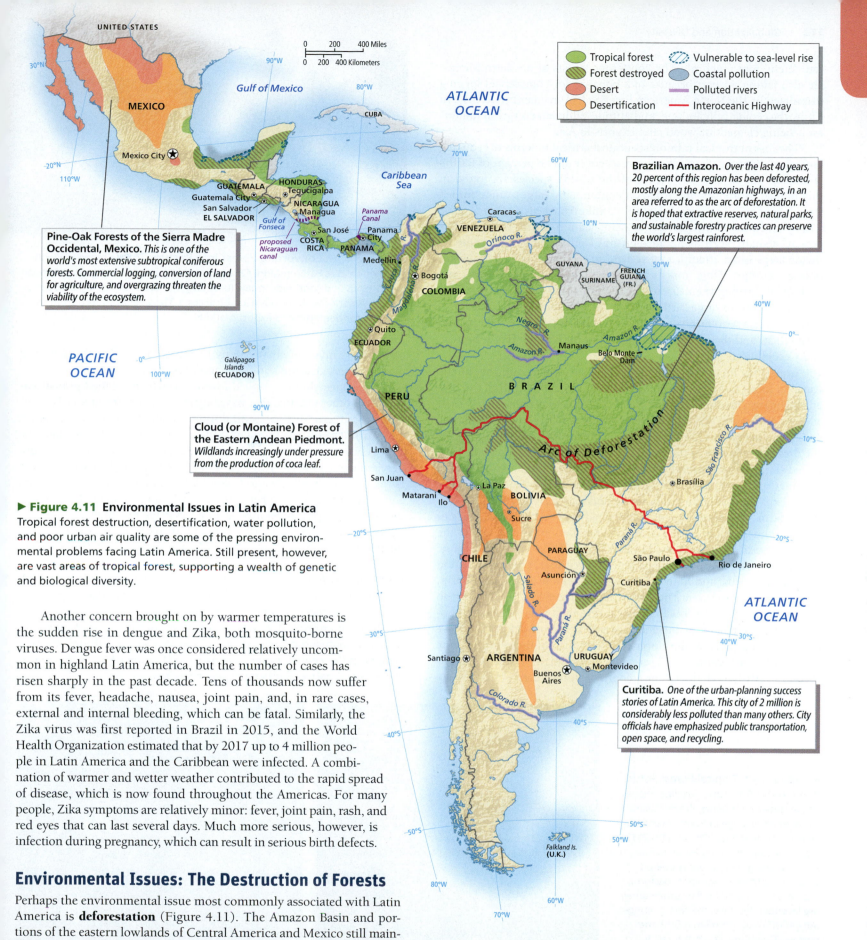

Pine-Oak Forests of the Sierra Madre Occidental, Mexico. *This is one of the world's most extensive subtropical coniferous forests. Commercial logging, conversion of land for agriculture, and overgrazing threaten the viability of the ecosystem.*

Brazilian Amazon. *Over the last 40 years, 20 percent of this region has been deforested, mostly along the Amazonian highways, in an area referred to as the arc of deforestation. It is hoped that extractive reserves, natural parks, and sustainable forestry practices can preserve the world's largest rainforest.*

Cloud (or Montaine) Forest of the Eastern Andean Piedmont. *Wildlands increasingly under pressure from the production of coca leaf.*

Curitiba. *One of the urban-planning success stories of Latin America. This city of 2 million is considerably less polluted than many others. City officials have emphasized public transportation, open space, and recycling.*

Legend:
- Tropical forest
- Forest destroyed
- Desert
- Desertification
- Vulnerable to sea-level rise
- Coastal pollution
- Polluted rivers
- Interoceanic Highway

▶ **Figure 4.11 Environmental Issues in Latin America** Tropical forest destruction, desertification, water pollution, and poor urban air quality are some of the pressing environmental problems facing Latin America. Still present, however, are vast areas of tropical forest, supporting a wealth of genetic and biological diversity.

Another concern brought on by warmer temperatures is the sudden rise in dengue and Zika, both mosquito-borne viruses. Dengue fever was once considered relatively uncommon in highland Latin America, but the number of cases has risen sharply in the past decade. Tens of thousands now suffer from its fever, headache, nausea, joint pain, and, in rare cases, external and internal bleeding, which can be fatal. Similarly, the Zika virus was first reported in Brazil in 2015, and the World Health Organization estimated that by 2017 up to 4 million people in Latin America and the Caribbean were infected. A combination of warmer and wetter weather contributed to the rapid spread of disease, which is now found throughout the Americas. For many people, Zika symptoms are relatively minor: fever, joint pain, rash, and red eyes that can last several days. Much more serious, however, is infection during pregnancy, which can result in serious birth defects.

Environmental Issues: The Destruction of Forests

Perhaps the environmental issue most commonly associated with Latin America is **deforestation** (Figure 4.11). The Amazon Basin and portions of the eastern lowlands of Central America and Mexico still maintain unique and impressive stands of tropical forest. Other woodland areas, such as the Atlantic coastal forests of Brazil and the Pacific forests of Central America, have nearly disappeared as a result of agriculture, settlement, and ranching. The coniferous forests of northern Mexico are also falling, in part because of a bonanza for commercial logging stimulated by the North American Free Trade Agreement (NAFTA). In Chile, the ecologically unique, midlatitude Valdivian evergreen rainforest is being cleared for wood chip exports to Asia.

The loss of tropical rainforests is most critical in terms of reducing biological diversity. Tropical rainforests cover only 6 percent of Earth's landmass, but at least 50 percent of the world's species are found in this biome. Moreover, the Amazon contains the largest undisturbed stretches of rainforest in the world. Unlike Southeast Asian forests, where hardwood extraction drives deforestation, Latin American forests are usually seen as an agricultural frontier. State governments divide areas in an attempt to give land to the landless and to reward political cronies. Thus, forests are cut and burned, with settlers and politicians carving them up to create permanent settlements, slash-and-burn plots, or large cattle ranches. In addition, some tropical forest cutting has been motivated by the search for gold (Brazil, Peru, and Costa Rica) and the production of coca leaf for cocaine (Peru, Bolivia, and Colombia).

Brazil has incurred more criticism than other countries for its Amazon forest policies. During the past 40 years, one-fifth of the Brazilian Amazon has been cleared. Nearly 60 percent of some states, such as Rondônia, have been deforested (Figure 4.12). What most alarms environmentalists and forest dwellers (Indians and rubber tappers) is the dramatic increase in the rate of rainforest clearing since 2000, estimated at nearly 8000 square miles (20,000 square kilometers) per year. The increased rates of deforestation in the Brazilian Amazon are due to the expansion of industrial mining and logging, the growth of corporate farms, the development of new road networks, the incidence of human-ignited wildfires, and continued population growth. Under the Advance Brazil program started in 2000, some US$40 billion has gone to new gas lines, hydroelectric projects, power lines, river canalization projects, railroads, and highways traversing remote areas of the basin, such as the Interoceanic Highway (see Figure 4.11). In an effort to slow deforestation rates, the Brazilian government has created new conservation areas, many of them alongside the "arc of deforestation"— a swath of agricultural development along the southern edge of the Amazon Basin (again, see Figure 4.11). Yet Brazil's Forest Code, as revised in 2012, has reduced the amount of "forest reserve" that private

▲ **Figure 4.13 Converting Forest into Pasture**   Cattle graze in northern Guatemala's Petén region. Clearing of this tropical forest lowland began in the 1960s and continues today. Ranching is a status-conferring occupation in Latin America with serious ecological costs. The beef produced from this region is for domestic and export markets.

landholders must maintain. Many conservationists fear this will lead to more forest clearing and fragmentation.

The conversion of tropical forest into pasture, called **grassification**, is another practice that has contributed to deforestation. Particularly in southern Mexico, Central America, and the Brazilian Amazon, an assortment of development policies from the 1960s through the 1980s encouraged deforestation to make room for cattle in areas of frontier settlement (Figure 4.13). Although many natural grasslands such as the Llanos (Venezuela and Colombia), the Chaco (Bolivia, Paraguay, and Argentina), and the Pampas (Argentina) are suitable for grazing, the rush to convert forest into pasture has made ranching a scourge on the land. Even when domestic demand for beef has increased, ranching in remote tropical frontiers is seldom economically self-sustaining.

Problems on Agricultural Lands   The pressure to modernize agriculture has produced a series of environmental problems. As peasants were encouraged to adopt new hybrid varieties of corn, beans, and potatoes, an erosion of genetic diversity occurred. Efforts to preserve dozens of native domesticates are under way at agricultural research centers in the central Andes and Mexico (Figure 4.14). Nonetheless, many useful native plants may have been lost.

▶ **Figure 4.12 Tropical Forest Settlement in the Amazon**   Satellite images of the state of Rondônia, Brazil, illustrate the dramatic change in forest cover in just 16 years, between (a) 2001 and (b) 2017 along the main highway BR-364. Intact forest is dark green, whereas cleared areas are light green (crops or pasture), and recently cleared land or urban areas are lavender. Typically, the first clearings appear off roads, forming a fishbone pattern. Over time, as more forest is cleared and settlements grow, the fishbone pattern collapses into a mosaic of pasture, farmland, and forest fragments.

(a) Rondonia, 2001

(b) Rondonia, 2017

▲ **Figure 4.14 Andean Potatoes** With some 4000 edible varieties, this region is the hearth of potato domestication. Since 1971 the International Potato Center has been dedicated to protecting this vast diversity of tubers to ensure food security and the well-being of rural Andean farmers who maintain these valuable varieties.

Modern agriculture also depends on chemical fertilizers and pesticides that eventually run off into surface streams and groundwater. Consequently, many rural areas suffer from contamination of local water supplies. Even more troublesome is the direct exposure of farm workers to toxic agricultural chemicals. Mishandling of pesticides and fertilizers can result in rashes and burns. In some areas, such as Sinaloa, Mexico, a rise in serious birth defects parallels the widespread application of chemicals.

Protecting Resources for Future Generations

Latin America has more nationally protected lands than any other developing region. The areas designated as national parks, nature reserves, wildlife sanctuaries, and scientific reserves with limited public access went from 10 percent of the territory in 1990 to 21 percent in 2013, according to World Bank estimates. Brazil's protected land went from just 9 percent of the national territory to 26 percent in 20 years. Although conservationists complain that many of these areas are "paper parks" with limited real protection, many countries in the region have used the conservation of forests and other lands as a means to attract tourists. A regional leader in this regard is Costa Rica (see *Working Toward Sustainability: Ecotourism in Costa Rica*).

Urban Environmental Challenges

For most Latin Americans, air pollution, water availability and quality, and garbage removal are the pressing environmental problems of everyday life. Consequently, many environmental activists in the region focus their efforts on making urban environments cleaner by introducing "green" legislation and calling people to action. In this most urbanized region of the developing world, city dwellers do have better access to water, sewers, and electricity than their counterparts in South Asia and Africa. Moreover, the density of urban settlement seems to encourage the widespread use of mass transportation; both public and private bus and van routes make getting around cities fairly easy. However, the usual environmental problems that come from dense urban settings ultimately require expensive remedies, such as new power plants and modernized sewer and water lines. The money for such projects is never enough, due to currency devaluation, inflation, and foreign debt. Because many urban dwellers tend to reside in unplanned squatter settlements, servicing these communities with utilities after they are built is difficult and costly.

Air Pollution Most major cities, but especially Santiago and Mexico City, suffer from air pollution. A combination of geographical factors (basin settings) and meteorological factors (winter inversion layers), along with dense human settlement and automobile dependence, has led to these two capital cities having some of the highest recorded concentrations of particulate matter and ozone, major contributors to air pollution. Air pollution is not just an aesthetic issue—the health costs of breathing such contaminated air are significant, as elevated death rates due to heart disease, asthma, influenza, and pneumonia suggest. The burden of air pollution is not evenly distributed among city residents, as the elderly, the very young, and the poor are more likely to suffer the negative health effects of contaminated air. Fortunately, both cities have taken steps to address this vexing problem.

Mexico City's smog has been so bad that most visitors today have no idea that mountains surround them. Air quality has been a major issue since the 1960s, driven in part by the city's unusually high rate of growth (4.8 percent annually between 1950 and 1980). Steps were finally taken in the late 1980s to reduce emissions from factories and cars. Unleaded gas is now widely available, and cars manufactured for the Mexican market must have catalytic converters. In addition, some of the worst polluting factories in the Valley of Mexico have closed. In the last few years, Mexico City has expanded a low-emissions bus system, eliminating thousands of tons of carbon monoxide. In 2007, the decision was made to close the elegant Paseo de la Reforma to traffic on Sunday mornings and open it to bike riders. This change was so popular that now bike lanes have been introduced to some downtown areas in an effort to encourage bike ridership. For longer commutes, a suburban train system that will complement the existing subway system is being expanded. The payoff is real: Mexico City no longer ranks among the most polluted cities in the world and seems to have cut most of its pollutants by at least half (Figure 4.15).

Water Providing access to clean and reliable freshwater is also a challenge for Latin America's large cities. In many cities, tap water is not considered safe to drink, so people purchase water, which can be

▼ **Figure 4.15 Air Pollution in Mexico City** The city's high elevation and immense size make air quality management challenging, but fortunately metropolitan and federal efforts have helped to reduce the most polluting factories and automobiles, and has led to overall improvements.

Ecotourism in Costa Rica

Costa Rica, a small tropical country straddling the Pacific and the Caribbean Sea, boasts impressive biodiversity in both its highland and lowland areas. It is the one Latin American country that has become nearly synonymous with the term **ecotourism**, because most of its tourism sector is oriented toward enjoying the natural environment. Whether you want to ride a zip-line through the cloud forest, bird-watch, hike to the rim of a volcano, or explore two distinctive coastal environments, the country offers a wide variety of ecologically oriented activities. Costa Rica received over 2.5 million tourists in 2015, nearly half of those from the United States. Tourism is the country's leading source of foreign exchange and the fastest growing sector of the economy.

National Parks Protect Land This approach to tourism did not happen overnight but developed over 40 years. Long known for agricultural commodities, especially coffee and bananas, Costa Rica was like many of its Central American neighbors in the 1960s. In the 1970s, Costa Rican conservationist Mario Boza successfully lobbied for the creation of national parks in response to the rampant forest destruction occurring to expand coffee production, banana plantations, and cattle pasture. Importantly, his first park was created near the capital city of San Jose, with the intent to attract Costa Ricans to enjoy the natural environment. Today Costa Ricans still visit the country's many parks, such as Manuel Antonio (Figure 4.1.1), and about 20 percent of the national territory is protected from development. International tourists also visit the parks but pay higher entrance fees that go to supporting conservation and park maintenance.

The business of ecotourism requires a protected environment as well as infrastructure: hotels for accommodations, knowledgeable guides and park rangers, roads and trails to access places. Ecotourism has grown through a mix of private and public investment—the private sector creating tourist accommodations and adventure opportunities, and the public sector investing in protected lands (Figure 4.1.2). Like any business, there is a delicate balance, as too many tourists

▲ **Figure 4.1.2 Eco-lodge in Costa Rica** International tourists view wildlife from the deck of an eco-lodge on the Osa Peninsula, one of the most biodiverse locations in Central America.

at a particular location can overwhelm the flora and fauna that is being protected.

Of course, there was resistance to protecting lands, especially from traditional agricultural and ranching interests. Yet today, Costa Ricans have come to embrace their status as a leading ecotourism destination. *Pura vida* is a widely used Costa Rican expression for enjoying life, happiness, and satisfaction. It literally means "pure life," but for eco-conscious Costa Ricans, *pura vida* is surely reinforced by the robust national park system they have created and maintained.

1. **Explain why Costa Rican ecotourism is more developed than that of other Central American countries.**

2. **Go online to search for ecotourism-oriented places or businesses in your home state or where you go to school. What components are necessary to foster ecotourism?**

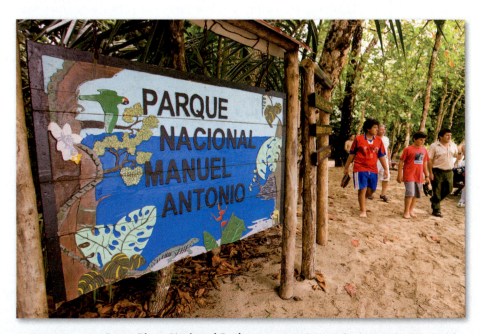

▲ **Figure 4.1.1 Costa Rican National Park** Hugging the Pacific Coast, the tropical forest and beaches of Manuel Antonio National Park make it a popular destination for Costa Ricans as well as international tourists. The pressures to develop tropical coasts are real, which makes creating protected areas an urgent need.

GOOGLE EARTH Virtual Tour Video
http://goo.gl/u8BvAw

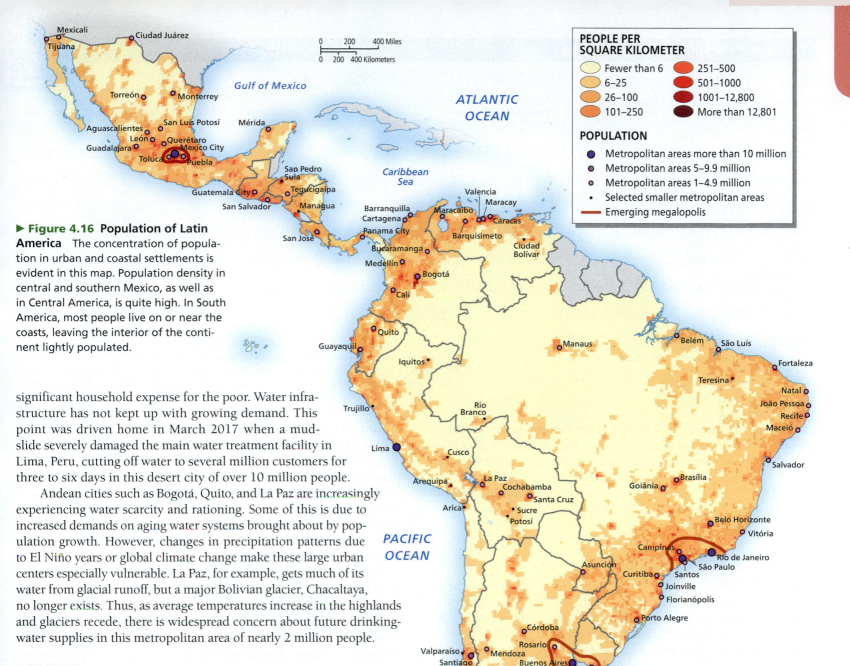

▶ **Figure 4.16 Population of Latin America** The concentration of population in urban and coastal settlements is evident in this map. Population density in central and southern Mexico, as well as in Central America, is quite high. In South America, most people live on or near the coasts, leaving the interior of the continent lightly populated.

PEOPLE PER SQUARE KILOMETER

Fewer than 6	251–500
6–25	501–1000
26–100	1001–12,800
101–250	More than 12,801

POPULATION

- ● Metropolitan areas more than 10 million
- ● Metropolitan areas 5–9.9 million
- ● Metropolitan areas 1–4.9 million
- • Selected smaller metropolitan areas
- ▬ Emerging megalopolis

significant household expense for the poor. Water infrastructure has not kept up with growing demand. This point was driven home in March 2017 when a mudslide severely damaged the main water treatment facility in Lima, Peru, cutting off water to several million customers for three to six days in this desert city of over 10 million people.

Andean cities such as Bogotá, Quito, and La Paz are increasingly experiencing water scarcity and rationing. Some of this is due to increased demands on aging water systems brought about by population growth. However, changes in precipitation patterns due to El Niño years or global climate change make these large urban centers especially vulnerable. La Paz, for example, gets much of its water from glacial runoff, but a major Bolivian glacier, Chacaltaya, no longer exists. Thus, as average temperatures increase in the highlands and glaciers recede, there is widespread concern about future drinking-water supplies in this metropolitan area of nearly 2 million people.

REVIEW

4.1 Describe the major ecosystems in Latin America and how humans have adapted to and modified these different ecosystems.

4.2 Summarize some of the major environmental issues impacting this region and how different countries have tried to address them.

KEY TERMS neotropics, sustainable development, Altiplano, shield, altitudinal zonation, environmental lapse rate, El Niño, deforestation, grassification, ecotourism

Population and Settlement: The Dominance of Cities

Historically, the highland areas of Latin America supported most of this region's population during the pre-Hispanic and colonial eras,

especially in Mexico, Central America, and the Andes. In the 20th century, population growth and migration to the Atlantic lowlands of Argentina and Brazil, along with continued growth of Andean coastal cities such as Guayaquil, Barranquilla, and Maracaibo, have reduced the demographic importance of the highlands. Major highland cities such as Mexico City, Guatemala City, Bogotá, and La Paz still dominate their national economies, but most large cities are on or near the coasts (Figure 4.16).

Like the rest of the developing world, Latin America has experienced dramatic population growth. In 1950, its population totaled 150 million people, which equaled the population of the United States at that time. By 1995, the population had tripled to 450 million; in comparison, the U.S. population only reached 300 million in 2006.

Country	Population (millions) 2018	Population Density (per square kilometer)[1]	Rate of Natural Increase (RNI)	Total Fertility Rate	Life Expectancy		Percent Urban	Percent <15	Percent >65	Net Migration (rate per 1000)
					Male	Female				
Argentina	44.5	16	1.0	2.3	74	80	92	25	11	0
Bolivia	11.3	10	1.6	2.9	67	72	69	32	7	−1
Brazil	209.4	25	0.8	1.7	72	79	86	22	8	0
Chile	18.6	24	0.8	1.8	77	82	87	21	11	2
Colombia	49.8	44	0.9	2.0	73	79	77	26	8	−1
Costa Rica	5.0	96	0.9	1.7	78	83	73	22	8	2
Ecuador	17.0	67	0.0	2.5	74	79	64	29	7	0
El Salvador	6.5	308	1.3	2.3	69	78	70	28	8	−7
Guatemala	17.2	158	1.9	2.8	69	76	51	40	5	−1
Honduras	9.0	83	1.7	2.5	71	76	54	34	5	0
Mexico	130.8	66	1.3	2.2	75	80	73	27	7	−1
Nicaragua	6.3	52	1.5	2.2	72	78	59	30	5	−4
Panama	4.2	55	1.4	2.4	75	81	69	27	8	1
Paraguay	6.9	17	−2.0	2.5	71	75	61	30	6	−2
Peru	32.2	25	−1.0	2.4	72	78	78	27	7	−1
Uruguay	3.5	20	−1.0	2.0	74	81	95	21	15	−1
Venezuela	31.8	36	−1.0	2.4	73	79	88	26	7	−1

Source: Population Reference Bureau, *World Population Data Sheet,* 2018.
[1] World Bank Open Data 2018.

Log in to Mastering Geography & access MapMaster to explore these data!
1) Which countries have the highest population densities, and where are they located within the region?
2) Which countries have total fertility rates below 2.1? How do you think this will impact their population growth over the next 30 years?

Latin America outpaced the United States because infant mortality fell and life expectancy soared, while birth rates remained higher than that of the United States. In 1950, Brazilian life expectancy was only 43 years; by the 1980s, it was 63, and now it is 75. Four countries account for 70 percent of the region's population: Brazil with 209 million, Mexico with 131 million, Colombia with 50 million, and Argentina with 45 million (Table 4.1).

Patterns of Rural Settlement

Although the majority of people live in cities, some 120 million people do not. Throughout the region, a distinct rural lifestyle exists, especially among peasant subsistence farmers. In Brazil alone, more than 30 million people live in rural areas. Interestingly, the absolute number of people living in rural areas today is roughly equal to the number in the 1960s. Yet rural life has definitely changed. Along with subsistence agriculture, highly mechanized, capital-intensive farming occurs in most rural areas. The links between rural and urban areas are much improved, making rural areas less isolated. Also, as international migration increases, many rural communities are directly connected to cities in North America and Europe, with immigrants sending back remittances and supporting hometown associations. This is especially evident in rural Mexico and Central America. The rural landscape is divided by extremes of poverty and wealth, and the root of social and economic tension in the countryside is the uneven distribution of arable land.

Rural Landholdings   Control of land has been the basis for political and economic power in Latin America. Colonial authorities granted large tracts of land to the colonists, who were also promised the services of Indian laborers. These large estates typically took up the best lands along the valley bottoms and coastal plains. The owners were often absentee landlords, spending most of their time in the city and relying on a mixture of hired and slave labor to run their rural operations. Passed down from one generation to the next, many estates can trace their ownership back a century or more. The establishment of large blocks of estate land meant that peasants were denied territory

of their own, so they were forced to work for the estates. This long-observed practice of maintaining large estates is called **latifundia**.

Although the pattern of estate ownership is well documented, peasants have always farmed small plots for their own use. This practice of **minifundia** can lead to permanent or shifting cultivation. Small farmers typically plant a mixture of crops for subsistence as well as for trade. Peasant farmers in Colombia or Costa Rica, for example, grow corn, fruits, and various vegetables alongside coffee bushes that produce beans for export. Strains on the minifundia system occur when rural populations grow and land becomes scarce, forcing farmers to divide their properties into smaller and less productive parcels or seek out new parcels on steep slopes.

Much of the turmoil in 20th-century Latin America surrounded the issue of land ownership, with peasants demanding its redistribution through the process of **agrarian reform**. Governments have addressed these concerns in different ways. The Mexican Revolution in 1910 led to a system of communally held lands called *ejidos*. In the 1950s, Bolivia crafted agrarian reform policies that led to the government appropriating estate lands and redistributing them to small farmers. As part of the Sandinista revolution in Nicaragua in 1979, lands were taken from the political elite and converted into collective farms. In 2000, President Hugo Chavez ushered in a new era of agrarian reform in Venezuela, and Bolivian President Evo Morales introduced an agrarian reform program in 2006 to give land title to indigenous communities in the eastern lowlands. Each of these programs has met with resistance and proved to be politically and economically difficult to implement. Eventually, the path chosen by most governments has been to make frontier lands available to land-hungry peasants.

Agricultural Frontiers   The expansion into agricultural frontiers serves several purposes: providing peasants with land, tapping unused resources, and filling in blank spots on the map with settlers. Several frontier colonization efforts are noteworthy. Peru developed its *Carretera Marginal* (Perimeter Highway) in an effort to lure colonists into the cloud forests and rainforests of eastern Peru some four decades ago. Bolivia, Colombia, and Venezuela have all devised agricultural frontier schemes in their interior lowland tropical plains that attracted more large-scale investors than peasant farmers. Guatemala built roads to open the Petén region. Most recently, the Interoceanic Highway, completed in 2013, links ports in Peru with those in Brazil and opens up new settlement areas in the Amazonian territories of both countries (Figure 4.17).

The opening of the Brazilian Amazon for settlement was the region's most ambitious frontier colonization scheme. In the 1960s, Brazil began its frontier expansion by constructing several Amazonian highways, a new capital (Brasília), and state-sponsored mining operations. The Brazilian military directed the opening of the Amazon to provide an outlet for landless peasants and to extract the region's many resources. However, the generals' plans did not deliver as intended. Government-promised land titles, agricultural subsidies, and credits were slow to reach small farmers. Instead, too much money went to subsidizing large cattle ranches through tax breaks and improvement deals in which "improved" meant cleared forestland. Today five times more people live in the Amazon than in the 1960s; thus, continued human modification of this region is inevitable (see the purple arrows in Figure 4.18). Yet most of the people in the Brazilian Amazon live in large cities such as Manaus and Belém.

▼ **Figure 4.17 Puerto Maldonado, Peru** The recently constructed Puerto Maldonado Bridge in the Peruvian Amazon spans the Madre de Dios River. The bridge is a major infrastructural feature of the Interoceanic Highway, which connects Atlantic ports in Brazil with Pacific ports in Peru by traversing the Andes and the Amazon.

Explore the SIGHTS of Puerto Maldonado Bridge
https://goo.gl/sPx8f3

▲ **Figure 4.18 Major Latin American Migration Flows** Internal, intraregional, and international migrations have opened frontier zones (purple arrows) and created transnational communities (red and blue arrows). Over the past three decades, the flow of Latin Americans to the United States has grown. In 2016, the U.S. Census Bureau estimated there were 59 million people of Hispanic ancestry in the United States. Most of these people either were born in Latin America or have ancestral ties to Latin America.

The Latin American City

A quick glance at the population map of Latin America shows a concentration of people in cities (see Figure 4.16). One of the most significant demographic shifts has been the movement out of rural areas to cities, which began in earnest in the 1950s. Just one-quarter of the region's population was urban in 1950; the rest lived in small villages and the countryside. Today the pattern is reversed, with three-quarters of the population living in cities. In the most urbanized countries, such as Argentina, Chile, Uruguay, and Venezuela, more than 87 percent of the population lives in cities (see Table 4.1). This preference for urban life is attributed to cultural as well as economic factors; under Iberian rule, people residing in cities had higher social status and greater economic opportunity. Initially, only Europeans were allowed to live in the colonial cities, but this exclusivity was not strictly enforced. Over the centuries, colonial cities became the hubs for transportation and communication, making them the primary centers for economic and social activities.

Latin America is noted for high levels of **urban primacy**, a condition in which a country has a **primate city** three to four times larger than any other city in the country. Examples of primate cities are Mexico City, Lima, Caracas, Guatemala City, Panama City, Santiago, and Buenos Aires (Figure 4.19). Primacy is often viewed as a problem because too many national resources are concentrated into one urban center. In three cases, urban growth has led to the emergence of a megalopolis: the Mexico City–Puebla–Toluca–Cuernavaca area on Mexico's Mesa Central; the Niterói–Rio de Janeiro–Santos–São Paulo–Campinas axis in southern Brazil; and the Rosario–Buenos Aires–Montevideo–San Nicolás corridor in Argentina and Uruguay's lower Rio Plata Basin (see Figure 4.16).

Urban Form   Latin American cities have a distinct urban form that reflects both their colonial origins and their present-day growth (Figure 4.20). Usually, a clear central business district (CBD) exists in the old colonial core. Radiating out from the CBD is older middle- and lower-class housing found in the zones of maturity and

in situ (natural) accretion (areas of mixed levels of housing and services). In this model, residential quality declines moving from the center to the periphery. The exception is the elite "spine," a newer commercial and business strip that extends from the colonial core to newer parts of the city. Along the spine are superior services, roads, and transportation. The city's best residential zones and shopping malls are usually on either side of the spine. Close to the elite residential sector, a limited area of middle-class housing is typically found. Most major urban centers also have a *periférico* (a ring road or beltway highway) encircling the city. Industry is located in isolated areas of the inner city and in larger industrial parks outside the ring road.

Straddling the *periférico* is a zone of peripheral **squatter settlements**, where many of the urban poor live in self-built housing on land that does not belong to them. Services and infrastructure are uneven: Roads are unpaved, water and sewer systems are irregular, and the neighborhoods may not appear on city maps (see *Humanitarian Geography: Putting Squatter Settlements on the Map*). The dense ring of squatter settlements that encircles Latin American cities reflects the speed and intensity with which these zones were created. In some cities, more than one-third of residents live in self-built homes of marginal or poor quality. These kinds of dwellings are found throughout the developing world, yet the practice of building homes on the "urban frontier" has a longer history in Latin America than in most Asian and African cities. The combination of a rapid inflow of migrants, the inability of governments to meet pressing housing needs, and the eventual official recognition of many of these neighborhoods with land titles and utilities meant that this housing strategy was seldom discouraged. Each successful squatter settlement on the urban edge encouraged more.

Population Growth and Mobility

Latin America's high growth rates throughout the 20th century are attributed to natural increases as well as to immigration in the early

▼ **Figure 4.19 Primacy in Buenos Aires**   This capital city with a metropolitan population of over 13 million is the economic and cultural hub of Argentina. The bustling ceremonial boulevard, 9 de Julio Avenue, cuts through the downtown and commemorates Argentine Independence Day (July 9). The obelisk built in the early 20th century is an iconic structure for the city. Buenos Aires is both a primate city and a megacity (over 10 million people).

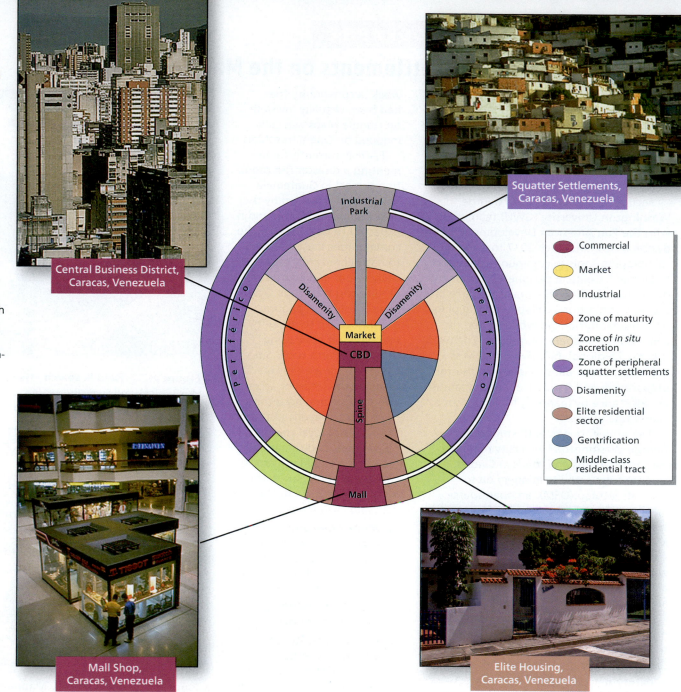

► **Figure 4.20 Latin American City Model** This urban model highlights the growth of Latin American cities and the class divisions within them. Although the central business district, elite spine, and residential sectors may have excellent access to services and utilities, life in the zone of peripheral squatter settlements (variously called *ranchos*, *favelas*, *barrios jóvenes*, or *pueblos nuevos*) is much more difficult. In many Latin American cities, one-third of the population resides in squatter settlements. **Q: How does the Latin American city model compare with that of North America in terms of where the rich and the poor live? What are the factors that drive urban growth?**

Central Business District, Caracas, Venezuela

Squatter Settlements, Caracas, Venezuela

Mall Shop, Caracas, Venezuela

Elite Housing, Caracas, Venezuela

Legend:
- Commercial
- Market
- Industrial
- Zone of maturity
- Zone of *in situ* accretion
- Zone of peripheral squatter settlements
- Disamenity
- Elite residential sector
- Gentrification
- Middle-class residential tract

part of the century. The 1960s and 1970s were decades of tremendous growth, resulting from high fertility rates and increasing life expectancy. In the 1960s, for example, a Latin American woman typically had six or seven children. By the 1980s, family sizes were half as large. Today the total fertility rate (TFR) for the region is 2.1, which is replacement value (see Table 4.1). Several factors explain this: more urban families, which tend to be smaller than rural ones; increased participation of women in the workforce; higher education levels of women; state support of family planning; and better access to birth control.

Even with family sizes shrinking—and in Brazil, Chile, Costa Rica, and Uruguay falling below replacement value—built-in potential for continued growth exists because of the relative demographic youth of these countries. The average percentage of the population below age 15 is 26 percent. In North America, that same group is 19 percent of the population, and in Europe, it is just 16 percent. This means that a proportionally larger segment of Latin Americans has yet to enter the childbearing years.

The population pyramids of two countries, Uruguay and Guatemala, contrast the profile of a country that has a stable population size with that of a demographically growing state (Figure 4.21). Uruguay is a small but prosperous country with a high human development index (HDI) ranking and relatively little poverty. Uruguayan women average two children, which is slightly below replacement level. Life expectancy is also high, but most population projections have the country growing very slowly between now and 2050. In contrast, Guatemala has a wider-based population pyramid and is considerably poorer. Total fertility rates have declined steadily in Guatemala and the average woman now has three children, but due to its youthful population and increased life expectancy, Guatemala's population is expected to increase to 27 million by 2050.

In addition to natural increase, waves of immigrants into Latin America and migrant streams within Latin America have influenced population size and patterns of settlement. Beginning in the late

Putting Squatter Settlements on the Map

Many settlements on Latin America's urban periphery are not mapped, so when emergencies happen, the residents are hard to assist. Mapping and surveying these areas improve the ability of communities and authorities to respond to hazards and improve services.

Faculty and students from George Washington University (GWU) teamed up with the Pan American Development Foundation (PADF) in March 2017 to bring digital mapping tools to community members in Ciudad Satélite, a peripheral settlement of Guatemala City. The goal is to make life-saving information available to local planners, humanitarian aid workers, and the community members themselves. The work is part of a global initiative called Missing Maps, which aims to improve disaster preparation and the delivery of humanitarian assistance by mapping the world's most vulnerable places.

The project began with students at George Washington University remotely tracing buildings and roads in Ciudad Satélite from satellite imagery on to OpenStreetMap (OSM), an open source platform. Then in Guatemala, the research team of geographers Nuala Cowan and Marie Price, along with master's students Andrii Berdnyk and Sudie Brown, worked with community members to validate and correct the maps using Field Papers, another open source tool (Figure 4.2.1). Community residents conducted surveys on the status of buildings as well as talked to household members to understand the threats that residents face and the available resources to combat these threats (Figure 4.2.2). The maps and information were then shared with the community for planning purposes. By the end of the

week, a community that had been virtually invisible on Google Maps was fully mapped on OpenStreetMap.

PADF is currently implementing a disaster risk reduction project in Guatemala. The project is funded by the government of Taiwan and is expected to help identify the communities most vulnerable to natural hazards, especially floods and landslides. Faculty and staff from GWU were involved in piloting this innovative community mapping initiative using the open source tools. Since these maps are accessible to anyone with Internet access, open-data mapping technologies can be readily put into the hands of local students, researchers, and community members to build grassroots capacity and prepare settlement residents for future risks and challenges.

1. **What environmental and natural disaster risks are common in informal urban settlements? What other problems challenge these communities?**

2. **Go to OpenStreetMap and see how your neighborhood is mapped. What could be added to the map?**

▲ **Figure 4.2.1 Field Research** Geography master's students, Andrii Berdnyk and Sudie Brown, lead an open source mapping project in Ciudad Satélite in Guatemala.

▲ **Figure 4.2.2 Participatory Mapping** Guatemalan community members and staff from the PADF review draft maps and discuss edits as neighborhood streets and buildings are added to OpenStreetMap.

GOOGLE EARTH
Virtual Tour Video
https://goo.gl/rE8vXE

19th century, new immigrants from Europe and Asia added to the region's size and ethnic complexity. Important population shifts within countries have also occurred in recent decades, as illustrated by the growth of Mexican border towns and the demographic expansion of the Bolivian plains. In an increasingly globalized economy, even more Latin Americans live and work outside the region, especially in the United States and Europe.

European Migration After Latin American countries gained independence from Spain and Portugal in the 19th century, their new leaders

sought to develop economically through immigration. Firmly believing that "to govern is to populate," many countries set up immigration offices in Europe to attract hard-working peasants to till the soils and "whiten" the *mestizo* (people of mixed European and Indian ancestry) population. The Southern Cone countries of Argentina, Chile, Uruguay, Paraguay, and southern Brazil were the most successful in attracting European immigrants from the 1870s until the depression of the 1930s. During this period, some 8 million Europeans arrived (more than the number who came during the entire colonial period), with Italians, Portuguese, Spaniards, and Germans being the most numerous.

▶ **Figure 4.21** Population Structure of Uruguay and Guatemala The two pyramids contrast the population structure of (a) the more developed and demographically stable Uruguay with (b) the youthful and growing Guatemala. The average Uruguayan woman has two children, whereas Guatemalan women average three children. Due in part to this difference, Uruguay is projected to have about the same population size in 2050, but Guatemala is projected to grow to 27 million people.

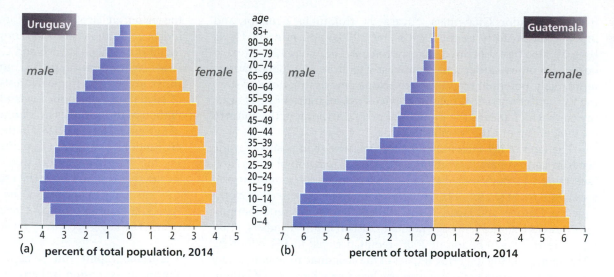

(a) percent of total population, 2014

(b) percent of total population, 2014

Asian Migration Less well known than the European immigrants to Latin America are the Asian immigrants, who also arrived during the late 19th and 20th centuries. Although considerably fewer, over time they established an important presence in the large cities of Brazil, Peru, Argentina, and Paraguay. Beginning in the mid-19th century, Chinese and Japanese laborers were contracted to work on the coffee estates in southern Brazil and the sugar estates and coastal mines of Peru. Over time, these Asian immigrants became prominent members of society; for example, Alberto Fujimori, a son of Japanese immigrants, was president of Peru from 1990 to 2000.

Between 1908 and 1978, one-quarter of a million Japanese immigrated to Brazil; today the country is home to 1.5 million people of Japanese descent (Figure 4.22). As a group, the Japanese have been closely associated with the expansion of soybean and orange production. Increasingly, second- and third-generation Japanese have taken professional and commercial jobs in Brazilian cities; many have married outside their ethnic group and are losing their fluency in Japanese. South America's economic turmoil in the 1990s encouraged many ethnic Japanese to emigrate to Japan in search of better wages. Nearly one-quarter of a million ethnic Japanese, mostly from Brazil and Peru, left South America in the 1990s and now work in Japan.

Latino Migration and Hemispheric Change Migration within Latin America and between Latin America and North America has significantly impacted both sending and receiving communities. Within Latin America, international migration is shaped by shifting economic and political realities. For decades Venezuela's oil wealth attracted Colombian immigrants, yet now waves of Venezuelan migrants are fleeing economic turmoil and entering Colombia, Brazil, and Panama. Argentina has long been a destination for Bolivian and Paraguayan laborers. Chile's booming economy has become an attractive destination for Peruvian migrants. Nicaraguans seek employment in Costa Rica. And farmers in the United States have depended on Mexican laborers for more than a century.

Political turmoil has also sparked international migration. The bloody civil wars in El Salvador and Guatemala in the 1980s, for example, sent waves of refugees into Mexico and the United States. Violence in northern Central America is again driving people toward Mexico and the United States. In 2014 and 2016, there were spikes in the numbers of unaccompanied minors from Guatemala, El Salvador, and Honduras crossing into the United States and turning themselves over to authorities as they sought to reunite with family in the United States and to find refuge from gang-driven violence in Central America. Although some of these youth were able to seek asylum status, many were returned to Central America. Moreover, Central American countries actively dissuaded people from sending their youth to the north because of the many dangers involved in the crossing.

Most U.S. Hispanics have ancestral ties with peoples from Latin America and the Caribbean (see Chapter 5 on Caribbean migration). Presently, Mexico is the country of origin for most documented Hispanic immigrants to the United States; two-thirds of the 58 million U.S. Hispanics claim Mexican ancestry, including approximately 12 million who were born in Mexico. Mexican labor migration to the United States dates back to the late 1800s, when relatively unskilled labor was recruited to work in agriculture, mining, and railroads. These immigrants are concentrated in California and Texas but increasingly are found throughout the country. Although Mexicans continue to have the greatest presence among Latinos in the United States, the number of migrants

▼ **Figure 4.22** Japanese Brazilians In 1908, the first Japanese immigrants arrived as agricultural workers, choosing Brazil as a destination after the United States and Canada had banned Japanese immigration. Today there are over 1.5 million ethnic Japanese in Brazil, mainly in the states of São Paulo and Paraná. Here people dance the samba at an event where the Japanese Brazilian community gathered to watch a soccer match between Brazil and Japan in São Paulo.

from El Salvador, Guatemala, Nicaragua, Colombia, Ecuador, and Brazil grew steadily in the 1990s and 2000s, and migrant flows from Mexico have declined sharply since 2010. As the U.S.–Mexico border becomes harder and more costly to cross, and deportations more likely, potential migrants are seeking new destinations or staying in their countries of origin. Mexico's net migration rate is nearing zero, which means the number of people arriving in the country are cancelling out the number leaving.

The rates of net migration are more negative than positive for the states in Latin America, which means it is still a region of emigration (see Table 4.1). Both skilled and unskilled immigrants from the region are an important source of labor within Latin America, as well as North America, Europe, and Japan. Many of these immigrants send monthly **remittances** (monies sent back home) to sustain family members. Peaking at nearly US$70 billion in 2008, remittances fell to $61 billion by 2013, indicating the lingering impact of the economic recession. Yet by 2017, remittances to Latin America reached a new high of $79 billion, with $27 billion going to Mexico alone. The economic significance of remittances will be discussed again at the end of the chapter.

REVIEW

4.3 What are the historical and economic explanations for urban dominance and urban primacy in Latin America?

4.4 How have policies such as agrarian reform and frontier colonization impacted the patterns of settlement and primary resource extraction in the region?

4.5 Demographically, Latin America has grown much faster than North America. What factors contribute to its faster growth, and is this growth likely to continue?

KEY TERMS latifundia, minifundia, agrarian reform, urban primacy, primate city, squatter settlement, *mestizo*, remittance

Cultural Coherence and Diversity: Repopulating a Continent

The Iberian colonial experience (1492 to the 1800s) imposed a political and cultural coherence on Latin America that makes it recognizable today as a world region. Yet this was not a simple transplanting of Iberia across the Atlantic. Instead, a process unfolded in which European and Indian traditions blended as indigenous groups were added into either the Spanish or the Portuguese empires. Indian cultures have shown remarkable resilience in some areas, as evidenced by the survival of Amerindian languages. However, the prevailing pattern was one of forced assimilation in which European religion, languages, and political organization were imposed on surviving Amerindian societies. Later, other cultures—especially more than 10 million enslaved Africans—added to the cultural mix of Latin America, the Caribbean, and North America. The legacy of the African slave trade is examined in greater detail in Chapters 5 and 6.

The Decline of Native Populations

It is difficult to grasp the enormity of cultural change and human loss due to this encounter between the Americas and Europe. Throughout the region, archaeological sites are reminders of the complexity of Amerindian civilizations prior to European contact. Dozens of stone temples found throughout Mexico and Central America, where the Mayan and Aztec civilizations flourished, attest to the ability of these societies to thrive in the area's tropical forests and upland plateaus. The Mayan city of Tikal flourished in the lowland forests of Guatemala, supporting tens of thousands, before its mysterious collapse centuries before the arrival of Europeans (Figure 4.23). In the Andes, the complexity of Amerindian civilizations can be seen in political centers such as Cuzco and Machu Picchu. The Spanish, too, were impressed by the sophistication and wealth they saw around them, especially in Tenochtitlán, where Mexico City sits today. Tenochtitlán, the political and ceremonial center of the Aztecs, supported a complex metropolitan area with some 300,000 residents. The largest city in Spain at the time was considerably smaller.

The Demographic Toll The most telling indicators of the impact of European expansion in Latin America are demographic. It is widely believed that the precontact Americas had 54 million inhabitants; by comparison, western Europe in 1500 had approximately 42 million people. Of the 54 million, about 47 million were in what is now Latin America, and the rest were in North America and the Caribbean. By 1650, after a century and a half of colonization, the indigenous population was one-tenth its precontact size. The human tragedy of this population loss is hard to comprehend. The relentless elimination of 90 percent of the native population was largely caused by epidemics

▼ **Figure 4.23 Tikal, Guatemala** This ancient Mayan city, located in the lowland forests of the Petén, was part of a complex network of cities located in the Yucatan and northern Guatemala. At its height, Tikal supported over 100,000 people before its collapse in the late 10th century. Today it is a major tourist destination.

of influenza and smallpox, but warfare, forced labor, and starvation due to the collapse of food-production systems also contributed to the rapid population decline.

The Columbian Exchange and Global Food Systems

Historian Alfred Crosby likens the contact period between the Old World (Europe, Africa, and Asia) and the New World (the Americas) to an immense biological swap, which he terms the **Columbian Exchange**. According to Crosby, Europeans benefited greatly from this exchange, and Amerindian peoples suffered the most from it. On both sides of the Atlantic, however, the introduction of new diseases, peoples, plants, and animals forever changed the human ecology, especially the foods we eat.

Consider, for example, the introduction of Old World crops. The Spanish brought their staples of wheat, olives, and grapes to plant in the Americas. Wheat did surprisingly well in the highland tropics and became a widely consumed grain over time. Grapes and olive trees did not fare as well, but eventually grapes were produced commercially in the temperate zones of the Americas. The Spanish grew to appreciate the domestication skills of Indian agriculturalists, who had developed valuable starch crops such as corn, potatoes, and bitter manioc, as well as condiments such as hot peppers, tomatoes, pineapple, cacao, and avocados. Corn never became a popular food for direct consumption by Europeans, but many African peoples adopted it as a vital staple. And corn as an animal grain and a sweetener (corn syrup) is an integral part of most foods today. Yet for many Mexicans and Central Americans, corn tortillas are considered a basic and daily food item (Figure 4.24).

The movement of Old World animals across the Atlantic had a profound impact on the Americas. Initially, these animals hastened Indian decline by introducing animal-borne diseases and by producing feral offspring that consumed everything in their paths. However, the utility of domesticated swine, sheep, cattle, and horses was eventually appreciated by native survivors. Draft animals were adopted, as was the plow, which facilitated the preparation of soil for planting. Wool became a very important fiber for indigenous communities in the uplands. Slowly, pork, chicken, and eggs added protein and diversity to the staple diets of corn, potatoes, and cassava. With the major exception of disease, many transfers of plants and animals ultimately benefited both worlds. Still, it is clear that the ecological and material basis for life in Latin America was completely reworked through the exchange process initiated by Columbus.

Indian Survival Presently, Mexico, Guatemala, Ecuador, Peru, and Bolivia have the largest indigenous populations. Not surprisingly, these areas had the densest native populations at the time of European contact. Indigenous survival also occurs in isolated settings where the workings of national and global economies are slow to break through, such as in eastern Panama, the Miskito Coast of Honduras, and the roadless sections of western Amazonia.

In many cases, Indian survival comes down to one key resource—land. Indigenous peoples who are able to maintain a territorial home, formally through land title or informally through long-term occupancy, are more likely to preserve a distinct ethnic identity. Because of this close association between identity and territory, native peoples are increasingly insisting on a recognized space within their countries. Some of Panama's indigenous groups have organized territories called *comarcas*, where they assert local authority and have limited autonomy. The *comarca* of Guna Yala, on Panama's Caribbean coast, is the recognized territory of some 40,000 Guna. These efforts to define indigenous territory are seldom welcomed by the state, but they are occurring throughout the region.

Patterns of Ethnicity and Culture

The Amerindian demographic collapse enabled Spain and Portugal to reshape Latin America into a European likeness. However, instead of a neo-Europe rising in the tropics, a complex ethnic blend evolved. Within the first years of contact, unions between European sailors and Indian women began the process of racial mixing that over time became a defining feature of the region. The courts of Spain and Portugal officially discouraged racial mixing, but such positions could not be realistically enforced in the colonial territories.

Generations of intermarriage led to four broad racial categories: *blanco* (European ancestry), *mestizo* (mixed ancestry), *indio* (Indian ancestry), and *negro* (African ancestry). The *blancos* (or Europeans) continue to be well represented among the elites, yet the vast majority of Latin Americans are of mixed racial ancestry. *Dia de la Raza*, the region's observance of Columbus Day, recognizes the emergence of a new *mestizo* race as the legacy of European conquest. Throughout Latin America, more than other regions of the world, miscegenation (or racial mixing) is the norm, making the process of mapping racial or ethnic groups especially difficult.

Languages Roughly two-thirds of Latin Americans are Spanish speakers, and one-third speak Portuguese. These colonial languages were so widespread by the 19th century that they were the unquestioned languages of government and education for the newly independent Latin American republics. In fact, until recently, many countries actively discouraged, and even repressed, Indian tongues. It took a constitutional amendment in Bolivia in the 1990s to legalize native-language instruction in primary schools and to recognize the country's multiethnic heritage (more than half the population is Indian; and Quechua, Aymara, and Guaraní are widely spoken) (Figure 4.25).

Because Spanish and Portuguese dominate, there is a tendency to overlook the influence of indigenous languages in the region.

▼ **Figure 4.24 Corn Tortillas** This basic food is consumed daily by many Mexicans and northern Central Americans. These women prepare tortillas for sale on the streets of Antigua, Guatemala. Corn was first domesticated in Mexico but is grown throughout the world as a food and animal grain.

Explore the TASTES of Tortillas
http://goo.gl/p7XnNY

Mapping the use of native languages, however, reveals important areas of Indian resistance and survival. In the Central Andes of Peru, Bolivia, and southern Ecuador, more than 10 million people still speak Quechua and Aymara, along with Spanish. In Paraguay and

Explore the SOUNDS of Quechua
http://goo.gl/9sNLBG

lowland Bolivia, there are 4 million Guaraní speakers, and in southern Mexico and Guatemala, at least 6 to 8 million speak Mayan languages. Small groups of native-language speakers are found scattered throughout the sparsely settled interior of South America and the more isolated forests of Central America. However, many of these languages have fewer than 10,000 speakers.

Explore the SOUNDS of Mayan
http://goo.gl/5UuHDJ

▼ **Figure 4.25** **Language Map of Latin America** The dominant languages of Latin America are Spanish and Portuguese. Nevertheless, there are significant areas in which native languages still exist and, in some cases, are recognized as official languages. Smaller language groups exist in Central America, the Amazon Basin, and southern Chile. **Q: What does this language map tell us about the patterns of Amerindian survival and endurance in Latin America?**

DOMINANT/OFFICIAL* LANGUAGES
- Spanish
- Portuguese

INDIGENOUS LANGUAGES
1. Aymara
2. Embera
3. Garifuna
4. Guaraní
5. Quechua
6. Guna
7. Mapuche
8. Mayan
9. Miskitu
10. Mixtec
11. Nawan/Spanish
12. Pemon
13. Zapotec
14. Wahiro
15. Yamomani
- Dispersed indigenous-language communities

*Multiple Official Languages:
*Bolivia: Spanish, Quechua, Aymara, Guaraní
*Peru: Spanish, Quechua

Christian Religions Like language, the Roman Catholic faith appears to have been imposed on Latin America without challenge. Most countries report 90 percent or more of their population as Catholic. Every major city has dozens of churches, and even the smallest hamlet maintains a graceful church on its central square. Throughout Central America, Brazil, and Uruguay, however, a sizable portion of the population practice Protestant evangelical faiths.

The demographic core of the Catholic Church is in Latin America, not Europe. Worldwide, Brazil has the largest Catholic population (150 million) followed by Mexico (106 million). In 2013, a new pope was selected, and for the first time the spiritual leader of nearly 1.2 billion Roman Catholics was from Latin America. Pope Francis, formerly Bishop Jorge Mario Bergoglio, is the son of Italian immigrants and was born in Buenos Aires, Argentina. As a religious leader and member of the Jesuit Order, he earned a reputation for his humility and devotion to the poor. Now, as leader of the Catholic Church, he oversees a vast global network of churches, schools, missions, and clergy that look to his guidance from Rome.

Throughout Latin America, **syncretic religions**—blends of different belief systems—enabled animist practices to be included in Christian worship. The syncretic blend of Catholicism with African traditions is most obvious in the celebration of carnival, Brazil's most popular festival and one of the major components of Brazilian national identity. The three days of carnival, known as the Reign of Momo, combine Christian Lenten beliefs with pagan influences and feature African musical traditions epitomized by the rhythmic samba bands. Although the street festival was banned for part of the 19th century, Afro-Brazilians in Rio de Janeiro resurrected it in the 1880s with nightly parades, music, and dancing. Within 50 years, the street festival had given rise to formalized samba schools and helped break down racial barriers. By the 1960s, carnival became an important symbol for Brazil's multiracial national identity. Today, the festival—which is most closely associated with Rio de Janeiro—draws tens of thousands of participants from all over the world (Figure 4.26).

Explore the SOUNDS of the Samba
http://goo.gl/LWiOpK

The Zumba Sensation

Go to a local gym anytime or to an urban park in the summer, and you may hear the pulsing sounds of merengue or reggaeton and see an enthusiastic, well-toned instructor leading a class in a fast-paced hip-swinging exercise dance called zumba. Zumba began in Colombia. The story goes that a popular fitness teacher, Alberto Pérez, was leading an aerobics class in the 1990s but forgot his music. He rushed to his car and grabbed cassettes featuring cumbia and salsa. As he improvised the steps to this Latin music, the students loved it. Perez eventually moved to Miami, trademarked the class style, and called it zumba, a word with no meaning but easy to remember. Zumba took off in 2001 and hasn't slowed down.

Zumba's appeal is its vast sample of Latin music including salsa, samba, mambo, and soca. The dance moves are a mix of Latin American–inspired steps mixed with aerobics. There are even Zumba performing artists, such as pop star Cláudia Leitte from Brazil, who performs some of the most popular Zumba songs. The combination of fast tempo, upbeat music, and fun moves has made Zumba a global sensation.

While most Zumba enthusiasts are women, men are welcome to shake it as well. There are Zumba classes for all ages and ability levels. In cities all over the world, people will exercise if it's fun—and Zumba fits the bill. In 2015, 13,000 Filipinos gathered for an outdoor Zumba class, breaking the Guinness World Record for the largest class size (Figure 4.3.1).

▲ **Figure 4.3.1 Zumba in the Philippines** What began in Colombia as an aerobics alternative with Latin music has become a global exercise craze. Nearly 13,000 Filipinos in the town of Mandaluyong participated in a zumba session in July 2015, earning a place in the Guinness World Records for the largest zumba class.

1. **Identify factors that make this type of group exercise globally appealing.**

2. **Have you ever taken a Zumba class and, if so, where?**

The Global Reach of Latino Culture

Latin American culture, vivid and diverse as it is, is widely recognized throughout the world. Whether it is the sultry pulse of the tango or the fanaticism with which Latinos embrace soccer as an art form, aspects of Latin American culture have been absorbed into the global culture. In the arts, Latin American writers such as Jorge Luis Borges, Gabriel García Márquez, and Isabel Allende have obtained worldwide recognition. In terms of popular culture, musical artists such as Colombia's Shakira and Brazil's hip-hop samba singer Max de Castro have international audiences. Through music, literature, art, and even sport, Latino culture is being transmitted to an eager worldwide audience (see *Globalization in Our Lives: The Zumba Sensation*).

Soccer Perhaps the quintessential global sport, soccer has a fanatical following throughout much of the world. Yet it is Latin America, and especially South America, where *fútbol* is considered a cultural necessity. Still largely a male game, young boys and men are constantly seen on fields, beaches, and blacktops playing soccer, especially in late afternoons and on weekends. This is beginning to change in some countries as women take up the sport, especially in Brazil. The great soccer stadiums of Buenos Aires (La Bombonera) and Rio de Janeiro (Maracaña) are regarded as shrines to the game. (Maracaña Stadium was also the site for the opening ceremony of the 2016 Olympics). Many fans use the victories and losses of their national soccer teams as important chronological markers of their lives. And as Latin Americans emigrate (both as players and laborers), they bring their enthusiasm for the sport with them.

▲ **Figure 4.26 Carnival in Rio de Janeiro** Samba schools, such as this one, compete each year during Carnival for the best costumes and music. Rio de Janeiro's Carnival is a spectacle that draws thousands of revelers.

REVIEW

4.6 What factors contributed to racial mixing in Latin America, and where are the areas of strongest Amerindian survival?

4.7 What are the cultural legacies of Iberia in Latin America, and how are they expressed?

KEY TERMS Columbian Exchange, syncretic religion

Geopolitical Framework: Redrawing the Map

Latin America's colonial history, more than its current condition, unifies this region geopolitically. For the first 300 years after the arrival of Columbus, Latin America was a territorial prize sought by various European countries, but the contest was effectively won by Spain and Portugal. By the 19th century, the independent states of Latin America had formed but continued to experience foreign influence and sometimes overt political pressure, especially from the United States. Such a situation can be termed **neocolonialism**. At various times, a more neutral hemispheric vision of American relations and cooperation has held sway, represented by the formation of the **Organization of American States (OAS)**. The present OAS was officially formed in 1948, but its origins date to 1889. Yet there is no doubt that U.S. policies toward trade, economic assistance, political development, and sometimes military intervention are often seen as undermining the independence of these states.

Today the geopolitical influence of the United States in the region is declining, especially in South America. For many South American countries, trade with the European Union, China, and Japan is as important as, if not more important than, trade with the United States. For example, Brazil's largest trading partner is now China. And as Latin America's largest economy and host of the 2016 Summer Olympics,

Brazil's own influence in the region and the world is rising. Trade blocs such as the Pacific Alliance, Mercosur, and UNASUR are reshaping patterns of trade and political engagement.

Iberian Conquest and Territorial Division

When Christopher Columbus claimed the Americas for Spain, the Spanish became the first active colonial agents in the Western Hemisphere. In contrast, the Portuguese presence in the Americas was the result of the **Treaty of Tordesillas** in 1493–1494. By that time, Portuguese navigators had charted much of the coast of Africa in an attempt to find a water route to the Spice Islands (Moluccas) in Southeast Asia. With the help of Columbus, Spain sought a western route to the Far East. When Columbus discovered the Americas, Spain and Portugal asked the Pope to settle how these new territories should be divided. Without consulting other European powers, the Pope divided the Atlantic world in half: The eastern half, containing the African continent, was awarded to Portugal; the western half, with most of the Americas, was given to Spain. The line of division established by the treaty actually cut through the eastern part of South America, placing it under Portuguese rule. The treaty was never recognized by the French, English, or Dutch, who also claimed territory in the Americas, but it did provide the legal justification for the creation of Portuguese Brazil, which would later become the largest and most populous country in Latin America (Figure 4.27a).

◀ **Figure 4.27 Shifting Political Boundaries** (a) The evolution of Latin American political boundaries began with the 1494 Treaty of Tordesillas, which gave much of the Americas to Spain and a slice of South America (Brazil) to Portugal. The larger Spanish territory was gradually divided into viceroyalties and *audiencias*, which formed the basis for many modern national boundaries. (b) The 1830 borders of these newly independent states were far from fixed. Bolivia lost its access to the coast, Peru gained much of Ecuador's Amazon, and Mexico was stripped of its northern territory by the United States.

Six years after the treaty was signed, Portuguese navigator Pedro Álvares Cabral accidentally reached the coast of Brazil on a voyage to southern Africa. The Portuguese soon realized that this territory was on their side of the Tordesillas line. Initially, they were unimpressed by what Brazil had to offer; there were no spices or major native settlements. Over time, they came to appreciate the utility of the coast as a provisioning site as well as a source for brazilwood, used to produce a valuable dye. Portuguese interest in the territory intensified in the late 16th century, with the development of sugar estates and the expansion of the slave trade, and in the 17th century, with the discovery of gold in the Brazilian interior.

Spain, in contrast, aggressively pursued the conquest and settlement of its new American territories from the very start. After discovering little gold in the Caribbean, by the mid-16th century Spain directed its energy toward developing the silver resources of Central Mexico and the Central Andes (most notably Potosí in Bolivia). Gradually, the economy diversified to include some agricultural exports, such as cacao (for chocolate) and sugar, as well as a variety of livestock. In terms of foodstuffs, the colonies were virtually self-sufficient. Manufacturing, however, was forbidden in the Spanish-American colonies in order to keep them dependent on Spain.

Revolution and Independence   Not until the rise of revolutionary movements between 1810 and 1826 was Spanish authority on the mainland challenged. Ultimately, European elites born in the Americas gained control, displacing leaders loyal to the crown. In Brazil, the evolution from Portuguese colony to independent republic was a slower and less violent process that spanned eight decades (1808–1889). Brazil was declared a separate kingdom from Portugal, with its own king, and later became a republic.

The territorial division of Spanish and Portuguese America into administrative units provided the legal basis for the modern states of Latin America (Figure 4.27b). The Spanish colonies were first divided into two viceroyalties (New Spain, which included what is now Mexico and most of Central America; and Peru, which included all of Spanish South America), and later subdivided into smaller viceroyalties that became the basis for the modern states. Unlike Brazil, which evolved from a colony into a single republic, the former Spanish colonies experienced fragmentation in the 19th century. For a brief period in the early 19th century, five Central American states formed the United Provinces of Central America, and Gran Colombia included present-day Venezuela, Colombia, Ecuador, and Panama (see Figure 4.27b).

As the colonial administrative units turned into states, it became clear that the territories were not clearly delimited, especially the borders that stretched into the sparsely populated interior of South America. This would later become a source of conflict as the new states struggled to demarcate their boundaries. Numerous border wars erupted in the 19th and 20th centuries, and the map of Latin America has been redrawn many times. Some notable conflicts were the War of the Pacific (1879–1882), in which Chile expanded to the north and Bolivia lost its access to the Pacific; warfare between Mexico and the United States in the 1840s, which resulted in the present border under the Treaty of Hidalgo (1848); and the War of the Triple Alliance (1864–1870), the bloodiest war of the postcolonial period, in which Argentina, Brazil, and Uruguay allied themselves to defeat Paraguay in its claim to control the upper Paraná River Basin. It is estimated that 90 percent of Paraguay's adult males died in this conflict. Sixty years later, the Chaco War (1932–1935) resulted in a territorial loss for Bolivia in its eastern lowlands and a gain for Paraguay. In the 1980s, Argentina lost a war with Great Britain over control of the Falkland, or Malvinas, Islands in the South Atlantic. As recently as 1998, Peru and Ecuador skirmished over a disputed boundary in the Amazon Basin.

The Trend Toward Democracy   Most of Latin America's 17 countries have or will soon celebrate their bicentennials. Compared with most of the rest of the developing world, Latin Americans have been independent for a long time. Yet political stability is not a characteristic of the region. Among these countries, some 250 constitutions have been written since independence, and military takeovers have been alarmingly frequent. Since the 1980s, however, the trend has been toward democratically elected governments, the opening of markets, and broader public participation in the political process. Where dictators once outnumbered elected leaders, by the 1990s, each country in the region had a democratically elected president. Yet documented electoral fraud in the presidential elections in Venezuela (2013 and 2018) and Honduras (2017) have undermined the democratic process in these countries.

Democracy may not be enough for the millions frustrated by the slow pace of political economic reform rising crime and corruption. In survey after survey, Latin Americans reveal their dissatisfaction with corrupt politicians and nontransparent governments. Many of the democratic leaders have been free-market reformers who are quick to eliminate state-backed social safety nets such as food subsidies, government jobs, and pensions, resulting in serious hardship for the poor. Corruption charges led to the impeachment and ouster of Brazil's first female president, Dilma Rousseff, in 2016. Due to the turmoil in Venezuela, President Madura promised to hold elections in 2018 but also banned the most popular opposition candidate, Leopold López, from running, and placed him under house arrest.

Democratic institutions are also being challenged by a rise in crime and concerns for personal security. A dangerous combination of the narcotics trade, gangs, youth unemployment, and readily available firearms has driven up homicide rates in the region. Moreover, only a small fraction of those responsible for the deaths are ever tried in court. According to InSight Crime, the Latin American countries with the highest homicide rate in 2017 were Venezuela at 89 per 100,000 and El Salvador at 60 per 100,000. Brazil (29.7) and Mexico (22.5) had far lower rates, but given their much larger populations, these rates accounted for a staggering 61,000 and 29,000 homicides respectively. By comparison, the U.S. homicide rate in 2016 was estimated at 5 per 100,000.

The threat of violence also silences political opposition and the press, which undermines democracy. In Mexico, journalists are regularly threatened, and 11 were killed in 2016. In March 2018 a popular Rio de Janeiro Councilor, Marielle Franco, was gunned down in her car along with her driver in what appeared to be a targeted assassination (Figure 4.28). Protests across Brazil were held to honor Franco, who was a voice for the disadvantaged in Brazil's favelas and for LGBTQ rights.

Regional Trade and Crime

At the same time that governments struggle to address the pressing needs of their countries, regional trade groups and crime organizations pose unique opportunities and challenges to the authority of Latin American leaders. Various regional groups have formed to support intraregional trade and development, as well as to shore up opportunities outside the region. Similarly, various insurgencies and organized crime groups have created instability in the region. Some of the more destabilizing groups have been the drug cartels.

Trade Blocs In the 1990s, **Mercosur (the Southern Cone Common Market)** and NAFTA emerged as supranational structures that could influence development (Figure 4.29). For Latin America, lessons learned from Mercosur in particular led to the creation of **UNASUR (Union of South American Nations)** in 2008, uniting virtually all of South America, and the **Pacific Alliance** in 2011 to expand trade with Asia.

NAFTA took effect in 1994 as a free trade area that would gradually eliminate tariffs and ease the movement of goods among the member countries (Mexico, the United States, and Canada). NAFTA increased intraregional trade, but it has provoked considerable controversy and was renegotiated in 2018 (see Chapter 3).

▲ **Figure 4.29 Geopolitics and Trade Blocs in Latin America**
Of the five economic trade blocs shown, Mercosur and NAFTA are the most dynamic. As UNASUR develops, it could unite all of South America into a single common market, but that seems unlikely at this point. Members of the Central American Common Market signed an agreement in 2004 to form CAFTA (Central American Free Trade Agreement), which also includes the Dominican Republic. **Q: How could the growth and strength of trade blocs impact how Latin America functions as a region?**

Guatemala/Mexico Border. *Increased border enforcement to deter Central Americans from migrating north.*

Crisis in Venezuela. *In 2017 Venezuela had the region's highest homicide rate. Political instability led to the country's 2016 suspension from Mercosur and a growing flow of refugees to neighboring states.*

Brazil. *President Rousseff was removed from office in 2016. In October 2018, far-right candidate Jair Bolsonaro was elected president on an anti-crime platform.*

U.S. – Mexican Border. *Intensified border security since 1996 over contraband trade and undocumented migration.*

Colombia. *A fragile peace agreement between the FARC and President Santos was signed in 2016 after nearly 50 years of fighting and 220,000 deaths.*

ECONOMIC TRADE BLOCS
- CAFTA
- Mercosur
- USMCA (formerly NAFTA)
- Pacific Alliance
- UNASUR

Falkland/Malvinas Islands. *Territorial dispute between the United Kingdom and Argentina.*

▲ **Figure 4.28 Marielle Franco** The Brazilian politician, human rights activist, and feminist in her office in Rio de Janeiro, where she served on the city council. Franco was outspoken about the high levels of violence and police brutality, especially in the *favelas*. In 2018 she was assassinated in her car along with her driver.

NAFTA did prove, however, that a free trade area combining industrialized and developing states was possible. In 2004, the United States, five Central American countries—Guatemala, El Salvador, Nicaragua,

Honduras, and Costa Rica—and the Dominican Republic signed **CAFTA (Central American Free Trade Agreement)**. Like NAFTA (renamed the United States-Mexico-Canada Agreement [USMCA] in 2018), CAFTA aims to increase trade and reduce tariffs among member countries. The treaty was fully ratified in 2009, but whether it will lead to more economic development in Central America is a much-debated question.

Mercosur was formed in 1991 with Brazil and Argentina—the two largest economies in South America—and the smaller states of Uruguay and Paraguay as members. Since its formation, trade among these countries has grown so much that Venezuela joined as a full member in 2012. In 2016, Mercosur suspended Venezuela's membership indefinitely to put pressure on the Maduro government to restore democracy. Mercosur's success is significant in two ways: It reflects the growth of these economies and their willingness to put aside old rivalries (especially long-standing antagonisms between Argentina and Brazil) for the economic benefits of cooperation, and it recognizes free movement of peoples within member countries for work and legal residence, which is something that NAFTA (USMCA) or CAFTA do not.

In 2008, Brazil initiated the formation of UNASUR, which includes all countries in South America except for French Guiana, a French territory. UNASUR was formally organized with a permanent secretariat and has responded to political crises in Bolivia (2008), Ecuador (2010), and Paraguay (2012). Significantly, the organization was a Brazilian-led effort, not one led by the United States. As Latin America's most populous country and largest economy, Brazil's role in UNASUR underscores its clout in South American development and its larger geopolitical ambitions to secure a permanent seat on the United Nations Security Council.

In contrast, the newer Pacific Alliance has already made many concrete moves to promote cooperation and trade among the member states and to increase bargaining power when creating trade deals with Asia, especially China (see Figure 4.29). This alliance is also seen as a geopolitical counterweight to Brazil's influence in the region. Led by Mexico, Colombia, Peru, and Chile, the members of the Pacific Alliance represent nearly half of the region's world trade.

Insurgencies and Drug Cartels   Guerrilla groups such as the Revolutionary Armed Forces of Columbia (FARC) have controlled large territories of their countries through the support of those loyal to the cause, along with theft, kidnapping, and violence. In Columbia, the FARC, along with the ELN (National Liberation Army), gained wealth and weapons through the drug trade. The level of violence in Colombia escalated further with the rise of paramilitary groups—armed private groups that terrorize those sympathetic to insurgency. The paramilitary groups have been blamed for hundreds of politically motivated murders each year. As many as 2.5 million Colombians have been internally displaced by violence since the late 1980s, most fleeing rural areas for towns and cities. Fortunately, after more than a decade of negotiations, a 2014 agreement for unilateral cessation of FARC hostilities was implemented, and the levels of violence declined. Interestingly, in 2016, Colombian voters narrowly rejected the peace agreement because they felt the terms were too lenient toward the FARC, and after the stunning vote, President Santos ushered in a revised peace agreement. In 2017, Colombia's homicide rate of 27 per 100,000 was the lowest level in 40 years. Yet Colombia remains the world's largest cocaine producer, followed by Peru and Bolivia.

Drug cartels and gangs in states as diverse as Mexico, Guatemala, El Salvador, Honduras, and Brazil have been blamed for increases in violence and lawlessness. The spike in violence and corruption in Mexico has been especially destabilizing. Profiting from the illegal production and/or shipment of cocaine, marijuana, methamphetamine, and heroin, the cartels generate billions of dollars. Beginning in 2006, Mexico's government brought in the army to quell the violence, kidnapping, and intimidation brought on by cartel groups, especially in the border region, but now extending throughout Mexico and into Central America (Figure 4.30). Some cartel leaders were captured, including the notorious leader of the Sinaloa Cartel, El Chapo (Joaquín Guzmán), who was extradited to the United States in 2017 to face criminal charges.

The Mexican government reported some 151,000 murders from 2006 to 2015 and more than 20,000 additional "disappeared" persons; many of these murders are at the hands of the cartels. Finding ways to stem the violence (rather than the flow of drugs) was one of the biggest issues in the 2012 Mexican election. Enrique Peña Nieto promised a shift toward police work and judicial reform to reduce violence, but accusations of corruption and a lack of progress on crime marred his

ORGANIZED CRIME GROUPS*

🟠 **Sinaloa**
Sinaloa Federation
Beltran Leyva Organization
Los Mazatlecos
El Chapo Trini/El Cadete

🟣 **Tierra Caliente**
Cartel de Jalisco Nueva Generación
Knights Templar
La Familia Michoacana
Guerreros Unidos
Los Rojos
Independent Cartel of Apapulco (CIDA)
Los Viagra

🟢 **Tamaulipas**
Los Zetas
Gulf Cartel (Velazquez network)
Gulf Cartel gangs

*Some organized crime groups overlap

DRUG ROUTES
→ All drug traffic
┈┈▸ Marijuana and methamphetamine traffic
→ Methamphetamine precursor supply lines
┄·▸ Cocaine traffic

▲ **Figure 4.30 Mapping the Influence of Drug Cartels in Mexico** This map suggests the major areas of influence of Mexican cartel groups in early 2015. It also shows some of the major entry points for drugs into Mexico (from Asia and South America) and into the United States.

presidency. Mexico continues to be a major transshipment zone for illegal drugs from Asia and South America (see Figure 4.30). The profit and violence associated with this drug trade is destabilizing for much of the region.

REVIEW

4.8 How did Iberian colonization lead to the formation of the modern states of Latin America?

4.9 How are trade blocs reshaping the region's geopolitics and development?

KEY TERMS neocolonialism, Organization of American States (OAS), Treaty of Tordesillas, Mercosur (Southern Cone Common Market), UNASUR (Union of South American Nations), Pacific Alliance, Central American Free Trade Agreement (CAFTA)

Economic and Social Development: Focusing on Neoliberalism

Most Latin American economies fit into the broad middle-income category set by the World Bank. Clearly part of the developing world, Latin American people are much better off than those in Sub-Saharan Africa, South Asia, and much of China. Still, the economic contrasts are sharp, both between states and within them (Table 4.2). Although

per capita incomes in Latin America are well below levels of developed countries, the region has witnessed steady improvements in various social indicators, such as life expectancy, child mortality, and literacy. Also, some small states, such as Costa Rica and Panama, do very well in the human development index.

The economic engines of Latin America are its two largest countries, Brazil and Mexico. According to the International Monetary Fund, in 2017, Brazil was the world's eighth largest economy and Mexico was the 15th largest, based on gross domestic product (GDP). The region has also seen reductions in extreme poverty as the percentage of people living on less than $3.10 per day dropped from roughly one in five people to one in ten from 1999 to 2017. The poorest Latin American countries, such as Nicaragua, Honduras, Guatemala, and now Venezuela, have one-quarter of their populations living in poverty (see Table 4.2).

The path toward economic development in Latin America has been a volatile one. In the 1960s, Brazil, Mexico, and Argentina all seemed poised to enter the ranks of the developed world. Multilateral agencies such as the World Bank and the Inter-American Development Bank loaned money for big development projects: continental highways, dams, mechanized agriculture, and power plants. All sectors of the economy were radically transformed. Agricultural production increased with the application of "green revolution" technology and mechanization (see Chapter 12). State-run industries reduced the need for imported goods, and the service sector ballooned due to new government and private-sector jobs. In the end, most Latin American countries made the transition from predominantly rural and agrarian economies, dependent on one or two commodities, to more

TABLE 4.2 Development Indicators

Explore these data in MapMaster 2.0 https://goo.gl/9E4oKV

Country	GNI per Capita, PPP 2017[1]	GDP Average Annual Growth 2009–2015[2]	Human Development Index (2016)[3]	Percent Population Living Below $3.10 a Day[2]	Under Age 5 Mortality Rate (per 1000 live births), 1990[1]	Under Age 5 Mortality Rate (per 1000 live births), 2016[1]	Secondary School Enrollment Ratios[4] Male (2009–2016)	Female (2009–2016)	Gender Inequality Index (2016)[3,6]	Freedom Rating (2018)[5]
Argentina	20,787	2.2	0.827	4	29	11	103	110	0.362	2.0
Bolivia	7,560	5.4	0.674	13	124	37	87	86	0.446	3.0
Brazil	15,484	2.2	0.754	9	64	15	97	102	0.414	2.0
Chile	24,085	4.3	0.847	3	19	8	100	101	0.322	1.0
Colombia	14,552	4.6	0.727	14	35	15	95	102	0.393	3.0
Costa Rica	17,044	3.7	0.776	4	17	9	121	126	0.308	1.0
Ecuador	11,617	4.8	0.739	12	57	21	105	109	0.391	3.0
El Salvador	8006	1.9	0.680	12	60	15	79	80	0.384	2.5
Guatemala	8150	3.7	0.640	27	82	29	68	63	0.494	4.0
Honduras	4986	3.5	0.625	35	58	19	65	77	0.461	4.0
Mexico	18,149	3.1	0.762	10	46	15	88	93	0.345	3.0
Nicaragua	5842	5.0	0.645	25	68	20	70	79	0.462	4.5
Panama	24,446	7.8	0.788	8	31	16	73	78	0.457	2.0
Paraguay	9691	5.9	0.693	6	47	20	74	79	0.464	3.0
Peru	13,434	5.3	0.740	10	80	15	96	96	0.385	2.5
Uruguay	22,563	4.2	0.795	–	23	9	90	100	0.284	1.0
Venezuela	17,640*	1.9	0.767	24	30	16	86	93	0.461	5.5

[1] World Bank Open Data, 2018. * 2014 Figure.
[2] World Bank—*World Development Indicators*, 2017.
[3] United Nations, *Human Development Report*, 2016.
[4] Population Reference Bureau, *World Population Data Sheet*, 2017.
[5] Freedom House, Freedom in the World 2018. See Ch.1, pp. 33 –34, for more info. on this scale (1–7, with 7 representing states with the least freedom).
[6] See Ch. 1, p. 39, for more info. on this scale (0–1, with higher values representing less gender equality).

Log in to Mastering Geography & access MapMaster to explore these data!

1) Rank the countries in this region by GNI per capita PPP and their Human Development Index figure. Compare the maps and describe the relationship between these two measures.

2) Compare the figures for secondary school enrollment with Gender Inequality. What patterns do you see?

economically diversified and urbanized countries with mixed levels of industrialization.

The modernization dreams of Latin American countries were trampled in the 1980s, when debt, currency devaluation, hyperinflation, and falling commodity prices undermined the region's aspirations. By the 1990s, most Latin American governments had radically changed their economic development strategies. State-run national industries and tariffs were jettisoned for policy reforms that emphasized privatization, direct foreign investment, and free trade, collectively labeled **neoliberalism**. Through tough fiscal policy, increased trade, privatization, and reduced government spending, most countries saw their economies grow and poverty decline. In aggregate, the economies of the region averaged an annual growth rate of 2.7 percent from 2009 to 2015. Much of the economic growth in the 2000s (average growth was an impressive 3.6 percent in 2000–2009) was attributed to a boom in primary exports, driven by increased demand from China. With that demand softening, commodity prices have dropped and growth rates have declined. This decline and other sporadic economic downturns have made neoliberal policies highly unpopular with the masses, at times causing major political and economic turmoil.

Primary Export Dependency

Historically, Latin America's abundant natural resources were its wealth. In the colonial period, silver, gold, and sugar generated great wealth for the colonists. With independence in the 19th century, the region began a series of export booms to an expanding world market, including commodities such as bananas, coffee, cacao, grains, tin, rubber, copper, wool, and petroleum. One of the legacies of this export-led development was a tendency for countries to specialize in one or two major commodities, a pattern that continued into the 1950s. During that decade, 90 percent of Costa Rica's export earnings came from bananas and coffee, 70 percent of Nicaragua's came from coffee and cotton, 85 percent of Chile's came from copper, and half of Uruguay's came from wood. Even Brazil generated 60 percent of its export earnings from coffee in 1955; by 2000, coffee accounted for less than 5 percent of the country's exports, even though Brazil remained the world leader in coffee production.

Agricultural Production   Since the 1960s, the trend in Latin America has been to diversify and mechanize agriculture. Nowhere is this more evident than in the Plata Basin, which includes southern Brazil, Uruguay, northern Argentina, Paraguay, and eastern Bolivia. Soybeans, used for oil and animal feed, transformed these lowlands in the 1980s and early 1990s. Brazil is now the world's second largest producer of soy (following the United States) and the world's largest soy exporter. Argentina is the third largest, and production is still increasing: between the late 1990s and 2010, soy production tripled. The speed with which the Plata and Amazon basins are being converted into soy fields alarms many, as forests and savannas are eliminated, negatively impacting biodiversity and increasing greenhouse gas emissions. But with soy prices high, the rush to plant continues (Figure 4.31). In addition to soy, acres of rice, cotton, and orange trees, as well as the traditional wheat and sugar, continue to be planted in the Plata Basin.

▲ **Figure 4.31 Soy Production in Brazil**   Fartura Farm in the state of Mato Grosso, Brazil, embodies the large-scale industrial agriculture that has transformed much of South America into one of the world's largest producers and exporters of soy products.

Explore the **SIGHTS** of Brazilian Farming
http://goo.gl/Am2ekd

Coffee looms large in Latin America's connection to global trade. For over a century, Brazil has been the world leader in coffee production. But with over 200 million people, it is also a major coffee consumer as well. Up until 1990, Colombia was the second largest global producer; but over the past two decades, Colombia's production has declined, and the output of two Southeast Asian countries—Vietnam and Indonesia—has surpassed it. Today, Latin America accounts for 58 percent of the world's coffee production, with Brazil, Colombia, and Mexico being the leading regional producers; most is bound for Europe, North America, and Japan. Over 145 million 60-kilogram bags were produced in the 2015–2016 harvest.

Similar large-scale agricultural frontiers exist along the Pacific slope of Central America (cotton and some tropical fruits) and in the Central Valley of Chile and the foothills of Argentina (wine and fruit production). In northern Mexico, water supplied from dams along the Sierra Madre Occidental has turned the valleys in Sinaloa into intensive agricultural centers of fruits and vegetables for U.S. consumers. Northern Mexico's relatively mild winters allow growers to produce strawberries, tomatoes, zucchinis, and peppers during the winter months.

In each of these cases, the agricultural sector is capital-intensive. By using machinery, hybrid crops, chemical fertilizers, and pesticides, many corporate farms are extremely productive and profitable. What these operations fail to do is employ many rural people, which is especially problematic in countries where one-third or more of the population depends on agriculture for its livelihood. Interestingly, a few traditional Amerindian foods, such as quinoa, are gaining consumers thanks to a growing appetite for organic and healthful foods. Peru, Bolivia, and Ecuador have experienced a recent boom in quinoa production and exports, with much of the crop being grown in small and medium-sized highland farms. The United Nations declared 2013 the "year of quinoa" in an effort to promote this nutritional native food; exports of quinoa bound for North America and Europe doubled from 2013 to 2016.

Mining and Forestry   The mining of silver, zinc, copper, iron ore, bauxite, and gold is an economic mainstay for many countries in the region (see *Exploring Global Connections: South America's Lithium Triangle*). Moreover, many commodity prices reached record levels in

South America's Lithium Triangle

High in a remote corner of the Andes, where Bolivia, Argentina, and Chile meet, is the largest known reserve of lithium in the world. This soft, silver-white metal is an essential element in lightweight batteries, like those that power cell phones and laptops. It is also a key metal for electric vehicle batteries and photovoltaic cells. Companies such as Tesla, Samsung, and Apple are keenly aware of the cost and scarcity of lithium, which could greatly benefit these developing economies. Yet possessing more than half of the world's lithium is only step one—being able to extract it for global markets has been the challenge.

Lithium is found under salt flats in South America's Altiplano region, at elevations of up to 13,000 ft (Figure 4.4.1). Miners must extract the lithium-bearing brine from wells sunk deep below the salt crust and then deposit the liquid into evaporation ponds to let the sun do its work. Once sun-baked, the concentrate is taken for processing into lithium carbonate. South America's lithium boom thus far has been hindered by a lack of technology and capital, as well as national laws that designate lithium a strategic metal and therefore limit investment from foreign companies. Bolivia and Argentina have the largest known reserves, but Australia is the leading producer, followed by Chile. China rounds out the top five lithium source countries.

For decades, Chile has been the region's export leader, sending lithium carbonate primarily to manufacturers in South Korea, China, and Japan. The Atacama salt flats have the highest quality reserves, and ports such as Antofagasta are relatively close (Figure 4.4.2). Moreover, Chile's neoliberal policies have been more open to foreign investment in mining. Argentina is trying to catch up through increased foreign investment in lithium extraction around Jujuy province. In 2016, it produced about half as much lithium as Chile. Bolivia, which may have the largest reserves under the Salar de Uyuni salt flat, has yet to become a significant producer. This is partly due to the state's tight control of the resource and the wariness of foreign investors to engage in this country, which is noted for nationalizing key resources such as natural gas. As far as Bolivians are concerned, they need only look to the nearby mountain of Potosí, whose silver financed Spain's colonization of the Americas, to understand that owning a resource does not mean profiting from it.

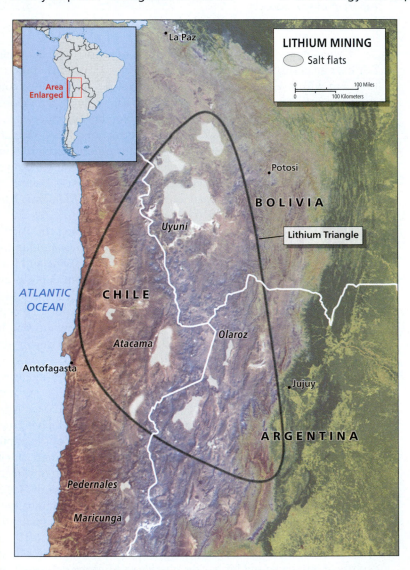

▲ **Figure 4.4.1 Lithium Mining in South America** The largest lithium deposits in the world are found where Bolivia, Chile, and Argentina converge. Lithium is a critical metal for lightweight batteries used in cell phones and laptops.

LITHIUM MINING
Salt flats

▲ **Figure 4.4.2 Lithium Processing in San Pedro de Atacama, Chile** In the high Atacama desert, lithium-laden brine is pumped out of the ground and into evaporation ponds. Once dried, the powdery substance is shipped and processed into lithium.

1. **What are the factors that make Chile the leading South American exporter of lithium?**

2. **What are the products that you use that need lithium to function?**

GOOGLE EARTH
Virtual Tour Video
https://goo.gl/5NLv7s

the last decade, boosting foreign exchange earnings, but prices began to fall in 2013 and have yet to recover fully. Chile is the world leader in copper production, far surpassing the next two largest producers, Peru and the United States. Mexico and Peru were top silver producers in 2015, but Chile and Bolivia were also top-ten producers. Peru was also Latin America's top gold producer in 2015, followed by Mexico.

Logging is another important, and controversial, resource-based activity. Several countries rely on plantation forests of introduced species of pine, teak, and eucalyptus to supply domestic fuel wood, pulp, and board lumber. These plantation forests grow single species and fall far short of the complex ecosystems occurring in natural forests. Still, growing trees for paper or fuel reduces the pressure on other forested areas. Brazil, Venezuela, Chile, and Argentina are the leaders in plantation forestry. Considered one of Latin America's economic stars, Chile exports timber and wood chips to add to its mineral export earnings. Thousands of hectares of exotics (especially pine) have been planted, systematically harvested, and cut into boards or chipped for wood pulp. East Asian capital is heavily invested in this sector of the Chilean economy. Expansion of the wood chip business, however, has led to a dramatic increase in the logging of native forests.

The Energy Sector   Oil-rich Mexico, Venezuela, Ecuador, and now Brazil can meet most of their own fuel needs and also earn vital revenue from oil exports. Although Latin American oil producers do not receive as much attention as those in Southwest Asia, they are significant. Venezuela was one of the five founding members of the Organization of the Petroleum Exporting Countries (OPEC), and Ecuador joined the group in 1973.

In 2014, Brazil was the world's 9th largest oil producer, Mexico was 10th, and Venezuela was 12th. These three countries account for 75 percent of the region's oil production. The largest new oil discovery in recent years has been off the coast of Brazil, and production has soared in the past decade, making Brazil the region's top oil producer and drawing increased foreign investment. Venezuela's overall production has declined, yet it still earns up to 90 percent of its foreign exchange from petroleum and natural gas products. As the board of the Venezuelan state oil company, PDVSA, became highly politicized in the early 2000s, basic investment in extractive infrastructure deteriorated, and production fell sharply. The unstable political climate under President Maduro detoured any foreign investment, adding to the country's economic collapse.

Natural gas production is also on the rise. Venezuela and Bolivia have the largest proven reserves of natural gas in Latin America, but Mexico and Argentina are by far the largest producers. In recent years, Argentina's natural gas production has been boosted by new finds in Patagonia. In 2012, Argentina made headlines when President Christina Fernandez de Kirchner seized majority control of YPF, the country's major producer of oil and natural gas, from a Spanish-owned company, claiming frustration over unnecessary declines in output. Argentina is especially dependent on natural gas for its urban markets and exports, and declining production led to shortages and price increases in the domestic liquefied natural gas market that required action.

In the area of biofuels, Brazil offers a story of sweet success. When oil prices skyrocketed in the 1970s, then oil-poor Brazil decided to convert its abundant sugarcane harvest into ethanol. Brazil continued to invest in ethanol even in years when oil prices plummeted, building mills and a distribution system that delivered ethanol to gas stations. A major technological success was inventing flex-fuel cars that run on any combination of ethanol and gasoline. At the time, Brazil was motivated by its limited oil reserves, but today, with interest in biofuels growing as a way to reduce CO_2 emissions, Brazil's support of its ethanol program looks visionary.

Thanks to these innovations, Latin America's energy mix has changed considerably since the 1970s (Figure 4.32). Fifty years ago, 60 percent of the region's energy was supplied by oil and 20 percent by wood fuel. By 2010, the energy supply was more diverse and cleaner, most notably through the growth in natural gas, hydroelectricity, and biofuels (or bagasse). That said, Latin America's energy consumption has also increased fivefold from 1970 to 2010 due to population increase, urban growth, improved transportation, and greater economic activity.

Latin America in the Global Economy

During the 1990s, Latin American governments and the World Bank became champions of neoliberalism as a sure path to economic development. Neoliberal policies accentuate the forces of globalization by reducing state intervention and self-sufficiency. Most Latin American political leaders have embraced neoliberalism and the benefits that come with it, such as increased trade, greater direct foreign investment, and more favorable terms for debt repayment. However, there are signs of discontent with neoliberalism throughout the region. Recent protests in Brazil and Chile reflect popular anger against trade policies that seem to benefit only the elite.

Maquiladoras and Foreign Investment   Growth in foreign investment and the presence of foreign-owned factories are examples of neoliberalism. **Maquiladoras**, the Mexican assembly plants that line the border with the United States, are characteristic of manufacturing systems in an increasingly globalized economy. Construction of these plants began in the 1960s as part of a border industrialization program. Today there are some 3,000 factories employing about 1 million workers, both along the border and throughout Mexico. With NAFTA, foreign-owned manufacturing plants are no longer restricted to the border zone and are increasingly being built near the population centers of Monterrey, Puebla, and Veracruz. Aguascalientes, in central Mexico, has emerged as the country's auto city, although most of the cars are produced by foreign companies and are destined for export. Chihuahua, a four-hour drive from the border, is Mexico's aerospace center (Figure 4.33). In the last decade, over three dozen aerospace plants have opened there, providing parts

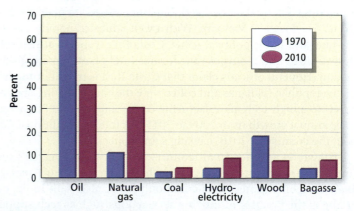

▲ **Figure 4.32 Latin America's Energy Mix, 1970 and 2010**   As the region's energy sources have increased and diversified, there is much less reliance on wood and more reliance on natural gas. **Q: What could explain the decline in wood fuel for this region?**

▲ **Figure 4.33 Chihuahua's Aerospace Industry** Mexican workers at the Hawker Beechcraft plant in Chihuahua assemble jet airplane parts for export to the United States. Chihuahua has the largest concentration of aerospace engineers and technicians in Mexico.

for the booming U.S. airplane manufacturing business. Labor costs average $6 per hour in Chihuahua, far cheaper than U.S. labor. Even though Mexican wages are higher than those in China, northern Mexico is still an attractive location because of its proximity to U.S. firms and its membership in NAFTA.

Considerable controversy on both sides of the border surrounds this form of industrialization. Organized labor in the United States complains that well-paying manufacturing jobs are being lost to low-cost competitors. Mexicans worry that these plants are poorly integrated into the rest of the economy and that many factories choose to hire young, unmarried women because they are viewed as docile laborers.

Other Latin American states are attracting foreign companies through tax incentives and low labor costs. Assembly plants in Honduras, Guatemala, and El Salvador draw foreign investors, especially in the apparel industry. A recent report from El Salvador claims that not one of its apparel factories has a union. Making goods for major American labels, many Salvadoran garment workers complain that they do not make a living wage, work 80-hour weeks, and will lose their jobs if they become pregnant. Costa Rica had been a major computer chip manufacturer for Intel since 1998, but Intel closed its plant in 2014, deciding to relocate to Vietnam. With a well-educated population, low crime rate, and stable political scene, Costa Rica wants to attract other high-tech firms and repurpose the former Intel plant into a high-tech incubator. Hopeful officials claim that Costa Rica is transitioning from a banana republic (bananas and coffee were the country's long-standing exports) to a high-tech manufacturing center and boasted 3.7 percent average annual growth from 2009 to 2015. Yet Intel cut additional R&D jobs in the country in 2018, causing some concerns.

Uruguay is another small country with a well-educated population that has recently emerged as the leader in Latin American **outsourcing** operations. Most commonly associated with India, outsourcing is the practice of moving service jobs such as tech support, data entry, and programming to cheaper locations. Partnered with the Indian multinational company Tata, in the last few years, TCS Iberoamerica in Uruguay has created the largest outsourcing operation in the region. Uruguay takes advantage of being in a time zone

close to that of the eastern United States. While India's top engineers sleep, Uruguayan engineers and programmers can serve their customers from Montevideo.

Latin America is linked to the world economy in ways other than trade. Figure 4.34 shows the changes in foreign direct investment (FDI) as a percentage of gross domestic product (GDP) from 1990 and 2016. For nearly every country in the region, the value of FDI in terms of the percentage of GDP went up. In 1995, Brazil's FDI was less than $5 billion, and Mexico's was $9.5 billion. By 2016, FDI in Brazil reached US$60 billion, and Mexico's was $26 billion. Much of this foreign investment was from Europe and Asia. In 2012, China became Brazil's largest trading partner. China and Brazil—the so-called rapidly developing BRICS nations, along with Russia, India and South Africa—have recognized their global strategic partnership. Not only is Brazil exporting grains, minerals, and energy resources to China, but it has also signed an agreement that will allow Brazilian airplane maker Embraer to manufacture and sell its regional jets in China.

Remittances Another important indicator that reflects the integration of Latin American workers into global labor markets is remittances. Scholars debate whether this flow of capital can actually lead to sustained development or is simply a survival strategy of last resort. World Bank research shows that remittances sharply dropped during the global economic recession that began in 2008 but recovered in less than a decade, topping $79 billion in 2017. Many economists project remittances will continue to be a major source of capital for the region.

The economic impact of remittances on a per capita basis is real (see Figure 4.34). Mexico is the regional leader, receiving over US$28.7 billion in remittance income in 2016 (equivalent to over $225 per capita). But for smaller countries such as El Salvador and Honduras, remittances contribute far more to the domestic economy. El Salvador, a country of about 6.3 million people, received nearly $4.6 billion in remittances in 2016, or about $730 per capita. For many Latinos, remittances are the surest way to alleviate poverty, even though they depend on an international migration system that is constantly changing and includes both legal and unauthorized channels of movement.

Dollarization During the 1990s, as Latin American governments faced various financial crises, many began to consider the economic benefits of **dollarization**, a process by which a country adopts—in whole or in part—the U.S. dollar as its official currency. In a totally dollarized economy, the U.S. dollar becomes the only medium of exchange, and the country's national currency ceases to exist. Dollarization is not new; Panama dollarized its economy in 1904, the year after it gained independence from Colombia. However, Panama was the only fully dollarized state in Latin America until 2000, when Ecuador adopted dollarization to address the dual problems of currency devaluation and hyperinflation rates of more than 1000 percent annually. El Salvador followed suit in 2001 as a means to reduce the cost of borrowing money. Some economists are recommending that Venezuela dollarize as a way to address its current economic crisis, but as of 2018, it has not.

A more common strategy in Latin America is limited dollarization, in which U.S. dollars circulate and are used alongside the country's national currency. Many banks in Latin America, for example, allow customers to maintain accounts in dollars to avoid the problem of capital flight should a local currency be devalued. Other countries keep their national currency but peg its value

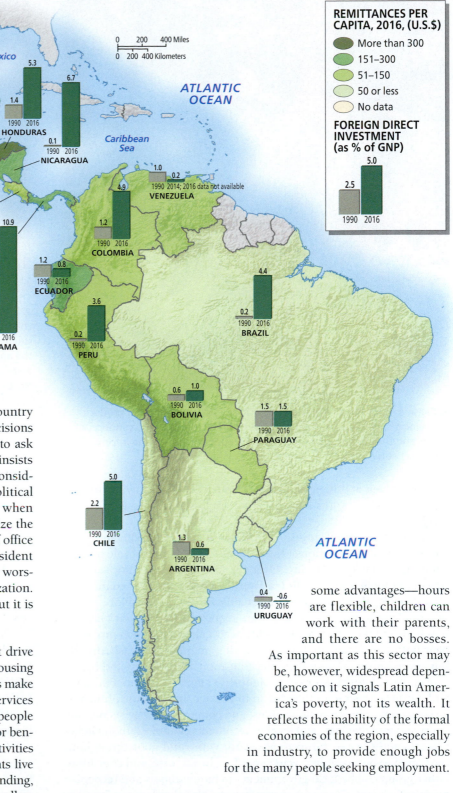

▶ **Figure 4.34 Global Linkages: Foreign Investment and Remittances** Foreign investors and immigrants are responsible for significant increases in the amount of capital flowing into Latin America. As the map indicates, most countries saw increases in foreign direct investment between 1990 and 2016. Immigrants working abroad sent $79 billion to the region in 2017, providing much-needed capital to many poor households. For Guatemala, El Salvador, and Honduras, remittances totaled more than $300 per capita.

one-for-one to the dollar; this was the innovative strategy adopted by Argentina in 1991, although it led to a serious financial crisis in 2001 and was eventually stopped. Dollarization, partial or full, tends to reduce inflation, eliminate fears of currency devaluation, and reduce the cost of trade by eliminating currency conversion costs.

Dollarization has its drawbacks. The obvious one is that a country no longer controls its monetary policy, instead relying on the decisions of the U.S. Federal Reserve. Foreign governments do not have to ask permission to dollarize their economies, but the United States insists that its monetary policies be based exclusively on domestic considerations, regardless of the impact on foreign countries. The political impact of eliminating a national currency is serious; in 1999, when Ecuador's President Jamil Mahuad announced his plan to dollarize the economy to head off hyperinflation, he was quickly forced out of office by a coalition of military and Indian activists. When Vice President Gustavo Noboa became president and the economic situation worsened, the country's political leadership went ahead with dollarization. In short, dollarization may help in a time of economic duress, but it is not a popular policy.

The Informal Sector Even in prosperous capital cities, a short drive to the urban periphery shows large neighborhoods of self-built housing filled with street traders and family-run workshops. Such activities make up the **informal sector**, which is the provision of goods and services without government regulation, registration, or taxation. Most people in the informal economy are self-employed and receive no wages or benefits except the profits they clear. The most common informal activities are housing construction (in many cities, one-third of all residents live in self-built housing), manufacturing in small workshops, street vending, transportation services (messenger services, bicycle delivery, and collective taxis), garbage picking, street performing, and even paid waiting in line (Figure 4.35).

No one is sure how big this economy is, in part because it is difficult to separate formal activities from informal ones. Visit Lima, Belém, Guatemala City, or Guayaquil, and it is easy to get the impression that the informal economy *is* the economy. From self-help housing that dominates the landscape to hundreds of street vendors who crowd the sidewalks, it is impossible to avoid. The informal sector has

some advantages—hours are flexible, children can work with their parents, and there are no bosses. As important as this sector may be, however, widespread dependence on it signals Latin America's poverty, not its wealth. It reflects the inability of the formal economies of the region, especially in industry, to provide enough jobs for the many people seeking employment.

Social Development

Over the past three decades, Latin America has experienced marked improvements in life expectancy, child survival, and educational equity. One telling indicator is the steady decline between 1990 and 2015 in mortality rates for children below age five (see Table 4.2). This indicator is important because an increase in survival of children younger than five suggests that basic nutritional and health care needs are being met. We can also conclude that resources are being used to

▲ **Figure 4.36 Schoolchildren in Panama** Uniformed public schoolchildren walk to school in Panama City. Latin American states have seen steady improvements in youth literacy, with 97 percent of youth (between the ages of 15 and 24) being literate. Per capita expenditures on education and access to postsecondary education still lag behind levels in Europe and North America.

▲ **Figure 4.35 Peruvian Street Vendors** Street vendors sell produce in Huancayo, Peru. Street vending plays a critical role in the distribution of goods and the generation of income. It is representative of the informal sector in Latin America.

sustain women and their children. Despite economic downturns, the region's social networks have been able to lessen the negative effects on children.

A combination of government policies and grassroots and nongovernmental organizations (NGOs) plays a fundamental role in contributing to social well-being. Over the last decade, conditional cash transfer programs, such as **Bolsa Familia**, have reduced extreme poverty. Poor Brazilian families who qualify for Bolsa Familia receive a monthly check from the state but are required to keep their children in school and take them to clinics for health checkups. Such programs have both the immediate impact of giving poor families cash and the long-term impact of improving the educational attainment and health care of their children. Mexico has adopted a similar program. For states with far fewer resources than Brazil or Mexico, international humanitarian groups, church organizations, and community activists provide many services that state and local governments cannot. Catholic Relief Services and Caritas, for example, work with rural poor throughout the region to improve their water supplies, health care, and education. Other groups lobby local governments to build schools and recognize squatters' claims. Grassroots organizations also develop cooperatives that market everything from sweaters to cheeses.

Other important indicators for social development are gender equity, secondary education enrollment, and overall political freedom (Table 4.2). In general, youth literacy is high in the region, with boys and girls having equal access to education (Figure 4.36) In terms of gender inequality, which considers reproductive health, empowerment, and labor market participation of women, there is room for improvement, with all states showing moderate levels of inequality. In terms of

political freedom, not surprisingly, Venezuela had the lowest score in 2017 and is the only state of the region classified as "unfree."

Race and Inequality There is much to admire about race relations in Latin America. The complex racial and ethnic mix that was created in the region fostered tolerance for diversity. That said, Indians and blacks are more likely to be counted among the region's poorest groups. More than ever before, racial discrimination is a major political issue in Brazil; reports of organized killings of street children, most of them Afro-Brazilian, make headlines. For decades, Brazil championed its vision of a color-blind racial democracy. Residential segregation by race is rare, and interracial marriage is common, but certain patterns of social and economic inequality seem best explained by racial discrimination.

Assessing racial inequalities in Brazil is problematic, because the Brazilian census asks few racial questions, and all are based on self-classification. In the 2000 census, less than 11 percent of the population called itself black. Some Brazilian sociologists, however, claim that more than half the population is of African ancestry, making Brazil the second largest "African state" after Nigeria. Racial classification is always highly subjective and relative, but some patterns support the existence of racism. In northeastern Brazil, where Afro-Brazilians are the majority, death rates approach those of some of the world's poorest countries, and throughout Brazil, blacks suffer higher rates of homelessness, landlessness, illiteracy, and unemployment. To address this problem, various affirmative action measures have been implemented (along with the Bolsa Familia program). From federal ministries to public universities, quota systems are being tried to improve the condition of Afro-Brazilians.

In areas of Latin America where Indian cultures are strong, indicators of low socioeconomic position are also present. In most countries, areas where native languages are widely spoken regularly are also areas of persistent poverty. In Mexico, the Indian south lags behind the booming north and Mexico City. Prejudice is embedded in the language. To call someone an *indio* (Indian) is an insult in Mexico. In Bolivia, women who dress in the Indian style of full, pleated skirts and bowler hats are called *cholas*, a descriptive term referring to the rural *mestizo* population that suggests a backwardness and even cowardice. No one of high social standing, regardless of skin color, would ever be called a *chola* or *cholo*.

It is difficult to separate status divisions based on class from those based on race. From the days of conquest, being European meant an immediate elevation in status over the Indian, African, and *mestizo* populations. Race does not necessarily determine one's economic standing, but it certainly influences it. Amerindian people, however, are politically asserting themselves. For example, Evo Morales was inaugurated as president of Bolivia in 2006, making him the first Indian leader in that country.

The Status of Women Many contradictions exist with regard to the status of women in Latin America. Many Latina women work outside the home. In most countries, the formal figures are between 30 and 40 percent of the workforce, not far off from many European countries, but lower than in the United States. Legally speaking, women can vote, own property, and sign for loans, although they are less likely to do so than men, reflecting the society's patriarchal tendencies. Even though Latin America is predominantly Catholic, divorce is legal and family planning is promoted. In most countries, however, abortion remains illegal.

Compared to other developing regions, access to education in Latin America is good, and thus illiteracy rates tend to be low. Rates of adult illiteracy are slightly higher for women than for men, but usually by only a few percentage points. Male and female students are equally represented today in higher education, and consequently, women are regularly employed in the fields of education, medicine, and law.

The biggest changes for women are the trends toward smaller families, urban living, and educational parity with men. These factors have greatly improved the participation of women in the labor force. In the countryside, however, serious inequalities remain. Rural women are less likely to be educated and tend to have larger families. In addition, they are often left to care for their families alone, as husbands leave in search of seasonal employment. In most cases, the conditions facing rural women have been slow to improve.

Women are increasingly playing an active role in politics. In 1990, Nicaragua elected Latin America's first woman president, Violeta Chamorro, the owner of an opposition newspaper. Nine years later, Panamanians voted Mireya Moscoso into power. In 2005, South America had its first woman president: Dr. Michelle Bachelet, a pediatrician and single mother, took the oath of office in Chile. She was reelected president for a second term in 2013. Brazilian President Dilma Rousseff took office in 2011, but her second term was cut short as she was removed from office following impeachment proceedings. As shown in Figure 4.37, many Latin American states have larger percentages of women in seats of national parliaments than does the United States.

Across the region, women and indigenous groups are active organizers and participants in cooperatives, small businesses, and unions

PERCENT OF SEATS HELD BY WOMEN IN NATIONAL PARLIAMENTS, 2015

- 5–14.9
- 15–24.9
- 25–34.9
- 35–44.9
- 45–54.9
- Countries that have or have had women presidents

In Latin America 29% of seats are held by women in national parliaments. By comparison, in the United States 19% of congressional seats are held by women. In Canada the figure is 26%.

▲ **Figure 4.37 Women's Participation in National Politics** Women are active players in Latin American politics. On average, 29% of seats in national parliaments are held by women. In Brazil, only 11% of the seats are held by women, but in Mexico, 43% and Argentina, 39% were held by women in 2017. Moreover, six countries have had or have female presidents. *Source:* World Bank Data, 2018

and are elected to national office. Over a relatively short period, they have won a formal place in the economy and a political voice. Moreover, evidence suggests that this trend will continue.

REVIEW

4.10 How has the export of primary products (food, fiber, and energy) shaped the economies of Latin America?

4.11 What explains some of the positive indicators of social development in Latin America?

KEY TERMS neoliberalism, maquiladora, outsourcing, dollarization, informal sector, Bolsa Familia

Latin America

REVIEW, REFLECT, & APPLY

Summary

- Latin America is still rich in natural resources and relatively lightly populated. Yet as populations continue to grow and trade in natural resources increases, environmental problems multiply. Tropical forest clearing and dam construction are particular concerns.

- Unlike in other developing areas, 80 percent of Latin Americans live in cities. This shift started early and reflects a cultural bias toward urban living with roots in the colonial past.

- Cities are large and combine aspects of the formal industrial economy with the informal one. Grinding poverty in rural areas drives people to cities or to overseas employment.

- Perhaps 90 percent of Latin America's native population died from disease, cruelty, and forced resettlement when colonized by Europe. The slow demographic recovery of native peoples and the continual arrival of Europeans and Africans resulted in an unprecedented level of racial and cultural mixing. Today Amerindian activism is on the rise as indigenous groups seek territorial and political recognition.

- Most Latin American states have been independent for 200 years yet remained politically and economically dependent on Europe and North America. Today Latin America, especially Brazil, is exerting more geopolitical influence. New political actors—from indigenous groups to women—are challenging old ways of doing things.

- Latin American governments were early adopters of neoliberal economic policies. Some states prospered whereas others faltered, sparking popular protests against the effects of neoliberalism and globalization. Extreme poverty has declined, and social indicators of development are improving.

Review Questions

1. How do different countries in the region address environmental issues, and why might approaches differ?

2. Give examples of major environmental issues facing the region?

3. How does Latin America's history and geography explain why the region is much more urbanized than other developing world regions? With its vast lightly inhabited areas, why does Latin America have so many megacities?

4. In what ways do Latin American countries influence popular culture around the world? Which countries are the most influential, and why?

5. NAFTA and Mercosur are established trade blocs in the region, whereas the Pacific Alliance is relatively new. What are the benefits of forming trade blocs? How have the trade blocs in Latin America impacted the region's economic growth and geopolitics?

6. How has the export of primary products (food, fiber, and energy) shaped the economies of Latin America, and why does the region continue to produce so many raw materials for the world?

Image Analysis

1. Two important sources of foreign capital in Latin America are foreign direct investment (FDI) and remittances. Remittances are often a sign of a country's dependence on emigration as a livelihood strategy. Which countries in the region are most dependent on remittances? What generalizations can you make about these countries?

2. The largest recipient of FDI in absolute dollars is Brazil, and yet FDI accounts for a relatively small portion of Brazil's economy. Which countries have seen the largest increase in FDI? What might explain that increase?

▶ **Figure IA4** Latin American Foreign Direct Investment and Remittances

REMITTANCES PER CAPITA, 2016, (U.S.$)
- More than 300
- 151–300
- 51–150
- 50 or less
- No data

FOREIGN DIRECT INVESTMENT (as % of GNP)

Join the Debate

Dams are necessary for flood control, irrigation, and energy production but are often controversial, especially large dams built in wilderness areas. Many dams are planned or under construction in the Amazon Basin. Considering the fact that the basin contains the world's largest tropical forest with immense biodiversity and thousands of acres of undisturbed land, do dams benefit or hurt Amazonian peoples?

Dam construction benefits Amazonian peoples!

- With the growing demand for energy in the region, dam construction creates a renewable and clean source of electricity.

- The Amazon Basin is vast, and for many South American countries it represents a resource and settlement frontier. By building roads and constructing dams, the countries of this region can better benefit from their natural resources.

- Dam construction creates jobs, advances in engineering abilities, and enhances overall infrastructural development. It can be done safely with limited environmental impact.

Dams construction is a poor choice for the Amazon Basin!

- Dams are expensive projects that have limited life spans, especially in a rainforest environment. The costs of construction can drive countries into debt with limited long-term benefits.

- The ecological costs of building dams in remote tropical forests are serious; many species can be lost, and the negative impact of the dam goes far beyond the footprint of the actual project.

- For indigenous people in the region, dam construction inevitably results in displacement.

▲ **Figure D4** Construction Site of the Belo Monte Hydroelectric Dam on the Xingu River in Brazil

Key Terms

agrarian reform (p. 116)
Altiplano (p. 106)
altitudinal zonation (p. 108)
Bolsa Familia (p. 136)
Central American Free Trade Agreement (CAFTA) (p. 129)
Columbian Exchange (p. 123)
deforestation (p. 111)
dollarization (p. 134)
ecotourism (p. 114)
El Niño (p. 110)

environmental lapse rate (p. 108)
grassification (p. 112)
informal sector (p. 135)
latifundia (p. 116)
maquiladora (p. 133)
Mercosur (Southern Cone Common Market) (p. 128)
mestizo (p. 120)
minifundia (p. 116)
neocolonialism (p. 126)
neoliberalism (p. 131)
neotropics (p. 105)
Organization of American States (OAS) (p. 126)

outsourcing (p. 134)
Pacific Alliance (p. 128)
primate city (p. 118)
remittance (p. 122)
shield (p. 107)
squatter settlement (p. 118)
sustainable development (p. 106)
syncretic religion (p. 124)
Treaty of Tordesillas (p. 126)
UNASUR (Union of South American Nations) (p. 128)
urban primacy (p. 118)

Mastering Geography

Looking for additional review and test prep materials? Visit the Study Area in Mastering Geography to enhance your geographic literacy, spatial reasoning skills, and understanding of this chapter's content by accessing a variety of resources, including MapMaster interactive maps, videos, flashcards, web links, self-study quizzes, and an eText version of Globalization and Diversity.

Scan to read about Geographer at Work
Corrie Drummond and her work with the U.S. Agency for International Development (USAID)
https://goo.gl/T5BZos

GeoSpatial Data Analysis

Changes in Coffee Exports Coffee production is extremely important for many Latin American countries, and the region produces more than half of the world's coffee on large estates and small farms in the tropics. For many rural families, the coffee bean provides access to international markets and income. The International Coffee Organization (ICO) maintains statistics on coffee production worldwide.

Open MapMaster 2.0 in the Mastering Geography Study Area. Then consult the ICO's website (http://www.ico.org) and go to Historical Data under the Statistics heading. Next, click on Trade Statistics Tables, then download the Excel table for total production figures by all exporting countries, and prepare and import the data from 2009/2010 and 2017/2018 to create two maps. Then answer the following questions.

International Coffee Organization
https://goo.gl/neyyeX

1. Which countries were the top producers in 2009/2010? Which were the top producers in 2017/2018? What political or economic factors might explain the changes?

2. Which world regions have seen the biggest increase in coffee production? Which have seen the biggest decline?

3. Explain how significant increases or decreases in coffee production could impact the physical environment and economic development potential of these regions.

The Caribbean

5

Physical Geography and Environmental Issues

Climate change threatens the Caribbean, with the potential for stronger and more frequent hurricanes, loss of land due to rising sea levels, and destruction of coral reefs.

Population and Settlement

Having experienced its demographic transition, the region now has slow population growth. In addition, large numbers of Caribbean peoples have emigrated from the region, leaving in search of economic opportunity and sending back billions of dollars.

Cultural Coherence and Diversity

Creolization—the blending of African, European, and Amerindian elements—led to unique Caribbean expressions of culture, such as rara, reggae, and steel drum bands. Caribbean-styled celebrations of carnival have diffused with the movement of the region's people to Europe and North America.

Geopolitical Framework

The first area of the Americas to be extensively explored and colonized by Europeans, the region has seen many rival European claims and, since the early 20th century, has experienced strong U.S. influence.

Economic and Social Development

Environmental, locational, and economic factors make tourism a vital component of this region's economy. Offshore manufacturing and banking are also significant in the region's modern economic development.

▶ A vendor prepares fish for sale in a Georgetown, Guyana, market. Fish is a regular part of the Caribbean diet as most people live near the sea, so residents are concerned about local fisheries as offshore oil is developed off Guyana's coast.

Georgetown, Guyana

THE CARIBBEAN

Guyana is an English-speaking country of less than one million inhabitants on the northern coast of South America. Part of the Caribbean rimland, most Guyanese reside near Georgetown, the capital, on the coast. The country is heavily forested and rich in minerals, but with limited infrastructure and a gross national income per capita (based on purchasing power parity) of $8100. Much is about to change for Guyana, however, as the government has signed a deal with Exxon-Mobil to extract some 3 billion barrels of light crude oil. The oil is about 120 miles offshore, and Guyana is not in the hurricane belt, making it an attractive place to invest. As offshore drilling begins, the big question is: Can Guyana benefit from its new wealth, or will a few elite reap most of the profits?

Oil production at the scale anticipated would catapult Guyana into the ranks of top oil exporters. The gross domestic product of Guyana is expected to triple in just five years! Yet this is a small country with weak institutions and a history of ethnic tensions. With no plans to build a refinery in the country, there will be few jobs directly tied to oil production. Still, Guyana's government will receive half the revenue once ExxonMobil has recovered its costs. The government's challenge will be to invest this money wisely, improving infrastructure, education, and job opportunities for this diverse country while avoiding the environmental costs of a poorly managed oil industry.

Defining the Region

The Caribbean was the first region of the Americas to be extensively explored and colonized by Europeans. Yet its modern identity is unclear—often merged with Latin America, but also viewed as apart from it. Today the region is home to 45 million inhabitants scattered across 26 countries and dependent territories, ranging from the small British dependency of Turks and Caicos, with 50,000 people, to the island of Hispaniola with 22 million. In addition to the Caribbean islands, Belize of Central America and the three Guianas—Guyana, Suriname, and French Guiana—of South America are considered part of the Caribbean. For historical and cultural reasons, these mainland peoples identify with the island nations and are thus included in this chapter (Figure 5.1).

The basis for treating the Caribbean as a distinct world region lies within its particular economic and cultural history. As in many developing areas, external control of the Caribbean produced highly dependent and inequitable economies, accentuated by a reliance on slave labor, plantation agriculture, and the prosperity that sugar promised in the 18th century. Culturally, the region can be distinguished from the largely Iberian-influenced mainland of Latin America by its more diverse European colonial history and strong African imprint.

Still a developing area, most Caribbean states have achieved life expectancies in the 70s, low child mortality rates, and high literacy rates. Millions of tourists view the Caribbean as an international playground for sun, sand, and fun. However, there is another Caribbean, far poorer and economically more dependent than the one portrayed on travel posters. Haiti is by most measures the Western Hemisphere's poorest country. In fact, the majority of the Caribbean people, living in the shadow of North America's vast wealth, suffer from serious economic problems and poverty.

Nevertheless, the Caribbean has evolved as a distinct but economically marginal world region. This status is expressed today as workers leave the region in search of better wages, while foreign companies are drawn to the Caribbean's cheaper labor costs and natural resources. The economic well-being of most Caribbean countries is precarious. Despite such uncertainty, an enduring cultural richness and an attachment to place are present here, which may explain a growing countercurrent of migration back to the region.

> The Caribbean has evolved as a distinct but economically marginal world region.

LEARNING OBJECTIVES

After reading this chapter you should be able to:

5.1 Differentiate between island and rimland environments, and the environmental issues that affect each.

5.2 Summarize demographic shifts in the Caribbean as population growth slows, settlement in cities intensifies, emigration abroad continues, and a return migration begins.

5.3 Identify the demographic and cultural implications of the transfer of African peoples to the Caribbean and the creation of a neo-African society in the Americas.

5.4 Explain why European colonists so aggressively sought control of the Caribbean and why independence in the region came about more gradually than in neighboring Latin America.

5.5 Describe how the Caribbean is linked to the global economy through offshore banking, emigration, and tourism.

5.6 Suggest reasons why the Caribbean does better in social indicators of development than in economic indicators.

Physical Geography and Environmental Issues: Paradise Undone

Tucked between the Tropic of Cancer and the equator, with year-round temperatures averaging in the high 70s, the hundreds of islands and picturesque waters of the Caribbean have often inspired comparisons to paradise. Columbus began the tradition by describing the islands of the New World as the most marvelous, beautiful, and fertile lands he had ever known, filled with flocks of parrots, exotic plants, and friendly natives. Writers today are still lured by the sea, sands, and swaying palms of the Caribbean.

Ecologically speaking, it is difficult to picture a landscape more completely altered than that of the Caribbean. For over five centuries, the destruction of forests and the unrelenting cultivation of soils

▲ **Figure 5.1 The Caribbean** Containing 26 states and dependent territories, the Caribbean today is a product of a long and complex history of colonialism and independence. More than 45 million residents live in the region, but most of the population is found on the four largest islands: Cuba, Hispaniola, Jamaica, and Puerto Rico.

resulted in the extinction of many endemic (native) Caribbean plants and animals, including various shrubs and trees, songbirds, large mammals, and monkeys. This severe depletion of biological resources helps explain some of the region's current economic and social instability. Most Caribbean environmental problems are associated with agricultural practices, soil erosion, excessive reliance on wood and charcoal for fuel, and the threat of global climate change. The devastating effects of Hurricanes Irma and Maria in 2017 demonstrated how vulnerable this region is, especially as tropical storms increase in intensity due to climate change. Likewise, the 2010 Port-au-Prince earthquake underscored how quickly a place such as Haiti can become undone due to a major natural disaster. However, because many countries rely on tourism as a vital source of income, the region has also experienced a growth in protected areas, both on the land and in maritime areas.

Island and Rimland Landscapes

The Caribbean Sea itself—the body of water enclosed between the Antillean islands (the arc of islands that begins with Cuba and ends with Trinidad) and the mainland of Central and South America—links the states of the region. Historically, the sea connected people through its trade routes and sustained them with its marine resources of fish, green turtle, manatee, lobster, and crab. The Caribbean Sea is noted for its clarity and biological diversity, but it has never supported large-scale commercial fishing because the quantities of any one species are not large. Sea surface temperatures range from 73° to 84°F (23° to 29°C), over which forms a warm tropical marine air mass that influences daily weather patterns. This warm water and tropical setting continue to be key resources for the region, as millions of tourists visit the Caribbean each year (Figure 5.2).

The arc of islands that stretches across the Caribbean Sea is its most distinguishing physical feature. The Antillean islands are divided into the Greater and Lesser Antilles. The majority of the region's people

▼ **Figure 5.2 Caribbean Sea**   Noted for its calm turquoise waters, steady breezes, and treacherous shallows, the Caribbean Sea has both sheltered and challenged sailors for centuries. Recreational sailors enjoy the waters around Tabago Cays in the Grenadines.

live on these islands. The **rimland** (the Caribbean coastal zone of the mainland) includes Belize and the Guianas as well as the Caribbean coast of Central and South America (see Figure 5.1). In contrast to the islands, the rimland has low population densities.

Most of the islands, with the exception of Cuba, are on the Caribbean tectonic plate, wedged between the South American and North American plates. Generally, this is not one of the most tectonically active zones, although earthquakes and volcanic eruptions do happen. In January 2010, a magnitude 7.0 earthquake leveled Port-au-Prince in one of the most tragic natural disasters to strike the Caribbean.

The Enriquillo Fault, near the densely settled and extremely poor capital city of Haiti, had been inactive for more than a century. When it violently shifted in 2010, the epicenter of the resulting earthquake was just a few miles from the city. The disaster affected nearly 3 million people, as homes were rendered unsafe and water and electricity supplies were disrupted. Shockingly, the earthquake left over 200,000 people dead and 1 million homeless. Many government agencies that would have assisted in the relief response were also destroyed. The tragedy was compounded by the state's poverty and corruption, as most buildings were not built to standards that could withstand an earthquake of this magnitude. The international community, as well as the large diaspora of Haitians living abroad, immediately offered financial aid and assistance. More than $12 billion in humanitarian and development aid and debt relief were pledged in the wake of the crisis. Rebuilding has slowly progressed, including a new two-lane highway between Port-au-Prince and Gonaives, a new airport, and hundreds of new schools. Neighborhoods such as Jalousie (where many earthquake survivors resettled) received a new paint job from the government, but residents in this neighborhood would have benefited more from investments in sanitation, electricity, and schools (Figure 5.3).

Greater Antilles   The four large islands of Cuba, Jamaica, Hispaniola (shared by Haiti and the Dominican Republic), and Puerto Rico make up the **Greater Antilles**. These islands contain the bulk of the region's population, arable lands, and large mountain ranges. Given the popular interest in the Caribbean coasts, it still surprises many people that Pico Duarte in the Cordillera Central of the Dominican Republic is more than 10,000 feet (3000 meters) tall, Jamaica's Blue Mountains top 7000 feet (2100 meters), and Cuba's Sierra Maestra is more than 6000 feet (1800 meters) tall. The mountains of the Greater Antilles were of little economic interest to plantation owners, who preferred the coastal plains and valleys, but were an important refuge for runaway slaves and subsistence farmers and thus figure prominently in the cultural history of the region.

The best farmlands are found in the Greater Antilles, especially in the central and western valleys of Cuba, where limestone contributes to the formation of a fertile red clay soil (locally called *mantanzas*), and in Jamaica, where a gray or black soil type called *rendzinas* is found. Surprisingly, given the region's agricultural history, many other soils on these islands are nutrient-poor, heavily leached, and acidic.

Lesser Antilles   The **Lesser Antilles** form a double arc of small islands stretching from the Virgin Islands

▲ **Figure 5.3 Housing in Haiti** The brightly colored houses of Jalousie, a Port-au-Prince informal settlement, look inviting from afar. Yet, this crowded neighborhood of 50,000 has no power grid or sanitation. It also is prone to mudslides and is located on a fault line.

Explore the SIGHTS of Jalousie, Port-au-Prince
https://goo.gl/U9koZQ

▲ **Figure 5.4 European Space Center** An Ariane 5 rocket carrying telecommunications and meteorology satellites launches from Kourou, French Guiana. The European Space Agency uses this French territory as it is near the equator, on the coast, and lightly populated.

to Trinidad. Smaller in size and population than the Greater Antilles, they were important early footholds for rival European colonial powers. The islands from St. Kitts to Grenada form the inner arc of the Lesser Antilles. These mountainous islands, with peaks ranging from 4000 to 5000 feet (1200 to 1500 meters), have volcanic origins. Erosion of the island peaks and accumulation of ash from volcanic eruptions formed small pockets of arable soils, although the steep terrain limits agricultural development.

Just east of this volcanic arc are the low-lying islands of Barbados, Antigua, Barbuda, and the eastern half of Guadeloupe. Covered in limestone that overlays volcanic rock, these lands were much more inviting for agriculture. Such soils were ideal for growing sugarcane, making these islands important early settings for the plantation economy that diffused throughout the region.

Rimland States Unlike the rest of the Caribbean, the rimland states of Belize and the Guianas still contain significant amounts of forest cover. As on the islands, agriculture in these states is closely tied to local geology and soils. Much of low-lying Belize is limestone. Sugarcane is the dominant crop in the drier north, whereas citrus is produced in the wetter central portion of the state. The Guianas are characterized by the rolling hills of the Guiana Shield, whose crystalline rock is responsible for the area's overall poor soil quality. Thus, most agriculture in the Guianas occurs on the narrow coastal plain, with sugarcane and rice as the primary crops. French Guiana, an overseas territory of France, relies mostly on French subsidies but exports shrimp and timber. It is also home to the European Space Center at Kourou (Figure 5.4).

Caribbean Climate and Climate Change

Much of the Antillean islands and rimland receives more than 80 inches (200 centimeters) of rainfall annually, which is enough to support tropical forests. Average temperatures are typically highs of 80°F degrees and lows of 70°F (27°C and 21°C) (Figure 5.5). Seasonality in the Caribbean is defined by changes in rainfall rather than temperature. Although rain falls throughout the year, for much of the region the rainy season is from July to October. This is when the Atlantic high-pressure cell is farthest north and easterly winds generate moisture-laden air and unstable atmospheric conditions that sometimes yield hurricanes. During the slightly cooler months of December through March, rainfall declines (see the climographs in Figure 5.5 for Havana, Port-au-Prince, and Bridgetown). This time of year corresponds with the peak tourist season.

The Guianas have a different rainfall cycle. On average, these territories receive more rain than the Antillean islands. In Cayenne, French Guiana, an average of 126 inches (320 centimeters) falls each year (see the climograph in Figure 5.5). Unlike the Antilles, the Guianas experience a brief dry period from September to October, while January tends to be a wet period. The Guianas also differ from the rest of the region because they are not affected by hurricanes.

Hurricanes Each year, **hurricanes** hit the Caribbean, as well as Central and North America, with heavy rains and fierce winds. Beginning in July, westward-moving low-pressure disturbances form off the coast of West Africa, picking up moisture and speed as they move across the Atlantic. Usually no more than 100 miles across, these disturbances achieve hurricane status when wind speeds reach 74 miles per hour. Hurricanes may take several paths through the region, but they typically enter through the Lesser Antilles. They then curve north or northwest and collide with the Greater Antilles, Central America, Mexico, or southern North America before moving to the northeast and dissipating in the Atlantic Ocean. The hurricane zone lies just north of the equator on both the Pacific and Atlantic sides of the Americas. Typically, a half-dozen to a dozen hurricanes form each season and move through the region, causing limited damage.

There are, of course, exceptions, and most longtime residents of the Caribbean have felt the full force of at least one major hurricane in their lifetimes. The destruction caused by these storms is not only from the high winds (which blow off roofs, take out power, and destroy crops), but also from the heavy downpours, which can cause severe flooding and deadly coastal storm surges. Modern tracking equipment has improved hurricane forecasting and reduced the number of fatalities, primarily through early evacuation of areas in a hurricane's path. This has saved lives but cannot reduce the damage to crops, forests, or infrastructure.

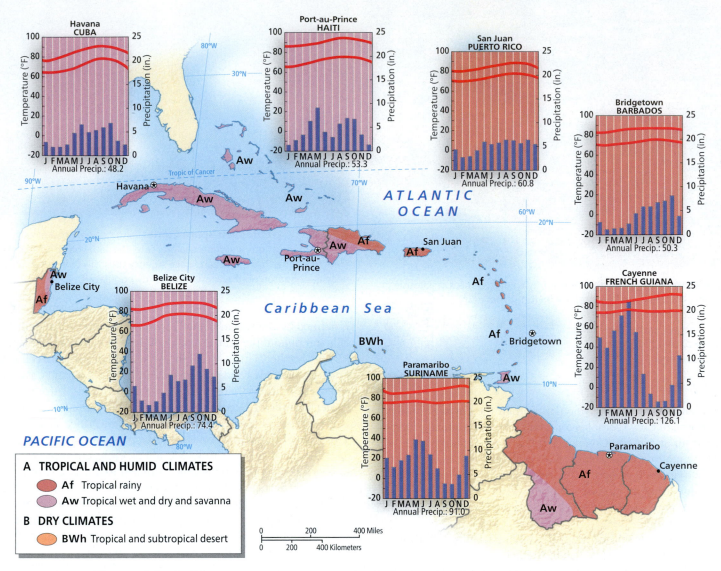

▲ **Figure 5.5** **Climate of the Caribbean** Most of the region is classified as having either a tropical wet (Af) or a tropical savanna (Aw) climate. Temperature varies little across the region, as shown by the relatively straight temperature lines. Important differences in total rainfall and the timing of the dry season distinguish different localities. **Q: What are the wettest and driest months for each of these Caribbean cities? Is there a general pattern between the northern vs. the southern Caribbean?**

Back-to-back hurricanes (Irma and Maria) in September 2017 devastated the northeastern Caribbean. Irma, a Category 5 storm (sustained winds of 175 mph), hit Puerto Rico, Hispaniola, Cuba, and The Bahamas before striking the Florida Keys. Days later, Hurricane Maria, another Category 5 hurricane, slammed into Dominica, inflicting damage on nearly every structure on the island. By the time Maria reached Puerto Rico, it had slowed to a Category 4 hurricane (sustained winds of 155 mph), but it was still the strongest hurricane to directly hit the island in 80 years. The storm destroyed homes, bridges, roads, and crops, and knocked out electricity across the entire island of over 3 million residents (Figure 5.6). Initial death tolls attributed to the storm were quite low. Since then, however, the U.S. government has acknowledged that some 1400 people perished due to lack of medical attention and the months required to restore electricity throughout the island; estimates by island officials are much higher. These two hurricanes were the most powerful and expensive storms in modern Caribbean history. Experts also fear that storms of such magnitude may become a more regular occurrence as a result of global climate change.

Climate Change Of all the issues facing the Caribbean, one of the most difficult to address is climate change. The Caribbean is not a major contributor of greenhouse gases (GHGs), but this maritime region is extremely vulnerable to the negative effects of climate change, including sea-level rise, increased intensity of storms, variable rainfall leading to both floods and droughts, and loss of biodiversity (both in forests and in coral reefs). The scientific consensus is that surface temperatures would increase between 1.2° and 2.3°C across the Caribbean in this century, and that rainfall will decrease by 5 to 6 percent. Perhaps most destructive of all, the latest climate change models show sea levels rising by 0.6 meters by the end of the century. In terms of land loss due to inundation, the low-lying Bahamas would be the region's most affected country. In terms of people affected by inundation, Suriname, French Guiana, Guyana, Belize, and The Bahamas would be the most

Throughout the Caribbean, protecting the environment and preparing for the effects of climate change are increasingly being recognized not as a luxury, but as a question of economic livelihood. In fact, Caribbean nations represent half of the 41 countries that make up the Small Island Developing States (SIDS), a group of developing countries that advocate for ecosystem-based adaptation and risk reduction strategies. Strategies such as conserving water resources, reducing GHG emissions, developing renewable energy, and protecting coastal reefs and mangroves are all efforts to reduce the risks associated with climate change. Keeping in line with the Paris Agreement (see Chapter 2), the government of Japan launched a $15 million partnership with eight Caribbean nations to support 50 coastal communities in adaptation efforts.

Environmental Issues

Climate change is both a medium- and a long-term concern for the Caribbean region. However, other environmental issues, such as soil erosion and deforestation, have preoccupied the region due to its long-standing dependence on agriculture (see Figure 5.7). Also, as the Caribbean has become more urbanized and more reliant on tourism, governments have come to realize that protection of local ecosystems is not just good for the environment but also good for the overall economy of the region.

Agriculture's Legacy of Deforestation   Prior to the arrival of Europeans, much of the Caribbean was covered in tropical forests. The great clearing of these forests began on European-owned plantations on the smaller islands of the eastern Caribbean in the 17th century and spread westward. The island forests were removed not only to make room for sugarcane but also to provide the fuel necessary to turn the cane juice into sugar, molasses, or rum. Lumber was needed for housing, fences, and ships. Primarily, however, tropical forests were removed because they were seen as unproductive; the European colonists valued cleared land. The newly exposed tropical soils easily eroded and ceased to be productive after several harvests, a situation that led to two distinct land-use strategies. On the larger islands of Cuba and Hispaniola, as well as on the mainland, new lands were constantly cleared and older ones abandoned or fallowed in an effort to keep up sugar production. On the smaller islands where land was limited, such as Barbados and Antigua, labor-intensive efforts to conserve soil and maintain fertility were employed. In either case, the island forests were replaced by a landscape devoted to crops for world markets.

Haiti's problems with deforestation were accentuated throughout the 20th century as a destructive cycle of environmental and economic impoverishment was established. Half of Haiti's people are peasants who work small hillside plots and seasonally labor on larger estates. As the population grew, people sought more land. They cleared the remaining hillsides, subdivided their plots into smaller units, and abandoned the practice of fallowing land in an effort to eke out an annual subsistence. When the heavy tropical rains came, the exposed and easily eroded mountain soils were washed away. As sediments collected in downstream irrigation ditches and behind dams, agriculture suffered, electricity production declined, and water supplies were degraded. Deforestation was further aggravated by reliance of many poor Haitians on charcoal (made from trees) for household needs. It is estimated that less than 3 percent of Haiti remains forested (Figure 5.8).

▲ **Figure 5.6 Hurricane Maria in Puerto Rico** The destructive capacity of hurricanes is due to high velocity winds, vast amounts of rain, and floods. This home in Utuado, in the mountains of central Puerto Rico, was destroyed by mudslides and flooding caused by Hurricane Maria in 2017.

Explore the SIGHTS Hurricane Maria's Aftermath
https://goo.gl/kPKkth

severely impacted: A 3-foot (1-meter) sea-level rise would be devastating because most of the population lives near the coast (Figure 5.7).

In addition to land loss and population displacement due to sea-level rise, other concerns include changes in rainfall patterns, leading to declines in agricultural yields and freshwater supplies; and more intense storms—especially hurricanes—that destroy infrastructure and cause other problems. All these changes would negatively affect tourism and thus the gross domestic income of Caribbean countries. Some of the worst-case scenarios are catastrophic.

In terms of biodiversity, continued warming of ocean temperatures will further damage the Caribbean's coral reefs. These reefs, particularly those of the rimland, are already threatened by water pollution and subsistence fishing practices. Fortunately, efforts by the Belizean government to protect mangrove trees and slow coastal development resulted in improvements in the country's barrier reef in 2018. Coral reefs are biologically diverse and productive ecosystems that function as nurseries for many marine species. Healthy reefs also serve as barriers to protect populated coastal zones as well as mangroves and wetlands.

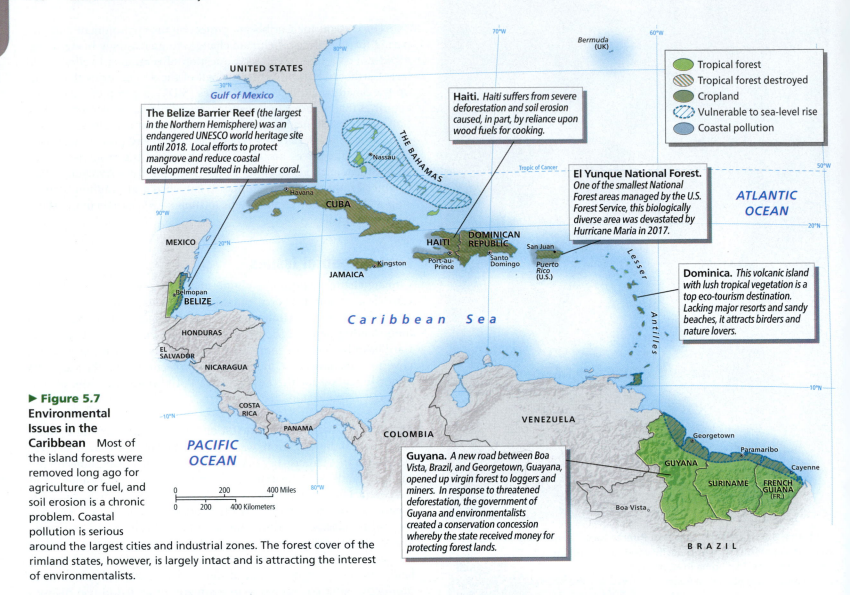

► **Figure 5.7**
Environmental Issues in the Caribbean Most of the island forests were removed long ago for agriculture or fuel, and soil erosion is a chronic problem. Coastal pollution is serious around the largest cities and industrial zones. The forest cover of the rimland states, however, is largely intact and is attracting the interest of environmentalists.

The Belize Barrier Reef (the largest in the Northern Hemisphere) was an endangered UNESCO world heritage site until 2018. Local efforts to protect mangrove and reduce coastal development resulted in healthier coral.

Haiti. Haiti suffers from severe deforestation and soil erosion caused, in part, by reliance upon wood fuels for cooking.

El Yunque National Forest. One of the smallest National Forest areas managed by the U.S. Forest Service, this biologically diverse area was devastated by Hurricane Maria in 2017.

Dominica. This volcanic island with lush tropical vegetation is a top eco-tourism destination. Lacking major resorts and sandy beaches, it attracts birders and nature lovers.

Guyana. A new road between Boa Vista, Brazil, and Georgetown, Guyana, opened up virgin forest to loggers and miners. In response to threatened deforestation, the government of Guyana and environmentalists created a conservation concession whereby the state received money for protecting forest lands.

Legend:
- Tropical forest
- Tropical forest destroyed
- Cropland
- Vulnerable to sea-level rise
- Coastal pollution

While Haiti has lost most of its forest cover, on Jamaica and Cuba nearly one-third of the land is still forested. In the case of the Dominican Republic, about 40 percent of the country is forested, and more than half of Puerto Rico is forested. On these more forested islands, the decline in agriculture has caused some fields to be abandoned and forests to recover. In the case of Puerto Rico, a territory of the United States, its national forests—such as El Yunque on the eastern side of the island—contribute greatly to the island's biological diversity (Figure 5.9).

◄ **Figure 5.8 Differences in Deforestation**
Hispaniola has serious deforestation problems. Hillsides are denuded to make room for agricultural production or to secure wood fuel. **Q: Which side of the orange border is Haiti? How would you compare land use in Haiti with that in the Dominican Republic?**

▲ **Figure 5.9 El Yunque National Forest** Puerto Rico's largest remaining rainforest is in El Yunque National Forest. Visitors enjoy the park's hiking trails and rich biological diversity, yet the forest sustained major damage by Hurricane Maria in 2017.

Energy Needs and Innovations
With the exception of Guyana's new offshore oil production and Trinidad and Tobago, which exports oil and liquefied natural gas, the Caribbean states are net importers of oil and highly dependent on foreign sources for their energy needs. The region has some oil refineries that process crude oil shipments into petroleum for domestic consumption and even some for export, but dependence on foreign energy and volatile oil pricing make the small economies of this region vulnerable.

Not surprisingly, Caribbean nations have a growing interest in renewable energy. In many ways, wind energy has long been important for the Caribbean economy. After all, the entire colonial enterprise depended on the trade winds to move commodities and people across the Atlantic. Commercial wind energy is gaining popularity. The Santa Isabel wind farm on Puerto Rico's southern coast near the city of Ponce is the largest wind generator in the Caribbean.

The region's potential for solar power is also excellent, and this sector has been growing. Nearly half the households in Barbados use solar water heaters, and solar panels are increasingly seen around government buildings, hospitals, businesses, and private homes. Some of the volcanically active Caribbean islands, especially in the Lesser Antilles, are tapping into clean **geothermal** energy (see *Working Toward Sustainability: Geothermal Energy for the Lesser Antilles*).

Conservation Efforts In general, biological diversity and stability are less threatened in the rimland states than in the rest of the Caribbean. Thus, current conservation efforts could produce important results. Even though much of Belize was selectively logged for mahogany in the 19th and 20th centuries, healthy forest cover still supports a diversity of mammals, birds, reptiles, and plants. Public awareness of the negative consequences of deforestation is also greater now, and Belize has established many protected areas. In the mid-1980s, villagers in Bermudian Landing, Belize, established a community-run sanctuary for black howler monkeys (locally referred to as baboons). The villagers banded together to maintain the habitat for the monkeys and commit to land management practices that accommodate this gregarious species. The success of the project led to tourists visiting the villages to see these indigenous primates up close. In 1986, a jaguar preserve was established in the Cockscomb Basin in southern Belize, the first of its kind in Central America.

Explore the SOUNDS of Howler Monkeys
http://goo.gl/KWDQLu

Slowly, the territorial waters surrounding the Caribbean nations have gained protection, although more could be done. The Caribbean island of Bonaire, which attracts large numbers of scuba divers, maintains the Bonaire Marine Park, recognized as one of the most effectively managed marine reserves in the region. Marine tourism in Dominica has also grown, especially the viewing sperm whales, which congregate off the deep waters of Dominica from November to June. Considered the whale viewing capital of the Caribbean, there are even exclusive permits to swim near the whales (Figure 5.10).

▼ **Figure 5.10 Swimming with Whales** The seafloor drops steeply off the west coast of Dominica, creating a sheltered habitat for a large resident population of sperm whales. For decades, visitors to the island have delighted in viewing the whales from boats. For those who want an up-close experience, limited and controlled in-water permits are available to researchers, film crews, and specialized underwater expedition leaders.

WORKING TOWARD SUSTAINABILITY

Geothermal Energy for the Lesser Antilles

As Caribbean states strive for greener and less expensive energy sources, geothermal energy has become an attractive alternative for the islands of the Lesser Antilles. Many of these islands lie in a subduction zone where the Caribbean plate collides with the North American or South American plates. Such locations are ideally positioned for drilling geothermal wells that tap into subsurface heat for electricity production. Geothermal energy is the intense heat deep within the Earth, felt at the surface in geysers, hot springs, and volcanoes. Now several Caribbean islands—with international grants and technological help from experts in Iceland, New Zealand, and Japan—are drilling geothermal wells and constructing the first power stations. Not many countries in the world can exploit geothermal energy, but those that do benefit from a reliable and renewable energy alternative.

For most of the residents in the Lesser Antilles, energy costs are three times that of wealthy mainland countries. Many of these islands burn oil or gas to generate electricity, which is expensive and produces greenhouse gases (GHGs). In addition to energy conservation, such as using LED light bulbs or highly efficient air conditioners, geothermal energy represents a clean, local, and renewable alternative for power generation. Seven island nations are exploring or developing geothermal energy sources: Guadeloupe, Dominica, Grenada, Montserrat, St. Lucia, St. Kitts and Nevis, and St. Vincent and the Grenadines. Currently Guadeloupe is the only Caribbean island with an active geothermal plant (Figure 5.1.1).

With the international focus on reducing GHG emissions, there are new sources of funding to support geothermal infrastructure in the region. After a grant from the European Union resulted in identifying drill sites, Dominica now has a loan from the World Bank to build two geothermal plants that will supply much of the island's energy needs and even energy for export. Guadeloupe, a French territory, has La Bouillante plant, which generates 15 megawatts of clean geothermal power (Figure 5.1.2).

Geothermal energy does have its risks. Tapping the wells can be costly and time consuming. Also, hot water from geothermal sources can contain trace amounts of toxic metals such as mercury and arsenic, which evaporate into the air as the water cools. Consequently, there have been local protests against geothermal plants, especially from nearby farmers and landowners. But countries such as Iceland, Japan, and New Zealand that have exploited geothermal energy for years demonstrate the enormous potential of this resource to produce electricity without using fossil fuels.

1. **What are the potential benefits of geothermal energy for the Lesser Antilles?**

2. **Consider Figure 2.3, which shows Earth's plate boundaries. Where else might geothermal energy be developed? Is this a possibility in your area?**

▲ **Figure 5.1.1 Powering the Lesser Antilles** The islands that arc from St. Kitts and Nevis to Grenada are part of the Volcanic Axis. On fault lines and with geothermal activity, these islands are seeking to develop renewable energy from the earth.

▲ **Figure 5.1.2 Green Energy for Guadeloupe** Workers tend to the turbines at the new geothermal plant in La Bouillante. Geothermal energy—widely used in Iceland, Japan, and New Zealand—is being developed in the Lesser Antilles, including the French territory of Guadeloupe and Dominica.

GOOGLE EARTH
Virtual Tour Video
https://goo.gl/4YgtUP

5.1 Describe the locational, environmental, and climate factors that together help make the Caribbean a major international tourist destination.

5.2 What environmental issues currently affect the Caribbean? Describe the risks and possible solutions.

KEY TERMS rimland, Greater Antilles, Lesser Antilles, hurricanes, geothermal

Population and Settlement: Densely Settled Islands and Rimland Frontiers

Caribbean population density is generally quite high and, as in neighboring Latin America, increasingly urban. Eighty-five percent of the region's population is concentrated on the four islands of the Greater Antilles (Figure 5.11). Add to this Trinidad's 1.4 million and Guyana's

800,000, and most of the population of the Caribbean is accounted for by six countries and one U.S. territory (Puerto Rico).

In terms of total population, few people inhabit the Lesser Antilles; nevertheless, some of these small island states are densely settled (Table 5.1). The island of Barbados is an extreme example. With only 166 square miles (430 square kilometers) of territory, it has over 1700 people per square mile (664 per square kilometer). Bermuda, which is one-third the size of the District of Columbia, has more than 3000 people per square mile (1300 per square kilometer). Population densities on St. Vincent, Martinique, and Grenada, while not as high, are still more than 700 people per square mile (280 per square kilometer). Because arable land is scarce on some islands, access to land is a basic resource problem for many island states. Population growth in the region, coupled with the scarcity of land, has forced many people into cities or abroad.

▼ **Figure 5.11 Population of the Caribbean** The major population centers are on the islands of the Greater Antilles. The pattern here, as in the rest of Latin America, is a tendency toward greater urbanism. The largest metropolitan areas in the region are San Juan, Santo Domingo, Port-au-Prince, and Havana; each has over 2 million residents. In comparison, the rimland states are lightly settled.

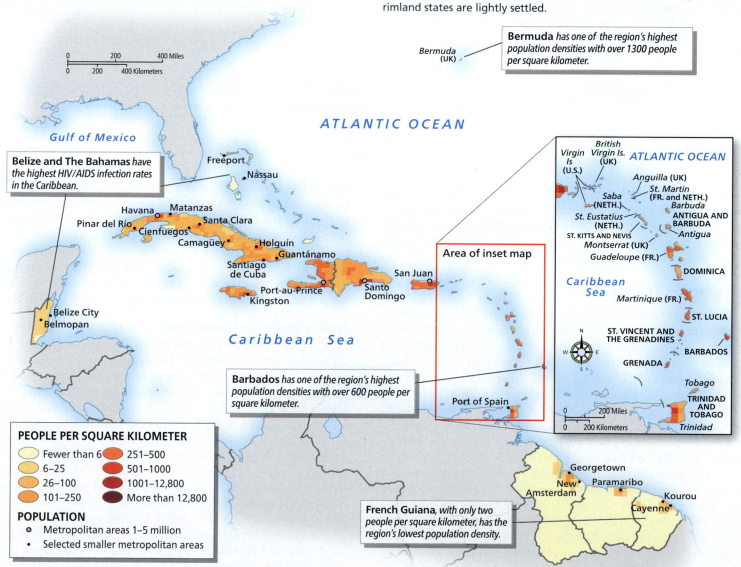

TABLE 5.1 Population Indicators

Explore these data in MapMaster 2.0 https://goo.gl/jphmih

Country	Population (millions) 2018	Population Density (per square kilometer)[1]	Rate of Natural Increase (RNI)	Total Fertility Rate	Life Expectancy		Percent Urban	Percent <15	Percent >65	Net Migration (rate per 1000)
					Male	Female				
Anguilla (UK)	0.02*	–	–	1.7*	–	–	100*	–	–	–
Antigua and Barbuda	0.1	232	0.9	1.9	74	79	25	24	7	0
Aruba (Netherlands)	0.1*	585	–	1.8*	–	–	43*	–	–	–
Bahamas	0.4	39	0.7	1.6	73	79	83	20	9	3
Barbados	0.3	664	0.1	1.6	73	78	31	19	15	1
Belize	0.4	16	1.8	2.6	71	77	45	36	4	4
Bermuda (UK)	0.07*	1309	–	1.9*	–	–	100*	–	–	–
Cayman Islands (UK)	0.06*	256	–	1.9*	–	–	100*	–	–	–
Cuba	11.1	110	0.2	1.6	76	81	77	17	15	–5
Curaçao (Netherlands)	0.2	363	0.1	1.5	75	82	89	18	17	–3
Dominica	0.07	99	0.4	1.8	73	78	71	22	11	1
Dominican Republic	10.8	223	1.4	2.5	71	77	80	30	7	–3
French Guiana (France)	0.3	–	2.3	3.6	75	81	85	33	5	4
Grenada	0.1	317	0.7	2.1	74	79	36	26	7	0
Guadeloupe (France)	0.4	–	0.1	1.7	77	84	99	19	18	–9
Guyana	0.8	4	1.2	2.5	64	69	77	29	5	–7
Haiti	10.8	398	1.7	3.0	62	67	55	33	4	–2
Jamaica	2.9	267	1.0	2.0	74	78	56	23	9	–6
Martinique (France)	0.4	–	0.0	1.7	78	84	89	17	20	–9
Montserrat (UK)	0.005*	–	–	1.3*	–	–	9.1*	–	–	–
Puerto Rico (U.S.)	3.3	376	–0.2	1.1	78	84	94	16	19	–17
St. Kitts and Nevis	0.05	213	0.4	1.8	73	78	31	21	8	3
St. Lucia	0.2	293	0.7	1.5	75	83	19	20	11	–3
St. Vincent and the Grenadines	0.1	282	0.7	2.1	70	75	52	24	8	–6
Suriname	0.6	4	1.1	2.4	68	75	66	27	7	–2
Trinidad and Tobago	1.4	267	0.5	1.6	70	76	53	19	10	–1
Turks and Caicos Islands (UK)	0.05*	37	–	1.7*	–	–	93*	–	–	–

Source: Population Reference Bureau, *World Population Data Sheet* 2018.
[1] World Bank Open Data 2018.
* Additional data from CIA World Factbook.

Log in to Mastering Geography & access MapMaster to explore these data!
1) Which countries in this region have a positive net migration rate, and which ones are negative? What might explain these differences?
2) Which countries have "older populations" (a higher percentage of the population over 65 years of age) and which have younger ones (a higher percentage of population under 15)?

In contrast to the islands, the larger rimland states are lightly populated: Guyana averages 10 people per square mile (4 per square kilometer), as does Suriname. These areas are sparsely settled in part because the relatively poor quality and accessibility of arable land made them less attractive to colonial enterprises.

Demographic Trends

Prior to European contact with the New World, diseases such as smallpox, influenza, and malaria did not exist in the Americas. As discussed in Chapter 4, these diseases contributed to the demographic collapse of Amerindian populations. Epidemics spread quickly in the Caribbean, and within 50 years of Columbus's arrival, the indigenous population was virtually gone. Only the name *Caribbean* suggests that a Carib people once inhabited the region. Initially, European planters experimented with white indentured labor to work on sugar plantations. However, newcomers from Europe were especially vulnerable to malaria in the lowland Caribbean; typically, half died during the first year of settlement. Those who survived were considered "seasoned." In contrast, Africans had prior exposure to malaria and thus some immunity. They, too, died from malaria, but at much lower rates. This is not to argue that malaria caused slavery in the region, but it did strengthen the economic rationale for it.

During the years of slave-based sugar production, mortality rates were extremely high due to disease, inhumane treatment, and malnutrition. Consequently, the only way to maintain population levels was to continue to import enslaved Africans. With the end of slavery in the mid- to late 19th century and the gradual improvement of health and sanitary conditions on the islands, natural population increase began to occur. In the 1950s and 1960s, many states achieved peak growth rates of 3.0 percent or higher, causing population totals and densities to soar. Over the past 30 years, however, growth rates have come down or stabilized. As noted earlier, the current population of the Caribbean is 45 million. The population is now growing at an annual rate of 0.9 percent, and the projected population for 2030 is 48 million.

Fertility Decline and Longer Lives The most significant demographic trends in the Caribbean are the decline in fertility, an increase in life expectancy, and out-migration. Most countries of the region have low rates of natural increase (RNI) and are below replacement values for total fertility. Puerto Rico's rate of natural increase is −0.2, and the average woman has 1.1 children. A weak economy, natural disasters, and migration to the U.S. mainland have had a significant demographic impact. In socialist Cuba, the RNI is only 0.2, and the average woman has 1.6 children. In general, educational improvements, urbanization, availability of contraception, and a preference for smaller families have contributed to slower growth rates in the region, regardless of political ideology. Even states with relatively high total fertility rates (TFR), such as Haiti, have seen a decline in family size. Haiti's TFR fell from 6.0 in 1980 to 3.0 in 2018. Currently French Guiana has the region's highest TFR at 3.6.

Figure 5.12 provides a stark contrast in the population profiles of Cuba and Haiti. Although both are poor Caribbean countries, Haiti has the classic broad-based pyramid of a developing country, where more

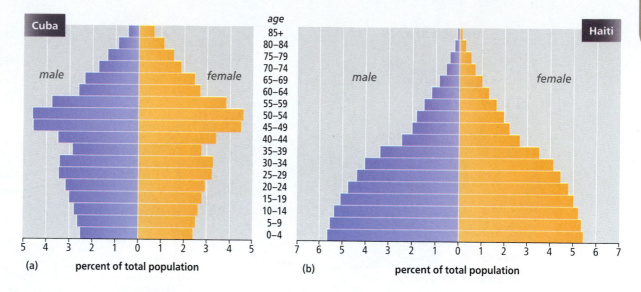

▶ **Figure 5.12** Population Pyramids of Cuba and Haiti Although neighbors, Cuba and Haiti have extremely different population profiles. (a) Cuba's population is stable and older, with a notable decline in family size. (b) Haiti's population is much younger and growing, which is reflected in its broad-based pyramid.

than one-third of the population is under the age of 15. Also, there are very few old people due to Haiti's relatively low life expectancy (64 years). In contrast, Cuba's population pyramid is more diamond-shaped, bulging in the 3 to 49-year-old age cohort. Here the impact of the Cuban revolution and socialism is evident. Family size fell sharply after education improved and modern contraception became readily available. With better health care, Cuba's population also lives longer, having the same life expectancy as people in the United States (76 years for men and 81 years for women). Cuba has 15 percent of its population over 65 and just 17 percent under 15; thus, it has an extremely low rate of natural increase, similar to many developed countries.

Emigration   Driven by the region's limited economic opportunities, a pattern of emigration to other Caribbean islands, North America, and Europe began in the 1950s. For more than 50 years, a Caribbean **diaspora**—the economic flight of Caribbean peoples across the globe—has defined existence and identity for much of the region (Figure 5.13). Barbadians generally choose England, most settling in the London suburb of Brixton with other Caribbean immigrants. In contrast, one out of every three Surinamese has moved to the Netherlands, with most residing in Amsterdam. As for Puerto Ricans, only slightly more live on the island than reside on the U.S. mainland. About 700,000 Jamaicans currently live in the United States, which is equal to one-quarter of the island's population. Cubans have made the city of Miami their destination of choice since the 1960s. Today they are a large percentage of that city's population.

Intraregional movements also are important. Perhaps one-fifth of all Haitians do not live in their country of birth. Their most common destination is the neighboring Dominican Republic, followed by the United States, Canada, and French Guiana. Dominicans are also on the move; the vast majority come to the United States, settling in New York City, where they are the single largest immigrant group. Others, however, simply cross the Mona Passage and settle in Puerto Rico. As a region, the Caribbean has one of the highest annual rates of net migration in the world at −4.0 per thousand. That means for every 1000 people in the region, 4 leave each year. Individual countries have much higher rates (see Table 5.1).

Crucial in this exchange of labor from south to north is the counterflow of cash **remittances**. Immigrants are expected to send something back, especially when immediate family members are left behind. Collectively, remittances add up; it is estimated that US$3.5 billion are sent annually to the Dominican Republic by immigrants in the United States, making remittances the country's second leading source of income. Jamaicans and Haitians remit nearly US$2 billion annually to their countries. Governments and individuals alike depend on these transnational family networks. Families carefully select the household member most likely to succeed abroad in the hope that money will flow back and a base for future immigrants will be established. A Caribbean nation in which no one left would soon face crisis.

Most migrants are part of a **circular migration** flow. In this type of migration, a man or woman typically leaves children behind with relatives in order to work hard, save money, and return home. Other times a **chain migration** begins, in which one family member at a time is brought over to the new country. In some cases, large numbers of residents from a Caribbean town or district send migrants to a particular locality in North America or Europe. Thus, chain migration can account for the formation of immigrant enclaves. Caribbean immigrants have increasingly practiced **transnational migration**—the straddling of livelihoods and households between two countries. Dominicans are probably the most transnational of all the Caribbean groups. They regularly move back and forth between two islands: Hispaniola and Manhattan. Former Dominican President Leonel Fernández was first elected in 1996 for a four-year term and was reelected in 2004 and in 2008. He grew up in New York City and for many years held a green card.

The Rural–Urban Continuum

Initially, plantation agriculture and subsistence farming shaped Caribbean settlement patterns. Low-lying arable lands were dedicated to export agriculture and controlled by the colonial elite. Only small amounts of land were set aside for subsistence production. Over time, villages of freed or runaway slaves were established, especially in remote island interiors. However, the vast majority of people lived on estates as owners, managers, or slaves. Cities were formed to serve the administrative and social needs of the colonizers, but most were small, containing just a fraction of a colony's population. The colonists who linked the Caribbean to the world economy saw no need to develop major urban centers.

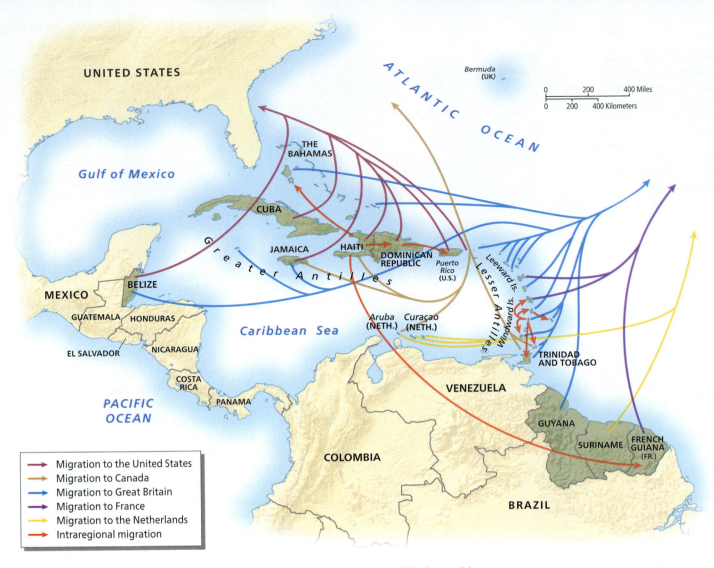

Migration to the United States
Migration to Canada
Migration to Great Britain
Migration to France
Migration to the Netherlands
Intraregional migration

▲ **Figure 5.13 Caribbean Diaspora** Emigration has long been a way of life for Caribbean peoples. With relatively high education levels but limited professional opportunities, migrants from the region head to North America, the United Kingdom, France, and the Netherlands. Intraregional migrations also occur between Haiti and the Dominican Republic and between the Dominican Republic and Puerto Rico. **Q: Compare this map with Figure 5.21 on Caribbean languages. What is the relationship between migration flows and the languages spoken in the origin and destination countries?**

Plantation America Anthropologist Charles Wagley coined the term **plantation America** to designate a cultural region that extends from midway up the coast of Brazil through the Guianas and the Caribbean into the southeastern United States. Ruled by a European elite dependent on enslaved African labor, this society was primarily coastal and produced agricultural exports. It relied on **mono-crop production** (production of a single commodity, such as sugar) under a plantation system that concentrated land in the hands of elite families. The term *plantation America* is meant to describe not a race-based division of the Americas but, rather, a production system that brought about specific ecological, social, and economic relations (Figure 5.14). Such a system created rigid class lines and formed a multiracial society in which people with lighter skin were privileged.

Even today the structure of Caribbean communities reflects this plantation legacy. Many of the region's subsistence farmers are descendants of enslaved Africans and continue to farm their small plots and work part-time as wage laborers on estates, especially in Haiti. The social and economic patterns generated by slavery still mark the landscape. Rural communities tend to be loosely organized, labor is temporary, and

▼ **Figure 5.14 Sugar Plantation** Commodities such as tobacco and sugar were profitable, but the work was arduous. Several million Africans were enslaved and forcibly relocated to the region to produce these commodities. This illustration from 1823 depicts slaves planting sugarcane in Antigua.

small farms are scattered on available pockets of land. Because men tend to leave home for seasonal labor, female-headed households are common.

Caribbean Cities   Since the 1960s, the mechanization of agriculture, offshore industrialization, and population growth have caused a surge in rural-to-urban migration. Cities have grown accordingly, and today two-thirds of the region is classified as urban. Caribbean cities are not large by world standards, as only five have 1 million or more residents: Santo Domingo, Havana, Port-au-Prince, San Juan, and Kingston.

Like their counterparts in Latin America, the Spanish Caribbean cities were laid out on a grid with a central plaza. Vulnerable to raids by rival European powers and pirates, these cities were usually walled and extensively fortified. The oldest continuously occupied European city in the Americas is Santo Domingo in the Dominican Republic, settled in 1496 (Figure 5.15). Today it is a metropolitan area of 3 million people. *Merengue*—fast-paced, highly danceable music that originated in the Dominican Republic—is the soundtrack that pulses through the metropolis day and night. As rural migrants poured into the city over the last four decades in search of employment and opportunity, the city steadily grew. In 2009, a high-speed Metro opened in Santo Domingo. It now has two lines and over 30 stations to help reduce the city's crushing traffic.

The region's second largest city is metropolitan San Juan, estimated at 2.6 million. It, too, has a renovated colonial core that is dwarfed by the modern sprawling city, which supports the island's largest port. San Juan is the financial, political, manufacturing, and tourism hub of Puerto Rico. With its highways, high-rises, shopping malls, and ever-present shoreline, it is an interesting blend of Latin American, North American, and Caribbean urbanism.

Strategically situated on Cuba's north coast at a narrow opening to a natural deep-water harbor, Havana emerged as the most important colonial city in the region, serving as a port for all incoming and outgoing Spanish ships. Consequently, Old Havana possesses a handsome collection of colonial architecture, especially from the 18th and 19th centuries, and is a UNESCO World Heritage Site. The modern city of 2 million is more sprawling, with a mix of Spanish colonial and Soviet-inspired

▲ **Figure 5.16 Caribbean Lifestyle**   A slower pace of life is one of the appeals of the region, along with a tropical climate and plenty of shoreline. This street scene in Vinales, a small town in western Cuba, illustrates a reliance on bicycles, horse-drawn vehicles, and time to chat with neighbors.

concrete apartment blocks. It is also a city that had to reinvent itself when subsidies from the former Soviet Union stopped flowing

Caribbean cities and towns do have their charms and reflect a blend of cultural influences. Throughout the region, houses are often simple structures made of wood, brick, or stucco, raised off the ground a few feet to avoid flooding, and painted in pastels. Most people still get around by foot, bicycle, motorbike, or public transportation; neighborhoods are filled with small shops and services within easy walking distance (Figure 5.16). Streets are narrow, and the pace of life is markedly slower than in North America and Europe. Even when space is tight in town, most settlements are near the sea and its cooling breezes. An afternoon or evening stroll along the waterfront is a common activity.

REVIEW

5.3   What are the major demographic trends for this region, and what factors explain these patterns?

5.4   How did the long-term reliance on a plantation economy influence settlement patterns in the Caribbean?

KEY TERMS   diaspora, remittances, circular migration, chain migration, transnational migration, plantation America, mono-crop production

▼ **Figure 5.15 Plaza Colon, Santo Domingo**   In the colonial core of Santo Domingo, a small crowd gathers in Plaza Colon below a statue of Christopher Columbus. This capital city of 3 million people in the Dominican Republic is the oldest continually occupied city in the Americas settled by Europeans.

Explore the SIGHTS of Santo Domingo's Ciudad Colonial
https://goo.gl/Y8VQps

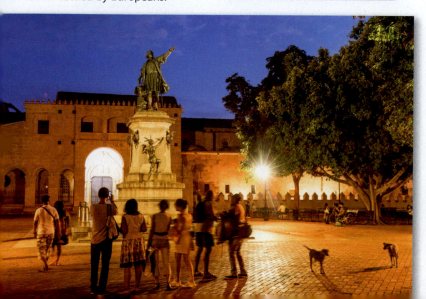

Cultural Coherence and Diversity: A Neo-Africa in the Americas

Linguistic, religious, and ethnic differences abound in the Caribbean. The presence of several former European colonies, millions of descendants of ethnically distinct Africans and indentured workers from India and China, and isolated Amerindian communities on the mainland challenge any notion of cultural coherence.

Common historical and cultural processes hold this region together. In particular, this section focuses on three cultural influences shared throughout the Caribbean: the European colonial presence, African influences, and the mix of European and African cultures termed *creolization*.

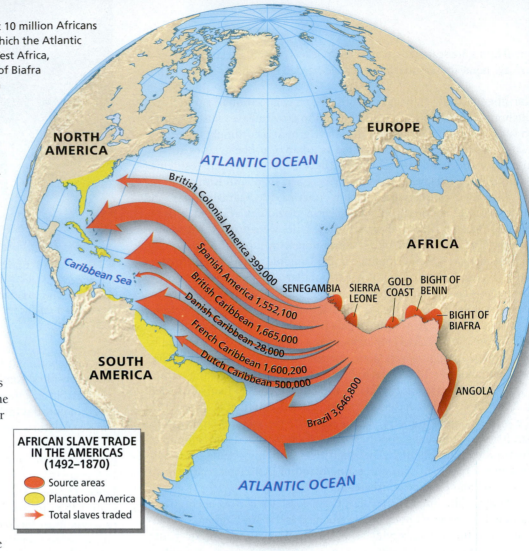

► **Figure 5.17** **Transatlantic Slave Trade** At least 10 million Africans landed in the Americas during the four centuries in which the Atlantic slave trade operated. Most of the slaves came from West Africa, especially the Gold Coast (now Ghana) and the Bight of Biafra (now Nigeria). Angola, in southern Africa, was also an important source area.

NORTH AMERICA

EUROPE

ATLANTIC OCEAN

British Colonial America 399,000

AFRICA

Spanish America 1,552,100   SENEGAMBIA   SIERRA LEONE   GOLD COAST   BIGHT OF BENIN

British Caribbean 1,665,000

Danish Caribbean 28,000

BIGHT OF BIAFRA

Caribbean Sea

French Caribbean 1,600,200

SOUTH AMERICA

Dutch Caribbean 500,000

ANGOLA

Brazil 3,646,800

AFRICAN SLAVE TRADE IN THE AMERICAS (1492–1870)

■ Source areas
■ Plantation America
→ Total slaves traded

ATLANTIC OCEAN

The Cultural Impact of Colonialism

The arrival of Columbus in 1492 triggered a devastating chain of events that depopulated much of the Caribbean islands within 50 years. A combination of Spanish brutality, enslavement, warfare, and disease changed the densely settled islands, which supported up to 3 million Caribs and Arawaks, into an uninhabited territory ready for the colonizer's hand. By the mid-16th century, as rival European states competed for Caribbean territory, the lands they fought for were virtually empty. In many ways, this simplified their task, as they did not have to recognize native land claims or work amid Amerindian societies. Instead, the colonizers reorganized the Caribbean territories to serve a plantation-based production system. The critical missing element was labor. Once slave labor from Africa and, later, contract labor from Asia were secured, the small Caribbean colonies became extremely profitable.

Creating a Neo-Africa The introduction of enslaved Africans to the Americas began in the 16th century and continued into the 19th century. This forced migration of Africans to the Americas was only part of a much more complex African diaspora— the forced removal of Africans from their native areas. The best-documented slave route was the transatlantic one—at least 10 million Africans landed in the Americas, and it is estimated that another 2 million died en route. More than half of these enslaved peoples were sent to the Caribbean (Figure 5.17).

The slave trade, combined with the elimination of nearly all the native inhabitants, remade the Caribbean as the area with the greatest concentration of relocated African people in the Americas. The African source areas extended from Senegal to Angola, and slave purchasers intentionally mixed tribal groups in order to weaken ethnic identities. Consequently, intact transfer of religion and languages into the Caribbean rarely occurred; instead, languages, customs, and beliefs were blended.

Maroon Societies Communities of runaway slaves—termed **maroons** in English, *palenques* in Spanish, and *quilombos* in Portuguese—offer interesting examples of African cultural diffusion across the Atlantic. Hidden settlements of escaped slaves existed wherever slavery was practiced. While many of these settlements were short-lived, others have endured and allowed for the survival of African traditions, especially farming practices, house designs, community organization, and language.

Whereas other maroon societies gradually blended into local populations, to this day the Suriname maroons, the largest maroon population in the Western Hemisphere, maintain a distinct identity.

Six distinct maroon tribes formed, ranging in size from a few hundred to 20,000. Clear connections to West African cultural traditions persist—including religious practices, crafts, patterns of social organization, agricultural systems, and even dress (Figure 5.18). Living relatively undisturbed for 200 years, these rainforest inhabitants fashioned

▼ **Figure 5.18** **Maroons in Suriname** Maroons (descendants of runaway slaves) dance in Paramaribo at the annual Black People's Day celebration, the first Sunday in January. Living in Suriname for over two centuries, Maroons maintain many West African cultural traditions.

Explore the SIGHTS of Maroon Communities
https://goo.gl/EBNSbc

The legacy of these indentured arrangements is clearest in Suriname, Guyana, and Trinidad and Tobago. In Suriname, a former Dutch colony, more than one-third of the population is of South Asian descent, and 16 percent is Javanese (from Indonesia, another former Dutch colony). Guyana and Trinidad were British colonies, and most of their contract labor came from India. Today half of Guyana's population, and 40 percent of Trinidad and Tobago's, claim South Asian ancestry. Hindu temples are found in the cities and villages, and many families speak Hindi at home. In 2015, Moses Nagamootoo, of Indian ancestry, was elected Prime Minister of Guyana (Figure 5.20).

Most of the former English colonies also have Chinese populations of not more than 2 percent. Once these East Asian immigrants fulfilled their agricultural contracts, they often became merchants and small-business owners, positions they still hold in Caribbean society. Cuba and Suriname have the largest ethnic Chinese populations in the region. Moreover, Suriname has experienced a substantial surge in Chinese immigration, with some reports suggesting that recent Chinese arrivals may account for 10 percent of the country's population.

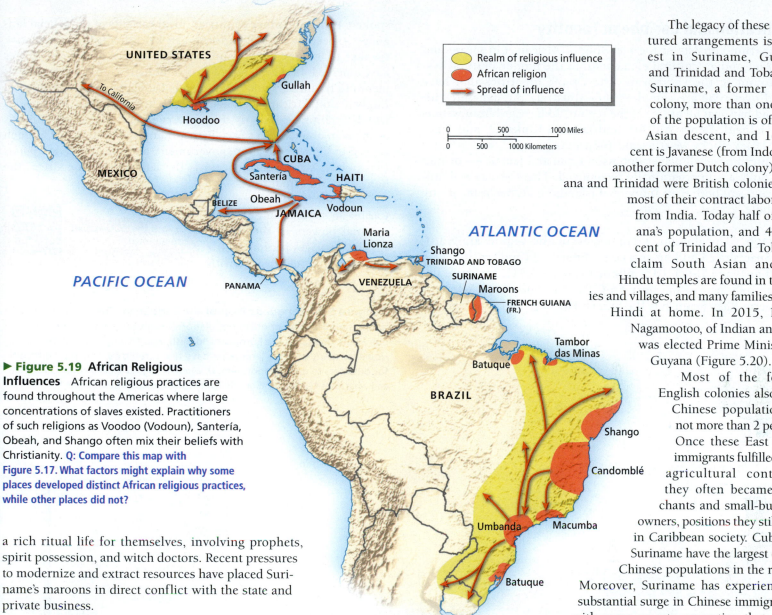

▶ **Figure 5.19 African Religious Influences**   African religious practices are found throughout the Americas where large concentrations of slaves existed. Practitioners of such religions as Voodoo (Vodoun), Santería, Obeah, and Shango often mix their beliefs with Christianity. **Q: Compare this map with Figure 5.17. What factors might explain why some places developed distinct African religious practices, while other places did not?**

a rich ritual life for themselves, involving prophets, spirit possession, and witch doctors. Recent pressures to modernize and extract resources have placed Suriname's maroons in direct conflict with the state and private business.

African Religions
Linked to maroon societies, but more widely diffused, was the transfer of African religious and magical systems to the Caribbean. These patterns, another reflection of neo-Africa in the Americas, are most closely associated with northeastern Brazil and the Caribbean. Millions of Brazilians practice the African-based religions of Umbanda, Macuba, and Candomblé, along with Catholicism. Likewise, Afro-religious traditions in the Caribbean have evolved into unique forms that have clear ties to West Africa. The most widely practiced are Voodoo (also Vodoun) in Haiti, Santería in Cuba, and Obeah in Jamaica. Each of these religions has its own priesthood and unique pattern of worship. Their impact is considerable; the father-and-son dictators of Haiti, the Duvaliers, were known to hire Voodoo priests to scare off government opposition (Figure 5.19).

Indentured Labor from Asia
By the mid-19th century, most colonial governments in the Caribbean had begun to free their enslaved populations. Fearful of labor shortages, they sought **indentured labor** (workers contracted to labor on estates for a set period of time, often several years) from South, Southeast, and East Asia.

▼ **Figure 5.20 South Asians in the Caribbean**   Guyana's Prime Minister, Moses Nagamootoo (in white), congratulates Veerasammy Permaul, a cricket player on the Guyana Amazon Warriors team, for being player of the match. Both men share South Asian ancestry.

Creolization and Caribbean Identity

Creolization consists of the blending of African, European, and some Amerindian cultural elements into the unique cultural systems found in the Caribbean. The Creole identities that have formed over time are complex; they illustrate the dynamic cultural and national identities of the region. Today Caribbean writers (V. S. Naipaul, Derek Walcott, and Jamaica Kinkaid), musicians (Bob Marley, Ricky Martin, and Juan Luis Guerra), and athletes (Dominican baseball player Robinson Cano and Jamaican sprinter Usain Bolt) are internationally recognized. These artists and athletes represent both their individual islands and Caribbean culture as a whole.

Language The dominant languages in the region are European: Spanish (25 million speakers), French (11 million), English (7 million), and Dutch (0.5 million) (Figure 5.21). Yet these figures tell only part of the story. In Cuba, the Dominican Republic, and Puerto Rico, Spanish is the official language and is universally spoken. As for the other countries, local variants of the official language, especially in spoken form, can be difficult for a nonnative speaker to understand. In some cases, completely new languages have emerged. In the islands of Aruba, Bonaire, and Curaçao, Papiamento (a trading language that blends Dutch, Spanish, Portuguese, English, and African languages) is the lingua franca, with use of Dutch declining. Similarly, French Creole, or *patois*, in Haiti has been given official status as a distinct language. In practice, French is used in higher education, government, and the courts, but *patois* (with clear African influences) is the language of the street, the home, and oral tradition.

With the independence of the Caribbean states from European colonial powers in the 1960s, Creole languages became politically and culturally charged with

Explore the SOUNDS of Papiamento
https://goo.gl/8cDbS9

▼ **Figure 5.21 Language Map of the Caribbean** Because this region has no significant Amerindian population (except on the mainland), the dominant languages are European: Spanish, French, English, and Dutch. However, many of these languages have been creolized, making it difficult for outsiders to understand them. Q: Where is English spoken in the Caribbean, and what does this tell us about the early colonization of this region?

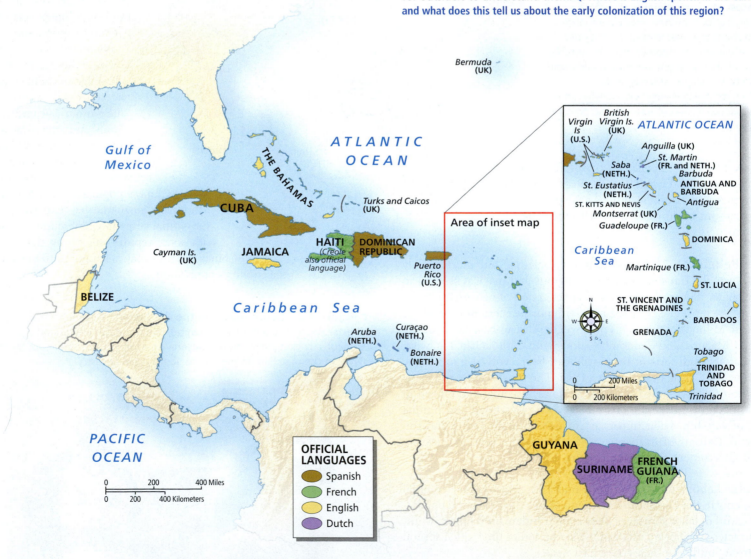

OFFICIAL LANGUAGES
- Spanish
- French
- English
- Dutch

national meaning. While most formal education is taught using standard language forms, the richness of vernacular expression and its capacity to instill a sense of identity are appreciated. As linguists began to study these languages, they found that, although the vocabulary came from Europe, the syntax or sematic structure had other origins, notably from African language families. Locals rely on their ability to switch from standard to vernacular forms of speech. Thus, a Jamaican can converse with a tourist in standard English and then switch to a Creole variant when a friend walks by, effectively excluding the outsider from the conversation. This code-switching ability is evident in many cultures, but it is widely used in the Caribbean.

Music   The rhythmic beats of the Caribbean might be the region's best-known product. This small area is the home of reggae, calypso, merengue, rumba, zouk, and scores of other musical forms. The roots of modern Caribbean music reflect a combination of African rhythms with European forms of melody and verse. These diverse influences, coupled with a long period of relative isolation, sparked distinct local sounds. As movement among the Caribbean population increased, especially during the 20th century, musical traditions were blended, but characteristic sounds remained.

Explore the Sounds of Garifuna Music
https://goo.gl/Zrrshk

The famed steel pan drums of Trinidad were created from oil drums discarded from a U.S. military base there in the 1940s. The bottoms of the pans are pounded with a sledgehammer to create a concave surface that produces different tones. During carnival, racks of steel pans are pushed through the streets by dancers while drummers play. So skilled are these musicians that they even perform classical music, and government agencies encourage troubled teens to learn steel pan (Figure 5.22).

Today carnival is celebrated on nearly every Caribbean island as a national street party that can go on for weeks and attract thousands of tourists. Many official carnivals are no longer held prior to Lent (the 40 days leading up to Easter) but are scheduled at other times of the year. Carnival celebrations have also followed the Caribbean diaspora to new settings in North America and Europe. One of North America's biggest carnivals is in Toronto, where a large Caribbean immigrant population maintains the tradition every July. In London, Birmingham, and Leicester, Caribbean carnivals are celebrated annually.

The distinct sound and the ingenious rhythms have made Caribbean music and carnival celebrations popular. When Jamaican Bob Marley and the Wailers crashed the international music scene with their soulful reggae sound, it was the lyrics about poverty, injustice, and freedom that resonated with the world. More than good dance beats, Caribbean music can be closely tied to political protest. In Haiti, rara music mixes percussion instruments, saxophones, and bamboo trumpets, weaving in funk and reggae bass lines. The songs are always performed in French Creole and typically celebrate Haiti's African ancestry and the use of Voodoo. The lyrics speak of difficult issues, such as political oppression and poverty.

Sports: From Baseball to Olympic Glory   Latin Americans are known for their love of soccer, but baseball is the dominant sport for much of the Caribbean. The popularity of baseball in is attributed to U.S. influence in the region in the early 20th century as American businesses and military personnel introduced the game to the region. The Dominican

▲ **Figure 5.22 Trinidadian Drummer**   A steel pan drummer performs while his drum cart is pushed through the streets during carnival in Port of Spain, Trinidad. Steel drums were created in Trinidad in the 1940s from discarded oil drums from a U.S. military base. They have become an iconic sound for the region.

Republic sends more players to Major League Baseball than any other Caribbean country, accounting for 10 percent of all players in 2016 (see *Exploring Global Connections: From Baseball to Béisbol*).

Two Caribbean countries dominate in producing Olympic-caliber athletes: Jamaica and Cuba. Each country brought home 11 medals during the 2016 Olympic Games in Rio de Janeiro. While other Caribbean countries direct athletes toward baseball or cricket, Jamaicans prefer track and field, especially sprinting. Jamaica's 11 medals were all in track and field, which is particularly impressive considering it is a nation of less than 3 million. An important individual in their success is Usain Bolt, a sprinter who has dominated his events since exploding onto the scene at the 2008 Beijing Olympics. In the last three Olympics, Bolt won gold in the 100- and 200-meter sprints and the 4 × 100 meter relay. His charismatic personality on the track has made him an international sensation and a much-admired figure at home (Figure 5.23). Jamaican women sprinters are also world class; at the 2016 games Elaine Thompson won gold in the 100-meter and 200-meter sprints and was part of the 4 × 100 women's team that took the silver medal.

For Cubans, however, Olympic glory has come through boxing. As in many socialist countries, promising athletes are nurtured from an early age. In the Rio de Janeiro games, six of Cuba's 11 medals were earned in the boxing ring. Cubans also won three medals in wrestling, one in judo, and one in track.

REVIEW

5.5   What kinds of African influences exist in the Caribbean, and how do they express themselves?

5.6   What is creolization, and how does it explain different cultural influences and patterns found in the Caribbean?

KEY TERMS   maroon, indentured labor, creolization

From Baseball to *Béisbol*

Cubans, Puerto Ricans, Venezuelans, and Dominicans adore *béisbol* (Figure 5.2.1). Several Cuban baseball stars defected to the United States in recent years; players such as José Abreu, Alex Guerrero, and Aroldis Chapman landed multimillion-dollar contracts and appreciative fans. Over one-quarter of Major League Baseball (MLB) players are born outside the United States, mostly from the Caribbean or Latin America. After the Dominican Republic, the top sources for players are Venezuela, Cuba, and Puerto Rico (a U.S. commonwealth).

The Dominican Republic became an MLB talent pipeline due to a complex mix of boyhood dreams, economic inequality,

▲ **Figure 5.2.2 Dominican Training Camps**   Major League hopefuls train at the Washington Nationals Dominican baseball training complex in Boca Chica. Many MLB teams have training facilities in the Dominican Republic, anchored near the city of San Pedro de Macorís.

▲ **Figure 5.2.1 Caribbean Baseball**   A young Cuban batter takes aim during a pickup game in rural Cuba. Several Caribbean islands have adopted baseball as their national sport—most notably the Dominican Republic, Cuba, and Puerto Rico.

and greed. This small country has produced many baseball legends, and over the decades franchises have invested millions in training camps there. In the past two decades, however, Dominican pride in its baseball prowess has been tinged by the realities of a merciless feeder system that depends on impoverished kids, performance-enhancing drugs, fake documents, and scouts who skim a percentage of the signing bonuses. Yet the reality is that more and more young boys, who can sign contracts at

age 16, see their future in baseball rather than in schooling. Even a modest signing bonus of $10,000 to $20,000 can build a nice home for a boy's family.

San Pedro de Macorís, not far from Santo Domingo, epitomizes this field of dreams (Figure 5.2.2). It is a place of cane fields, kids on bicycles with bats and gloves, sugarcane factories, dusty baseball diamonds, and large homes of former players such as George Bell, Pedro Guerrero, and Sammy Sosa. These houses are silent testaments to what is possible through baseball. In an effort to clean up baseball's image, MLB has officials in the Dominican Republic investigating drug use and fraudulent papers. However, as long as there are families pushing their teenage boys and a talent pool that delivers, this transnational system is self-perpetuating.

1. **What has attributed to the popularity of baseball in the Caribbean?**

2. **In what other countries is baseball popular?**

GOOGLE EARTH
Virtual Tour Video
https://goo.gl/llvstr

▲ **Figure 5.23 Jamaican Sprinter** Usain Bolt is considered one of greatest sprinters in history. Wearing Jamaican yellow, he celebrates gold after winning the 200-meter sprint at the Olympics in Rio de Janeiro. Track and field is a national obsession in Jamaica, and the country produces many world-class athletes.

Geopolitical Framework: Colonialism, Neocolonialism, and Independence

Caribbean colonial history is a patchwork of competing European powers fighting over profitable tropical territories. By the 17th century, the Caribbean had become an important proving ground for European ambitions. Spain's grip on the region was slipping, and rival European nations felt confident that they could gain territory by gradually pushing Spain out. Many territories, especially islands in the Lesser Antilles, changed European rulers several times.

Europeans viewed the Caribbean as a strategically located, profitable region in which to produce sugar, rum, and spices. Geopolitically, rival European powers also felt that their presence in the Caribbean limited Spanish authority there. However, Europe's geopolitical dominance in the Caribbean began to diminish by the mid-19th century, just as the U.S. presence increased.

Inspired by the **Monroe Doctrine**, which claimed that the United States would not tolerate European military involvement in the Western Hemisphere, the U.S. government made it clear that it considered the Caribbean to be within its sphere of influence. This view was highlighted during the Spanish–American War in 1898. Even though several English, Dutch, and French colonies remained after this date, the United States indirectly (and sometimes directly) asserted its control over the region, bringing in a period of **neocolonialism**. Neocolonialism is the indirect control of one country or region by another through economic and cultural domination, rather than through direct military or political control as occurs under colonialism.

In an increasingly global age, however, even neocolonial interests can be short-lived or sporadic. The Caribbean has not attracted the level of private foreign investment seen in other regions, and as its strategic importance in a post–Cold War era fades, new geopolitical forces are shaping the region. Taiwan began wooing small Caribbean islands in the 1990s with strategic investments, in the hope of winning United Nations votes for its cause. Not surprisingly, China responded by investing still more money in the region, in part to convince nations that once supported Taiwan, such as the Dominican Republic and Grenada, to switch their support to China—which they eventually did.

Life in "America's Backyard"

To this day, the United States exerts considerable influence in the Caribbean, which was commonly referred to as "America's backyard" in the early 20th century. The stated foreign policy objectives were to free the region from European authority and encourage democratic governance. Yet time and again, U.S. political and economic ambitions undermined those goals. President Theodore Roosevelt made his priorities clear with imperialistic policies that extended the influence of the United States beyond its borders. Policies and projects such as the construction of the Panama Canal and the maintenance of open sea lanes benefited the United States but did not necessarily support social, economic, or political gains for the Caribbean people. The United States later offered benign-sounding development packages, such as the Good Neighbor Policy (1930s), the Alliance for Progress (1960s), and the Caribbean Basin Initiative (1980s). The Caribbean view of these initiatives has been wary at best. Rather than feeling liberated, many residents believe that one kind of political dependence was traded for another—colonialism for neocolonialism.

In the early 1900s, the U.S. role in the Caribbean was overtly military and political. The Spanish–American War (1898) secured Cuba's freedom from Spain and also resulted in Spain giving up the Philippines, Puerto Rico, and Guam to the United States; the latter two are still U.S. territories. The U.S. government also purchased the Danish Virgin Islands in 1917, renaming them the U.S. Virgin Islands and developing the harbor of St. Thomas. French, English, and Dutch colonies were tolerated as long as these allies recognized U.S. supremacy in the region. Outwardly anticolonial, the United States had become much like an imperial force.

When a Caribbean state refused to abide by U.S. trade rules, U.S. Navy vessels would block its ports. These were not short-term engagements: U.S. troops have occupied the Dominican Republic (1916–1924), Haiti (1913–1934), and Cuba (1906–1909 and 1917–1922) (Figure 5.24). Even today, the United States maintains several important military bases in the region, including Guantánamo in eastern Cuba.

Critics of U.S. policy in the Caribbean complain that business interests overwhelm democratic principles when foreign policy is determined. The U.S. banana companies that settled the coastal plain of the Caribbean rimland operated as if they were independent states. Meanwhile, truly democratic institutions remained weak. True, exports increased, railroads were built, and port facilities were improved, but levels of income, education, and health remained dreadfully low throughout the first half of the 20th century.

The Commonwealth of Puerto Rico
Puerto Rico is both within the Caribbean and apart from it because of its status as a commonwealth of the United States. Throughout the 20th century, various Puerto Rican independence movements sought to separate the island from

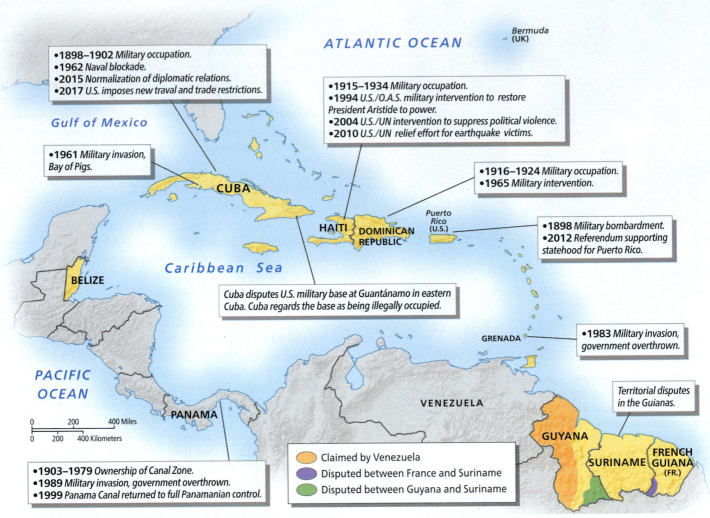

•1898–1902 *Military occupation.*
•1962 *Naval blockade.*
•2015 *Normalization of diplomatic relations.*
•2017 *U.S. imposes new traval and trade restrictions.*

ATLANTIC OCEAN

Bermuda
(UK)

Gulf of Mexico

•1915–1934 *Military occupation.*
•1994 *U.S./O.A.S. military intervention to restore President Aristide to power.*
•2004 *U.S./UN intervention to suppress political violence.*
•2010 *U.S./UN relief effort for earthquake victims.*

•1961 *Military invasion, Bay of Pigs.*

CUBA

•1916–1924 *Military occupation.*
•1965 *Military intervention.*

HAITI   DOMINICAN REPUBLIC

Puerto Rico (U.S.)

•1898 *Military bombardment.*
•2012 *Referendum supporting statehood for Puerto Rico.*

Caribbean Sea

BELIZE

Cuba disputes U.S. military base at Guantánamo in eastern Cuba. Cuba regards the base as being illegally occupied.

PACIFIC OCEAN

GRENADA

•1983 *Military invasion, government overthrown.*

PANAMA

VENEZUELA

Territorial disputes in the Guianas.

0   200   400 Miles
0   200   400 Kilometers

•1903–1979 *Ownership of Canal Zone.*
•1989 *Military invasion, government overthrown.*
•1999 *Panama Canal returned to full Panamanian control.*

GUYANA

SURINAME   FRENCH GUIANA (FR.)

Claimed by Venezuela
Disputed between France and Suriname
Disputed between Guyana and Suriname

▲ **Figure 5.24 Geopolitical Issues in the Caribbean: U.S. Military Involvement and Regional Disputes** The Caribbean was regarded as the geopolitical backyard of the United States, and U.S. military occupation was a common occurrence in the first half of the 20th century. Border and ethnic conflicts also exist—most notably in the Guianas.

the United States. Even today Puerto Ricans are divided about their island's political future. At the same time, Puerto Rico depends on U.S. investment and welfare programs; U.S. food stamps are a major source of income for many Puerto Rican families. Commonwealth status also means that Puerto Ricans can freely move between the island and the U.S. mainland, a right they actively assert. In other ways, Puerto Ricans symbolically display their independence. For example, they support their own "national" sports teams and send a Miss Puerto Rico to international beauty pageants.

In 2012, a controversial island referendum resulted in the majority of residents voting for a change in political status, with their preference being for statehood. In 2018, a Puerto Rican delegation presented a bill to the U.S. Congress asking to transition the island from a territory to a state. The formal push for statehood began a year after Hurricane Maria devastated the island and support from the mainland was viewed as slow and inadequate. Presently, there is little indication that the bill will pass (Figure 5.25).

Puerto Rico led the Caribbean in the transition from an agrarian economy to an industrial one, beginning in the 1950s. By the 1990s, Puerto Rico was one of the most industrialized places in the region,

with a significantly higher per capita income than its neighbors. By the 2000s, many of the U.S. tax benefits associated with establishing factories in Puerto Rico had ceased to exist, and manufacturing declined. Consequently, there are growing signs of underdevelopment, including widespread poverty, extensive out-migration, low rates of education, and serious debt problems, all of which were made worse by the hurricanes of 2017 (see *Humanitarian Geography: The Puerto Rican Exodus*).

Cuba and Geopolitics The most profound challenge to U.S. authority in the region came from Cuba and its superpower ally, the former Soviet Union. A revolutionary effort led by Fidel Castro against the pro-American Batista government began in the 1950s. Cuba's economic productivity had soared, but its people were still poor, uneducated, and increasingly angry. The contrast between the lives of average sugarcane workers and the foreign elite was sharp. Castro, who took power in 1959, had tapped a deep vein of Cuban resentment against six decades of American neocolonialism.

After Castro's government nationalized U.S. industries and took ownership of all foreign-owned properties, the United States responded by

Independence and Integration

Given the repressive colonial history of the Caribbean, it is no wonder that the struggle for political independence began more than 200 years ago. Haiti was the second colony in the Americas to gain independence, in 1804, after the United States in 1776. However, political independence in the region has not guaranteed economic independence. Many Caribbean states struggle to meet the basic needs of their people. Today some Caribbean territories maintain their colonial status as an economic asset. For example, the French territories of Martinique, Guadeloupe, and French Guiana are overseas departments of France; residents have full French citizenship and social welfare benefits.

Independence Movements Haiti's revolutionary war began in 1791 and ended in 1804. During this conflict, the island's population was cut in half by casualties and emigration; ultimately, the formerly enslaved became the rulers. Independence, however, did not allow this crown jewel of the French Caribbean to prosper. Slowed by economic and political problems, Haiti was ignored by the European powers and never fully accepted by the mainland Latin American countries once they gained their own political independence in the 1820s.

Several revolutionary periods followed in the 19th century. In the Greater Antilles, the Dominican Republic finally gained independence in 1844, after taking control of the territory from Spain and Haiti. Cuba and Puerto Rico were freed from Spanish colonialism in 1898, but their independence was weakened by greater U.S. involvement. The British colonies also faced revolts, especially in the 1930s; it was not until the 1960s that independent states emerged from the English Caribbean. First, the larger colonies of Jamaica, Trinidad and Tobago, Guyana, and Barbados gained independence. Other British colonies followed throughout the 1970s and early 1980s. Suriname, the only Dutch colony on the rimland, became a self-governing territory in 1954 but remained part of the Netherlands until 1975, when it declared itself an independent republic. Today the territories in the region are doing well in terms of their overall Freedom Rating, with the exception of Cuba and Haiti (see Table 5.2).

Limited Regional Integration Perhaps the most difficult task facing the Caribbean is increasing economic integration. Scattered islands, a divided rimland, different languages, and limited economic resources hinder the formation of a meaningful regional trade bloc. Economic cooperation is more common between groups of islands with a shared colonial background than between, for example, former French and English colonies.

During the 1960s, the Caribbean began to experiment with regional trade associations as a means to improve its economic competitiveness. The goal of regional cooperation was to raise employment, increase intraregional trade, and ultimately reduce economic dependence. The countries of the English Caribbean took the lead in this development strategy. In 1963, Guyana proposed an economic integration plan with Barbados and Antigua. In 1972, the integration process intensified with the formation of the **Caribbean Community and Common Market (CARICOM)**. Representing the former English colonies, CARICOM proposed an ambitious regional

▼ **Figure 5.25 Puerto Rican Statehood** Puerto Rican Congresswoman Jenniffer González-Colón and Governor Ricardo Rosselló celebrate a statehood referendum in San Juan, Puerto Rico on June 11, 2017. The U.S. territory overwhelmingly choose statehood in this non-binding vote with low voter turnout. The island's economic crisis and the exodus to the mainland may have driven voter sentiment. In order to become the 51st state, the U.S. Congress would have to approve statehood, which does not seem likely.

refusing to buy Cuban sugar and ultimately ending diplomatic relations with the state. Various U.S. embargoes (laws forbidding trade with a particular country) against Cuba have existed for five decades. When Cuba established strong diplomatic relations with the Soviet Union in 1960, at the height of the Cold War, the island state became a geopolitical enemy of the United States. With the Soviet Union financially and militarily backing Castro, a direct U.S. invasion of Cuba was too risky. The fall of 1962 produced one of the most dangerous episodes of the Cold War, when Soviet missiles were discovered on Cuban soil. Ultimately, the Soviet Union removed its weapons; in return, the United States promised not to invade Cuba.

Even with the loss of Soviet financial support once the Cold War ended, Cuba managed to reinvent itself by growing its tourism sector and courting foreign investment, especially from Spain, and importing oil from Venezuela.

In many ways, Cuba is entering a new political era. Fidel Castro stepped down from power in 2008 and died eight years later. In 2015, President Obama and President Raúl Castro, Fidel's younger brother, agreed to restore full diplomatic relations between the two countries. The first regular commercial flights from the United States to Cuba resumed in 2016, and the number of American tourists in Cuba increased. Unlike Fidel, Raúl encouraged private enterprise in Cuba and expanded the availability of licenses permitting self-employed individuals and small businesses to operate. Consequently, privately employed workers in Cuba increased steadily. In 2018, the orchestrated transition of Miguel Díaz-Canel as president of Cuba was complete, ending nearly six decades of Castro family leadership. Yet as this transition was occurring, President Trump reintroduced travel restrictions for individuals going to Cuba and bans on business investment as well. By 2018, the number of Americans visiting Cuba declined.

The Puerto Rican Exodus

Even before Hurricanes Irma and Maria struck Puerto Rico in 2017, young Puerto Ricans were leaving for the U.S. mainland due to economic stagnation and lack of jobs. As U.S. citizens, Puerto Ricans can settle throughout the United States as they please. In fact, nearly half a million Puerto Ricans left for the mainland between 2006 and 2016, representing some 10 percent of the island's population. In the aftermath of the Hurricane Maria, many more Puerto Ricans are leaving, in some cases through the help of Federal Emergency Management Agency (FEMA) resettlement activities or through networks of friends and family. More than 7000 families participated in a FEMA voucher program that placed them in modest hotels on the U.S. mainland, mostly in Florida. That program ended in June 2018. After eight months of support, displaced Puerto Ricans had to relocate with family or friends, find places to rent, or return to the island. A recent study estimates that the two-year, post–Hurricane Maria exodus could total 400,000 people (Figure 5.3.1). Florida is currently the top destination, closely followed by Texas and Pennsylvania.

Impacts of Mass Migration Understandably, people are moving in response to the island's serious economic problems and the collapse of basic infrastructure caused by the storm. Yet the scale of this outflow will have serious repercussions for the long-term health of the island. Most of these migrants are young adults with children, which has greatly affected the education sector. Over 243 schools have closed throughout the island from 2006 to 2017, and another 467 are predicted to close by 2022 as a result of the exodus. In addition, the number of teachers has dropped by 50 percent since 2006.

In 2005, there were about as many Puerto Ricans living on the island as there were on the mainland. By 2017, the demographic balance shifted to the north, with nearly 5.5 million Puerto Ricans on the mainland and 3.4 million on the island. Many fear that an entire generation of Puerto Ricans, the future of the island, may be lost. Others hope that as the island's economy and infrastructure recover, many will decide to return to their Caribbean homeland.

1. **What are the advantages of commonwealth status with regard to migration?**

2. **What factors might explain where Puerto Ricans are settling on the mainland? Research a large immigrant population in your community to explain why they settled there.**

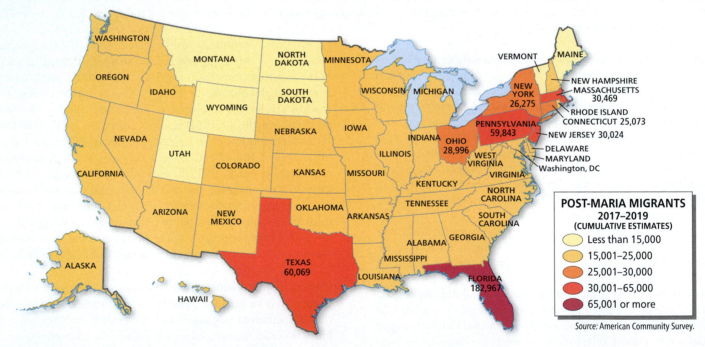

POST-MARIA MIGRANTS
2017–2019
(CUMULATIVE ESTIMATES)

- Less than 15,000
- 15,001–25,000
- 25,001–30,000
- 30,001–65,000
- 65,001 or more

Source: American Community Survey.

▲ **Figure 5.3.1 Puerto Rican Exodus** Puerto Ricans leaving the island after Hurricane Maria have settled throughout the United States. The top destinations of the latest arrivals are Florida, Texas, Pennsylvania, Massachusetts, and New Jersey.

GOOGLE EARTH
Virtual Tour Video
https://goo.gl/UhZs9B

industrialization plan and the creation of the Caribbean Development Bank to assist the poorer states. As important as CARICOM is as an economic symbol of regional identity, it has produced limited improvements in intraregional trade. It has 15 full member states—all of the English Caribbean, French-speaking Haiti, and Dutch-speaking Suriname. Other dependencies, such as Anguilla, Turks and Caicos, Bermuda, and the British Virgin Islands, are associate members. In 2005, CARICOM passports began to be issued to facilitate travel between member nations, but creating a unified economic development policy has been challenging. CARICOM has been, however, an important voice in global climate negotiation treaties to reduce GHG emissions and to respond to the worst impacts of climate change.

The dream of regional integration as a way to produce a more stable, self-sufficient Caribbean has never been realized. One scholar of the region argues that a limiting factor is a "small-islandist ideology." Islanders tend to keep their backs to the sea, oblivious to the needs of neighbors. At times, such isolationism results in suspicion, distrust, and even hostility toward nearby states. Yet economic necessity dictates engagement with partners outside the region. This peculiar status of isolated proximity expresses itself in the Caribbean's uneven social and economic development trends.

REVIEW

5.7 Which countries have had colonial or neocolonial influences in the Caribbean, and why have they engaged with the region?

5.8 Describe the obstacles to Caribbean political or economic integration.

KEY TERMS Monroe Doctrine, neocolonialism, Caribbean Community and Common Market (CARICOM)

Economic and Social Development: from Cane Fields to Cruise Ships

Collectively, Caribbean peoples, although poor by U.S. standards, are economically better off than most residents of Sub-Saharan Africa, South Asia, and even China. Despite periods of economic stagnation, social gains in education, health, and life expectancy are significant (Table 5.2). Historically, the Caribbean's links to the world economy were through tropical agricultural exports, but several specialized industries—such as tourism, offshore banking, and assembly plants—have challenged agriculture's dominance. These industries grew because of the region's proximity to North America and Europe, abundant cheap labor, and policies that created a nearly tax-free environment for foreign-owned companies. Unfortunately, growth in these industries does not employ all the region's displaced rural laborers, so the lure of jobs outside the region is still strong.

From Fields to Factories and Resorts

Agriculture once dominated Caribbean economic life. However, decades of unstable crop prices and a decline in special trade agreements with former colonial states have produced more hardship than prosperity. Ecologically, the soils are overworked, and there are no frontier areas into which to expand production except for areas of the rimland. Moreover, agricultural prices have not kept pace with rising production costs, so wages and profits remain low. With the exception of a few mineral-rich territories—such as Trinidad, Guyana, Suriname, and Jamaica—most countries have tried to diversify their economies, relying less on their soils and more on manufacturing and services.

Comparing export figures over time illustrates the shift away from mono-crop dependence. In 1955, Haiti earned more than 70 percent of its foreign exchange through coffee exports; by 1990, coffee accounted for only 11 percent of its export earnings. Similarly, in 1955, the Dominican Republic earned close to 60 percent of its foreign exchange through sugar, but 35 years later, sugar accounted for less than 20 percent of the country's foreign exchange, and today it earns about 2 percent.

Sugar   The economic history of the Caribbean cannot be separated from the production of sugarcane. Even relatively small territories such as Antigua and Barbados yielded fabulous profits because there was no limit to the demand for sugar in the 18th century. Once considered a luxury crop, it became a necessity for European and North American laborers by the 1750s. It sweetened tea and coffee and made jams a popular spread for stale bread. In short, it made the meager and bland diets of ordinary people tolerable and also boosted caloric intake. Distilled into rum, sugar produced a popular intoxicant. Though it is difficult to imagine today, consumption of a pint of rum a day was not uncommon in the 1800s. The Caribbean and Latin America still produce the majority of the world's rum (see *Globalization in Our Lives: Yo-Ho-Ho and a Bottle of Rum*).

Sugarcane is still grown throughout the region for domestic consumption and export. Its declining economic importance is mostly due to increased competition from corn and sugar beets grown in the midlatitudes. The Caribbean and Brazil are the world's major sugar exporters. Until 1990, Cuba alone accounted for more than 60 percent of the value of world sugar exports and earned 80 percent of its foreign exchange through sugar production. However, Cuba's dominance in sugar exports had more to do with its subsidized and guaranteed markets in Eastern Europe and the Soviet Union than with exceptional productivity. Since the breakup of the Soviet Union in 1991, the value and volume of the Cuban sugar harvest have plummeted.

Assembly-Plant Industrialization   Another important Caribbean development strategy has been to invite foreign investors to set up assembly plants and thus create jobs. This was first tried successfully in Puerto Rico in the 1950s and was copied throughout the region. Today the main driver of the Puerto Rican economy is manufacturing, especially pharmaceuticals, textiles, petrochemicals, and electronics. However, since the signing of NAFTA (renegotiated and renamed the United States-Mexico-Canada Agreement, or USMCA, in late 2018) and CAFTA (see Chapter 4), Puerto Rico has faced increased competition from other states with even lower wages. The 1996 decision of the U.S. Congress to phase out many of the tax exemptions also undercut Puerto Rico's ability to maintain its specialized industrial base. Consequently, since 2006, the Puerto Rican economy has been in recession, although manufacturing still comprises 46 percent of the island's economy.

Through the creation of **free trade zones (FTZs)**—duty-free and tax-exempt industrial parks for foreign corporations—the Caribbean has become an increasingly attractive location for assembling goods for North American consumers. The Dominican Republic took advantage of tax incentives and guaranteed access to the U.S. market offered through the Caribbean Basin Initiative, and the country now has 50 FTZs. The majority of them are clustered around the outskirts of Santo Domingo and Santiago, the two largest cities (Figure 5.26). The most frequent investors in these zones are U.S. and Canadian firms, followed by Dominican, South Korean, and Taiwanese firms. Traditional manufacturing on the island was tied to sugar refining, whereas production

TABLE 5.2   Development Indicators

Explore these data in MapMaster 2.0   https://goo.gl/nTafYh

Country	GNI per Capita, PPP 2017[1]	GDP Average Annual Growth 2009–2015[2]	Human Development Index (2016)[3]	Percent Population Living Below $3.10 a Day[2]	Under Age 5 Mortality Rate (per 1000 live births), 1990[1]	Under Age 5 Mortality Rate (per 1000 live births), 2016[1]	Secondary School Enrollment Ratios[4] Male (2009–2016)	Secondary School Enrollment Ratios[4] Female (2009–2016)	Gender Inequality Index (2016)[3,6]	Freedom Rating (2018)[5]
Anguilla (UK)	12,200*	–	–	–	–	–	–	–	–	–
Antigua and Barbuda	23,594	0.9	0.786	–	26	9	102	104	–	2.0
Aruba (Netherlands)	25,300*	–	–	–	–	–	–	–	–	–
Bahamas	30,430	0.7	0.792	–	24	11	90	95	0.362	1.0
Barbados	18,640	0.3	0.795	–	18	12	108	111	0.291	1.0
Belize	8590	2.6	0.706	33	39	15	80	82	0.375	1.5
Bermuda (UK)	52,436	–3.4	–	–	–	–	–	–	–	–
Cayman Islands (UK)	43,800*	–	–	–	–	–	–	–	–	–
Cuba	12,300*	2.6	0.775	–	13	6	98	103	0.304	6.5
Curaçao (Netherlands)	15,000*	–	–	–	–	–	86	91	–	–
Dominica	10,620	0.5	0.726	–	17	34	101	100	–	1.0
Dominican Republic	16,030	5.2	0.722	10	60	31	74	82	0.470	3.0
Grenada	14,924	2.3	0.754	–	22	16	99	99	–	1.5
Guyana	8163	4.6	0.638	59	60	32	90	89	0.508	2.5
Haiti	1815	2.5	0.493	73	145	67	–	–	0.593	5.0
Jamaica	8995	0.4	0.730	11	30	15	79	85	0.422	2.5
Montserrat (UK)	34,000*	–	–	–	–	–	–	–	–	–
Puerto Rico (U.S.)	37,741	–2.0	–	–	–	–	79	84	–	–
St. Kitts and Nevis	27,067	2.8	0.765	–	32	9	88	93	–	1.0
St. Lucia	14,219	–0.1	0.735	–	21	13	85	85	0.354	1.0
St. Vincent and the Grenadines	11,777	0.6	0.722	–	24	17	108	105	–	1.0
Suriname	15,114	3.0	0.725	–	46	20	72	91	0.448	2.0
Trinidad and Tobago	31,578	1.0	0.780	8	30	19	–	–	0.324	2.0
Turks and Caicos Islands (UK)	29,100*	–	–	–	–	–	–	–	–	–

[1] World Bank Open Data, 2018. *Additional Data from CIA World Factbook. No data available for French Guiana, Guadaloupe, Martinique.
[2] World Bank—*World Development Indicators,* 2017.
[3] United Nations, *Human Development Report,* 2016.
[4] Population Reference Bureau, *World Population Data Sheet,* 2017.
[5] Freedom House, Freedom in the World 2018. See Ch.1, pp. 33 –34, for more info. on this scale (1–7, with 7 representing states with the least freedom).
[6] See Ch. 1, p. 39, for more info. on this scale (0–1, with higher values representing less gender equality).

Log in to Mastering Geography & access MapMaster to explore these data!
1) Identify some of the fastest growing economies in the Caribbean. What might attribute to their growth?
2) What generalizations can you make about educational attainment in this region? Which countries are lagging in this indicator?

in the FTZs focuses on garments and textiles. These manufacturing centers now account for three-quarters of the country's exports.

Growth in manufacturing depends on national and international policies supporting export-led development through foreign investment. Certainly, new jobs are being created, and national economies are diversifying in the process, but critics believe that foreign investors gain more than the host countries. Most goods are assembled from imported materials, so there is little development of national suppliers. Wages may be higher than local averages, but they are still low compared to those in the developed world—sometimes just a few dollars a day.

Offshore Banking and Online Gambling   The rise of offshore banking in the Caribbean began with The Bahamas in the 1920s. **Offshore banking** centers appeal to foreign banks and corporations by offering specialized services that are confidential and tax exempt. These offshore banks make money through registration fees, not taxes. The Bahamas was so successful in developing this sector that by 1976 the country was the world's third largest banking center. Its dominance began to decline due to corruption concerns linked to money laundering and competition from the Caribbean, Hong Kong, and Singapore. In the 1990s, the Cayman Islands emerged as the

region's leader in financial services—and remains one of the world's leading financial centers. With a population of 60,000, this crown colony of Britain has some 50,000 registered companies and a per capita purchasing power parity of nearly $44,000. Yet with growing competition from Asia, the Pacific islands, and even Europe, the Caymans has slipped from one of the top 5 banking centers to one of the top 50 (Figure 5.27).

An estimated $20–$30 trillion is hidden in offshore tax havens all over the world; the Caribbean is just one location where corporations and rich individuals park their money. Each Caribbean offshore banking center tries to develop specialized financial services to attract clients, such as functional operations, insurance, and trusts. Bermuda, for example, is a global leader in the reinsurance business, which makes money from underwriting part of the risk of other insurance companies. The Caribbean is an attractive location for such services because of proximity to the United States (home of many of the registered firms), client demand for these services in different countries, and advances in telecommunications that make this industry possible. The resource-poor islands of the region see financial services as a way to bring foreign capital to state treasuries. Envious of the economic success of The Bahamas, Bermuda, and the Cayman Islands,

GLOBALIZATION IN OUR LIVES

Yo-Ho-Ho and a Bottle of Rum

Closely linked to the production of sugar was the distillation of rum, usually from molasses. Rum is produced throughout the tropical world, but it is still strongly associated with the Caribbean and Latin America, where it is popularly consumed. Colorful cocktails such as piña coladas, mojitos, and daquiris are cliché drinks at beachside resorts. And one can't image a pirate without his bottle of rum.

Rum is part of the popular imagination for this region because it was an early product of globalization. Its creation depended on a blend of materials, technologies, and people that crisscrossed the Atlantic in economically powerful and oppressive ways, as demonstrated by the sugar–rum–slave triangle, and its production and distribution fundamentally shaped the Caribbean region.

By the 17th century, the British Royal Navy gave its sailors a daily ration of rum, which, when served with limes, reduced scurvy. People from all walks of life consumed it, but it was most likely sailors who ensured rum's diffusion throughout the world. Usually considered a crude but strong beverage, it is still the alcohol of choice in the Caribbean. Today, many quality rums are produced in the Caribbean

▲ **Figure 5.4.1 Rum production** A worker in the Havana Club distillery in Cuba checks bottles in the production process. Quality rum aged in oak barrels is a sought-after export around the world.

Explore the TASTES of Rum
https://goo.gl/9fS7Fs

and Latin America for domestic consumption and export (Figure 5.4.1).

1. **How is rum production an early expression of globalization?**

2. **What factors might explain the continued popularity of rum in the Caribbean? Is there a beverage associated with your own region?**

► **Figure 5.26 Free Trade Zones in the Dominican Republic** One sign of globalization is the increase in duty-free and tax-exempt industrial parks in the Caribbean. The Dominican Republic, also a member of the Central American Free Trade Association, has 50 FTZs with foreign investors from the United States, Canada, South Korea, and Taiwan. Q: Where are FTZs clustered in the Dominican Republic, and what might explain this pattern?

▲ **Figure 5.27 Financial Services in the Cayman Islands** Upland House has the offices of Maples and Calder, the largest law firm in the Cayman Islands specializing in financial services. It is also the official address of nearly 13,000 companies. Offshore financial services are vital to the Cayman Islands economy.

countries such as Antigua, Aruba, Barbados, and Belize have also established international banking services. Barbados, for example, is the preferred tax haven for Canadians who desire to bank offshore.

Online gambling is the newest industry for the microstates of the Caribbean. Antigua and St. Kitts were the region's leaders, beginning legal online gambling services in 1999. Other states soon followed; as of 2003, Dominica, Grenada, Belize, and the Cayman Islands had gambling domain sites. In 2007, the World Trade Organization deemed U.S. restrictions imposed on overseas Internet gambling sites to be illegal. The tiny nation of Antigua is currently seeking $3 billion in compensation from the United States for lost revenue due to illegal restrictions placed on Antigua's business.

Meanwhile, drawn to a lucrative business opportunity, U.S. efforts to legalize Internet gambling moved into high gear. By 2013, the governors of Delaware, New Jersey, and Nevada had signed laws to allow online poker in their states. Other cash-strapped states have followed suit, seeing the taxation of online gambling as an attractive revenue source. For the Caribbean, however, this suggests that its era as an online gambling destination may soon be eclipsed.

Tourism Environmental, locational, and economic factors converge to support tourism in the Caribbean. The earliest visitors to this tropical sea admired its clear and sparkling turquoise waters. By the 19th century, wealthy North Americans were fleeing winter to enjoy the healing warmth of the Caribbean during the dry season. Developers later realized that the simultaneous occurrence of the Caribbean's dry season and the Northern Hemisphere's winter was ideal for beach

resorts. By the 20th century, tourism was well established, with both destination resorts and cruise lines. By the 1950s, the leader in tourism was Cuba, with The Bahamas a distant second. Castro's rise to power, however, eliminated this sector of the Cuba's economy for nearly three decades and opened the door for other islands to develop tourism economies.

Six countries or territories hosted two-thirds of the 21.5 million international tourists who came to the Caribbean in 2014: the Dominican Republic, Puerto Rico, Cuba, The Bahamas, Jamaica, and Aruba (Figure 5.28). Puerto Rico's tourist sector began to grow with commonwealth status in 1952. San Juan is now the largest home port for cruise lines and the second largest cruise-ship port in the world in terms of total visitors. The Dominican Republic is the region's largest tourist destination, receiving 5 million visitors in 2014, many of them Dominican nationals who live overseas. Since 1980, tourist receipts have increased 20-fold, making tourism a leading foreign exchange earner at more than $5.6 billion.

After years of neglect, Cuba has revived tourism in an attempt to earn badly needed foreign currency. Tourism represented less than 1 percent of the national economy in the early 1980s. By 2014, 2.9 million tourists (mostly Canadians and Europeans) poured onto the island, bringing in $2.5 billion in tourism receipts. Conspicuous in their absence were U.S. visitors, forbidden to travel to Cuba as "tourists" because of the U.S.-imposed sanctions. But with the normalization of U.S.–Cuban relations in 2015, there has been a growth in authorized tours to Cuba. Even with the 2017 sanctions imposed by U.S. government, record numbers of American tourists traveled to Cuba.

As important as tourism is for the larger states, it is often the principal source of income for smaller ones. The Virgin Islands, Barbados, Turks and Caicos, and Belize all significantly depend on international tourists. To show how quickly this sector can grow, consider this example: When Belize began promoting tourism in the early 1980s, it had just 30,000 arrivals per year. An English-speaking country close to North America, Belize specialized in ecotourism that showcased its interior tropical forests and coastal barrier reef. By the 1990s, the number of tourists topped 300,000, and tourism employed one-fifth of the workforce. Belize City became a port of call for day visitors from cruise ships in 2000, making it the fastest-growing tourist port in the Caribbean. Yet the influx of day visitors to this impoverished coastal town of 60,000 has done little to improve the city's infrastructure or high unemployment.

For more than four decades, tourism has been the foundation of the Caribbean economy, but the industry has grown more slowly in recent years, compared to other tourist destinations in the Middle East, southern Europe, and even Central America. Americans seem to favor domestic destinations—such as Hawaii, Florida, and Las Vegas—or more "exotic" settings—such as Costa Rica. European tourists also seem to be either sticking closer to home or venturing to new locations, such as Dubai on the Persian Gulf and Goa in India. Increasingly, tourists are opting to experience the Caribbean from the decks of cruise ships rather than land-based resorts. This trend undermines the local benefits of tourism, directing capital to large cruise lines rather than island economies (Figure 5.29).

Tourism-led growth has other drawbacks. It heavily depends on the overall health of the world economy and current political affairs. Thus, if North America experiences a recession or international

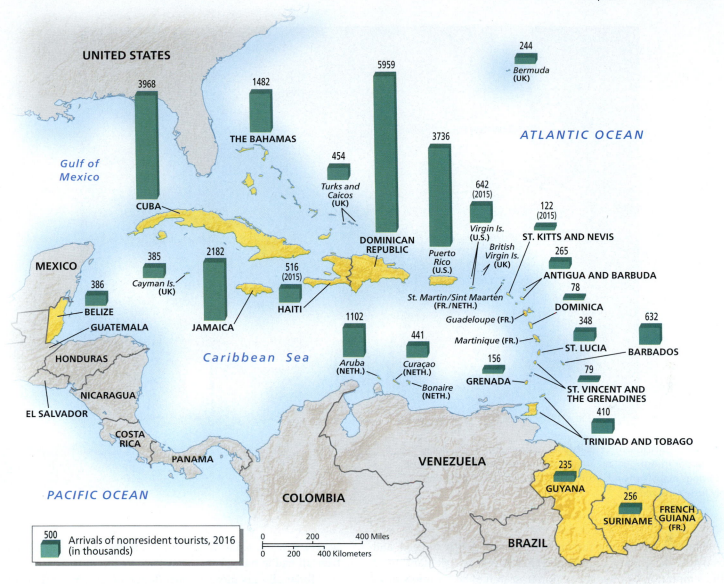

▲ **Figure 5.28** **Global Linkages: International Tourism**   The Caribbean is directly linked to the global economy through tourism. Each year more than 21 million tourists come to the islands, mostly from North America, Latin America, and Europe. The most popular destinations are the Dominican Republic, Puerto Rico, Cuba, Jamaica, The Bahamas, and Aruba. **Q: What might explain the lower levels of tourism in the Guianas compared to the rest of the region?**

tourism declines due to terrorism concerns, the flow of tourist dollars to the Caribbean dries up. Natural disasters like Hurricane Maria can negatively impact tourist flows to those areas damaged by storms. Where tourism is on the rise, local resentment may build as residents confront the differences between their own lives and those of the tourists. There is also a serious problem of **capital leakage**, which is the huge gap between gross income and the total tourist dollars that remain in the Caribbean. Because many guests stay in hotel chains or cruise ships with corporate headquarters outside the region, leakage of profits is inevitable. On the plus side, tourism tends to promote stronger environmental laws and regulations. Countries quickly learn that their physical environment is the foundation for success. And though tourism does have its costs (higher energy and water consumption and demand for more imports), it is environmentally less destructive than traditional export agriculture—and currently more profitable.

Social Development

The Caribbean's record for economic growth is inconsistent, but measures of social development are generally strong. For example, most Caribbean peoples have an average life expectancy of more than 73 years (see Table 5.2). Literacy levels are high, with near parity in terms of school enrollment by gender. And child mortality is also at low rates, with the exception of Haiti.

These demographic and social indicators explain why Caribbean nations fare well on the Human Development Index. Most fall within the high and medium human development categories. Many of these well-ranked states—especially Jamaica, Guyana, and St. Kitts and Nevis—have significant annual per capita inflows of remittances from the Caribbean diaspora. Many argue that remittances have become extremely important in boosting the overall level of social and economic development in the region. Yet despite real social gains,

▲ **Figure 5.29 Caribbean Cruising** A massive cruise ship docks in St. John's harbor in Antigua, an island of the Lesser Antilles. The cruise business brings many visitors to these small islands, but they spend relatively little time there.

Explore the **SIGHTS** of **St. Johns Harbor**
https://goo.gl/X1HoOC

many inhabitants are chronically underemployed, poorly housed, and perhaps overly dependent on foreign remittances. For rich and poor alike, the temptation to leave the region in search of better opportunities remains.

Gender, Politics, and Culture

The matriarchal (female-dominated) basis of Caribbean households is often singled out as a distinguishing characteristic of the region. The rural custom of men leaving home for seasonal employment tends to nurture strong, self-sufficient female networks. Women typically run the local street markets. With men absent for long stretches, women tend to make household and community decisions. Although this gives women local power, it does not always imply higher status. In rural areas, female status is often undermined by the relative exclusion of women from the cash economy—men earn wages, while women provide subsistence.

As Caribbean society urbanizes, more women are being employed in assembly plants (the garment industry, in particular, prefers to hire women), in data-entry firms, and in tourism. With new employment opportunities, female participation in the labor force has surged; in countries such as Barbados, Haiti, and Jamaica, more than 45 percent of the workforce is female. Increasingly, women are the principal earners of cash and are more likely to complete secondary education than men. There are also signs of greater political involvement by women.

In recent years, Jamaica, Dominica, Trinidad and Tobago, and Guyana have all had female prime ministers.

Today slightly more women than men migrate to the United States, and many seek employment in the globalized care economy as health care workers, nannies, and eldercare aides. Many feminist scholars argue that the care industry has increasingly become the domain of immigrant women of color throughout North America and Europe. This segmentation of the labor market is driven by a complex mix of income inequalities, racial and gender preferences, and the demands and relatively low status of care work in general.

An estimated 300,000 to 500,000 Caribbean-born health care professionals, mostly women, work in the United States. More than half are home health care aides, and another third are registered nurses or technicians (Figure 5.30). Many of these women are reliable senders of remittances. Yet it is also true that these female migrants may leave their own families and children behind due to work demands and visa stipulations. Many Caribbean governments worry about this trend, but it is not one that can be easily addressed.

Education Many Caribbean states have excelled in educating their citizens. Literacy is the norm, and most people receive at least a high school diploma (see Table 5.2). When considering gender differences in secondary school enrollment, in most cases the gap between males and females is not large—and in The Bahamas, Cuba, Jamaica, and Puerto Rico, female students are enrolled at higher rates

▲ **Figure 5.30 Caribbean Health Care Worker**   A Jamaican nurse with her colleagues in a Massachusetts hospital. Up to half a million Caribbean healthcare professionals work in the United States.

than males. Two demographically large countries underperform in this measure; only 74 percent of young men are enrolled in secondary school in the Dominican Republic, while there are no estimates for Haiti, a country that has suffered from chronic illiteracy for decades.

Education is expensive for these nations, but it is considered essential for development. Ironically, many states express frustration about training professionals for the benefit of developed countries in a phenomenon called **brain drain**. Brain drain occurs throughout the developing world, especially between former colonies and the mother countries. In the early 1980s, the prime minister of Jamaica complained that 60 percent of his country's newly university-trained workers left for the United States, Canada, and Britain, representing

a subsidy to these economies far greater than the foreign aid Jamaica received from them. A World Bank study of skilled migrants revealed that 40 percent of Caribbean migrants living abroad were college educated. For countries such as Guyana, Grenada, Jamaica, St. Vincent and the Grenadines, and Haiti, over 80 percent of the college-educated population will emigrate. No other world region has proportionately this many educated people leaving. Given the small population of many Caribbean territories, each professional person lost to emigration can negatively impact local health care, education, and enterprise. However, despite the high outflow of professionals, many countries are more recently experiencing a return migration of Caribbean peoples from North America and Europe as a **brain gain**. This term refers to the potential of returnees to contribute to the social and economic development of their home country with the experiences they have gained abroad.

Many Caribbean countries are also reaching out to their diasporas in an effort to forge transnational linkages that can foster greater investment in development for the region. In countries as diverse as Haiti and Barbados, the combination of brain gain and the potential of the diaspora to stimulate social and economic development are strategies being exploited to improve the region's future.

REVIEW

5.9 As the Caribbean has shifted out of dependence on agriculture, what other economic sectors have emerged in the region?

5.10 What explains the relatively high levels of social development in the Caribbean, given the region's relative poverty?

KEY TERMS free trade zone (FTZ), offshore banking, capital leakage, brain drain, brain gain

The Caribbean

REVIEW, REFLECT, & APPLY

Summary

- This tropical region was exploited to produce export commodities such as sugar, tobacco, and bananas. The region's warm waters and mild climate attract millions of tourists, yet serious problems with deforestation and soil erosion have degraded urban and rural environments. Global warming poses a serious threat to the region, with the likelihood of more-intense hurricanes and sea-level rise.

- Population growth in the Caribbean has slowed over the past two decades; the average woman now has two to three children. Life expectancy is quite high, as are literacy rates. Most Caribbean people live in cities, but Caribbean cities are not large by world standards. The Caribbean region is noted for high rates of emigration, especially among the highly skilled who settle in North America and Europe.

- The Caribbean was forged through European colonialism and the labor of millions of Africans. The blending of African and European elements, termed creolization, has resulted in many unique cultural expressions in music, language, and religion. Others view the Caribbean as a neo-Africa, in which African peoples, cultures, and even some agricultural practices dominate, especially in isolated maroon communities.

- Today the region contains 26 independent countries and territories. The end of Fidel Castro's government has brought some political and economic change to the Caribbean's largest country, but U.S. relations with Cuba remain troubled. In addition to the United States, other countries such as Brazil, China, and the United Kingdom also exert influence in the Caribbean.

- The Caribbean has gradually shifted from being an exporter of primary agricultural resources (especially sugar) to a service and manufacturing economy. Employment opportunities in assembly plants, tourism, and offshore banking have replaced jobs in agriculture. The region's strides in social development—especially in education, health, and the status of women—distinguish it from other developing areas.

Review Questions

1. What are the historical, cultural, and resource differences between the Caribbean islands and the rimland, and what environmental issues affect these distinct areas?

2. What is the relationship between the shift from agrarian to urban settlement and the reliance on emigration in assessing the long-term population growth of this region?

3. What are the demographic and cultural implications of the forced transfer of African people to the Caribbean and the creation of a neo-African society in the Americas?

4. Why did European colonists so aggressively seek control of the Caribbean, and why did independence in the region come about more gradually than in neighboring Latin America?

5. How is the contemporary Caribbean linked to the world economy through emigration, offshore banking, free trade zones, and tourism? What changes might occur over the next 20 years?

Image Analysis

1. Look at the population pyramid for Barbados. Based on the pyramid, what percentage of the population is under 15? Over 65?

2. When did the TFR begin to decline? Is the population growing, declining, or stable? What factors might account for demographic trends in Barbados?

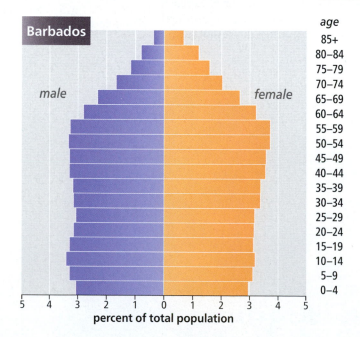

▶ **Figure IA5** Barbados Population Pyramid

Join the Debate

For over five decades, the United States had a trade embargo with Cuba in an effort to isolate this socialist country and to force political change. In 2015, relations between the two countries began to normalize due to the executive action of the Obama administration. Still a socialist country, some private-sector activities and foreign investments are allowed by the government of Cuba, which has a new president. In 2017, however, the Trump administration reimposed travel and investment restrictions. Does the economic isolation of Cuba support the status quo or help to bring about change?

▲ **Figure D5** **U.S.–Cuba Relations** This Cuban gives a thumbs-up from his Havana balcony following the beginning of normalization of U.S.–Cuba relations in 2015. The administration of President Trump has since reimposed travel and financial restrictions.

The embargo has a positive impact overall.

- The trade embargo kept pressure on the Cuban government and discouraged other Caribbean and Latin American countries from following Cuba's path.

- By forcing Cuba to develop other trade partners and alternative sources of revenue, the country became an innovator in health care and biomedical research.

- Limited interaction with the United States over 50 years allowed for a distinct Cuban way of life and identity to form.

The embargo serves neither U.S. citizens nor the citizens of Cuba.

- The embargo resulted in needless suffering of the Cuban people and did not change the political system.

- The United States would have had more influence on Cuba earlier if it had stayed economically and politically engaged in the country, as the United States did with communist China.

- By maintaining the embargo for so long, foreign investors from Europe and Asia have more economic influence in the country.

Key Terms

brain drain (p. 171)
brain gain (p. 171)
capital leakage (p. 169)
Caribbean Community and Common Market (CARICOM) (p. 163)
chain migration (p. 153)
circular migration (p. 153)

creolization (p. 158)
diaspora (p. 153)
free trade zones (FTZs) (p. 165)
geothermal (p. 149)
Greater Antilles (p. 144)
hurricanes (p. 145)
indentured labor (p. 157)
Lesser Antilles (p. 144)
maroons (p. 156)

mono-crop production (p. 154)
Monroe Doctrine (p. 161)
neocolonialism (p. 161)
offshore banking (p. 166)
plantation America (p. 154)
remittances (p. 153)
rimland (p. 144)
transnational migration (p. 153)

Mastering Geography

Looking for additional review and test prep materials? Visit the Study Area in Mastering Geography to enhance your geographic literacy, spatial reasoning skills, and understanding of this chapter's content by accessing a variety of resources, including MapMaster interactive maps, geoscience animations, videos, flashcards, web links, self-study quizzes, and an eText version of *Globalization and Diversity*.

Scan to read about Geographer at Work
Sarah Blue and her company Candela Cuba Tours
https://goo.gl/ogfUCZ

GeoSpatial Data Analysis

Remittances and Development Remittances are an important source of revenue for many Caribbean countries and may contribute to their overall high levels of social development. This analysis allows you to explore the relationship between the Human Development Index (HDI) and remittances.

Begin by logging into MapMaster, and create a map of HDI data for the Caribbean, found in Table 5.2. Then go to the World Bank website (www.worldbank.org) and scroll down to Data, then click on Browse Data by Indicator. Search for Personal Remittances, Received in the Economy & Growth category, then download the .csv file and prepare the data for Caribbean countries. Create a map of remittances received, and use the split screen feature to compare your maps and answer the following:

World Bank Personal Remittances, Received
https://goo.gl/RxBt2v

1. Which countries receive the most remittances, and why? Keeping in mind the relative population sizes of each country, which countries receive proportionally higher levels of remittances?

2. Describe the relationship between HDI rankings and the inflow of remittances for the countries mapped. Suggest reasons for the pattern you see, and explain the importance of remittances for developing economies.

Sub-Saharan Africa

Physical Geography and Environmental Issues

The Sahel is chronically prone to drought, but an El Niño–induced drought in 2015–2016 gripped southern and eastern Africa. Drought creates food and water shortages and, in the worst cases, famine. Tree planting through the Greenbelt Movement addresses fuel shortages and improves resilience.

Population and Settlement

Sub-Saharan Africa is demographically young and growing. With over 1 billion people, its rate of natural increase is 2.7, making it the fastest-growing world region in terms of population. It is also the region hit hardest by HIV/AIDS, which has lowered overall life expectancies in many countries.

Cultural Coherence and Diversity

The region is culturally complex, with dozens of languages spoken in some countries. Religious life is also important, with large and growing numbers of Muslims and Christians. Religious diversity and tolerance have been distinctive features of this region; but religious conflict, especially in the Sahel, is on the rise.

Geopolitical Framework

Most countries won independence in the 1960s. Since then, many ethnic conflicts have occurred as governments struggle for national unity within boundaries drawn by European colonialists. Muslim extremist groups, especially Boko Haram and the Shabab, have increased violence and instability. The number of internally displaced people and refugees in the region is large and growing.

Economic and Social Development

Extreme poverty is all too common in this region. The UN's Sustainable Development Goals to reduce extreme poverty by 2030 are the new benchmarks to measure progress. Fortunately, significant improvements in education, life expectancy, and economic growth are occurring.

▶ **The Sahel** A view of the Niger River from the Dune of Koima near the city of Gao, Mali. Many major settlements of the Sahel are along the Niger River, along with much of the region's agriculture.

Gao, Mali

SUB-SAHARAN AFRICA

The Sahel is a vast arid region that stretches across the African continent, with the Saharan desert to the north and the wetter forest and savanna zones to the south. Long an important area for farmers, pastoralists, and traders, the Sahel has been settled for centuries. The average woman in Mali and Niger has six to seven children, which means this area is demographically growing, but also vulnerable.

The people of the Sahel all too often face drought, famine, and conflict. Much recent interest in the region has focused on the western Sahel, in the countries of Mali, Mauritania, Burkina Faso, Niger, northern Nigeria, and Chad. The two principle water sources are the Niger River and Lake Chad. On the banks of the Niger River are the capitals of Mali (Bamako) and Niger (Niamey), as well as the historical city of Tombouctou (Timbuktu). Four countries border on Lake Chad, which is rapidly shrinking in a process called desiccation. Yet the lake is still the main water source for some 30 million people.

Vulnerability in the Sahel has been exacerbated by climate change and violent attacks led by Muslim extremists. Boko Haram has been especially deadly in northern Nigeria, Niger, and Chad. In Mali, an Al Qaeda affiliate and the Movement for Unity and Jihad in West Africa have been responsible for sporadic attacks. Concern over Islamic extremism has led religious leaders and civil society to call for religious tolerance in the wake of armed conflict; in 2014, five Sahelian countries formed the G5 Sahel to coordinate economic development and security.

> The impact of outsiders has shaped the region's identity.

Some positive trends are evident. In particular, Niger has increased its tree cover due to policy changes that led to greater conservation. France announced a major project in 2018 to support sound agroecological practices to enhance food security. Yet at the same time, the U.S. Air Force established a major drone base costing over $100 million in Agadez, Niger, to better monitor the region and fight terrorism.

Defining Sub-Saharan Africa

Sub-Saharan Africa—that portion of the African continent lying south of the Sahara Desert—is a commonly accepted world region, sharing similar livelihood systems and colonial experiences (Figure 6.1). No common religion, language, philosophy, or political system ever united the region.

Instead, its unity stems from loose cultural bonds developed from a variety of lifestyles and idea systems that evolved here. The impact of outsiders also shaped the region's identity. Slave traders from Europe, North Africa, and Southwest Asia treated Africans as chattel; until the mid-1800s, millions of Africans were abducted and sold into slavery. In the late 1800s, the entire continent was divided by European colonial powers, imposing political boundaries that, for the most part, remain to this day. In the postcolonial period, Sub-Saharan African countries shared many of the same economic and political challenges.

The major question is whether to include North Africa and treat the entire continent as one world region. However, North Africa is more closely linked, both culturally and physically, to Southwest Asia, as Arabic is the dominant language and Islam the dominant religion. Consequently, North Africa is discussed along with Southwest Asia in Chapter 7. In this chapter, the new country of South Sudan, along with the Sahelian states of Mauritania, Mali, Niger, and Chad, form the region's northern boundary.

The people of Sub-Saharan Africa are generally poorer, more rural, and much younger than those in Latin America and the Caribbean. Over 1 billion people reside in this region, which includes 48 states and one territory (Reunion, off the coast of Madagascar). Demographically, this is the fastest-growing world region; in most countries, over 40 percent of the population is younger than 15 years. By 2050, the region is expected to reach 2 billion people. Income levels are low: 41 percent of the people live in extreme poverty (less than $1.90 per day). Such statistics and the all-too-frequent negative headlines about violence, disease, and poverty might lead to despair. Yet many Africans are optimistic about the future. Local and international nongovernmental organizations (NGOs), diaspora-led groups, and government agencies have worked to reduce infant mortality, expand basic education, and increase food production in the past two decades. Most national economies are growing faster than their populations, another positive indicator for the region. One transformational change is the rapid diffusion of cell phones, along with innovative applications that improve communication, commerce, and information sharing. Sub-Saharan Africa is a vast region with abundant natural resources and a young population that needs better education. Sustained investments and infrastructural improvements are also needed. To better appreciate these opportunities and challenges, we must first consider the region's physical geography.

LEARNING OBJECTIVES

After reading this chapter you should be able to:

6.1 Describe the region's major ecosystems and how humans have adapted to living in them.

6.2 Outline the environmental issues that challenge Africa south of the Sahara.

6.3 Explain the region's rapid demographic growth and describe the differential impact of diseases such as HIV/AIDS.

6.4 Connect ethnicity and religion to conflicts in this region and identify strategies for maintaining peace.

6.5 Summarize the various cultural influences of African peoples within the region and globally.

6.6 Trace the colonial history of Sub-Saharan Africa and link colonial policies to postindependence challenges.

6.7 Assess the roots of African poverty and explain why many of the world's fastest-growing economies are in Sub-Saharan Africa.

6.8 List the region's major resources, especially metals and fossil fuels, and describe how they are being developed.

▲ Figure 6.1 Sub-Saharan Africa This vast world region, which includes 48 states and one territory, is often categorized into the western, central, eastern, and southern subregions (see Table 6.1). The rainforests, tropical savannas, and deserts of Sub-Saharan Africa are home to over 1 billion people. Much of the region consists of broad plateaus ranging in elevation from 1600 to 6500 feet (500 to 2000 meters). Although the population is growing rapidly, overall population density is low. Considered one of the least developed world regions, Africa south of the Sahara remains an area rich in natural resources.

Physical Geography and Environmental Issues: The Plateau Continent

Sub-Saharan Africa is the largest landmass straddling the equator (Figure 6.2). Its physical environment is remarkably beautiful, dominated by extensive elevated plateaus. The highest elevations are found on the continent's eastern edge, where the **Great Rift Valley** forms a complex upland of lakes, volcanoes, and deep valleys. In contrast, lowlands prevail in West Africa (see Figure 6.1). Despite this region's immense biodiversity, vast water resources, and wealth of precious minerals, it also has relatively poor soils, widespread disease, and vulnerability to drought.

Plateaus and Basins

A series of plateaus and elevated basins dominates the African interior and forms much of the region's unique physical geography (see Figure 6.1). Generally, elevations increase toward the south and east; most of southern and eastern Africa lies well above 2000 feet (600 meters), and sizable areas sit above 5000 feet (1500 meters). These areas are typically referred to as High Africa; Low Africa includes western Africa and much of central Africa. Steep escarpments form where plateaus abruptly end, as illustrated by the majestic Victoria Falls on the Zambezi River or the dramatic backdrop of Table Mountain in Cape Town, South Africa (Figure 6.3). Much of southern Africa is rimmed by the **Great Escarpment** (a high cliff separating two comparatively level

▶ **Figure 6.2 Africa's Relative Size**
Commonly used map projections tend to understate the vast size of the African continent. This graphic shows how much of the United States and Europe, as well as Japan, China, and India, would fit within the boundaries of Africa.

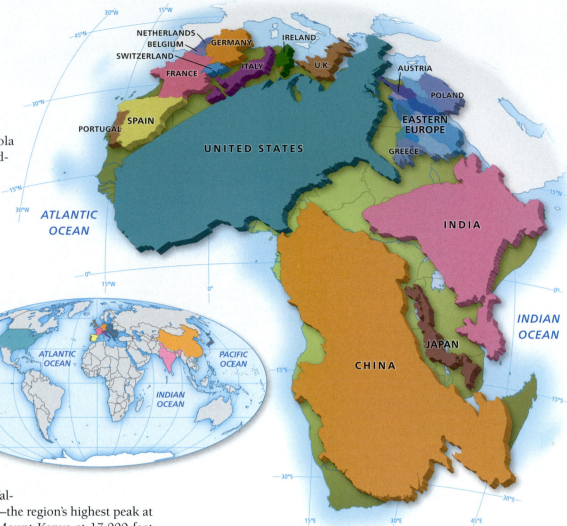

areas) stretching from southwestern Angola to northeastern South Africa. Such landforms proved to be barriers to European colonial settlement in the interior of the continent, explaining in part the prolonged colonization process.

Though Sub-Saharan Africa is an elevated landmass, the high plateaus are dominated by deep valleys rather than dramatic mountain ranges. The one extensive area of mountainous topography is in Ethiopia, in the northern portion of the Rift Valley zone. Receiving heavy rains in the wet season, the Ethiopian Plateau forms the headwaters of several important rivers—most notably the Blue Nile, which joins the White Nile at Khartoum, Sudan. Volcanic mountains, some quite tall, are located in the southern half of the Rift Valley, such as Tanzania's Mount Kilimanjaro—the region's highest peak at 19,000 feet (5900 meters)—and Kenya's Mount Kenya at 17,000 feet (5200 meters) (Figure 6.4).

Watersheds   Africa south of the Sahara does not have the broad, alluvial lowlands that influence patterns of settlement throughout other regions. The four major river systems are the Congo, Niger, Nile, and

▼ **Figure 6.3 Cape Town**   Table Mountain is the dramatic backdrop for this South African city, an early site of European settlement and trade. The mountains around Cape Town are a system of folds extending to the Great Escarpment to the north where the plateau begins.

Zambezi (the Nile is discussed in Chapter 7). Smaller rivers—such as the Orange in South Africa; the Senegal, which divides Mauritania and Senegal; and the Limpopo in Mozambique—are locally important but drain much smaller areas. Ironically, most people believe that Sub-Saharan Africa suffers from water scarcity and tend to discount the size and importance of the watersheds (or catchment areas) that these river systems drain.

The Congo (or Zaire) River is the region's largest watershed in terms of both drainage area and water volume. It is second only to South America's Amazon River in terms of annual flow. The Congo flows across a relatively flat basin more than 1000 feet (300 meters) above sea level, meandering through Africa's largest tropical forest, the Ituri (Figure 6.5). Entry from the Atlantic into the Congo Basin is prevented by a series of rapids and falls, making the river only partially navigable. Despite these limitations, the Congo River has been the major corridor for travel within the Republic of the Congo and the Democratic Republic of the Congo (formerly Zaire); the capitals of these countries, Brazzaville and Kinshasa, lie on opposite sides of the river to form a metropolitan area of nearly 13 million people.

The Niger River is the critical water source for all of West Africa, but especially for the arid countries of Mali and Niger. Beginning in the humid Guinea

▲ **Figure 6.4 Mount Kenya** Africa's second highest mountain after Kilimanjaro, Mount Kenya supports a unique tropical alpine ecosystem. Kenya's tallest mountain at 5200 m (17,000 ft) was first summitted in 1899 by British geographer Sir Halford Mackinder.

▲ **Figure 6.6 Settlement in the Sudd** South Sudan's Sudd is one of the region's largest wetlands, and during the rainy season, the White Nile River can expand the wetlands to four times its normal size. This is a traditional village near the Bor, South Sudan.

Explore the SIGHTS of the Sudd
http://goo.gl/hqVsra

highlands, the Niger flows to the northeast and then spreads out to form a huge inland delta in Mali before making a great bend southward at the margins of the Sahara near Gao. After flowing through the Sahel, the Niger returns to the humid lowlands of Nigeria, where the Kainji Reservoir temporarily blocks its flow to produce electricity for Africa's most populous state. At the river's end lies the Niger Delta, which is also the center of Nigeria's oil industry. This fertile delta region, home to ethnic groups such as the Igbo and Ogoni, is extremely poor. For decades, conflict in the region has centered on who benefits from the region's oil and on the serious environmental degradation resulting from oil and gas extraction.

The Nile River, the world's longest, is the lifeblood of Egypt and Sudan. Yet this river originates in the highlands of the Rift Valley zone and is an important link between North and Sub-Saharan Africa. The Nile begins in the lakes of the Rift Valley (Victoria and Edward) before descending into a vast wetland in South Sudan known as the Sudd. One

of the world's great wetlands, the Sudd averages over 30,000 square kilometers but can expand to four times that size during the rainy season. Navigating the Sudd is tricky, but settlers have long been drawn to this area's abundant water supply and aquatic life (Figure 6.6). Development projects in the 1970s increased the Sudd's agricultural potential, especially its peanut crop. Unfortunately, three decades of civil war in what is now South Sudan ravaged this area, turning farmers and herders into refugees and undermining the productive capacity of this important ecosystem.

Soils With a few major exceptions, Sub-Saharan Africa's soils are relatively infertile. Generally speaking, fertile soils are young soils, deposited in recent geologic time by rivers, volcanoes, glaciers, or windstorms. In older soils—especially those found in moist tropical environments—natural processes tend to wash out most plant nutrients over time. Over most of the region, the agents of soil renewal are largely absent.

Portions of Sub-Saharan Africa are, however, noted for their natural soil fertility; and not surprisingly, these areas support denser settlement. Some of the most fertile soils are in the Great Rift Valley, made productive by the area's volcanic activity. The population densities of rural Rwanda and Burundi, for example, are partially explained by the highly productive soils. The same is true for highland Ethiopia, which supports the region's second largest population after Nigeria. The Lake Victoria lowlands and central highlands of Kenya are also noted for their sizable populations and productive agriculture.

The drier grassland and semidesert areas feature a soil type called *alfisols*. High in aluminum and iron content, these red soils have greater fertility than comparable soils found in wetter zones. This helps to explain why farmers tend to plant in drier areas, such as the Sahel, despite the risk of drought. Many agronomists suggest that with irrigation, the southern African countries of Zambia and Zimbabwe could greatly increase commercial grain production on these soils.

▼ **Figure 6.5 Congo River** Africa's largest river by volume, the mighty Congo flows through the Ituri rainforest in the Democratic Republic of the Congo.

Climate and Vegetation

Most of Sub-Saharan Africa lies in the tropical latitudes, with only the far south extending into the subtropical and temperate belts. Much of the region averages high temperatures from 70°F to 80°F (22°C to 28°C)

year-round (Figure 6.7). The seasonality and amount of rainfall, more than temperature, determines the different vegetation belts characterizing the region—tropical forests, savannas, and deserts. Addis Ababa, Ethiopia, and Walvis Bay, Namibia, have similar average temperatures (see the climographs in Figure 6.7); but Addis Ababa is in the moist

▼ **Figure 6.7 Climates of Sub-Saharan Africa** Much of the region lies within the tropical humid and tropical dry climatic zones; thus, seasonal temperature changes are not great. Precipitation, however, varies significantly from month to month. Compare the distinct rainy seasons in Lusaka and Lagos: Lagos is wettest in June, and Lusaka receives most of its rain in January. Although important tropical forests are found in West and Central Africa (coinciding with the tropical wet and monsoon climate zones), much of the region's vegetation is tropical savanna.

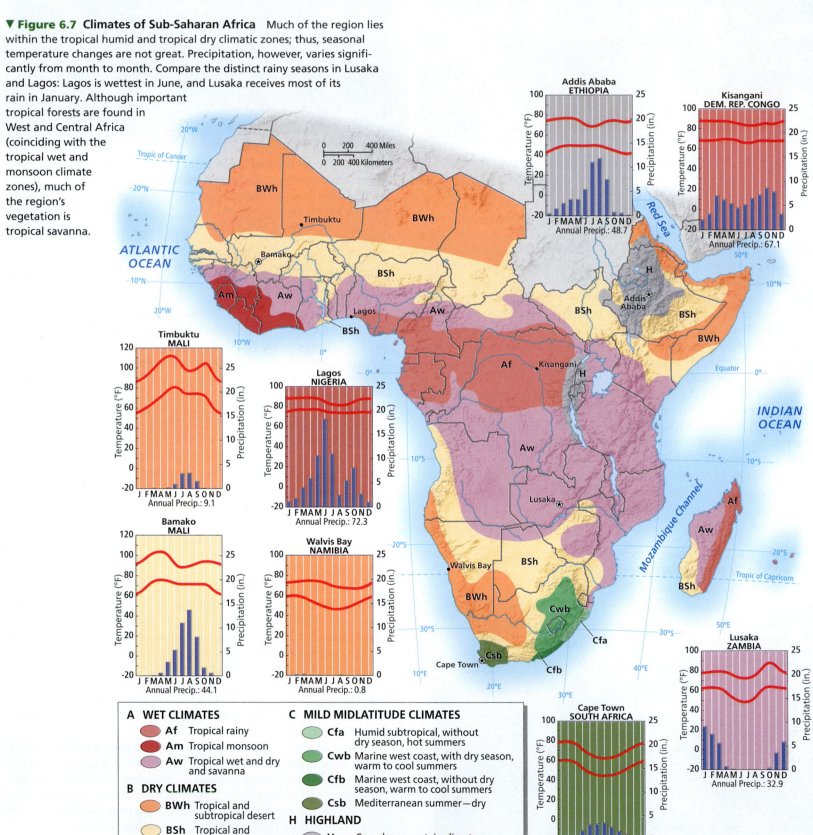

A WET CLIMATES

	Af	Tropical rainy
	Am	Tropical monsoon
	Aw	Tropical wet and dry and savanna

B DRY CLIMATES

| | **BWh** | Tropical and subtropical desert |
| | **BSh** | Tropical and subtropical steppe |

C MILD MIDLATITUDE CLIMATES

	Cfa	Humid subtropical, without dry season, hot summers
	Cwb	Marine west coast, with dry season, warm to cool summers
	Cfb	Marine west coast, without dry season, warm to cool summers
	Csb	Mediterranean summer—dry

H HIGHLAND

| | **H** | Complex mountain climates |

highlands and receives nearly 50 inches (127 cm) of rainfall annually, whereas Walvis Bay is in the Namibian Desert and receives less than 1 inch (2.5 cm).

Tropical Forests   Sub-Saharan Africa core is remarkably moist. The world's second largest expanse of equatorial rainforest, the Ituri, lies in the Congo Basin, extending from the Atlantic Coast of Gabon two-thirds of the way across the continent, including the northern portions of the Republic of the Congo and the Democratic Republic of the Congo. Conditions here are constantly warm to hot, and precipitation falls year-round (see the climograph for Kisangani in Figure 6.7).

Commercial logging and agricultural clearing have degraded the western and southern fringes of the Ituri, but much of the northeastern section is still intact. The Ituri has, so far, fared much better than Southeast Asia and Latin America in terms of tropical deforestation. Major national parks such as Okapi and Virunga have been created in the Democratic Republic of the Congo. Virunga National Park—one of the oldest in Africa—sits on the eastern fringe of the Ituri and is home to the endangered mountain gorillas. Poor infrastructure and political chaos in the Democratic Republic of the Congo over the past 25 years have made large-scale logging impossible but have also made conservation difficult. Parks such as Virunga have been repeatedly taken over by rebel groups, and park rangers have been killed; poaching has become a means of survival for people in the area. It seems likely that Central Africa's rainforest, and the wildlife within it, could suffer the kind of degradation experienced in other tropical forests, especially with illegal logging on the rise.

Savannas   Wrapped around the Central African rainforest belt in a great arc lie Africa's vast tropical wet and dry savannas. Savannas are dominated by a mix of trees and tall grasses in the wetter zones immediately adjacent to the forest belt, and by shorter grasses with fewer trees in the drier zones. North of the equatorial belt, rain generally falls only from May to October. The farther north you travel, the less the total rainfall and the longer the dry season. Climatic conditions south of the equator are similar, only reversed, with the wet season occurring between October and May and precipitation generally decreasing toward the south (see the climograph for Lusaka in Figure 6.7). A larger area of wet savanna exists south of the equator, with extensive woodlands located in southern portions of the Democratic Republic of the Congo, Zambia, Zimbabwe, and eastern Angola. These savannas are also a critical habitat for the region's large fauna (Figure 6.8).

Deserts   Major deserts exist in the southern and northern boundaries of the region. The Sahara, the world's largest desert and one of its driest, crosses the landmass from the Atlantic coast of Mauritania all the way to the Red Sea coast of Sudan. A narrow belt of desert extends to the south and east of the Sahara, wrapping around the **Horn of Africa** (the northeastern corner that includes Somalia, Ethiopia, Djibouti, and Eritrea) and pushing as far as eastern and northern Kenya. An even drier zone is found in southwestern Africa. In the striking red dunes of the Namib Desert of coastal Namibia, rainfall is a

▲ **Figure 6.8** **Savanna Wildlife**   A herd of elephants gather at a riverbank in South Africa's Krueger National Park. The savannas of southern Africa are a noted habitat for the region's larger mammals—including buffalo, impala, wildebeest, zebra, lions, and elephants.

rare event, although temperatures are usually mild (Figure 6.9). Inland from the Namib lies the Kalahari Desert. Because it receives slightly more than 10 inches (25 cm) of rain a year, most of the Kalahari is not dry enough to be classified as a true desert. Its rainy season, however, is brief, and most of the rainfall is immediately absorbed by the underlying sands. Surface water is thus scarce, giving the Kalahari a desert-like appearance for most of the year.

Africa's Environmental Issues

The prevailing perception of Africa south of the Sahara is one of environmental scarcity and degradation, no doubt reinforced by televised images of drought-ravaged regions and starving children. Single explanations such as rapid population growth and colonial exploitation cannot fully capture the complexity of Africa's environmental issues or the

▼ **Figure 6.9** **Namib Desert**   Arid Namibia has one of the region's lowest population densities, yet its coastal dunes attract foreign tourists. These hikers are scaling the orange-colored dunes in Namib-Naukluft National Park.

ways that people have adapted to living in marginal ecosystems with debilitating tropical diseases. Because most of Sub-Saharan Africa's population is rural, earning its livelihood directly from the land, sudden environmental changes can have devastating effects on household income and food consumption.

As Figure 6.10 illustrates, deforestation and **desertification**—the expansion of desert-like conditions due to human-induced degradation—is commonplace. Sub-Saharan Africa is also vulnerable to drought, most notably in the Horn of Africa, parts of southern Africa, and the Sahel. Many scientists fear that drought will become more frequent and prolonged under global climate change. Rainfall levels were abnormally low in southern and eastern Africa during the El Niño event of 2015–2016, contributing to water and food scarcity. Yet the region is also home to some of the most impressive wildlife populations in the world, which is a source of pride and revenue for many African states.

The Sahel and Desertification   The **Sahel** is a zone of ecological transition between the Sahara to the north and the wetter savannas and forest in the south (see Figure 6.10). In the 1970s, the Sahel was a symbol for the dangers of unchecked population growth and human-induced environmental degradation when a relatively wet period came to an abrupt end. Six years of drought (1968–1974) were followed by a second prolonged drought in the mid-1980s, ravaging the land. During these droughts, area rivers diminished and desert-like conditions expanded southward. Unfortunately, tens of millions of people lived in this area, and farmers and pastoralists whose livelihoods had come to depend on the more abundant precipitation of the relatively wet period were temporarily forced out.

Life in the Sahel depends on a delicate balance of limited rain, drought-resistant plants, and a pattern of **transhumance** (the movement of livestock between wet-season and dry-season pasture). What

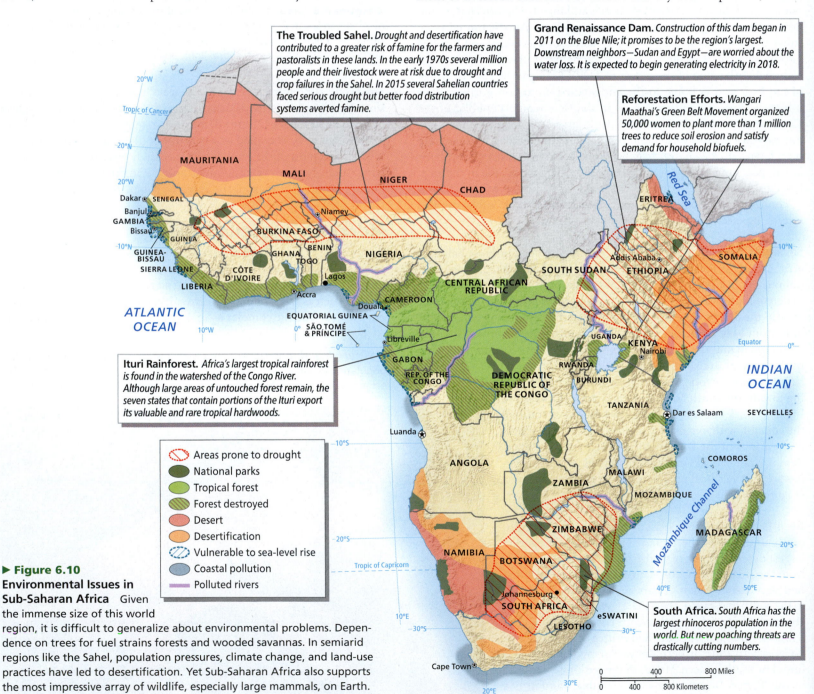

The Troubled Sahel. *Drought and desertification have contributed to a greater risk of famine for the farmers and pastoralists in these lands. In the early 1970s several million people and their livestock were at risk due to drought and crop failures in the Sahel. In 2015 several Sahelian countries faced serious drought but better food distribution systems averted famine.*

Grand Renaissance Dam. *Construction of this dam began in 2011 on the Blue Nile; it promises to be the region's largest. Downstream neighbors—Sudan and Egypt—are worried about the water loss. It is expected to begin generating electricity in 2018.*

Reforestation Efforts. *Wangari Maathai's Green Belt Movement organized 50,000 women to plant more than 1 million trees to reduce soil erosion and satisfy demand for household biofuels.*

Ituri Rainforest. *Africa's largest tropical rainforest is found in the watershed of the Congo River. Although large areas of untouched forest remain, the seven states that contain portions of the Ituri export its valuable and rare tropical hardwoods.*

South Africa. *South Africa has the largest rhinoceros population in the world. But new poaching threats are drastically cutting numbers.*

Legend:
- Areas prone to drought
- National parks
- Tropical forest
- Forest destroyed
- Desert
- Desertification
- Vulnerable to sea-level rise
- Coastal pollution
- Polluted rivers

▶ **Figure 6.10**
Environmental Issues in Sub-Saharan Africa   Given the immense size of this world region, it is difficult to generalize about environmental problems. Dependence on trees for fuel strains forests and wooded savannas. In semiarid regions like the Sahel, population pressures, climate change, and land-use practices have led to desertification. Yet Sub-Saharan Africa also supports the most impressive array of wildlife, especially large mammals, on Earth.

appears to be desert wasteland in April or May is transformed into a lush garden of millet, sorghum, and peanuts after the drenching rains of June. The fertile Sahelian soils, and relative freedom from the tropical diseases found in the wetter zones to the south, explain why people continue to live there despite unreliable rainfall patterns.

The main practices cited in Sahel desertification are the expansion of agriculture and overgrazing, leading to the loss of natural vegetation and declines in soil fertility. For example, French colonial authorities forced villagers to grow peanuts as an export crop, a policy continued by the newly independent Sahelian states. However, peanuts tend to deplete several key soil nutrients, which means that peanut fields are often abandoned after a few years as cultivators move on to fresh sites. Peanut harvesting also turns up the soil at the onset of the dry season, leading to accelerated wind erosion of valuable topsoil.

Overgrazing by livestock, another traditional product of the region, is also implicated in Sahelian desertification (Figure 6.11). Development agencies hoping to increase livestock production dug deep wells in areas previously unused by herders through most of the year. The new water supplies allowed year-round grazing in places that, over time, could not support it. Large barren circles around each new well began to appear even on satellite images.

Some Sahelian areas are seeing vegetation recovery thanks to simple actions taken by farmers, changes in government policy, and better rainfall. In the Sahelian portion of Niger, local agronomists have documented an unanticipated increase in tree cover over the past 35 years. More interesting still, increases in tree cover have occurred in some of the most densely populated rural areas. After the 1984 drought, farmers began to actively protect trees rather than clear them from their fields, including the nitrogen-fixing *goa* tree, which had disappeared from many villages. During the rainy season, the goa tree loses its leaves, so it does not compete with crops for water or sun. The leaves themselves fertilize the soil. Sahelian farmers also use branches, pods, and leaves from the trees for fuel and for animal fodder.

Until the 1990s, all trees were considered property of the state of Niger, thus giving farmers little incentive to protect them. Since then, the government has recognized the value of allowing individuals to own trees. Villages that protect their trees are much greener and more resilient during droughts than villages that do not. The Sahel is still poor and prone to drought; but as the case of Niger shows, relatively simple conservation practices can have a positive impact.

▼ **Figure 6.11 Sahel Goat Herder** A herder in the Ferlo region of Senegal tends his goats. Pastoralism is critical for survival in this part of the Sahel, where the dry season may last for nine months. Yet too many animals can lead to overgrazing and accelerated desertification.

Deforestation Although Sub-Saharan Africa still contains extensive forests, much of the region is either grasslands or deforested agricultural lands. Lush forests that once existed in places such as highland Ethiopia were long ago reduced to a few remnant patches. Throughout history, local populations have relied on such woodlands for their daily needs. Tropical savannas, which cover large portions of the region to the north and south of the tropical rainforest zone, are dotted with woodlands. For many in the region, savannah deforestation is of greater local concern than the commercial logging of the rainforest, due to the importance of **biofuels**—wood or wood-derived charcoal used for household energy needs, especially cooking—as the leading source of energy for many rural settlements. Loss of woody vegetation causes in extensive hardship, especially for women and children who must spend many hours a day looking for wood.

In some countries, village women have organized into community-based NGOs to plant trees and create green belts to meet ongoing fuel needs. One of the most successful efforts is the Green Belt Movement in Kenya, started by Wangari Maathai. The Green Belt Movement has 15,000 members, mostly women, who planted half a million trees in 2015 alone. Since the group's beginning in 1977, millions of trees have been planted. In Green Belt areas, village women now spend less time collecting fuel, and local environments have improved. Kenya's success has drawn interest from other African countries, spurring a Pan-African Green Belt Movement largely organized through NGOs interested in biofuel generation, conservation, and empowerment of women. In 2004, Maathai was awarded a Nobel Peace Prize for her contribution to sustainable development, democracy, and peace. She died in 2011, but the Green Belt Movement—now led by her daughter, Wanjira Maathai—remains a powerful force in the region (see *Working Toward Sustainability: Reforesting a Continent*).

Destruction of tropical rainforests through logging is most evident in the fringes of the Ituri (see Figure 6.10). Given its vast size and the relatively small number of inhabitants, however, the Ituri is less threatened than other forest areas. Two smaller rainforests—one along the Atlantic coast from Sierra Leone to western Ghana, and the other along the eastern coast of the island of Madagascar—have nearly disappeared due to commercial logging and agricultural clearing. Deforestation in Madagascar is especially worrisome because the island forms a unique environment with a large number of native species—most notably, the charismatic lemurs.

Energy Issues The people of Sub-Saharan Africa suffer from serious energy shortages; one of the justifications for the Great Renaissance Dam project is that, when completed, it will meet Ethiopia's electricity needs. At the same time, foreign investors are actively developing the region's oil and natural gas supplies, mostly for export. Many Sub-Saharan states extract oil and natural gas; major producers such as Nigeria and Angola are even members of OPEC (Figure 6.12). More recently, countries such as the Ivory Coast, Tanzania, and Mozambique have developed their natural gas reserves for domestic consumption and export. Yet for most Africans, wood and charcoal account for the majority of total energy production. Figure 6.12 shows the 20 states for which the World Bank estimates the percentage that biofuels (labeled as combustible renewables) contribute to national energy production. Even though Angola is a major oil producer, biofuels supply half of the country's energy, and biofuels account for 80 percent of oil-rich Nigeria's energy supply. More than 90 percent of that figure is in large countries such as Ethiopia and the Democratic Republic of the Congo. This is why energy production places tremendous strain on forests and vegetation, and why

WORKING TOWARD SUSTAINABILITY

Reforesting a Continent

A significant outcome of the 2015 Paris Agreement was a pledge by a dozen Sub-Saharan African nations to restore their natural forests, even though the developing region is the least responsible for greenhouse gas (GHG) emissions. The African Forest Landscape Restoration Initiative, or AFR100, is the first regional effort at forest restoration, with the goal of replanting 100 million hectares (about 386,000 square miles) of forest by 2030. The World Bank, the German government, and private funders committed $1.5 billion toward restoration, and the World Resources Institute, a U.S.-based NGO, will monitor the effort.

A New Green Belt Movement Reforestation has many benefits for rural Africa, including reduced soil erosion, improved soil fertility, and greater food security. Trees can slow desertification and supply animal feed and household cooking needs. If successful, such a project will increase biodiversity, create jobs, reduce food insecurity, and increase capacity for climate change resilience and adaptation. Growing trees also absorbs large amounts of carbon dioxide. Wanjira Mathai, president of the Green Belt Movement, supports this initiative: "I have seen restoration in communities both large and small across Africa.... Restoring landscapes will empower and enrich rural communities while providing downstream benefits to those in cities" (Figure 6.1.1).

Ethiopia, Kenya, Uganda, Madagascar, Burundi, and Rwanda have pledged millions of acres to the project. The Democratic Republic of the Congo alone has pledged 8 million hectares (20 million acres) to forest restoration. Sahelian states have also pledged to plant more trees to slow desertification.

▲ **Figure 6.1.1 Reforestation Efforts**   Women tend seedlings in a tree nursery in Meru, Kenya. Beginning with the Green Belt Movement, Kenyan women have played a leadership role in planting trees for fuel needs and ecological restoration.

Turning Talk into Action Countries pledge to implement these programs because they hope to get funding. In the worst cases, however, the promised funds go to government agencies and never reach the intended communities. Tree-planting programs have been tried in the region before, with mixed results. The difference this time is the scale of the AFR100 program and the financial support. Yet the challenges of having people on the ground with the training and the tools to do the work are real, and there must be local community buy-in to the benefits of the program—something that the Green Belt Movement has worked to build. Finally, any reforestation program, just like the seedlings it plants, takes years to mature.

The need for wood fuel, the demand for farmland, and the abundance of free-grazing animals put enormous pressure on the forests of Sub-Saharan Africa. At the same time, forests provide valuable ecosystem services such as reducing erosion and absorbing carbon. It is hoped that AFR100 can meet its goals, and that millions of hectares of forest can be restored.

1. **What services do tropical forests provide to rural residents? What do forests in your area offer to your community?**

2. **What strategies should be employed to make this program work?**

GOOGLE EARTH
Virtual Tour Video
http://goo.gl/jLlISM

so many countries are developing alternatives such as hydroelectricity. Another environmental concern for the majority who rely on biofuels is the woodsmoke that fills homes, causing respiratory problems.

Developing oil and natural gas reserves is not a sure path to economic development, and some have even called it a curse for Sub-Saharan Africa. Nigerian politicians and oil executives have prospered from oil revenue, but in the Niger Delta, where oil was first extracted over 50 years ago, many places lack roads, electricity, and schools. Moreover, careless and unregulated oil extraction has grossly degraded the delta ecosystem. As geographer Michael Watts has observed about the delta, oil has been "a dark tale of neglect and unremitting misery." Not all oil production leads to such misery, but Nigeria is a cautionary tale about the limits of oil's ability to foster development.

Wildlife Conservation   Sub-Saharan Africa is famed for its wildlife. No other world region has such abundance and diversity of large mammals. The survival of wildlife here reflects, to some extent, the historically low human population density and the fact that sleeping sickness

▶ **Figure 6.12** **Energy Production** Many Sub-Saharan African countries produce oil and natural gas. Two of the largest producers, Nigeria and Angola, are OPEC members. Yet in many states, most of the total energy comes from burning wood and agricultural waste.

Q: A few countries in this region are less dependent on biofuels. Why might that be?

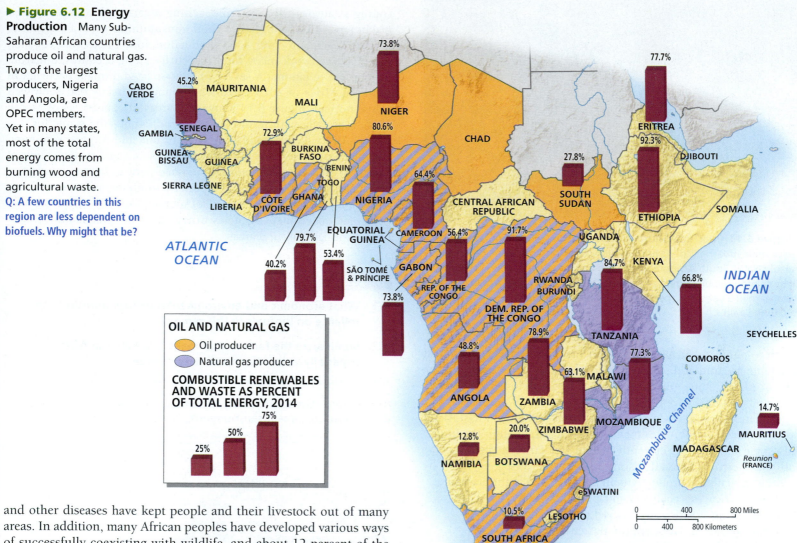

and other diseases have kept people and their livestock out of many areas. In addition, many African peoples have developed various ways of successfully coexisting with wildlife, and about 12 percent of the region is included in nationally protected lands.

However, as is true elsewhere in the world, wildlife is declining in much of the region. The most noted wildlife reserves are in East Africa (Kenya and Tanzania) and southern Africa (South Africa, Zimbabwe, Namibia, and Botswana). These reserves are vital for wildlife protection and are major tourist attractions. Wildlife reserves in southern Africa now seem to be the most secure, and in Zimbabwe, elephant populations are considered too large for the land to sustain. Yet throughout the region, population pressure, political instability, and poverty make the maintenance of large reserves difficult, even though many countries benefit from wildlife tourism.

In 1989, a worldwide ban on ivory trade was imposed as part of the Convention on International Trade in Endangered Species (CITES). Although several African states, such as Kenya, lobbied hard for the ban, others, including Zimbabwe, Namibia, and Botswana, complained that their herds were growing and the sale of ivory helped to fund conservation efforts. Conservationists feared that lifting the ban would bring a new wave of poaching and illegal trade, but in the late 1990s some southern African states were permitted to sell down their inventories of elephant ivory confiscated from poachers, and limited sales continued. The last legal auction of elephant ivory was in 2008; officials are reluctant to hold more auctions due to a corresponding increase in poaching.

Today the rhinoceros is especially threatened. The illegal market for rhino horn is lucrative, with most of the demand coming from Vietnam and China. Wildly valued for its questionable medicinal properties, ground rhino horn can fetch $65,000 per kilo in the black market. According to geographer Elizabeth Lunstrum, 80 percent of the region's rhinos are in South Africa, and half of those are in one place—Kruger National Park, which borders Mozambique. In 2008, rangers noted a spike in shot and dehorned rhinos, and by 2014, 1200 rhinos were reported killed in South Africa. The figures in 2017 were around 1000. Many poachers stage their attacks from Mozambique, making this an international as well as a domestic issue. South African officials are deploying drones and military personnel in an attempt to stop the slaughter. Even with these resources, protecting these endangered animals is difficult (Figure 6.13).

Climate Change and Vulnerability in Sub-Saharan Africa

Global climate change poses extreme risks for the region due to its poverty, recurrent droughts, and overdependence on rain-fed agriculture. Sub-Saharan Africa is the world's lowest emitter of greenhouse gases, but it is likely to experience greater-than-average human vulnerability to climate change because of the region's limited resources

▲ **Figure 6.13 Poaching South African Rhinos** The tragic remains of a black rhino, poached for its horn in Hluhluwe Umfolozi Game Reserve in Kwazulu Natal, South Africa, hunted with a high-powered rifle. The horn will be sold on the black market for more money than most people in the region earn in a year. This species of rhino is endangered, with less than 3,000 in the wild.

to both respond and adapt to environmental changes. The areas most vulnerable are arid and semiarid regions such as the Sahel and the Horn of Africa, some grasslands, and the coastal lowlands of West Africa and Angola.

Climate change models suggest that parts of highland East Africa and equatorial Central Africa may receive more rainfall in the future. Thus, some lands that are currently marginal for farming might become more productive. These effects are likely to be offset, however, by declining agricultural productivity in the Sahel and in the grasslands of southern Africa, especially in Zambia and Zimbabwe. Drier conditions in southern Africa are believed to be the reason for the death of baobab trees. These enormous trees can be as wide as a bus and live for over 2,000 years, yet researchers have documented a sudden die-off of the oldest and largest trees in the last decade. The impact of climate change was brought home to the 4 million residents of Cape Town, South Africa, in 2018 as water was severely rationed and even turned off in parts of the city due to dry reservoirs. Given the region's relatively high elevations, the negative consequences of a rising sea level would be felt mostly on the West African coast (Senegal, Gambia, Sierra Leone, Nigeria, Cameroon, and Gabon).

Even without the threat of climate change, famine stalks many areas of Africa. The Famine Early Warning Systems (FEWS) network monitors food insecurity throughout the developing world, but especially in Sub-Saharan Africa. **Food insecurity** measures both daily food intake and irreversible

▶ **Figure 6.14 Food Insecurity in Southern Africa** Anticipating areas of food insecurity, based upon the timing and amount of rainfall and changes in vegetation/crop cover, is the mission of the FEWS network. Since the late 1980s, FEWS has mapped areas of potential famine. The 2015–2016 drought in southern Africa was a major area of concern.

coping strategies—for example, selling assets such as livestock or machinery—that will lead to food consumption gaps. By tracking rainfall, vegetation cover, food production, food prices, and conflict, the FEWS network maps food insecurity along a continuum from food secure to famine. A severe El Niño–related drought occurred in southern Africa in 2015–2016 where rainfall already had been well below normal. Figure 6.14 shows the areas of concern as of March 2016, especially with regard to corn production—the staple for this part of Africa. While conditions in Tanzania and Angola were rated favorable, much of Zimbabwe, eSwatini, Lesotho, and even portions of South Africa and Mozambique were rated poor to failure in terms of basic food production. This is especially troubling as these areas are important production zones for all of southern Africa. Corn prices have risen, and government agencies worry about possible famine conditions, especially in rural areas.

REVIEW

6.1 What economic and environmental factors contribute to reliance on biofuels in this region?

6.2 Summarize the factors that make Sub-Saharan Africans especially vulnerable to climate change.

KEY TERMS Great Rift Valley, Great Escarpment, Horn of Africa, desertification, Sahel, transhumance, biofuels, food insecurity

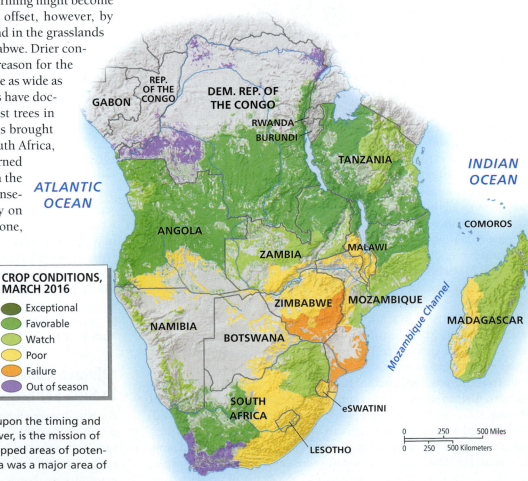

CROP CONDITIONS, MARCH 2016
- Exceptional
- Favorable
- Watch
- Poor
- Failure
- Out of season

Population and Settlement: Young and Restless

Sub-Saharan Africa's population is growing quickly. By 2050, the population is expected to reach 2 billion, double the current number. It is also a young population, with 41 percent of the people under age 15, compared to just 16 percent for more developed countries. Only 3 percent of the region's people are older than age 65. Families tend to be large, with the average woman having five children (Table 6.1). However, child and maternal mortality rates are also high, although child mortality rates have declined dramatically in the past two decades. The most troubling indicator is the region's low life expectancy, which dropped to 50 years in 2008 (in part due to the AIDS epidemic) and is currently estimated at 58 years for men and 62 for women. Life expectancy in developing nations such as India and China is much longer. Urban growth is also a major trend in this world region. In 1980, an estimated 23 percent of the population lived in cities; now the figure is 39 percent (Figure 6.15).

Behind these demographic facts lie complex differences in settlement patterns, livelihoods, belief systems, and health care access. Although the region is seeing rapid population growth, Sub-Saharan Africa is not densely populated. The entire region has about 1 billion people, whereas South Asia has almost twice as many people in a far smaller area. Just six states account for over half the region's population: Nigeria, Ethiopia, the Democratic Republic of the Congo, South Africa, Tanzania, and Kenya (see Table 6.1). Some states (such as Rwanda and Mauritius) have very high population densities, whereas others are sparsely settled (Namibia and Botswana). However, population density is not correlated with overall development. Mauritius is a densely settled island nation that is well governed and relatively prosperous, and the same could be said for the arid, sparsely settled state of Botswana. Population density, however, does provide an indicator of the relative population pressures of states in the region, so it is included in Table 6.1.

Demographic Trends

One positive change in the region's demographic profile is falling child mortality due to greater access to primary health care and new disease prevention efforts. Gone are the days when 1 child in 5 did not live past his or her fifth birthday; today that figure is closer to 1 in 10—still high by world standards but a considerable improvement. Also, life expectancy figures bottomed out in the 2000s due to the devastating impact of HIV/AIDS and are now on the rise. Finally, like people in other world regions, Africans are moving to cities, leading to fertility declines. Family size in South Africa, one of the more urbanized large countries, is half the regional average.

Figure 6.16 compares the population pyramids of Ethiopia and South Africa. Ethiopia exhibits the classic broad-based pyramid of a demographically growing and youthful country. Most Ethiopian women have four or five children, and the rate of natural increase is 2.6 percent. The numbers of men and women are nearly even, and only 4 percent of the people are over 65 (life expectancy is 64 for men and 67 for women). In contrast, the South African population pyramid tapers down, reflecting the country's smaller family size (the average woman has two or three children). One unusual aspect of this graphic is the smaller number of women in their 30s and early 40s compared to men the same age, due to the disproportionate impact of AIDS on women in Africa. South Africa has more people over the age of 65 (6 percent), but its life expectancy is 61 for men and 67 for women.

This, too, is attributable to the AIDS epidemic, which hit southern Africa with deadly force in the 1990s. Thankfully, infection rates are on the decline in many states, but still high by world standards.

Family Size   A continued preference for large families is the basis for the region's demographic growth. In the 1960s, many areas in the developing world had total fertility rates (TFRs) of 5.0 or higher. Today Sub-Saharan Africa, at 5.0, is the only region with such a high TFR. A combination of cultural practices, rural lifestyles, child mortality, and economic realities encourages large families (Figure 6.17). Yet average family sizes are coming down; as recently as 1996, the regional TFR was 6.0.

Throughout the region, large families guarantee a family's lineage and status. Even now, most women marry young, typically as teenagers, maximizing their childbearing years. Demographers often point to the limited formal education available to women as another factor contributing to high fertility. Religious affiliation has little bearing on the region's fertility rates; Muslim, Christian, and animist communities all have similarly high birth rates.

The everyday realities of rural life make large families an asset. Children are an important source of labor; from tending crops and livestock to gathering wood, they add more to the household economy than they take. Children are also seen as social security: When parents' health falters, they expect their grown children to care for them.

National policies began shifting in the 1980s. For the first time, government officials argued that smaller families and slower population growth were needed for social and economic development. Other factors are bringing down the rate of natural increase. As African states slowly become more urban, there is a corresponding decline in family size—a pattern seen throughout the world. Tragically, declines in natural increase were also occurring as a result of AIDS.

The Disease Factor: Malaria, HIV/AIDS, and Ebola

Historically, the hazards of malaria and other tropical diseases such as sleeping sickness limited European settlement in the tropical portions of Sub-Saharan Africa. It was only in the 1850s, when European doctors discovered that a daily dose of quinine could protect against malaria, that the balance of power in Africa radically shifted. Explorers immediately began to penetrate the interior of the continent, while merchants and expeditionary forces moved inland from the coast. The first imperial claims soon followed, culminating in colonial division of Africa in the 1870s (discussed later in this chapter).

Malaria   A scourge in this region for centuries, malaria is transmitted from infected individuals to others via the anopheles mosquito. Malaria causes high fever, severe headache, and, in the worst cases, death. The World Health Organization (WHO) estimates that 200 million people contract malaria each year, resulting in some 400,000 deaths, mostly young children. The majority of infections and deaths occur in Sub-Saharan Africa. Since 2000, African governments, NGOs, and foreign aid sources have increased spending to reduce the threat of infection. Presently, a malaria vaccine does not exist, but research in this area is promising. Medication helps in many cases but is not reliable over the long term, and insecticide use has led to mosquito resistance to the chemicals. The most effective tool to reduce infection has been the distribution of insecticide-treated mosquito nets to millions of African homes. That, along with rapid diagnostic tests and access to medication once infected, has cut infections and related deaths by 30 percent since 2010.

TABLE 6.1   Population Indicators

Explore these data in MapMaster 2.0 https://goo.gl/Xp3CBi

Country	Population (millions) 2018	Population Density (per square kilometer)[1]	Rate of Natural Increase (RNI)	Total Fertility Rate	Life Expectancy Male	Life Expectancy Female	Percent Urban	Percent <15	Percent > 65	Net Migration (rate per 1000)
Western Africa										
Benin	11.5	99	2.8	5.0	59	62	45	43	3	0
Burkina Faso	20.3	70	3.1	5.5	60	61	29	47	3	−1
Cape Verde	0.6	136	1.4	2.2	71	75	66	31	4	−3
Gambia	2.2	208	3.1	5.4	60	63	61	46	2	−1
Ghana	29.5	127	2.2	3.9	62	64	56	39	3	−1
Guinea	11.9	52	2.8	4.8	58	61	35	45	4	−2
Guinea-Bissau	1.9	66	2.6	4.6	56	59	43	42	3	−1
Ivory Coast	24.9	76	2.3	4.6	52	55	51	43	3	0
Liberia	4.9	49	2.5	4.2	62	64	51	42	3	−1
Mali	19.4	15	3.5	6.0	58	62	42	47	2	−4
Mauritania	4.5	4	2.6	4.6	62	65	54	40	3	1
Niger	22.2	17	3.8	7.2	59	61	16	50	3	0
Nigeria	195.9	210	2.6	5.5	53	54	50	44	3	0
Senegal	16.3	82	2.7	4.6	65	69	46	43	3	−1
Sierra Leone	7.7	105	2.0	4.2	51	52	41	42	3	−1
Togo	8.0	143	2.5	5.5	59	61	42	42	3	0
Eastern Africa										
Burundi	11.8	423	3.0	5.5	59	62	13	46	2	0
Comoros	0.8	437	2.5	4.3	62	65	29	40	3	−2
Djibouti	1.0	41	1.5	2.9	61	64	78	32	4	1
Eritrea	6.0	44	2.3	4.1	62	68	40	41	4	−15
Ethiopia	107.5	105	2.6	4.4	64	67	20	41	4	0
Kenya	51.0	87	2.6	3.9	65	69	32	41	3	0
Madagascar	26.3	44	2.6	4.1	64	68	37	41	3	0
Mauritius	1.3	623	0.3	1.4	71	78	41	19	11	−1
Rwanda	12.6	495	2.6	4.2	65	69	19	40	3	−1
Seychelles	0.1	208	1.0	2.4	68	78	57	22	8	−3
Somalia	15.2	24	3.2	6.3	55	58	45	47	3	−3
South Sudan	13.0	–	2.6	5.0	56	58	20	42	3	−4
Tanzania	59.1	65	3.3	5.2	64	67	34	45	3	−1
Uganda	44.1	214	3.2	5.4	62	64	24	48	3	−1
Central Africa										
Cameroon	25.6	51	2.6	4.7	57	60	56	43	3	0
Central African Republic	4.7	7	2.2	4.9	50	54	41	44	4	−10
Chad	15.4	12	3.1	6.4	52	54	23	47	2	1
Dem. Rep. of Congo (Kinshasa)	84.3	36	3.3	6.3	58	61	45	46	3	0
Rep of. Congo (Brazzaville)	5.4	15	2.3	4.4	58	61	67	42	3	−5
Equatorial Guinea	1.3	45	2.4	4.7	56	59	71	37	3	14
Gabon	2.1	8	2.2	3.8	65	68	89	36	4	3
São Tomé and Principe	0.2	213	2.6	4.4	65	69	73	42	4	−5
Southern Africa										
Angola	30.4	24	3.5	6.2	58	62	63	48	2	0
Botswana	2.2	4	1.2	2.6	61	65	70	32	5	3
eSwatini	1.4	79	1.9	3.3	54	61	24	37	3	−1
Lesotho	2.3	74	1.6	3.3	52	56	28	36	5	−2
Malawi	19.1	198	2.6	4.2	61	55	17	44	3	−1
Mozambique	30.5	38	2.9	5.3	56	60	32	45	3	0
Namibia	2.5	3	2.0	3.4	62	65	48	37	4	0
South Africa	57.7	47	1.2	2.4	61	67	66	30	6	4
Zambia	17.7	23	3.1	5.2	59	64	44	45	42	0
Zimbabwe	14.0	43	2.4	4.0	58	62	32	39	4	−10

Source: Population Reference Bureau, *World Population Data Sheet*, 2018.
[1] World Bank Open Data, 2018.

Log in to Mastering Geography & access MapMaster to explore these data!

1) Which countries have the highest RNI? What is the relationship between RNI and TFR? The doubling of a population in years can be calculated using the formula 70/RNI. Use this fraction to determine the countries with the shortest doubling time.

2) Which five countries have the highest population density in this region? Which countries have the lowest? What characteristics does each group have in common?

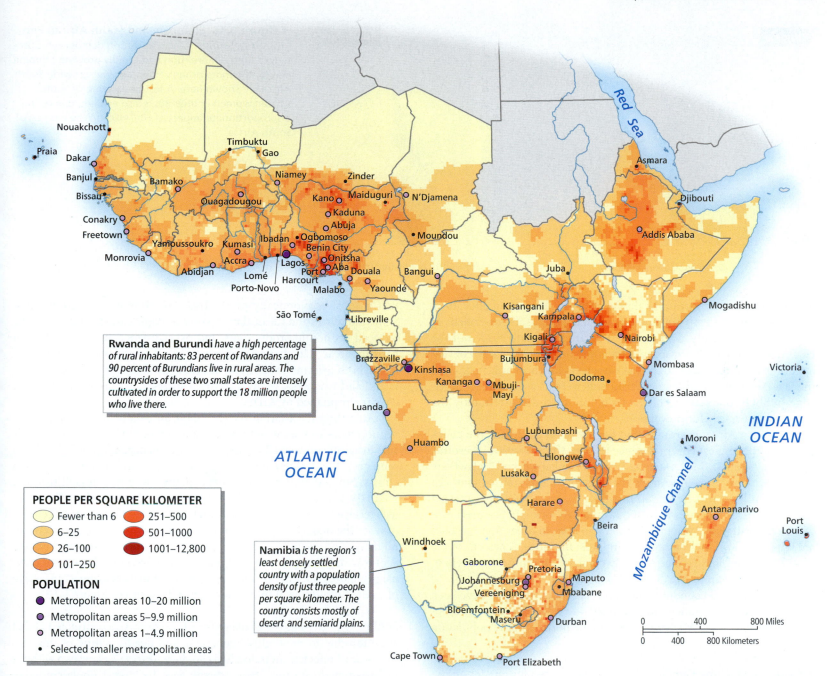

Rwanda and Burundi have a high percentage of rural inhabitants: 83 percent of Rwandans and 90 percent of Burundians live in rural areas. The countrysides of these two small states are intensely cultivated in order to support the 18 million people who live there.

Namibia is the region's least densely settled country with a population density of just three people per square kilometer. The country consists mostly of desert and semiarid plains.

PEOPLE PER SQUARE KILOMETER

- Fewer than 6
- 6–25
- 26–100
- 101–250
- 251–500
- 501–1000
- 1001–12,800

POPULATION

- Metropolitan areas 10–20 million
- Metropolitan areas 5–9.9 million
- Metropolitan areas 1–4.9 million
- Selected smaller metropolitan areas

▲ **Figure 6.15 Population of Sub-Saharan Africa**   The majority of the region's 1 billion people live in rural areas. However, some of these rural zones—such as West Africa and the East African highlands—are densely settled. Major urban centers, especially in South Africa and Nigeria, support millions. Lagos, Nigeria, is the one megacity in the region, but more than three dozen cities have more than 1 million residents. **Q: What factors contributes to the extremely low population density in the southwest corner of the continent?**

Malaria and poverty are closely related in Sub-Saharan Africa, with many of the poorest tropical countries experiencing higher infection rates. West and Central Africa are the areas hardest hit. The Democratic Republic of the Congo and Nigeria, along with India, account for 40 percent of worldwide malaria cases, and 40 percent of malaria deaths occur in these two African nations.

HIV/AIDS   Now in its fourth decade, HIV/AIDS has been one of the deadliest epidemics in modern human history, yet it is beginning to be

tamed. This is especially welcome news for Sub-Saharan Africa, home to 70 percent of the 37 million people living with HIV/AIDS. Human immunodeficiency virus (HIV) is the virus that can lead to acquired immunodeficiency syndrome (AIDS). The human body cannot get rid of HIV, but antiretroviral drugs can suppress it and keep it from becoming AIDS. As of 2016, nearly 20 million people were receiving life-prolonging antiretroviral drugs, with the goal of reaching 30 million people by 2020. Since first identified, AIDS has infected nearly 80 million people and killed 35 million.

The HIV/AIDS virus is thought to have originated in the forests of the Congo, possibly crossing over from chimpanzees to humans sometime in the late 1950s. Yet it was not until the 1980s that the impact of the disease was first felt. In Sub-Saharan Africa, as in much of the developing world, the virus is transmitted mostly by unprotected heterosexual activity or from mother to child during the birth process or through breastfeeding. A long-standing pattern of seasonal male labor

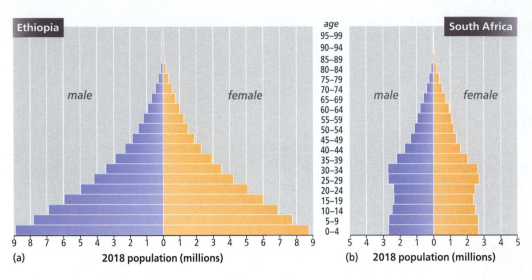

migration helped to spread the disease. So, too, did lack of education, inadequate testing early on in the epidemic, and disempowerment of women. Consequently, women bear a disproportionate burden of the HIV/AIDS epidemic. They account for approximately 60 percent of HIV infections, and they are usually the caregivers for those infected. Until the late 1990s, many African governments were unwilling to acknowledge publicly the severity of the situation or to discuss frankly the measures necessary for prevention.

Southern Africa is ground zero for the AIDS epidemic; the countries with the highest HIV prevalence (South Africa, eSwatini, Lesotho, Botswana, Namibia, Mozambique, Zambia, Zimbabwe, and Malawi) are all located there. In South Africa, the most populous state in southern Africa, 7 million (nearly one in five people ages 15–49) are infected with HIV/AIDS. The rates of infection in neighboring Botswana, Lesotho, and eSwatini are even higher. Infection rates in other African states are lower, but still high by world standards. In Kenya, 5 percent of the 15–49 age group has HIV or AIDS. By comparison, only 0.6 percent of the same age group in North America is infected.

The social and economic implications of this epidemic have been profound. Life expectancy rates tumbled—in a few places even dropping to the early 40s. AIDS typically hits the portion of the population that is most economically productive. The time lost to care for sick family members and the outlay of workers' compensation benefits have reduced economic productivity and overwhelmed public services in hard-hit areas. The disease makes no class distinctions: Countries are losing both peasant farmers and educated professionals (doctors, engineers, and teachers). Many countries struggle to care for millions of children orphaned by AIDS.

After three devastating decades, there are finally hopeful signs. Prevention measures are widely taught, and treatment with a mix of available drugs means that HIV infection is now manageable and not a death sentence. Due to international financial support and national outreach efforts, Sub-Saharan Africa now has many more health facilities that offer HIV testing and counseling. More than half of the infected individuals receive life-prolonging drugs. Prevention services in prenatal clinics provide the majority of pregnant HIV-positive women with antiretroviral drugs to prevent transmission of the virus to their babies. Political activism and changes in sexual practices—driven by educational campaigns, more condom use, and higher rates of male circumcision—have prevented hundreds of thousands of new HIV cases (Figure 6.18).

Ebola The 2014–2015 Ebola outbreak in West Africa attracted global attention due to the highly contagious and deadly nature of this disease—once infected, an individual can die within a week or two without intensive medical care. This outbreak was the largest Ebola epidemic in history, affecting multiple countries in West Africa and resulting in some 30,000 confirmed cases. A rare disease, Ebola was first identified in 1976 along the Ebola River in the Democratic Republic of the Congo. There had been isolated outbreaks before, but the 2014 outbreak was by far the largest and deadliest (Figure 6.19).

International organizations such as Médecins Sans Frontières (Doctors Without Borders) and various government agencies contributed money, medical personnel, and equipment to fight this epidemic, fearing that it could spread and potentially infect millions. This often meant quarantining infected people, developing stringent procedures for health care personnel, conducting education efforts, and abandoning traditional burial practices to reduce the spread of infection. International cooperation and the efforts on the ground in Sierra Leone, Liberia, and Guinea worked in the 2014–2015 outbreak, and worst-case scenarios were not realized. By 2018, when a new outbreak of Ebola occurred in the Democratic Republic of the Congo, the WHO was able to use a new vaccine, which offered hope and help in controlling the outbreak.

▼ **Figure 6.17** **Large Families** The average total fertility rate for Africa south of the Sahara is five children per woman. A Ugandan family with four children stands in front of their rural home.

▲ **Figure 6.18 Combating HIV/AIDS in South Africa** To address high infection rates, southern African countries have created explicit campaigns to educate the public on the need to use condoms to prevent sexually transmitted diseases such as AIDS. This billboard is in Johannesburg, South Africa's largest city.

Patterns of Settlement and Land Use

Because of the dominance of rural settlements in Sub-Saharan Africa, people are widely scattered throughout the region (see Figure 6.15). Population concentrations are highest in West Africa, highland East Africa, and the eastern half of South Africa. The first two areas have

▼ **Figure 6.19 Ebola Outbreak in West Africa** Liberia, Sierra Leone, and Guinea were the states most impacted by the recent Ebola outbreak. In Liberia, cases were found throughout the country but especially in the capital, Monrovia. Sierra Leone's cases were also more concentrated near Freetown and Kenema. Guinea, which had fewer deaths, had a concentration of rural cases near the Liberia border.

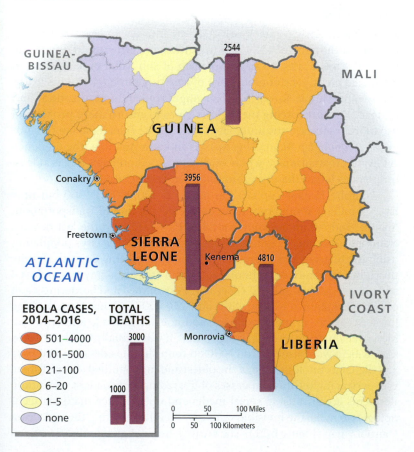

some of the region's best soils, and native systems of permanent agriculture developed there. In South Africa, the more densely settled east is a result of an urbanized economy based on mining as well as the forced concentration of black South Africans into eastern homelands.

As more Africans move to cities, settlement patterns are becoming more concentrated. Towns that were once small administrative centers for colonial elites have grown into major cities. The region even has its own megacity, Lagos, estimated at 13–17 million residents. Cities throughout the continent are growing faster than rural areas. Before examining the Sub-Saharan urban scene, however, a more detailed discussion of rural subsistence is needed.

Agricultural Subsistence and Foodways The staple crops over most of Sub-Saharan Africa are millet, sorghum, and corn (maize) as well as various tubers and root crops such as yams. Irrigated rice is widely grown in West Africa and Madagascar. Geographer Judith Carney, in her book *Black Rice*, documents how enslaved African introduced rice cultivation to the Americas. Corn, in contrast, was introduced to Africa from the Americas through the slave trade and quickly became a staple. In the higher elevations of Ethiopia and South Africa, wheat and barley are grown. Intermixed with subsistence foods are a variety of export crops—coffee, tea, rubber, bananas, cacao, cotton, and peanuts—grown in distinct ecological zones, often in some of the best soils.

In areas that support high annual crop yields, population densities are greater. In parts of humid West Africa, for example, the yam became the dominant subsistence crop. The mastery of yam production in earthen mounds throughout West Africa, and especially Nigeria, enhanced food supplies so that people could live in denser permanent settlements (Figure 6.20). Much of traditional Igbo culture is tied to the demanding tasks of clearing the fields, tending the delicate plants, and celebrating the harvest.

Over much of Sub-Saharan Africa, however, agriculture remains relatively unproductive, and population densities tend to be low. On poorer tropical soils, farming usually entails shifting cultivation (or **swidden**). This process involves burning the natural vegetation to release fertilizing ash and then planting crops such as maize, beans, sweet potatoes, bananas, papayas, manioc, yams, melons, and squash. Each plot is temporarily abandoned once its source of nutrients has been exhausted. Swidden cultivation is often a very finely tuned adaptation to local environmental conditions, but it cannot support high population densities. Women are often the subsistence farmers of the region, producing for their household needs as well as for local markets.

Export Agriculture Agricultural exports, whether from large estates or small producers, are critical to the economies of many Sub-Saharan states. If African countries are to import the modern goods and energy resources they require, they must sell their own products on the world market. Because the region has few competitive industries, the bulk of its exports are primary products derived from farming, mining, and forestry.

In densely settled Rwanda, most farms are small, but the highland volcanic soil is ideal for growing coffee. However, for decades, the country's coffee production languished, and farmers earned very little for their low-quality beans. Yet across Rwanda's hillsides were older varieties of coffee plants with high value in today's premium coffee market. Farmers just needed a better way to prepare and market the beans. In an effort to rebuild the country after the ethnic genocide in the 1990s, the government targeted improving the quality of Rwandan coffee by forming cooperatives that would bring ethnic groups together and raise

▲ **Figure 6.20 Harvesting Yams in Nigeria** Farmers harvest yams planted in traditional earthen mounds in the village of Barangoni, Nigeria. As Nigeria's population continues to grow, various NGOs support agricultural outreach programs to improve yam yields.

Explore the TASTES of Yams
http://goo.gl/SweKlx

farmers' incomes. Through community-run cooperatives, the quality of washed and sorted beans improved. So, too, did the farmers' ability to bargain with major buyers such as Starbucks and Green Mountain, leaving out the middleman. For many small farmers in Rwanda, their premium coffee is now a source of pride. The cooperatives have taken off, and farmers, many of them war widows, have seen their incomes rise.

Several African countries rely heavily on one or two export crops. Coffee, for example, is vital for Ethiopia, Kenya, Rwanda, Burundi, and Tanzania. Peanuts have historically been the primary source of income in the Sahel, whereas cotton is tremendously important for the Central African Republic and South Sudan. Ghana and the Ivory Coast have long been the world's main suppliers of cacao for chocolate. Liberia produces plantation rubber, and many farmers in Nigeria specialize in palm oil. The export of such products brings good money when commodity prices are high, but when prices collapse, as they periodically do, economic devastation may follow (see *Globalization in Our Lives: West Africa's Chocolate Fix*).

Nontraditional agricultural exports that depend on significant capital inputs and refrigerated air transport have emerged in past 20 years. One such industry is floriculture for the plant and cut flower industry. Here the highland tropical climates of Kenya, Ethiopia, and South Africa is advantageous. After Colombia in South America, Kenya is the world's largest exporter of cut flowers, with most of its exports going to Europe. Similarly, the European market for fresh vegetables and fruits in the winter is being met by some producers in West and East Africa.

Pastoralism Animal husbandry is extremely important in Sub-Saharan Africa, particularly in semiarid zones. Camels and goats are the principal livestock of the Sahel and the Horn of Africa, but cattle are more common farther south. Many African peoples have traditionally specialized in cattle raising and are often tied in mutually beneficial relationships with neighboring farmers. Such **pastoralists** typically graze their stock on the stubble of harvested fields during the dry season and then move them to drier uncultivated areas during the wet season, when the pastures turn green. Farmers thus have their fields fertilized by the manure of the pastoralists' stock, while the pastoralists find good dry-season grazing. At the same time, these nomads can trade their animal products for grain and other goods of the sedentary world. Several pastoral peoples of East Africa, particularly the Masai of the Tanzanian–Kenyan borderlands, are noted for their extreme reliance on cattle and general (but never complete) independence from agriculture.

Large expanses of Sub-Saharan Africa have been off-limits to cattle because of infestations of **tsetse flies**, which spread sleeping sickness to cattle, humans, and some wildlife. Environments containing brush or woodland necessary for tsetse fly survival can support large numbers of wild animals immune to sleeping sickness, but vulnerable cattle cannot thrive. Tsetse fly eradication programs are currently reducing the threat, and cattle raising is spreading into once-restricted areas. This benefits African herders but may threaten the continued survival of many wild animals. When people and their livestock move into new areas in large numbers, wildlife almost inevitably declines.

Urban Life

Sub-Saharan Africa and South Asia are the least urbanized world regions, although in both regions cities are growing at twice the national growth rates. Nearly 40 percent of Sub-Saharan people live in urbanized areas. One consequence of this surge in city living is urban sprawl and slums. Rural-to-urban migration, industrialization, and refugee flows are forcing the region's cities to absorb more people and use more resources, straining city services (see *Humanitarian Geographies: Cleaning Up Akure, Nigeria*). As in Latin America, the tendency is toward urban primacy, where one major city dominates and is at least three times larger than the country's next largest city.

Nairobi, Kenya's capital, was a city of 250,000 in 1960 but now has 3.5 million residents and is considered the hub of transportation, finance, and communication for all of East Africa. In the last decade, Nairobi has become a high-tech superstar, with half the city's population using the Internet and many start-up companies being created. Despite this robust embrace of technology, Nairobi still has many unemployed and impoverished residents. Close to the downtown and bordering a golf course is the slum of Kibera, with some 250,000 residents. Here, garbage lines the streets, crime is rampant, and housing is crude and crowded. Moreover, on many municipal maps, Kibera was virtually blank. In 2009, Map Kibera was formed as an NGO to train slum residents in mapping and digital technology. One of the oldest and most studied slums, Kibera is experiencing the dual processes of upgrading and resident relocation. Residents are also using social media and open-source mapping technologies to make the invisible visible by literally inserting their place on Nairobi's map (Figure 6.21). The creative use of digital tools and citizen

GLOBALIZATION IN OUR LIVES

West Africa's Chocolate Fix

People love chocolate, and many eat it every day. Chocolate comes from the cacao bean, first domesticated in Central America but now grown throughout the humid tropics. West African countries produce more than half of the world's cacao beans, so when you bite into a chocolate bar, you taste the flavors of West Africa.

Large yellow pods form on the trunk and branches of cacao trees; mature pods are harvested and cracked open, and the sticky beans are removed. The beans are then washed, dried, and sorted, often by hand (Figure 6.2.1). Ivory Coast, Ghana, Nigeria, and Cameroon produced over 3 million tons of cacao beans in 2016— 60 percent of global production. North American and European companies buy the beans, which are then roasted and ground to produce cocoa powder.

Cacao has been a valuable West African export for a century. The time-consuming tasks of tending the trees and processing the beans does not create wealth but generates income and improves subsistence. The value added occurs further up the commodity chain where multinational companies such as Mars, Nestlé, Hershey, and Cadbury add sugar and milk to the cocoa powder, extracting the delicate flavor while offsetting the natural bitterness, and then adding this to candy bars.

▲ **Figure 6.2.1 Cocoa Beans in Ivory Coast** Women sort cacao beans in Duekoue, Ivory Coast. In the background is a public notice against the use of child labor. Cocoa, used for chocolate, is a major export of this West African country.

Several cacao-growing nations have tried to enter the candy business but with limited impact, as multinational corporations tightly control retail markets. Niche Cocoa, a Ghanaian brand, is trying to change this cycle by producing cocoa powder, cocoa butter, and chocolate bars under the Niche label in a modern factory outside of Accra. This high-end chocolate has found its way into West African markets, and Niche hopes it will become a sweet success abroad.

1. **Why is West Africa an important region for cocoa production?**

2. **How much chocolate do you consume, and where does it come from?**

---

reporting and training allows the issues of slum dwellers to be recorded and widely shared. Logistically, Kibera and other mapped slums also receive better services, as police and firefighters now know where to go when needed.

European colonialism greatly influenced urban form and development in the region. Although a very small percentage of the population lived in cities, Africans did have an urban tradition prior to the colonial era. Ancient cities, such as Axum in Ethiopia, thrived 2000 years ago, while in the Sahel, major trans-Saharan trade centers such as Tombouctou (Timbuktu) and Gao have existed for more than a millennium. In East Africa, an urban trading culture rooted

▶ **Figure 6.21 Community Mapping in Kibera** A map of Kibera produced by its residents through Open Street Map, showing clinics and places of worship along with pathways through the neighborhood. Before the Map Kibera project, this huge slum was a blank spot on the city map.

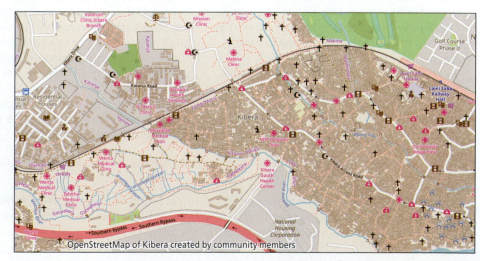

OpenStreetMap of Kibera created by community members

Cleaning up Akure, Nigeria

Garbage is a major environmental problem for the rapidly growing cities of Sub-Saharan Africa. When Temidayo Isaiah Oniosun moved to Akure, a city of 550,000 residents in Nigeria's Ondo state, he was alarmed at the vast piles of trash illegally dumped throughout the city. The garbage was not only unsightly, but also a potential source of water- and airborne disease. Temidayo, who studied Meteorology and majored in remote sensing and GIS at the Federal University of Technology, Akure (FUTA), turned to his colleagues in Youth-Mappers, an international organization of college students who collect and map data needed for disaster response or community development. "One of the things YouthMappers promotes is getting students to come up with innovative ideas for development issues using GIS," he explains. "I thought we could use mapping to solve this problem."

Mapping Service Needs Using satellite imagery, the students created a base map of Akure, then hit the streets on bicycles and motorbikes to locate illegal dumping sites, take photos, and record their locations using GPS (Figure 6.3.1). Temidayo recalls that initially, the students were just mapping the sites, "but we thought, instead of just analyzing the problem, why don't we create a solution." The students then proposed new disposal sites away from residential areas or water supplies, and took their plan to the Ondo State Ministry of the Environment to inform state-level planning. The FUTA YouthMappers also worked with area NGOs to address urban waste and to influence legislation banning illegal dumping in Akure. Temidayo hopes this will eventually lead to a similar project in Lagos. "What we did in Akure was like a pilot," he notes. "Let's solve this problem in other cities … at the end of the project, we want to have a cleaner country."

Formed in 2015, YouthMappers (http://www.youthmappers.org/) is funded by a grant from the U.S. Agency for International Development (USAID) and trains students to use open source geospatial data from platforms such as OpenStreetMap (https://www.openstreetmap.org) to solve problems and foster sustainable development. As of late 2018, the network has more than 140 chapters in over 42 countries; many are in geography departments in Sub-Saharan Africa, with Uganda boasting the most Youth-Mappers chapters. Student-run chapters are also active at U.S. universities, and members collaborate with mappers all over the world on international projects.

New Directions for GIS Temidayo chose to study remote sensing and GIS because he "was fascinated by the amount of problems you can solve with geospatial technologies … immediate problems in my society." In addition to extending the waste mapping project, his current interest is artificial intelligence (AI) applications in disaster monitoring. AI is being used for spatial temporal analysis and for Land Use Land Cover (LULC) classifications, using patterns instead of pixels and objects. "There are lots of ways to incorporate AI into remote sensing and GIS, and that excites me," he explains. "I'm already thinking of ways to solve problems and improve the technologies."

▲ **Figure 6.3.1 Locating Dumping Sites** Nigerian geographer Temidayo Oniosun collected data on illegal dumping sites in Akure using the KoBoToolbox software.

1. **What approach did the students take to solve the urban waste problem?**

2. **Is there a YouthMappers organization at your university? If so, what projects are members doing?**

GOOGLE EARTH Virtual Tour Video https://goo.gl/5pCwFJ

in Islam and the Swahili language emerged. West Africa, however, had the most developed precolonial urban network, reflecting both native and Islamic traditions. It also supports some of the region's largest cities today.

West African Urban Traditions The West African coastline is dotted with cities, from Dakar, Senegal, in the far north to Lagos, Nigeria, in the east. Half of Nigerians live in cities, and in 2018 the country had eight metropolitan areas with populations of more than 1 million. Historically,

▲ **Figure 6.23** **Elite Accra Neighborhood** The Legon area east of downtown houses Accra's nicest homes, private schools, and the University of Ghana.

the Yoruba cities in southwestern Nigeria have been the best documented. Developed in the 12th century, cities such as Ibadan were walled and gated, with a palace encircled by large rectangular courtyards at its core. An important center of trade for an extensive area, Ibadan was also a religious and political center. Lagos was another Yoruba settlement. Founded on a coastal island on the Bight of Benin, most of the modern city has spread onto the nearby mainland (Figure 6.22). Its coastal setting and natural harbor made this relatively small native city attractive to colonial powers. When the British took control in the mid-19th century, the city grew in size and importance. Following Nigerian independence in 1960, Lagos was a city of 1 million; today it is Sub-Saharan Africa's largest city with a metropolitan area of over 15 million people.

Most West African cities are hybrids—combining Islamic, European, and national elements such as mosques, Victorian architecture, and streets named after independence leaders. Accra, the capital of Ghana, is home to nearly 2.6 million people. Originally settled by the Ga people in the 16th century, it became a British colonial administrative center by the late 1800s. The modern city is being transformed through neoliberal policies introduced in the 1980s to attract international corporations. Like other cities in the region, Accra is rapidly changing due to an influx of foreign capital. Increased foreign investment in financial and producer services led to the creation of a "Global Central Business District (CBD)" on the city's east side, away from the "National CBD." Here foreign companies cluster in areas with secure land title, new roads, parking, and airport access. Upper-income gated communities have also formed near the Global CBD—with names such as Trasacco Valley, Airport Hills, Buena Vista, and Legon (Figure 6.23). The result is highly segregated urban spaces reflecting a global phase in urban development in which world market forces, rather than colonial or national ones, drive the change.

▼ **Figure 6.22** **Lagos, Nigeria** This satellite image shows sprawling Lagos, the largest city in Nigeria and Sub-Saharan Africa. The city's central market, Oshodi, is marked on the image to provide a sense of scale.

Explore the SIGHTS of Oshodi Market in Lagos http://goo.gl/gmcUzC

Oshodi Market

Urban Industrial South Africa

The major cities of southern Africa, unlike those of West Africa, are colonial in origin. Most of these cities, such as Lusaka, Zambia, and Harare, Zimbabwe, grew as administrative or mining centers. South Africa is one of the most urbanized states in the entire region, and it is certainly the most industrialized. The foundations of South Africa's urban economy rest largely on its incredibly rich mineral resources (diamonds, gold, chromium, platinum, tin, uranium, coal, iron ore, and manganese). Eight of its metropolitan areas have more than 1 million people; the largest are Johannesburg, Durban, and Cape Town.

The form of South African cities continues to be imprinted by the legacy of **apartheid**, an official policy of racial segregation that shaped social relations in South Africa for nearly 50 years. Even though apartheid was abolished in 1994, it is still evident in the landscape. Under apartheid laws, South African cities were divided into residential areas according to racial categories: white, **coloured** (a South African term describing people of mixed African and European ancestry), Indian (South Asian), and African (black). Whites occupied the largest and most desirable portions of the city. Blacks were crowded into the least desired areas, called **townships**, such as Soweto outside of Johannesburg and Gugulethu outside of Cape Town. Today blacks, coloureds, and Indians can legally live anywhere they want, but economic differences between racial groups, as well as deep-rooted animosity, hinder residential integration.

Johannesburg is the African metropolitan area that most consciously aspires to global city status, a dream underscored by its hosting of soccer's World Cup in 2010. However, as apartheid ended, Johannesburg became infamous for its high crime rate. Many white-owned businesses fled the CBD for the northern suburb of Sandton, and by the late 1990s, Sandton was the new financial and business hub for Johannesburg (Figure 6.24). It epitomizes the modern urban face of South Africa: Although racially mixed, affluent whites are overrepresented. When the province of Gauteng invested $3 billion in a new high-speed commuter rail called Gautrain, it was no accident that the first functioning line linked Sandton with Tambo International Airport.

6.3 How have infectious diseases impacted population trends, and what are governments and aid organizations doing to fight diseases such as HIV/AIDS and malaria?

6.4 What are the major rural livelihoods in this world region?

6.5 Explain the factors that contribute to the region's high population growth rates and rapid urbanization.

KEY TERMS swidden, pastoralist, tsetse fly, apartheid, coloured, township

Cultural Coherence and Diversity: Unity Through Adversity

No world region is culturally homogeneous, but most have been partially unified in the past through widespread systems of belief and communication. Traditional African religions, however, were largely limited to local areas, and the religions that did become widespread—namely Islam and Christianity—are primarily associated with other world regions. A handful of trade languages have long been understood over vast territories (Swahili in East Africa, Mandingo and Hausa in West Africa), but none spans the entire Sub-Saharan region. Sub-Saharan Africa also lacks a history of widespread political union or even an indigenous system of political relations. The powerful African kingdoms and empires of past centuries were limited to distinct subregions of the landmass.

The lack of traditional cultural and political coherence across Sub-Saharan Africa is not surprising if you consider the region's huge size—more than four times larger than Europe or South Asia (see Figure 6.2). Had foreign imperialism not impinged on the region, it is quite possible that West Africa and southern Africa would have developed into their own distinct world regions.

An African identity south of the Sahara was created through a common history of slavery and colonialism as well as through struggles for independence and development. More telling is the fact that the region's people often define themselves as African, especially to the

▲ **Figure 6.24 Sandton, Johannesburg** Just north of Johannesburg's central business district, Sandton has emerged as the financial and business center of the new South Africa. Many businesses relocated here, fleeing high crime rates in the old CBD. With world-class shopping, hotels, a convention center, and even Nelson Mandela Square, Sandton has become the part of Johannesburg that most tourists and businesspeople experience.

outside world. No one will deny that Sub-Saharan Africa is poor. Yet the cultural expressions of its people—its music, dance, and art—are joyous. Visitors often comment on the shared resilience and optimism of the region's residents. The cultural diversity is obvious, yet there is unity among the people, drawn from surviving many adversities.

Language Patterns

In most Sub-Saharan countries, as in other former colonies, people speak multiple languages that reflect tribal, ethnic, colonial, and national affiliations. Indigenous languages, many from the Bantu sub-family, are often localized to relatively small rural areas. More widely spoken languages, such as Swahili and Hausa, serve as lingua francas over broader areas. Overlaying native languages are Indo-European (French and English) and Afro-Asiatic (Arabic) languages. Figure 6.25 illustrates the complex pattern of language families and major languages found in the region today. A comparison of the larger map with the inset of current "official" languages shows that most African countries are multilingual, which can be a source of tension within states. In Nigeria, for example, the official language is English, yet there are millions of Hausa, Yoruba, Igbo (or Ibo), Ful (or Fulani), and Efik speakers as well as speakers of dozens of other languages.

African Language Groups Three of the six language groups mapped in Figure 6.25 are unique to the region (Niger-Congo, Nilo-Saharan, and Khoisan), while the other three (Afro-Asiatic, Austronesian, and Indo-European) are more closely associated with other parts of the world. Afro-Asiatic languages, especially Arabic, dominate North Africa and are understood in Islamic areas of Sub-Saharan Africa as well. Amharic in Ethiopia and Somali in Somalia are also Afro-Asiatic languages. The Austronesian language family is limited to the island of Madagascar, which many believe was first settled by seafarers from present-day Indonesia some 1500 years ago. Indo-European languages, especially French, English, Portuguese, and Afrikaans, are a legacy of colonialism and are widely spoken today.

Of the three language groups found exclusively in the region, the Niger-Congo family is by far the most important. This linguistic group originated in West Africa and includes Mandingo (one of the many Mande languages), Yoruba, Ful (Fulani), and Igbo (Ibo), among others. Around 3000 years ago, a people who spoke a Niger-Congo language began to expand out of western Africa into the equatorial zone (see arrows in Figure 6.25). This group, called the Bantu, commenced one of the most far-ranging migrations in human history, introducing agriculture into large areas in the central and southern Africa.

One Bantu tongue, Swahili, eventually became the most widely spoken Sub-Saharan language. Swahili originated on the East African coast, where merchants from Arabia established several colonies around 1100 CE. A hybrid society grew in a narrow coastal band of modern Kenya and Tanzania, speaking a language of Bantu structure enriched with many Arabic words. Swahili became the primary language only in this narrow coastal belt, but it spread far into the interior as the language of trade. After independence was achieved, Kenya, Tanzania, and Uganda adopted Swahili as an official language, along with English. With some 100 million speakers, Swahili remains a lingua franca for East Africa and parts of Central Africa. It has generated a fairly extensive literature and is often studied in other regions of the world.

Explore the SOUNDS of Swahili Language
http://goo.gl/oJo9kO

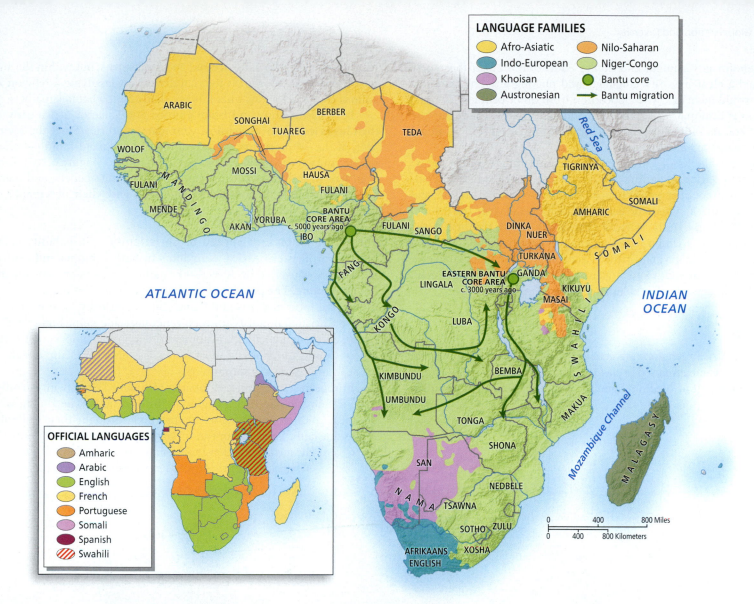

LANGUAGE FAMILIES
- Afro-Asiatic
- Indo-European
- Khoisan
- Austronesian
- Nilo-Saharan
- Niger-Congo
- Bantu core
- → Bantu migration

OFFICIAL LANGUAGES
- Amharic
- Arabic
- English
- French
- Portuguese
- Somali
- Spanish
- Swahili

▲ **Figure 6.25 Language Map of Sub-Saharan Africa** Mapping the region's languages is a complex task. Some languages, such as Swahili, have millions of speakers, while others are spoken by a few hundred people living in isolated areas. Six language families are represented in the region. Among these families are scores of individual languages (see the labels on the map). Because most modern states have many native languages, the colonial language often became the "official" language. English and French are the most common official languages in the region (see inset). **Q: What does this pattern of language families tell us about this region's interaction with peoples from other regions?**

Language and Identity Historically, ethnic identity and linguistic affiliation have been highly unstable over much of Sub-Saharan Africa. The tendency was for new language groups to form when people threatened by war fled to less-settled areas, where they often mixed with peoples from other places. In such situations, new languages arise quickly, and divisions between groups are blurred. Nevertheless, distinct **tribes**, initially consisting of a group of families with common kinship, language, and definable territory, formed. European colonial administrators were eager to establish a fixed indigenous social order to better control native peoples. During this process, a flawed cultural map of Sub-Saharan Africa evolved. Some tribes were artificially divided, meaningless names were applied, and cultural areas were often misinterpreted.

Social boundaries between different ethnic and linguistic groups have become more stable in recent years, and some individual languages have become particularly important for communication on a national scale. Wolof in Senegal; Mandingo in Mali; Mossi in Burkina

Faso; Yoruba, Hausa, and Igbo in Nigeria; Kikuyu in central Kenya; and Zulu, Xhosa, and Sesotho in South Africa are all nationally significant languages spoken by millions. None, however, has the status of official language for any country. With the end of apartheid, South Africa officially recognized 11 languages, although English is still the language of business and government. Indeed, a single language has a clear majority status in only a handful of countries. The more linguistically homogeneous states include Somalia (where virtually everyone speaks Somali) and the very small states of Rwanda, Burundi, eSwatini, and Lesotho.

European Languages In the colonial period, European countries used their own languages for administrative purposes in their African empires. Education in the colonial period also stressed literacy in the language of the imperial power. Postindependence, most Sub-Saharan African countries continued to use the languages of their former

197

colonizers for government and higher education. Few of these new states had a clear majority language, and picking any one minority tongue would have met with opposition from other peoples. The one exception is Ethiopia, which maintained its independence during the colonial era. Its official language is Amharic, although other indigenous languages are also spoken.

Two vast blocks of European language dominate Africa today: Francophone Africa, including the former colonies of France and Belgium, where French serves as the main language of administration, and Anglophone Africa, where the use of English prevails (see the inset to Figure 6.25). Early Dutch settlement in South Africa resulted in the use of Afrikaans (a Dutch-based language) by several million South Africans. In Mauritania and Eritrea, Arabic serves as a main language. Interestingly, when South Sudan gained its independence from Sudan in 2011, it changed its official language from Arabic to English.

Religion

Indigenous African religions are generally termed *animist*, a somewhat misleading catchall term used to classify all local faiths that do not fit into one of the handful of "world religions." Most practitioners of **animism**

focus on the worship of nature and ancestral spirits, but within the animist tradition there is great internal diversity. Classifying a religion as animist says more about what it is not than about what it actually is.

Both Christianity and Islam entered Sub-Saharan Africa early in their histories but advanced slowly for many centuries. Since the beginning of the 20th century, both religions have spread rapidly—more rapidly, in fact, than in any other part of the world (Figure 6.26). However, tens of millions of Africans still hold animist beliefs, and many others combine animist practices and ideas with their observance of Christianity or Islam.

The Introduction and Spread of Christianity Christianity came first to northeastern Africa. Kingdoms in both Ethiopia and central Sudan were converted by 300 CE—the earliest conversions outside the Roman Empire. The peoples of northern and central Ethiopia adopted the Coptic form of Christianity and thus historically have looked to Egypt's Christian minority for their religious leadership (Figure 6.27). At present, roughly half of the population in both Ethiopia and Eritrea follow Coptic Christianity; most of the rest practice Islam, but there are still some animist communities.

European settlers and missionaries introduced Christianity to other parts of Sub-Saharan Africa beginning in the 1600s. The Dutch, who began to colonize South Africa at that time, brought their Calvinist Protestant faith. Later, European immigrants to South Africa brought Anglicanism and other Protestant creeds as well as Catholicism. Most black South Africans eventually converted to one or

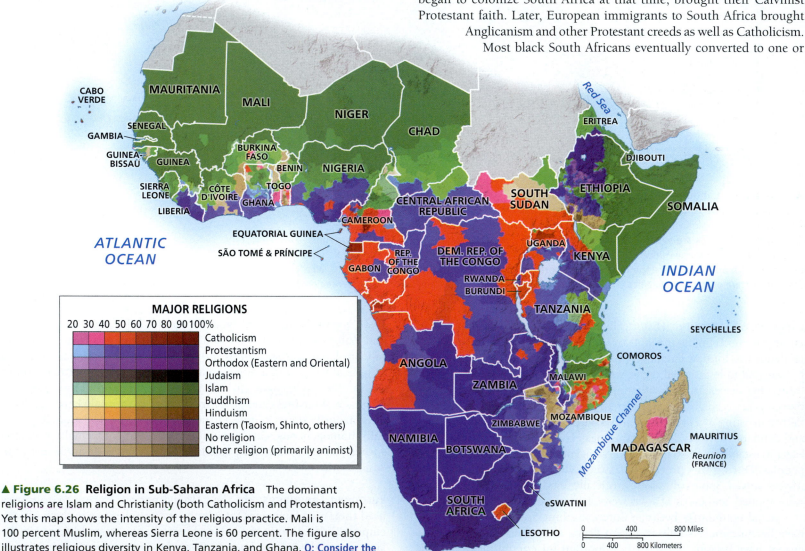

▲ **Figure 6.26 Religion in Sub-Saharan Africa** The dominant religions are Islam and Christianity (both Catholicism and Protestantism). Yet this map shows the intensity of the religious practice. Mali is 100 percent Muslim, whereas Sierra Leone is 60 percent. The figure also illustrates religious diversity in Kenya, Tanzania, and Ghana. **Q: Consider the distribution of religions in this region, especially areas of intense religious intermixing. Are these areas more prone or less prone to conflict?**

▲ **Figure 6.27 Coptic Christians** In Lalibela, Ethiopia, Coptic Christians worship at the walls of St. Michael's Church.

▲ **Figure 6.28 Grand Mosque of Touba** One of the largest mosques in Africa, Touba is a holy city for Muslims, and residents there dedicate themselves to devotion and scholarship. An annual pilgrimage, called the Grand Magal, brings 1 to 2 million faithful each year. Senegal, like much of the Sahel, converted to Islam more than six centuries ago.

Explore the SIGHTS of Islam in Senegal
http://goo.gl/KbU9S3

another form of Christianity as well. In fact, churches in South Africa were instrumental in the long fight against white racial supremacy. Religious leaders such as Bishop Desmond Tutu were outspoken critics of the injustices of apartheid and worked to bring down the system.

Elsewhere in Africa, Christianity came with European missionaries, most of whom arrived after the mid-1800s. As was true in the rest of the world, missionaries had little success where Islam had preceded them, but they eventually made numerous conversions in animist areas. As a general rule, Protestant Christianity prevails in areas of former British colonization, while Catholicism is more important in former French, Belgian, and Portuguese territories. There are nearly 150 million Catholics in the region. In the postcolonial era, African Christianity has spread out, at times taking on a life of its own, independent from foreign missionary efforts. Still active in the region are various Pentecostal, Evangelical, and Mormon missionary groups.

The Introduction and Spread of Islam Islam began to advance into Sub-Saharan Africa 1000 years ago. Berber traders from North Africa and the Sahara introduced the faith to the Sahel, and by 1050 the Kingdom of Tokolor in modern Senegal emerged as the first Sub-Saharan Muslim state (Figure 6.28). Somewhat later, the ruling class of the powerful Mande-speaking mercantile empires of Ghana and Mali converted as well. In the 14th century, the emperor of Mali astounded the Muslim world when he and his royal court made the pilgrimage to

Makkah (Mecca), bringing with them so much gold that they set off a brief period of high inflation throughout Southwest Asia.

Mande-speaking traders, whose networks spanned the Sahel to the Gulf of Guinea, gradually introduced the religion to other areas of West Africa. Many peoples remained committed to animism, however, and Islam made slow and unsteady progress. Today orthodox Islam prevails through most of the Sahel. Farther south, Muslims are mixed with Christians and animists, but their numbers continue to grow, and their practices tend to be orthodox as well (see Figure 6.26).

Interaction Between Religious Traditions The southward spread of Islam from the Sahel, coupled with the northward spread of Christianity from the port cities, created a complex religious frontier across much of West Africa. In Nigeria, the Hausa are firmly Muslim, while the southeastern Igbo are largely Christian. The Yoruba of the southwest are divided between Christian and Muslim. In the more remote parts of Nigeria, animist traditions remain strong. Despite this religious diversity, religious conflict in Nigeria has been relatively rare until recently. In 2000, seven northern Nigerian states imposed **sharia law** (Islamic religious law), which has triggered intermittent violence ever since, especially in the northern cities of Kano and Kaduna.

Around 2009, Boko Haram, an armed Islamic extremist group formed in northeastern Nigeria, escalated the violence through kidnapping and killing. In 2014, the group garnered international attention when it kidnapped 200 girls from a school in the village of Chibok. This launched an international campaign of #BringBackOurGirls (Figure 6.29). After several years in captivity, government officials brokered the release of half of the girls, but 100 are still held or were killed by Boko Haram. The terrorist group has been responsible for thousands of deaths in the border areas between Nigeria, Niger, and Chad. One tactic is to strap bombs on young girls and then detonate them in crowded areas. Over 2 million people have been displaced due to violence. In 2015, Boko Haram claimed a formal allegiance with the militant group ISIS—the Islamic State of Iraq and the Levant (also known as ISIL or IS; discussed in detail in Chapter 7). The Nigerian military has been sent to Borno Province in the northeast to defeat Boko Haram, with limited success. As mentioned at the beginning of the chapter, smaller and, so far, less deadly radical groups are active in Mali, leading to political instability in that Sahelian country.

Religious conflict historically has been far more acute in northeastern Africa, where Muslims and Christians have struggled against each other for centuries. Such a clash eventually led to the creation of the region's newest state when South Sudan separated from Sudan in 2011. Islam was introduced to Sudan in the 1300s by an invasion of Arabic-speaking pastoralists who destroyed the indigenous Coptic Christian kingdoms of the area. Within a few hundred years, northern and central Sudan had become completely Islamic. The southern equatorial province of Sudan, where tropical diseases and extensive wetlands prevented Arab advances, remained animist or converted to Christianity under British colonial rule.

In the 1970s, the Arabic-speaking Muslims of northern and central Sudan began to build an Islamic state. Experiencing both religious discrimination and economic exploitation, the Christian and animist peoples of the south launched a massive rebellion. Fighting became intense in the 1980s, with the government generally controlling the main towns and roads and the rebels maintaining power in the countryside. A peace was brokered in 2003, and as part of the peace agreement southern Sudan was promised an opportunity to vote on secession from the north in 2011. The vote took place, and the new nation was formed with Juba as its capital. Yet this landlocked territory is still not at peace, as discussed later in the chapter.

Sub-Saharan Africa is a land of religious vitality. Both Christianity and Islam are spreading rapidly, and devotional activities are part of the daily flow of life in cities and rural areas. Animism continues to have widespread appeal as well, so that new and syncretic (blended) forms of religious expression are also emerging. With such a diversity of faiths, it is fortunate that religion is not typically the cause of overt conflict.

Globalization and African Culture

The slave trade that linked Africa to the Americas and Europe set in motion paths of diffusion that transferred African peoples and cultures across the Atlantic. Tragically, slavery damaged the demographic and political strength of African societies, especially in West Africa, from which most slaves were taken. An estimated 12 million Africans were shipped to the Americas as slaves from the 1500s until 1870 (see Figure 5.17 in Chapter 5). Slavery impacted the entire region, sending Africans not just to the Americas, but also to Europe, North Africa, and Southwest Asia. The vast majority, however, worked on plantations in the Americas.

Out of this tragic displacement of people came a blending of African cultures with Amerindian and European ones. African rhythms are at the core of various American musical styles, from rumba to jazz, the blues, and rock and roll. Brazil, Latin America's largest country, is said to be the second largest "African state" (after Nigeria) because of its huge Afro-Brazilian population. Thus, the forced migration of enslaved Africans has had major cultural influence on many areas of the world.

So, too, have contemporary movements of Africans influenced the cultures of many world regions. Perhaps the most celebrated person of African ancestry today is former U.S. President Barack Obama, whose father was Kenyan. Obama's heritage and upbringing embody the forces of globalization. In Kenya, he is hailed as part of the modern African **diaspora**—young professionals (and their offspring) who leave the continent for work or education and make their mark elsewhere. In popular culture, South African comedian Trevor Noah was selected to take over Jon Stewart's *Daily Show* in 2015 when Stewart retired. This comedy "news" show's brand of political and social commentary has become more international with a South African at the helm. And Academy Award–winning actress Lupita Nyong'o, raised in Kenya, is also a global style icon. From film to fashion, the people of this region influence and respond to global cultural trends.

Nollywood   Africa's undisputed film capital is Lagos, Nigeria. Called Nollywood, the Nigerian movie industry makes more films than Hollywood, currently grinding out as many as 2500 films a year (50 per week) and employing more than 1 million people. Relying on relatively inexpensive digital video technology, most of these movies are shot in a few days and with budgets of $10,000 to $20,000. The typical themes of religion, ethnicity, corruption, witchcraft, the spirit world, violence, and injustice resonate with African audiences. The films are almost always shot on location—in city streets, office buildings, and homes or in the countryside. Nollywood films can be bloody and exploitive; they can also be overtly evangelical, promoting Christianity over indigenous faiths. Many are scripted in English, but there are also movies for Yoruba, Igbo, and Hausa speakers.

Rather than theatrical release, most Nollywood films go directly to DVD and are rarely viewed beyond Africa. The shelf life of these movies is rather short; production companies need to make their money back quickly, before pirated copies undermine profits. However, a $20,000 film can earn $500,000 in DVD sales in just a couple of weeks. Consequently, film distribution remains tightly controlled by Igbo businesspeople sometimes called the Alaba cartel for their distribution center on the outskirts of Lagos. But this is changing as middle-class Africans want to go to theaters and directors want to create higher-quality films for Nigeria and beyond.

▼ **Figure 6.29** **#BringBackOurGirls**   Activists gather at a 2014 "Bring Back Our Girls" rally in Abuja, Nigeria. The women rallied daily against the abduction of 200 schoolgirls in Chibok, Nigeria.

Music in West Africa Nigeria is West Africa's music center, with a well-developed and cosmopolitan recording industry. Modern Nigerian styles such as juju, highlife, and Afro-beat are influenced by jazz, rock, reggae, and gospel, but they are driven by an easily recognizable African sound. Yet one of the continent's biggest stars is Youssou N'Dour, a singer and percussionist from Senegal who performs to packed venues around the world. He is also a politician, having been appointed Senegal's Minister of Tourism and Culture in 2012.

Farther up the Niger River lies Mali; Bamako, the capital, is also a music center that has produced scores of recording artists. Many Malian musicians descend from a traditional caste of musical storytellers performing on either the traditional kora (a cross between a harp and a lute) or the guitar. The musical style is strikingly similar to that of the blues from the Mississippi Delta—so much so that the late Ali Farka Touré, from northern Mali, was known as the Bluesman of Africa because of his distinctive, yet familiar, guitar work (Figure 6.30).

Explore the SOUNDS of Kora Music
http://goo.gl/i0Er50

Contemporary African music can be both commercially and politically important. Nigerian singer Fela Kuti became a voice of political conscience for Nigerians struggling for true democracy. Born to an elite family and educated in England, Kuti borrowed from jazz, traditional, and popular music to produce the Afro-beat sound in the 1970s. The music was irresistible, but his searing lyrics also attracted attention. Acutely critical of the military government, he sang of police harassment, the inequities of the international economic order, and even Lagos's infamous traffic. Singing in English and Yoruba, his message was transmitted to a larger audience, and he became a target of state harassment. Kuti died in Lagos in 1997 from complications related to AIDS, yet his music and politics later became the subject of the awarding-winning Broadway musical *Fela!*

Pride in East African Runners Ethiopia and Kenya have produced many of the world's greatest distance runners. Abebe Bikila won Ethiopia's and Africa's first Olympic gold medal, running barefoot at the Rome games in 1960. Since then, nearly every Olympic Games has yielded medals for Ethiopia and Kenya. At the 2016 Rio Olympics, Kenyan runners won 12 medals and Ethiopians, eight.

Running is a national pastime in Ethiopia and Kenya, where elevation—Addis Ababa sits at 7300 feet (2200 meters) and Nairobi at 5300 feet (1600 meters)—increases oxygen-carrying capacity. Past medalists Haile Gebrselassie and Derartu Tulu are national celebrities in Ethiopia, where they are idolized by the country's youth. Tulu, the first black African

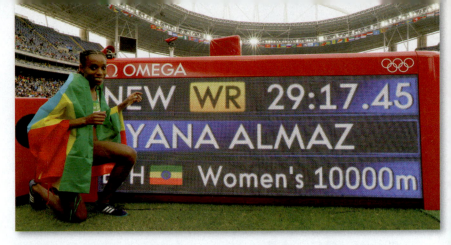

▲ **Figure 6.31 Ethiopian Running Star Almaz Ayana** Kneeling beside her world record time and draped in the Ethiopian flag, Almaz Ayana was one of several East African medalists at the 2016 Olympic Games in Rio de Janeiro.

woman to win a gold medal in distance running, is a forceful voice for women's rights in a country where women are discouraged from putting on running shorts. In the 2016 Olympics, Ethiopian Almaz Ayana won gold in the women's 10,000 meter race, setting a new world record (Figure 6.31).

African Fashion West Africa has long been a major center of textile production, but by the early 2000s, runway models were strutting in West African designs for a growing market. While European designers have taken inspiration from African fabrics, the region has not had major fashion houses of their own. But this is changing. Beginning in Dakar in Senegal, Lagos, Nairobi, Johannesburg, and Windhoek, Namibia now host Fashion Weeks that are equal to the glitz and serious business associated with Paris and Milan.

Known for daring use of color, geometric designs, and comfortable cottons, African fabrics are easily recognized. With investment in new designers and Fashion Week formats, the exuberant fashion industry is catching on, especially in West Africa (Figure 6.32). Celebrities such as Beyoncé, Lupita Nyong'o, and former First Lady Michelle Obama have worn designs by Nigerian Maki Oh. Adama Paris, a Senegalese designer, sells her line to the region's largest musical artists and to shops in France. Paris notes that African fashion is also about changing notions of beauty, and taking pride in one's ethnic community through design.

▼ **Figure 6.32 African Fashion** A model presents a creation by Ivorian designer Zak Kone during Dakar Fashion Week in the Senegalese capital. West Africa has a long textile tradition, but the area's designers have only recently begun to influence global fashion trends.

Explore the SIGHTS of Namibian Fashion
https://goo.gl/fk6f6H

▼ **Figure 6.30 Ali Farka Touré** Known as the Bluesman of Africa, Touré was an extremely popular and influential musician. Here he performs at the famous Desert Festival near Timbuktu, Mali. His musical style and guitar abilities earned worldwide praise.

REVIEW

6.6 What are the dominant religions of Sub-Saharan Africa, and how have they diffused throughout the region?

6.7 Describe the ways in which African peoples have influenced world regions beyond Africa.

KEY TERMS   tribes, animism, sharia law, diaspora

Geopolitical Framework: Legacies of Colonialism and Conflict

The duration of human settlement in Sub-Saharan Africa is unmatched by any other region. Evidence shows that humankind originated there, and many diverse ethnic groups have formed in the region over the past few thousand years. Although ethnic conflicts have occurred, cooperation and coexistence among different peoples have also continued over centuries.

Some 2000 years ago, the Kingdom of Axum arose in northern Ethiopia and Eritrea, strongly influenced by political models derived from Egypt and Arabia. The first wholly indigenous African states were founded in the Sahel around 700 CE. Over the next several centuries, various other states emerged in West Africa. By the 1600s, states located near the Gulf of Guinea took advantage of the opportunities presented by the slave trade—namely, selling slaves to Europeans (Figure 6.33).

Thus, prior to European colonization, Sub-Saharan Africa presented a complex mosaic of kingdoms, states, and tribal societies. The arrival of Europeans forever changed patterns of social organization and ethnic relations. As Europeans rushed to carve up the continent to serve their imperial ambitions, they set up various administrations that heightened ethnic tensions and promoted hostility. Many of the region's modern conflicts can trace their roots back to the colonial era, especially the drawing of political boundaries.

European Colonization

Unlike the relatively rapid colonization of the Americas, Europeans needed centuries to gain effective control of Sub-Saharan Africa. Portuguese traders arrived along the coast of West Africa

in the 1400s, and by the 1500s they were well established in East Africa. Initially, the Portuguese made large profits, converted a few local rulers to Christianity, established several defensive trading posts, and gained control over the Swahili trading cities of the east. They stretched themselves too thin, however, and failed in many of their colonizing activities. Only along the coasts of modern Angola and Mozambique, where a sizable population of mixed African and Portuguese peoples emerged, did Portugal maintain power. Along the Swahili, or eastern, coast, the Portuguese were eventually expelled by Arabs from Oman, who then established their own trade-based empire in the area.

One reason for the Portuguese failure was Sub-Saharan Africa's disease environment. With no resistance to malaria and other tropical diseases, roughly half of all Europeans who remained on the African mainland died within a year. Protected both by their armies and by the diseases of their native lands, African states were able to maintain an upper hand over European traders and adventurers well into the 1800s. Unlike in the Americas, where European conquest was facilitated by the introduction of Old World diseases that devastated native populations (see Chapters 4 and 5), in Sub-Saharan Africa endemic disease limited European settlement until the mid-19th century.

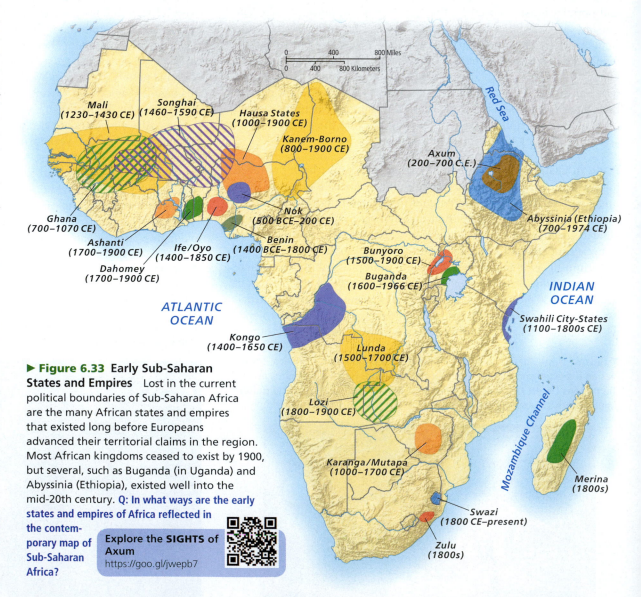

▶ **Figure 6.33 Early Sub-Saharan States and Empires**   Lost in the current political boundaries of Sub-Saharan Africa are the many African states and empires that existed long before Europeans advanced their territorial claims in the region. Most African kingdoms ceased to exist by 1900, but several, such as Buganda (in Uganda) and Abyssinia (Ethiopia), existed well into the mid-20th century. **Q: In what ways are the early states and empires of Africa reflected in the contemporary map of Sub-Saharan Africa?**

Explore the SIGHTS of Axum
https://goo.gl/jwepb7

▶ **Figure 6.34** European
Colonization in 1913 Before
1880, few areas of Africa were
under direct European
control. When the Berlin
Conference convened
in 1884, Africa was
carved up and traded
between European
powers. France and
Britain controlled the
most territory, but
Germany, Portugal, Belgium,
Spain, and Italy all had claims
as well. By 1913 the entire continent
except for Ethiopia, Liberia, and South
Africa, was under European colonial
control.

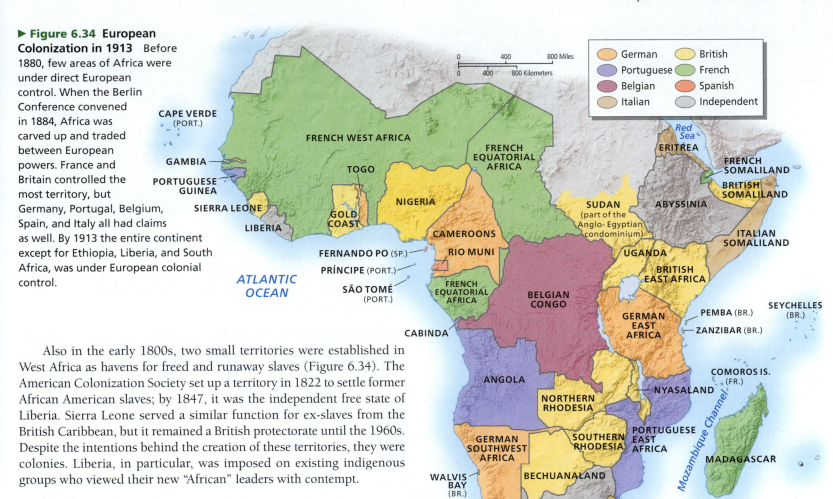

Also in the early 1800s, two small territories were established in
West Africa as havens for freed and runaway slaves (Figure 6.34). The
American Colonization Society set up a territory in 1822 to settle former
African American slaves; by 1847, it was the independent free state of
Liberia. Sierra Leone served a similar function for ex-slaves from the
British Caribbean, but it remained a British protectorate until the 1960s.
Despite the intentions behind the creation of these territories, they were
colonies. Liberia, in particular, was imposed on existing indigenous
groups who viewed their new "African" leaders with contempt.

The Scramble for Africa In the 1880s, European colonization quickly
accelerated, leading to the so-called scramble for Africa. By this time,
the colonists had practices to reduce malaria transmission, and after
the invention of the machine gun, no African state could long resist
European force.

As the colonization of Africa intensified, tensions among the colonizing forces of Britain, France, Belgium, Germany, Italy, Portugal, and
Spain mounted. Rather than risk war, 13 countries convened in Berlin
at the invitation of German Chancellor Bismarck in 1884, at a gathering known as the **Berlin Conference**. During the conference, in which
no African leaders were included, rules were established about what
determined "effective control" of a territory, and Sub-Saharan Africa
was carved into pieces that were traded like properties in a game of
Monopoly (see Figure 6.34).

Although European weapons in the 1880s were far superior to
anything found in Africa, several indigenous states organized effective
resistance campaigns. In South Africa, Zulu warriors resisted British
invasion into their lands in what have been termed the Anglo-Zulu
Wars (1879–1896). Eventually, European forces prevailed everywhere
except Ethiopia. The Italians had conquered the Red Sea coast and the
far northern highlands (modern Eritrea) by 1890 and quickly set their
sights on the large Ethiopian kingdom of Abyssinia, which had been
vigorously expanding for several decades. In 1896, however, Abyssinia
defeated the invading Italian army, earning the respect of the European
powers. In the 1930s, fascist Italy launched a major invasion of the
country, by this time renamed Ethiopia, to redeem its earlier defeat and
quickly prevailed with the help of poison gas and aerial bombardment.
However, Ethiopia had regained its freedom by 1942.

Although Germany was a principal instigator of the scramble for
Africa, it lost its own colonies after suffering defeat in World War I.
Britain and France then partitioned most of Germany's African empire
between themselves. Figure 6.34 shows the region's colonial status in
1913, prior to Germany's territorial loss.

While the Europeans were cementing their rule over Africa, South
Africa was inching toward political independence, at least for its white
population. One of the oldest colonies in Sub-Saharan Africa, in 1910,
South Africa became the first to obtain its political independence from
Europe. However, its formalized system of discrimination and racism
hardly made it a symbol of liberty. Ironically, as the Afrikaners tight-
ened their political and social control over the nonwhite population
through their policy of apartheid, introduced in 1948, the rest of the
continent was preparing for political independence from Europe.

Decolonization and Independence

Decolonization of Sub-Saharan Africa happened rather quickly and
peacefully, beginning in 1957. Independence movements, however, had
sprung up throughout the continent, some dating back to the early
1900s. Workers' unions and independent newspapers became voices
for African discontent and the hope for freedom.

By the late 1950s, political demands from within Sub-Saharan Africa and changing attitudes within Europe made it clear to leaders in Britain, France, and Belgium that they could no longer maintain their African empires. (Italy had already lost its colonies during World War II, and Britain gained Somalia and Eritrea.) Once started, decolonization progressed rapidly. By the mid-1960s, virtually the entire region had achieved independence. In most cases, the transition was relatively peaceful and smooth, with the exception of southern Africa.

Dynamic African leaders put their mark on the region during the early decades after independence. Kenya's Jomo Kenyatta, the Ivory Coast's Félix Houphouët-Boigny, Tanzania's Julius Nyerere, Ghana's Kwame Nkrumah, and others became powerful father figures who molded their new nations (Figure 6.35). President Nkrumah's vision for Africa was the most expansive. After helping to secure independence for Ghana in 1957, his ultimate aspiration was African political unity. Although never realized, his dream set the stage for the founding of the Organization of African Unity (OAU) in 1963, renamed the **African Union (AU)** in 2002. The AU is a continent-wide organization whose main role has been to mediate disputes between neighbors. Certainly in the 1970s and 1980s, the AU was a constant voice of opposition to South Africa's minority rule, and it continues to intervene in some of the region's ethnic conflicts and humanitarian emergencies.

Southern Africa's Independence Battles   Independence did not come easily to southern Africa. In Southern Rhodesia (modern-day Zimbabwe), the problem was the presence of some 250,000 white residents, most of whom owned large farms. Unwilling to see power pass to the country's black majority, then some 6 million strong, these settlers declared themselves the rulers of an independent, white-supremacist state in 1965. The black population continued to resist, however, and in 1978 the Rhodesian government was forced to give up power. Renamed Zimbabwe, the country was henceforth ruled by the black majority, although the remaining whites still form an economically privileged community. Since the mid-1990s, disputes over government land reform (splitting up the large commercial farms mostly owned by whites and giving the land to black farmers) and President Robert Mugabe's strongman politics have resulted in serious racial and political tensions as well as the collapse of the country's economy. The 93-year-old Mugabe was removed from office in 2017 in a bloodless military coup, but the effects of a contested election in 2018 remain to be seen.

In the former Portuguese colonies, independence came violently. Unlike the other imperial powers, Portugal refused to hand

▶ **Figure 6.35 Julius Nyerere Monument**   An independence leader from 1961 until he retired from the presidency in 1985, Julius Nyerere is Tanzania's founding father. This statue is located in Dodoma, the official capital since 1996. Dar es Salaam, the former capital, is still the largest city and has many government offices.

over its colonies in the 1960s, so the people of Angola and Mozambique turned to armed resistance. The most powerful rebel movements adopted a socialist orientation and received support from the Soviet Union and Cuba. When a new government came to power in 1974, however, Portugal withdrew suddenly from its African colonies, and Marxist regimes quickly took over in both Angola and Mozambique. The United States, and especially South Africa, responded to this perceived threat by supplying arms to rebel groups opposing the new governments. Fighting dragged on for three decades in Angola and Mozambique; their respective countrysides became so heavily laden with land mines that it was hard to farm. With the end of the Cold War, however, outsiders lost interest in continuing these conflicts, and sustained efforts to negotiate peace settlements began. Mozambique has been at peace since the mid-1990s. After several failed attempts, Angola's army signed a peace treaty with rebels in 2002, ending a 27-year conflict in which more than 300,000 Angolans died and 3 million more were displaced. With peace, Angola's impressive oil reserves have gone online, led by Chinese investment.

Apartheid's Demise in South Africa   While fighting continued in the former Portuguese zone, South Africa underwent a remarkable transformation. From 1948 through the 1980s, the ruling Afrikaners' National Party was firmly committed to white supremacy. Under apartheid, only whites enjoyed real political freedom, whereas blacks were denied even citizenship in their own country—technically, they were citizens of "homelands."

The first major change came in 1990 when South Africa withdrew from Namibia, which it had controlled as a protectorate since the end of World War I. South Africa now stood alone as Africa's single white-dominated state. A few years later, the leaders of the Afrikaner-dominated political party decided they could no longer resist internal and international pressure for change. In 1994, free elections were held, and Nelson Mandela, a black leader who had been imprisoned for 27 years by the old regime, emerged as the new president from the African National Congress (ANC) party. Black and white leaders pledged to put the past behind them and work together to build a new, multiracial South Africa. Since then, orderly elections have been held. Jacob Zuma, first elected president in 2009, resigned in 2018 due to mounting scandals and a troubled economy, and was replaced by Cyril Ramaphosa.

Unfortunately, the legacy of apartheid is not so easily erased. Residential segregation is illegal, but neighborhoods are still sharply divided along racial lines. Under the multiracial political system, a black middle class emerged, but most blacks remain extremely poor (and most whites remain prosperous). Violent crime has increased, and rural migrants and immigrants have poured into South African cities, producing a xenophobic anti-immigrant backlash, especially in greater Johannesburg. Because the political change was not matched by significant economic transformation, the hopes of many people have been frustrated.

Persistent Conflict

Although most Sub-Saharan countries made a relatively peaceful transition to independence, virtually all of them immediately faced a difficult set of institutional and political problems. In several cases, the old authorities had done almost nothing to prepare their colonies for independence. Lacking an institutional framework for independent government, countries such as the Democratic Republic of the Congo faced a chaotic situation from the beginning. Only a handful of Congolese had received higher education, let alone training for administrative posts. The indigenous African political framework had been essentially destroyed by colonization, and in most cases, very little had been built in its place.

Even more problematic in the long run was the political geography of the newly independent states. Civil servants could always be trained and administrative systems built, but little could be done to rework the region's basic political map. European colonial powers had essentially ignored indigenous cultural and political boundaries, both in dividing Africa among themselves and in creating administrative subdivisions within their own imperial territories. These imposed boundaries remained after independence.

The Tyranny of the Map All over Sub-Saharan Africa, different ethnic groups found themselves forced into the same state with peoples of different linguistic and religious backgrounds, many of whom had recently been enemies. At the same time, several larger ethnic groups found their territories split between two or more countries. The Hausa people of West Africa, for example, were divided between Niger (formerly French) and Nigeria (formerly British), sharing land with several former ethnic rivals.

Given these imposed borders, it is no wonder that many African countries struggled to generate a national identity or establish stable political institutions. **Tribalism**, or loyalty to the ethnic group rather than to the state, has become the bane of African political life. Especially in rural areas, tribal identities are usually more important than national ones. Because nearly all African countries inherited inappropriate colonial borders, some observers advocated drawing a new political map based on indigenous identities. However, such a strategy was impossible, as all the leaders of the newly independent states realized. Any new territorial divisions would have created winners and losers, thus resulting in even more conflict. Moreover, because ethnicity in Sub-Saharan Africa was traditionally fluid and because many ethnic groups were intermixed, it would have been difficult to generate clear-cut divisions. Finally, most African ethnic groups were considered too small to form viable countries. With such complications in mind, the new African leaders, meeting in 1963 to form the Organization of African Unity, agreed that colonial boundaries should remain. The violation of this principle, they argued, would lead to pointless wars between states and endless civil struggles within them.

Despite the determination of Africa's leaders to build their new nations within existing political boundaries, challenges to the states began soon after independence. Figure 6.36 maps the ethnic and

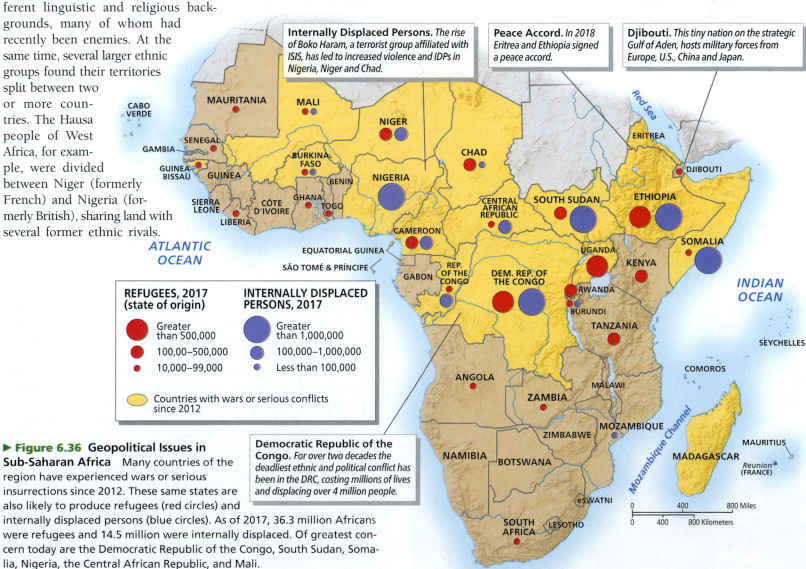

▶ **Figure 6.36 Geopolitical Issues in Sub-Saharan Africa** Many countries of the region have experienced wars or serious insurrections since 2012. These same states are also likely to produce refugees (red circles) and internally displaced persons (blue circles). As of 2017, 36.3 million Africans were refugees and 14.5 million were internally displaced. Of greatest concern today are the Democratic Republic of the Congo, South Sudan, Somalia, Nigeria, the Central African Republic, and Mali.

Internally Displaced Persons. *The rise of Boko Haram, a terrorist group affiliated with ISIS, has led to increased violence and IDPs in Nigeria, Niger and Chad.*

Peace Accord. *In 2018 Eritrea and Ethiopia signed a peace accord.*

Djibouti. *This tiny nation on the strategic Gulf of Aden, hosts military forces from Europe, U.S., China and Japan.*

Democratic Republic of the Congo. *For over two decades the deadliest ethnic and political conflict has been in the DRC, costing millions of lives and displacing over 4 million people.*

REFUGEES, 2017 (state of origin)
- Greater than 500,000
- 100,00–500,000
- 10,000–99,000

INTERNALLY DISPLACED PERSONS, 2017
- Greater than 1,000,000
- 100,000–1,000,000
- Less than 100,000

Countries with wars or serious conflicts since 2012

political conflicts that have disabled parts of Africa since 2010. The human cost of this turmoil is several million refugees and internally displaced persons. **Refugees** are people who flee their state because of well-founded fear of persecution based on race, ethnicity, religion, or political orientation. According to the United Nations, nearly 6.3 million Sub-Saharan Africans were considered refugees at the end of 2017, with South Sudan accounting for 2.4 million. Uganda, Sudan, and Ethiopia are the largest host countries for the region's refugees, each having around 1 million. Added to this figure are another 14.5 million **internally displaced persons (IDPs)**. IDPs have fled from conflict but still reside in their country of origin. The Democratic Republic of the Congo has the largest number of IDPs (4.5 million), followed by Somalia (2.1 million), South Sudan (1.9 million), Nigeria (1.7 million), and Ethiopia (1 million). These people are not technically considered refugees, but they may receive some assistance and/or protection from the UN High Commissioner for Refugees. Together refugees and IDPs in the region total over 20 million people, nearly double the counts in 2013 (Figure 6.37).

Political and Ethnic Conflicts

As Figure 6.36 shows, much of the conflict since 2012 has occurred in the Sahelian states, Central Africa, and the Horn of Africa. Fortunately, in the past two decades, peace has returned to Sierra Leone, Liberia, the Ivory Coast, and Angola, states that produced large numbers of refugees in the 1990s and early 2000s.

For two decades, ethnic and political conflict have ravaged the Democratic Republic of the Congo. The International Red Cross estimated that between 1998 and 2010, 5 million people died there, although many of the deaths were from war-induced starvation and disease rather than bullets or machetes. Current UN figures show half a million refugees are living outside the country and over 4 million IDPs within it. President Joseph Kabila is the son of the late Laurent Kabila, who led rebel forces across the country from Uganda and Rwanda in the late 1990s and took control of the Democratic Republic of the Congo until his assassination in 2001. The younger Kabila was elected in 2006 and has remained in power despite ongoing turmoil and violence. He postponed presidential elections in 2016 and promised them by December 2018, only after years of vigorous protest. Yet opposition candidates have been banned from running, and protestors have been gunned down in the streets of Kinshasa. Sub-Saharan Africa's largest state in terms of territory has only limited experience with

democracy, its civil service barely functions, corruption is rampant, and there are few roads and little working infrastructure for the over 80 million residents. In 2018, it was estimated that 70 armed groups roam the country, especially in the eastern region. UN peacekeepers are a stabilizing presence, but military groups such as Democratic Forces for the Liberation of Rwanda (FDLR), the Uganda Allied Democratic Froes (ADF), and the March 23 Movement (M23) terrorize communities and exploit natural resources. The massive and largely untapped mineral wealth in the region fuels the ongoing violence; the mineral trade reportedly finances groups that purchase and operate weapons.

A low-intensity ethnic conflict in northern Mali heated up in 2012 as a Touareg-based National Movement for the Liberation of Azawad (MNLA) proclaimed independence from the Bamako-based government in the south, largely controlled by Mande-speaking groups. The MNLA was formed in 2011, partly by armed Touareg fighters returning from Libya after the fall of Muammar al-Qaddafi's regime as part of the Arab Spring (see Chapter 7). As fighting intensified, a military coup in Bamako removed President Touré from office in 2012, citing dissatisfaction with the government's inability to fight Touareg rebels armed with Libyan weapons. In 2013, Ibrahim Keïta, an ethnic Touareg, was elected president and continues to run the country. With the help of the G5 Sahel (Mali, Burkina Faso, Chad, Mauritania, and Niger) and UN peacekeepers, radical Islamist groups have been pushed to the fringes of northern Mali or into the bordering Saharan countries. Although this latest conflict seems to be driven by political events in North Africa and ethnic rivalries, others point to ecological pressures brought on by drought and/or climate change that undermine traditional patterns of resource sharing between farmers and pastoralists. In addition, northern Mali is a hub for migrants from all over West Africa seeking to cross the Sahara into Algeria and Libya, eventually bound for Europe. Militant groups fund themselves by taxing human traffickers and smugglers.

Another area experiencing ethnic-religious conflict is northeast Nigeria near the Lake Chad basin, where almost 2 million Nigerians were displaced at the end of 2017. As discussed earlier in the chapter, violence in this area is led by Boko Haram, which has spilled into Niger and Chad, resulting in a steep rise in IDPs and refugees in all three states.

Secessionist Movements

Problematic African political boundaries have occasionally led to attempts by territories to secede and form new states. The Shaba (or Katanga) Province tried to leave what was then Zaire soon after independence. The rebellion was crushed a couple of years after it started, with the help of France and Belgium. Similarly, the Igbo in oil-rich southeastern Nigeria declared an independent state of Biafra in 1967. After a short but brutal war, during which Biafra was essentially starved into submission, Nigeria was reunited.

In 1991, the government of Somalia disintegrated, and the territory has been in civil war ever since. The lack of political control facilitated the rise of piracy; Somali pirates raid vessels and extort ransom from ships from the Gulf of Aden to the Indian Ocean. The territory has been ruled by warlords and their militias, who have informally divided the country into clan-based units. **Clans** are social units constituting branches of a tribe or an ethnic group larger than a family. Early in the conflict, the northern portion of Somalia declared its independence as a new country—Somaliland. Somaliland has a constitution, a functioning parliament, government ministries,

▼ **Figure 6.37 Refugee Camp** South Sudanese refugees collect water at a tap provided by the UNHCR in this Ugandan settlement. Over one million South Sudanese refugees crossed into Uganda in 2017.

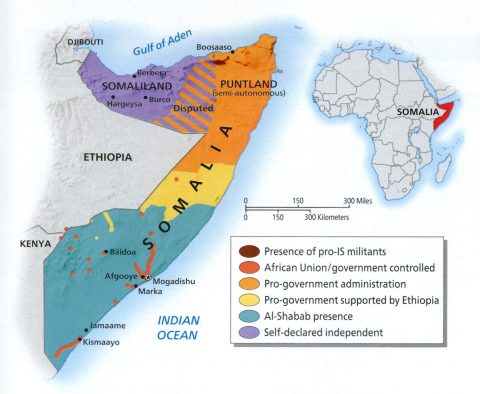

▲ Figure 6.38 Divided Somalia For three decades, conflict has raged in Somalia, challenging every aspect of governance. The territory is fractured with one area seeking independence; another seeking semi-autonomy; and pro-government and pro-Al Shabab groups, along with the Ethiopian military, vying for control.

Legend:
- Presence of pro-IS militants
- African Union/government controlled
- Pro-government administration
- Pro-government supported by Ethiopia
- Al-Shabab presence
- Self-declared independent

a police force, a judiciary, and a president and produces its own currency and passports. Yet no country has recognized this territory. In 1998, neighboring Puntland also declared autonomy but is not seeking independence. Meanwhile, well-armed Islamic insurgents (led by the Shabab) are present in the south. The political instability has been exacerbated by several years of drought, creating a humanitarian emergency (Figure 6.38).

Only two territories in the region have successfully seceded. In 1993, Eritrea gained independence from Ethiopia after two decades of civil conflict. This territorial secession is striking because Ethiopia gave up its access to the Red Sea, making it landlocked. Yet the creation of Eritrea still did not bring about peace. After years of fighting, the transition to Eritrean independence began remarkably well. Unfortunately, border disputes between the two countries erupted in 1998, resulting in the deaths of some 100,000 troops. In 2000, a peace accord was reached and the fighting stopped, but in 2016, border clashes erupted again. In 2018, Eritrea and Ethiopia signed another, and hopefully lasting, peace accord.

The second example is South Sudan, which gained its independence from Sudan in 2011 after some three decades of violent conflict between the largely Arab and Muslim north and the Christian and animist south. But peace has not come to this new territorial state either. A power struggle between rival factions of Dinka and Nuer began in 2013 and, despite intermittent violence over five years, the conflict has displaced nearly 2 million people and produced some 300,000 refugees. The difficulties experienced by the newly created states of South Sudan and Eritrea suggest that major changes to Africa's political map should not be expected.

REVIEW

6.8 What are the processes behind Sub-Saharan Africa's political map, and why have there been relatively few boundary changes since the 1960s?

6.9 What are the current major conflicts in this region, and where are they occurring?

KEY TERMS Berlin Conference, African Union (AU), tribalism, refugee, internally displaced person (IDP), clan

Economic and Social Development: The Struggle to Develop

By almost any measure, Sub-Saharan Africa is the poorest world region. According to World Bank estimates, 41 percent of the population lives in extreme poverty, surviving on less than $1.90 per day, although in 1993 the figure was 61 percent. Due to poverty and low life expectancy, nearly all the states in the region are ranked at the bottom of the Human Development Index. Some demographically small or resource-rich states—such as Botswana, Equatorial Guinea, Mauritius, the Seychelles, and South Africa—have much higher per capita gross national incomes adjusted for purchasing power parity (GNI-PPP), but the average for the region was about $3500 in 2017. By way of comparison, the figure for Southwest Asia and North Africa, the next poorest region, was $8000.

Since 2000, strong commodity prices, new infrastructure, and improved technology (mobile phone subscriptions cover most of the population) have brightened the region's economic prospects. The average annual growth rate for the region from 2000 to 2009 was an impressive 6.2 percent, although from 2009 to 2015 the growth rate slipped to 4.3 percent. Ethiopia's annual growth rate averaged a stunning 10.3 percent during 2009–2015. Over the past 20 years, real income per person grew (20–30 percent) whereas in the previous 20 years it actually decreased. The most optimistic views of the region see strengthened democracies, greater civic engagement, less violence, and growing investment (from within and outside the region). American economist Jeffrey Sachs argues that in order to get out of the poverty trap, the region will need substantial sums of new foreign aid and investment (Table 6.2).

Roots of African Poverty

In the past, observers often attributed Africa's poverty to its colonial history, poorly conceived development policies, and/or corrupt governance. Those who favored environmental explanations pointed to the region's infertile soils, erratic rainfall patterns, lack of navigable rivers, and virulent tropical diseases as reasons for underdevelopment. The best explanations for the region's poverty now point more to historical and institutional factors than to environmental circumstances.

Numerous scholars have singled out the slave trade for its debilitating effect on Sub-Saharan African economic life. Large areas of the region were depopulated, and many people were forced to flee into poor, inaccessible refuges. Colonization was another blow to Africa's economy. European powers invested little in infrastructure, education, and public health and were instead interested mainly in developing

TABLE 6.2   Development Indicators

Explore these data in MapMaster 2.0 https://goo.gl/RkzJsH

Country	GNI per Capita, PPP 2017[1]	GDP Average Annual Growth 2009–2015[2]	Human Development Index (2016)[3]	Percent Population Living Below $3.10 a Day[2]	Under Age 5 Mortality Rate (per 1000 live births), 1990[1]	Under Age 5 Mortality Rate (per 1000 live births), 2016[1]	Secondary School Enrollment Ratios[4] Male (2009–2016)	Secondary School Enrollment Ratios[4] Female (2009–2016)	Gender Inequality Index (2016)[3,6]	Freedom Rating (2018)[5]
Western Africa										
Benin	2266	4.7	0.485	77	178	98	67	47	0.613	2.0
Burkina Faso	1870	5.8	0.402	81	199	85	35	32	0.615	3.5
Cape Verde	6832	1.7	0.648	36	63	21	88	98	–	1.0
Gambia	1715	2.8	0.452	86	168	65	59	56	0.641	4.5
Ghana	4641	8.0	0.579	61	127	59	63	61	0.547	1.5
Guinea	2285	2.3	0.414	81	235	89	47	31	–	5.0
Guinea-Bissau	1700	2.8	0.424	81	219	88	–	–	–	5.0
Ivory Coast	3953	6.0	0.474	55	151	92	51	37	0.672	4.0
Liberia	827	5.8	0.427	–	258	67	42	33	0.649	3.0
Mali	2211	3.3	0.442	76	254	111	46	37	0.689	4.5
Mauritania	3950	5.2	0.513	33	117	81	32	29	0.626	5.5
Niger	1017	6.6	0.353	82	329	91	24	17	0.695	4.0
Nigeria	5861	5.2	0.527	79	213	104	58	53	–	4.0
Senegal	2712	3.9	0.494	66	140	47	50	49	0.521	2.0
Sierra Leone	1526	7.0	0.420	81	262	114	46	40	0.650	3.0
Togo	1570	4.8	0.487	77	145	76	–	–	0.556	4.0
Eastern Africa										
Burundi	771	3.4	0.404	95	170	72	44	41	0.474	6.5
Comoros	1552	2.6	0.497	–	126	73	58	62	–	3.5
Djibouti	2705	5.0	0.473	37	118	64	53	43	–	5.5
Eritrea	1511	–	0.420	–	151	45	33	28	–	7.0
Ethiopia	1899	10.3	0.448	76	203	58	36	34	0.499	6.5
Kenya	3286	5.7	0.555	46	98	49	63	57	0.565	4.0
Madagascar	1555	2.3	0.512	–	160	46	39	38	–	3.5
Mauritius	22,279	3.7	0.781	3	23	14	94	98	0.380	1.5
Rwanda	2036	7.1	0.498	81	151	39	35	38	0.383	6.0
Seychelles	28,964	5.7	0.782	<2	17	14	79	84	–	3.0
Somalia	–	–	–	–	181	133	–	–	–	7.0
South Sudan	1,693	–10.4	0.418	–	256	91	12	7	–	7.0
Tanzania	2946	6.7	0.531	78	179	57	34	31	0.544	4.0
Uganda	1864	5.3	0.493	69	175	53	24	22	0.522	5.0
Central Africa										
Cameroon	3694	4.9	0.518	54	143	80	63	54	0.568	6.0
Central African Republic	726	–7.1	0.352	84	174	124	23	12	0.648	7.0
Chad	1941	5.9	0.396	85	211	127	31	14	0.695	6.5
Dem. Rep. of Congo (Kinshasa)	887	7.7	0.435	98	184	94	54	33	0.663	6.5
Rep. of Congo (Brazzaville)	5359	4.6	0.592	72	91	54	58	51	0.592	6.0
Equatorial Guinea	26,058	–0.1	0.592	–	191	91	–	–	–	7.0
Gabon	18,183	5.6	0.697	–	92	47	–	–	0.542	6.0
São Tomé and Principe	3351	4.7	0.574	62	105	34	81	92	0.524	2.0
Southern Africa										
Angola	6389	4.8	0.533	54	221	83	35	23	–	6.0
Botswana	17,354	5.6	0.698	49	54	41	–	–	0.435	2.5
Eswatini	8496	2.9	0.541	70	66	70	66	66	0.566	6.5
Lesotho	3130	4.5	0.497	79	91	94	46	62	0.549	3.0
Malawi	1202	4.4	0.476	90	232	55	46	41	0.614	3.0
Mozambique	1247	7.1	0.418	92	248	71	34	31	0.574	4.0
Namibia	10,476	5.6	0.640	55	71	45	–	–	0.474	2.0
South Africa	13,498	2.3	0.666	36	57	43	88	112	0.394	2.0
Zambia	4050	5.9	0.579	77	182	63	–	–	0.526	4.0
Zimbabwe	2086	7.2	0.516	–	75	56	48	47	0.540	5.5

[1] World Bank Open Data, 2018.
[2] World Bank - *World Development Indicators*, 2017.
[3] United Nations, *Human Development Report*, 2016.
[4] Population Reference Bureau, *World Population Data Sheet*, 2017.
[5] Freedom House—Freedom in the World 2018. See Ch.1, pp. 33–34, for more info. on this scale (1–7, with 7 representing states with the least freedom).
[6] See Ch. 1, p. 39, for more info. on this scale (0–1, with higher values representing less gender equality).

Log in to Mastering Geography & access MapMaster to explore these data!

1) Which countries rank the highest with regard to the Human Development Index? What factors account for these relatively high rankings?

2) Look at the secondary school enrollments for the various subregions of Sub-Saharan Africa. Which subregion has the lowest enrollments, and what might explain it? What could be done to improve education in these areas?

mineral and agricultural resources for their own benefit. Several plantation and mining zones did achieve some prosperity under colonial regimes, but strong national economies failed to develop. In almost all cases, the basic transport and communication systems were designed to link administrative centers and zones of extraction directly to the colonial powers, rather than to their own surrounding areas. As a result, after achieving independence, Sub-Saharan African countries faced economic and infrastructural challenges that were as daunting as their political problems.

Failed Development Policies   The first decade or so of independence was a time of relative prosperity and optimism for many African countries. Most relied heavily on the export of mineral and agricultural products, and through the 1970s commodity prices generally remained high. The region attracted some foreign capital, and in many cases, the European economic presence actually increased after decolonization.

In the 1980s, as most commodity prices began to decline, foreign debt began to weigh down many Sub-Saharan countries. By the 1990s, most states were registering low or negative growth rates. Not only was the AIDS crisis raging, but an economic and debt crisis in the 1980s and 1990s prompted the introduction of **structural adjustment programs** by the International Monetary Fund (IMF) and the World Bank. These programs typically reduce government spending, cut food subsidies, and encourage private-sector initiatives. Yet these same policies caused immediate hardships for the poor, especially women and children, and led to social protest, most notably in cities. Although the region's debt was low compared to that of other developing regions (such as Latin America), as a percentage of economic output, Sub-Saharan Africa's debt was the highest in the world.

Many economists argue that the region's governments enacted counterproductive economic policies and thus brought some of their misery on themselves. Eager to build their own economies and reduce dependency on the former colonial powers, most African countries followed a course of economic nationalism. More specifically, they set about building steel mills and other forms of heavy industry that were simply not competitive. Local currencies were often maintained at artificially elevated levels, which benefited the elite who consumed imported products, but undercut exports.

Reflecting a neoliberal turn toward agricultural production, several Sub-Saharan states—Uganda, Tanzania, Somalia, and Mozambique—have experienced a 21st-century land grab driven by food-insecure governments. Dutch geographer Annelies Zoomers describes food-insecure governments, such as China and the Persian Gulf states, as those that seek to outsource their domestic food production by buying or leasing vast areas of farmland abroad for their own offshore food production. Ironically, the places that these countries invest in tend to be poor countries with their own food insecurity issues.

Corruption   Although prevalent throughout most of the world, corruption seems to have been particularly widespread in several African countries. Part of this is driven by a lack of transparent and representative governance and by a civil service class that lacks both resources and professionalism. According to a 2018 international poll, Nigeria ranks as the most corrupt country. (Skeptical observers, however, point out that several Asian nations with highly successful economies, such as China, are also noted for having high levels of corruption, so corruption alone may not be the problem.)

With millions of dollars in loans and aid pouring into the region, officials at various levels have been tempted to take something for themselves. Some African states, such as the Democratic Republic of the Congo, were dubbed *kleptocracies*. A **kleptocracy** is a state in which corruption is so institutionalized that most politicians and government bureaucrats siphon off a huge percentage of the country's wealth. President Mobutu, who ruled the Democratic Republic of the Congo (then Zaire) from 1965 to 1997, was a legendary kleptocrat. While his country was saddled with an enormous foreign debt, he reportedly skimmed several billion dollars from government funds and deposited them in Belgian banks.

Signs of Economic Growth

Most of the region's economies are growing at a rate of 4–5 percent, even with the recent dip in commodity prices. And there is a small but growing African middle class who aspire to own homes and cars and consume more than basic needs. This growth is not just tied to natural resources; for example, mobile telephones have transformed commerce throughout the region. The percentage of people living in extreme poverty has steadily declined over the last 20 years. In addition, more children are enrolled in school, and tremendous strides have been made in combating HIV/AIDS and malaria (Table 6.2).

One bright spot in the region's economy is the growth in cellular and digital technology. Admittedly, fixed telephone lines are scarce; the regional average is 1 line per 100 people. Cell-phone usage, however, has soared, so that there are 76 cell phones per 100 people. Multinational providers compete for mobile-phone customers. Development specialists and entrepreneurs are exploring new uses for cell phones and smartphones, with applications to secure not only microfinance, but also educational tools and updates about health issues or weather patterns. In Kenya, one of the most wired countries in the region, 46 percent of the population uses the Internet; the average for the region as a whole is 22 percent. And Kenya's cell phone–based money transfer service, M-Pesa, allows residents to pay for goods and services from street food to bus fare with their phone (Figure 6.39). Africans are developing applications for their needs. Peter Kariuki of Rwanda taught himself to code as a child. He has developed SafeMotos, an Uber-like hailing service for clients to catch a motorcycle ride in the city.

▼ **Figure 6.39 Mobile Phones for Africa**   Kenyan farmers share information through cell phone texting. Cell phones have been transformational in improving access to information and communication in the region.

A Surge in New Infrastructure Limited paved roads and railroads place limits on national economies, but large-scale infrastructural programs are in the works. South Africa is the only African state with a fully developed modern road network. Recently, Kenya inaugurated its first superhighway, an eight-lane, 42-kilometer road from Nairobi to Thika. Not only has the highway transformed many towns along its route; it is also an expression of growing Chinese investment in this region because a Chinese firm performed much of the engineering and construction work.

Two major projects to improve regional train networks also are under way. In East Africa, billions are being invested to renovate existing railroads, standardize gauges, and construct new lines to link East African cities to the port of Mombasa, Kenya. A new line from Kigali, Rwanda, is under construction and will connect to Uganda and then to Kenya. There are also plans to extend the East African rail network into South Sudan, Ethiopia, and the eastern Democratic Republic of the Congo. Much of the engineering and some of the financing for this 3000-mile (4800-km) network comes from China (Figure 6.40). In 2015, seven West African states (Ivory Coast, Ghana, Togo, Benin, Nigeria, Burkina Faso, and Niger) announced plans to renovate, build, and integrate 1860 miles (3000 km) of railroads to facilitate the export of minerals and other primary products. For landlocked Burkina Faso and Niger, this project could greatly reduce transportation costs.

Major water and energy projects are also under way. Ethiopia is constructing the Grand Renaissance Dam on the Blue Nile. When completed, it will be Africa's largest dam. A project of this scale is not without controversy; the downstream nation of Egypt is deeply concerned that the dam will reduce total water flow into the Nile River. Others complain that the scale of the dam is more than Ethiopia needs—annual energy production is estimated to be 15,000 gigawatt-hours—and that the dam's environmental impacts are not fully understood. The region also has great solar energy potential. Sub-Saharan Africa's largest photovoltaic solar power project opened near Kimberly, South Africa, in 2014. The Jasper plant can produce 180,000 megawatt-hours of electricity annually, enough to power 80,000 homes. Other states in the region are interested in following South Africa's lead in solar power, but interest in coal-powered plants is growing as well (see *Exploring Global Connections: Coal-Powered Energy Comes to Africa*).

▼ **Figure 6.40 East African Railroad Construction** Kenyan workers lay new standard-gauge railroad track as part of the Chinese-funded regional railway project that will modernize the rail link between Kenya's port of Mombasa and its capital, Nairobi. Plans to improve railroad links throughout East Africa are under way.

Explore the **SIGHTS** of Mombasa's Port Facilities
http://goo.gl/BzSe41

Links to the World Economy

The economies of Sub-Saharan Africa generate about 2 percent of world's GNI, yet the region contains 13 percent of the global population. Not surprisingly, the overall level of international trade is low, both within and outside the region. Traditionally, most exports went to the European Union (EU), especially the former colonial powers of England and France. The United States is the second most common destination. That pattern is changing rapidly; China is now the single largest trading partner for the region, although collective trade between the EU nations and Sub-Saharan Africa is greater. Throughout the decade of the 2000s, China's trade with Sub-Saharan Africa grew 30 percent per year on average. During the same decade, India's and Brazil's trade with Sub-Saharan Africa grew more than 20 percent annually.

The rise of China as the largest trading partner of and investor in Sub-Saharan Africa has generated much geopolitical discussion about China's influence in this resource-rich, but developing, world region. In 2013, China's President Xi Jinping visited Tanzania, South Africa, and the Democratic Republic of the Congo, promoting the mutual benefits of Sino–African relations. Estimates of the number of Chinese currently working in Sub-Saharan Africa vary. At the high end, Chinese sources claim over 1 million Chinese migrants have settled in the region; the low end is a quarter that number. Most agree that the top destination is South Africa, where Chinese migrants number somewhere between 200,000 and 400,000.

Aid Versus Investment In many ways, Sub-Saharan Africa is more tightly linked to the global economy through the flow of financial aid and loans than through the flow of goods. As Figure 6.41 shows, for several states (Central African Republic, Liberia, Malawi, Sierra Leone, and Somalia) foreign aid accounted for more than 20 percent of GNI in 2015. Most of the aid comes from a handful of developed regions (Europe, North America, and East Asia). In 2014, the total value of foreign assistance was $33 billion, whereas remittances to Sub-Saharan African countries from migrants were valued at $34 billion.

Although aid is critical for many African states, foreign direct investment in the region substantially increased from only $4.5 billion in 1995 to $42 billion in 2015. Yet the overall level of foreign investment remains low when compared to other developing regions. In 2017, the largest foreign investment recipients were Angola, Ghana, Nigeria, and Ethiopia. China has been the leading investor in the region at a time when the United States and the EU are more focused on offering aid or fighting terrorism. China wants to secure the oil and ore it needs for its massive industrial economy. In exchange, it offers Sub-Saharan nations money for roads, railways, housing, and schools, with relatively few strings attached. Some African leaders see China as a new kind of global partner, one that wants straight commercial relations without an ideological or political agenda. Angola, a country in which China has invested heavily, is now one of China's top suppliers of oil.

Economic Differentiation Within Africa

As in most other regions, considerable differences in levels of economic and social development persist in Sub-Saharan Africa. In many respects, the small island nations of Mauritius and the Seychelles have little in common with the mainland. With high levels of per capita GNI, life expectancies averaging in the low 70s, and economies built on tourism, they could more easily fit into the Caribbean were it not for their Indian Ocean location.

EXPLORING GLOBAL CONNECTIONS

Coal-Powered Energy Comes to Africa

While much of the world is turning away from coal production in an attempt to reduce CO_2 emissions, Sub-Saharan Africa is experiencing a boom in coal-fired power plants. This demographically growing region, still overly reliant on biofuels (namely wood), views modern power generation as crucial for its development.

Sub-Saharan Africa is turning to coal because it desperately needs more reliable power supplies as demand grows even faster than its population. The reality is that over half of the region's residents do not have electricity in their homes. Yes, there are investments in alternative energy, but coal is relatively cheap and the technology to build these plants is well known. Africa does have coal supplies, but imports are also relatively cheap and available from China, India, South Africa, and even the United States (Figure 6.4.1).

New Power Plant Projects As of 2018, there were 100 coal-generating power plants in various stages of planning and construction in 11 African countries, not counting South Africa, which has long relied on coal for its electricity needs. If all the projects are completed, the region's coal-powered energy will increase eightfold. Nearly all these projects are financed by foreign direct investment, with half of the money coming from China (Figure 6.4.2).

One of the more controversial power plant projects is on the coast of Kenya, near the UN World Heritage Site of Lamu, a Swahili-speaking community of fishers who still rely on sailboats and donkeys to ferry their goods. Residents would like to have electricity, but they also worry about the environmental side effects: air pollution, warm water discharged from the plant's cooling system, and seepage from the open-air ash pit, which could endanger fisheries and the mangrove logging industry and pollute the coast.

Energy Versus Environment The irony is that Africa's coal-powered infrastructure is going up when much of the world has put coal-fired plants on hold in an effort to reduce greenhouse gas emissions. Coal

▶ **Figure 6.4.1** **African Coal** A train in Mozambique carries coal bound for export.

is dirty, and regions that depend on these plants suffer from foul air and the residents from respiratory ailments. In 2016, China slowed its building of domestic coal projects due to excess capacity and an urban public weary of excessive air pollution. But China has been a major developer of fossil fuel–based energy in Africa since the early 2000s. The initial interest was to develop oil resources, especially in Angola. Now Africa is seen as a new and eager market for Chinese coal and coal-burning plants. The long-term consequences of this strategy for the local and global environment are worrisome.

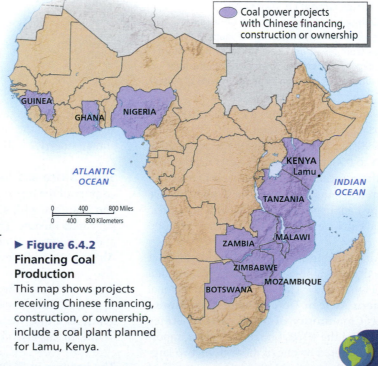

Coal power projects with Chinese financing, construction or ownership

GUINEA
GHANA
NIGERIA
ATLANTIC OCEAN
KENYA
Lamu
INDIAN OCEAN
TANZANIA
0   400   800 Miles
0   400   800 Kilometers
ZAMBIA
MALAWI
ZIMBABWE
MOZAMBIQUE
BOTSWANA

▶ **Figure 6.4.2** **Financing Coal Production** This map shows projects receiving Chinese financing, construction, or ownership, include a coal plant planned for Lamu, Kenya.

1. **Why is coal an appealing energy source for many Sub-Saharan African countries?**

2. **How important is coal for electricity production in your city, state, or country?**

GOOGLE EARTH
Virtual Tour Video
https://goo.gl/r6e7kd

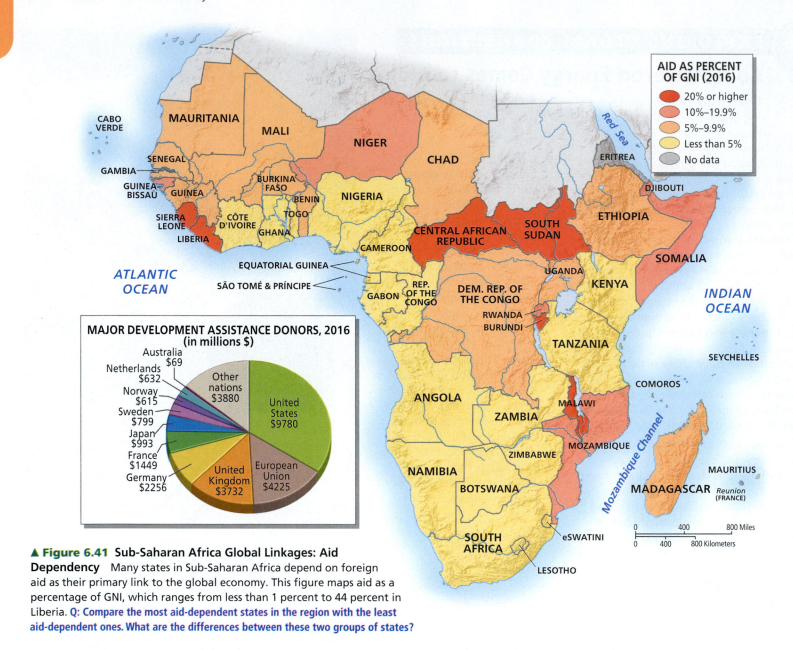

▲ Figure 6.41 Sub-Saharan Africa Global Linkages: Aid Dependency Many states in Sub-Saharan Africa depend on foreign aid as their primary link to the global economy. This figure maps aid as a percentage of GNI, which ranges from less than 1 percent to 44 percent in Liberia. **Q: Compare the most aid-dependent states in the region with the least aid-dependent ones. What are the differences between these two groups of states?**

Two small African states noted for oil wealth are Gabon (population 2 million) and Equatorial Guinea (less than 1 million), which began producing oil in the 1990s. In 2017, the GNI-PPP levels of these two states were $18,183 and $26,058, respectively. Yet after nearly three decades of oil production, these revenues have not been invested in the country's citizens; rather, as is often the case, they seem to have fallen into the pockets of a few members of the elite. Only a few states, mostly in southern Africa, have per capita GNI-PPP levels over $5000 (see Table 6.2).

Given the scale of the African continent, it is not surprising that groups of states have formed trade blocs to facilitate intraregional exchange and development (Figure 6.42). The two most active regional organizations are the **Southern African Development Community (SADC)** and the **Economic Community of West African States (ECOWAS)**. Both were founded in the 1970s but became more important in the 1990s, and each is anchored by one of the region's two largest economies: South Africa and Nigeria. In 2018, 44 African heads of state signed an agreement to create a new global trade bloc, the **African Continental Free Trade Area (AfCFTA)**. Though still in its infancy, if

this trade group is ratified by 22 national parliaments, it could eventually include all the countries on the continent (including North Africa) and another 1.2 billion people.

South Africa and SADC   South Africa is the most developed large country of the region, with a per capita GNI-PPP of $13,948. Botswana and Namibia, with strong mining economies, also do well in terms of per capita income. Through SADC, there have been efforts to integrate and improve the infrastructure of its member countries. Yet only South Africa has a well-developed and well-balanced industrial economy. It also boasts a healthy agricultural sector and, more important, it is one of the world's mining superpowers. South Africa remains unchallenged in gold production and is a leader in many other minerals and precious gems, including diamonds. Hosting the 2010 World Cup symbolized South Africa's arrival as a developed and modern nation.

The Leaders of ECOWAS   The largest economy and the most populous country in Africa, Nigeria is the core of ECOWAS. Nigeria has Sub-Saharan Africa's largest oil reserves, and it is an OPEC member. Yet

▶ **Figure 6.42 Regional Organizations of Sub-Saharan Africa** Political affiliations in Sub-Saharan Africa are both continental and regional. The AU includes all African countries, and the newly proposed AfCFTA could become a continental trade group. Smaller organizations, such as SADC, ECCAS, and ECOWAS, represent regional affiliations. Of these, SADC and ECOWAS show the most economic promise.

despite its natural resources, its per capita GNI-PPP was only $5861 in 2017. It has been argued that oil money has helped to make Nigeria notoriously inefficient and corrupt. A small minority of Nigerians have grown fantastically wealthy, more by manipulating the system than by engaging in productive activities. Eight out of ten Nigerians, however, live on less than $3.10 per day.

The second and third most populous states in ECOWAS, the Ivory Coast and Ghana, are also important commercial centers that rely on a mix of agricultural and mineral exports. In the mid-1990s, the Ivorian economy began to take off. Boosters within the country called it an emerging "African elephant" (comparing it to the successful "economic tigers" of eastern Asia). However, a destructive civil war that began in 2002 resulted in rebel forces controlling the northern half of the country and over half a million displaced Ivorians. A peace agreement was signed in 2007, but the economic growth of the 1990s has yet to return. Ghana, a former British colony, also began to see economic recovery in the 1990s. In 2001, Ghana negotiated with the IMF and the World Bank for debt relief to reduce its nearly $6 billion foreign debt. Between 2009 and 2015, Ghana maintained an average annual growth rate of 8 percent. In 2011, Ghana also became an emerging oil producer for Africa, with offshore wells being pumped near the city of Takoradi.

East Africa Long the commercial and communications center of East Africa, Kenya experienced economic decline and political tension throughout the 1990s. Yet from 2009 to 2015, its economy averaged 5.7 percent annual growth, and its per capita GNI-PPP was at $3,286. Kenya boasts good infrastructure by African standards, and over 1 million foreign tourists come each year to marvel at its wildlife and natural beauty. Traditional agricultural exports of coffee and tea, as well as nontraditional exports such as cut flowers, dominate the economy.

Kenya is also East Africa's technological leader. In 2009, the government launched plans for Konza City, a walkable technology-oriented development 60 kilometers (37 miles) southeast of Nairobi.

The intent is to capture the growing business in outsourced information technology services; Kenya is an English-speaking country where 82 percent of youth (ages 15–24) are literate. If Kenya can avoid political unrest due to ethnic rivalries, it could lead East Africa toward better economic integration.

The political and economic indicators for Kenya's neighbors, Uganda and Tanzania, are also improving, with average annual growth rates at 5.3 and 6.7 percent, respectively. Both countries rely heavily on agricultural exports and mining (especially gold), and both benefited from debt reduction agreements in the 2000s that redirected debt repayment funds to education and health care.

Measuring Social Development

By global standards, measures of social development in Sub-Saharan Africa are extremely low. Yet some positive trends, especially with regard to child survival and youth literacy, are cause for hope (see Table 6.2). Many governments in the region have reached out to the modern African diaspora now living in Europe and North America. In other parts of the world, African immigrant organizations have worked to improve schools and health care, and former emigrants have returned to invest in businesses and real estate. The economic impact of remittances in this region is small, but growing.

Child Mortality and Life Expectancy   Reductions in child mortality are a surrogate measure for improved social development, because if most children make it to their fifth birthday, it usually indicates adequate primary care and nutrition. A child mortality rate of 200 per 1000 means that 1 child out of 5 dies before his or her fifth birthday. As Table 6.2 shows, most states in the region saw modest to significant improvements in child survival between 1990 and 2016; Eritrea, Liberia, Madagascar, Malawi, and Rwanda actually experienced dramatic gains in child survival rates. However, countries with prolonged conflict (such as Somalia and the Democratic Republic of the Congo) have seen little improvement. In 1990, the regional child mortality rate was 175 per 1000; in 2015, it was down to 83. Thus, high child mortality is still a major concern, but steady reductions in this rate are significant for African families.

Life expectancy for Sub-Saharan Africa is 60 years. Yet there are indications that access to basic health care is improving and, eventually, so will life expectancy. Keep in mind that high infant- and child-mortality figures depress overall life expectancy figures; average life expectancies for people who make it to adulthood are much better.

Low life expectancies are generally related to extreme poverty, environmental hazards (such as drought), and various diseases (malaria, cholera, AIDS, and measles). Often these factors work in combination. Malaria, for example, kills 300,000 African children each year. The death rate is also affected by poverty, as undernourished children are the most vulnerable to the effects of high fevers. Tragically, preventable diseases such as measles occur when people have no access to or cannot afford vaccines. National and international health agencies, along with NGOs such as the Gates Foundation, are working to improve access to vaccines, bed nets (to prevent malaria), and primary health care. These efforts are making a difference.

Meeting Educational Needs   Basic education is another challenge for the region. The goal of universal access to primary education is a daunting one for a region where 43 percent of the population is under 15 years old. The UN estimates that fewer than half of all African children receive a secondary school (high school or its equivalent) education. Sub-Saharan Africa is home to one-sixth of the world's children under 15, but to half of the world's uneducated children. Girls are still less likely than boys to attend school. In West African countries such as Chad, Benin, Niger, and the Ivory Coast, girls are decidedly underrepresented.

A renewed focus on education since 2000 has been attributed to UN efforts to reduce extreme poverty by focusing on basic education, health care, and access to clean water through the **Sustainable Development Goals**. More government and nonprofit organizations' resources have been directed toward education. More schools are being built across the region, and more children are attending them (Figure 6.43).

Women and Development

Development gains cannot be achieved in Africa unless the economic contributions of African women are recognized. Women are the

▲ **Figure 6.43 Educating African Youth**   Shown here is a high school classroom in Hevie, Benin. One of the greatest challenges for this region is educating its youth beyond primary school.

invisible contributors to local and national economies. In agriculture, women account for 75 percent of the labor that produces more than half the food consumed in the region. Tending subsistence plots, taking in extra laundry, and selling surplus produce in local markets all contribute to household income. Yet because many of these activities are considered part of the informal sector, they are not counted. For many of Africa's poorest people, however, the informal sector is the economy, and within this sector, women dominate.

Status of Women   The social position of women is difficult to measure for Sub-Saharan Africa. Female traders in West Africa, for example, have considerable political and economic power. By such measures as female labor force participation, many Sub-Saharan African countries show relative gender equality. Also, women in most Sub-Saharan societies do not suffer the kinds of traditional social restrictions encountered in much of South Asia, Southwest Asia, and North Africa; in Sub-Saharan Africa, women work outside the home, conduct business, and own property. In 2006, Ellen Johnson-Sirleaf was sworn in as Liberia's president, making her Africa's first elected female leader. In 2012, she was joined by Joyce Banda, elected president of Malawi. In fact, throughout the region, women occupied 22 percent of all seats in national parliaments in 2014. In Rwanda, over half the parliamentary seats are filled by women (Figure 6.44).

By other measures, however, such as the prevalence of polygamy, the practice of the "bride-price," and the tendency for males to inherit property over females, African women do suffer discrimination. Perhaps the most controversial issue regarding women's status is the practice of female circumcision, or genital mutilation. In Ethiopia, Somalia, and Eritrea, as well as parts of West Africa, the majority of girls are subjected to this practice, which is extremely painful and can have serious health consequences. Yet because the practice is considered traditional, most African states are unwilling to ban it.

Regardless of social position, most African women still live in remote villages where educational and wage-earning opportunities remain limited and caring for large families is time-consuming and

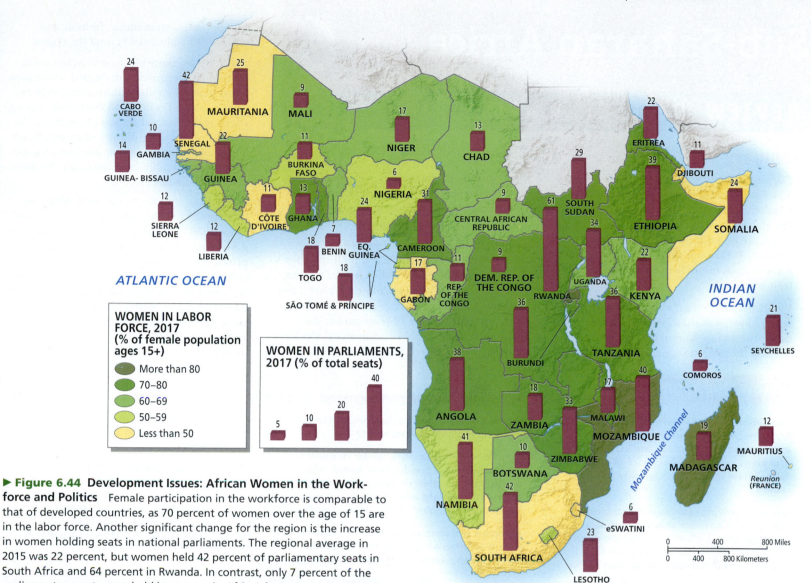

▶ **Figure 6.44** **Development Issues: African Women in the Workforce and Politics** Female participation in the workforce is comparable to that of developed countries, as 70 percent of women over the age of 15 are in the labor force. Another significant change for the region is the increase in women holding seats in national parliaments. The regional average in 2015 was 22 percent, but women held 42 percent of parliamentary seats in South Africa and 64 percent in Rwanda. In contrast, only 7 percent of the parliamentary seats were held by women in Africa's largest country, Nigeria.

demanding labor. As education levels increase and urban society expands—and as reduced infant mortality provides greater security—we can expect fertility in the region to gradually decrease. Governments can speed up the process by providing birth control information and cheap contraceptives—and by investing more money in women's health and education. As the economic importance of women receives greater attention from national and international organizations, more programs are being directed exclusively toward them.

Building from Within

Surveys reveal that the majority of people in Sub-Saharan Africa are optimistic about their future. Considering many of the real development hurdles the region faces, this surprises outside observers. Yet considering the levels of conflict, food insecurity, and neglect that Sub-Saharan Africa experienced during the 1990s, perhaps the developments of the past two decades are a cause for hope. Most African states have been independent for only half a century. During that time, these countries have shifted from one-party authoritarian states to multiparty democracies. Targeted aid projects to address particular critical indicators, such as mother and child mortality or

malaria prevention, have proved their effectiveness. Civil society is also vigorous, from raising the status of women to supporting small businesses with micro-credit loans. Even members of the African diaspora are beginning to return and invest in their countries. The rapid adoption of and adaptation to mobile phone technology in the region demonstrates Sub-Saharan Africans' desire to be better connected with each other and with the world.

REVIEW

6.10 **What are the historical, structural, and institutional reasons offered to explain poverty in the region?**

6.11 **What technological and infrastructural investments are affecting the region's development?**

KEY TERMS structural adjustment program, kleptocracy, Southern African Development Community (SADC), Economic Community of West African States (ECOWAS), African Continental Free Trade Area (AfCFTA), Sustainable Development Goals

REVIEW, REFLECT, & APPLY

Summary

- The largest landmass straddling the equator, Africa is called the plateau continent because it is dominated by extensive uplifted plains. Key environmental issues facing this tropical region are desertification, deforestation, and drought. At the same time, the region supports a tremendous diversity of wildlife, especially large mammals.

- With 1 billion people, Sub-Saharan Africa is the fastest-growing world region in terms of population; the average woman has five children. Yet it is also the poorest region, with 41 percent of the population living in extreme poverty (less than $1.90 per day). In addition, it has the lowest average life expectancy, at 60 years. HIV/AIDS has hit this region especially hard.

- Culturally, Sub-Saharan Africa is extremely diverse, where multiethnic and multireligious societies are the norm. With a few exceptions, religious diversity and tolerance have been distinctive features of the region. Most states have been independent for 50 years, and in that time, pluralistic, but distinct, national identities have been forged. Many African cultural expressions, such as music, fashion, and religion, are influential beyond the region.

- In the 1990s, many bloody ethnic and political conflicts occurred in the region. Peace now exists in many conflict-ridden areas, such as Angola, Sierra Leone, and Liberia, but ongoing ethnic and territorial disputes in Somalia, the Democratic Republic of the Congo, and South Sudan, and more recently in Mali and Nigeria, have produced millions of internally displaced persons and refugees.

- Widespread poverty and limited infrastructure are the region's most pressing concerns. Since 2000, Sub-Saharan economies have grown, led in part by higher commodity prices, greater investment, foreign assistance, and the end of some of the longest-running regional conflicts. Social indicators of development are also improving, due to greater attention from the international community and the formulation of Sustainable Development Goals.

Review Questions

1. Sub-Saharan Africa is noted for its wildlife, especially its large mammals. What environmental, historical, and institutional processes explain the existence of so much fauna? How important is wildlife for this region?

2. Consider how disease impacted colonization in Latin America and the Caribbean (demographic decline). How did the disease environment differ in Sub-Saharan Africa, and how did this delay colonization of this region? How do various diseases still influence the region's development?

3. Rates of urbanization are on the rise everywhere, including Sub-Saharan Africa. What are the urbanization challenges facing this region? How might a more urbanized region impact demographic and economic trends?

4. Compare and contrast the role of tribalism in Sub-Saharan Africa with that of nationalism in Europe. How might competing forms of social loyalties explain some of the development challenges and opportunities for this region?

5. Consider the development model put forward by the United States and Europe, with an emphasis on foreign assistance, and China's emphasis on foreign investment. Will Chinese influence in the region alter the course of development for Sub-Saharan Africa?

Image Analysis

Consider this image of per capita income for selected states in Africa South of the Sahara.

1. Compare the data for the three African subregions pictured, and describe the income patterns for each. What demographic, geopolitical, and socioeconomic factors explain these patterns?

2. How might this map appear if data were shown at the sub-state level? For example, look at Figure 6.1 and suggest which areas of South Africa have the highest and lowest incomes, and why.

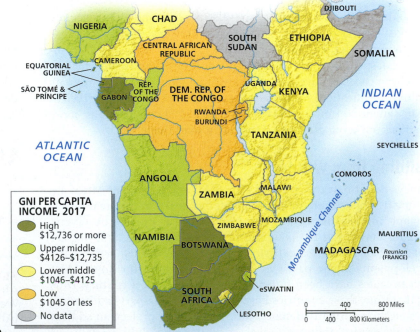

GNI PER CAPITA INCOME, 2017
- High $12,736 or more
- Upper middle $4126–$12,735
- Lower middle $1046–$4125
- Low $1045 or less
- No data

▶ **Figure IA6** Income Level by State

The United Nations recently implemented 17 Sustainable Development Goals that focus on ending poverty and addressing inequalities while promoting sustainable growth. SDGs replace Millennium Development Goals that targeted only developing countries, by covering more ground and calling on developed countries to help attain benchmarks. Can SDGs effectively promote sustainable development?

SDGs will reduce poverty in Sub-Saharan states.

- Since the creation of the Millennium Development Goals in 2000, more people have been lifted out of extreme poverty. The SDGs hope to continue that trend.

- Focusing on specific targets, such as decreasing maternal mortality or increasing youth literacy, will contribute to the region's overall social and economic development.

- The SDGs are broadly conceived benchmarks to support a holistic approach toward development, and do not focus exclusively on economic growth as a measure of success.

SDGs are unrealistic and do not impact development.

- Such metrics are often abused, and countries manipulate the data to look good rather than address development concerns.

- Much of the development that occurred globally in the 2000s was due to factors other than foreign assistance to support these goals.

- National-level data mask the complex patterns of poverty and underdevelopment, especially for large countries.

▲ **Figure D6** **Sustainable Development Goals** Pearson Education supports the Sustainable Development Goals

Key Terms

African Continental Free Trade Area (AfCFTA) (p. 212)
African Union (AU) (p. 204)
animism (p. 198)
apartheid (p. 195)
Berlin Conference (p. 203)
biofuels (p. 183)
clan (p. 206)
coloured (p. 195)
desertification (p. 182)
diaspora (p. 200)

Economic Community of West African States (ECOWAS) (p. 212)
food insecurity (p. 186)
Great Escarpment (p. 177)
Great Rift Valley (p. 177)
Horn of Africa (p. 181)
internally displaced person (IDP) (p. 206)
kleptocracy (p. 209)
pastoralist (p. 192)
refugee (p. 206)
Sahel (p. 182)
sharia law (p. 199)

Southern African Development Community (SADC) (p. 212)
structural adjustment program (p. 209)
Sustainable Development Goals (p. 214)
swidden (p. 191)
township (p. 195)
transhumance (p. 182)
tribalism (p. 205)
tribes (p. 197)
tsetse fly (p. 192)

Mastering Geography

Looking for additional review and test prep materials? Visit the Study Area in Mastering Geography to enhance your geographic literacy, spatial reasoning skills, and understanding of this chapter's content by accessing a variety of resources, including MapMaster interactive maps, geoscience animations, videos, flashcards, web links, self-study quizzes, and an eText version of *Globalization and Diversity*.

Scan to read about Geographer at Work
Fenda Akiwumi and her study of human–environment dynamics in West Africa.
https://goo.gl/T2hep6

GeoSpatial Data Analysis

HIV/AIDS Across Sub-Saharan Africa Sub-Saharan Africa has experienced more AIDS-related deaths and holds more people living with HIV/AIDS than any other world region. Country-specific data can shed light on the impact of this disease and its geography.

Open MapMaster 2.0 in the Mastering Geography Study Area. Now go to the United Nations AIDS information website (http://aidsinfo.unaids.org/) and access the data sheet that includes country data that can be downloaded as .xls files. Download and then prepare the data for people living with HIV; new HIV infections; and AIDS-related deaths. Now select only Sub-Saharan African countries, and map variables.

UNAIDS AIDSinfo
https://goo.gl/KPu5h3

1. Which countries have the largest population living with HIV/AIDS in absolute numbers, and how has this changed over time? Which countries are experiencing the highest rates of new HIV infections?

2. What does the map of AIDS-related deaths look like? Which states have been the most impacted?

7

Southwest Asia and North Africa

Physical Geography and Environmental Issues

The region's vulnerability to water shortages is increasing as growing populations, rapid urbanization, increasing demands for agricultural land, and climate change pressure limited supplies.

Population and Settlement

Many settings within the region continue to see rapid population growth. These demographic pressures are particularly visible in fragile, densely settled rural zones as well as in fast-growing large cities.

Cultural Coherence and Diversity

Islam continues to be a vital cultural and political force within the region, but increasing fragmentation within that world has led to more culturally defined political instability.

Geopolitical Framework

Widespread political instability dominates the region. While the threat of ISIS has lessened in Iraq and Syria, broad tensions exist between Sunni and Shiite political factions across the region. Particularly deadly conflicts still consume nations such as Syria and Yemen.

Economic and Social Development

Unstable world oil prices and unpredictable geopolitical conditions have discouraged investment and tourism in many countries. The pace of social change, especially for women, has quickened, stimulating diverse regional responses.

► Symbolic heart of the city, Mosul's Al Nuri Grand Mosque was all but destroyed in the recent conflict with ISIS. The Mosque is now being rebuilt, thanks to a generous donation from the United Arab Emirates.

Mosul, Iraq

NORTH AFRICA
SOUTHWEST ASIA

The legendary Al Nuri Grand Mosque in Iraq's city of Mosul became a tragic victim of war in 2017. Much of the centuries-old structure was destroyed by **ISIS** (Islamic State of Iraq and the Levant; also known as ISIL or IS) fighters as they lost their hold on Iraq's second largest city in a brutal street-by-street battle that ultimately drove the extremist Sunni group out of the area. Iraqi forces retaking Mosul mourned the rubble they found. Only the Grand Mosque's leaning minaret, built in 1172 CE and an enduring visual icon of the city's identity, partially survived the onslaught.

But things are looking up for Mosul as it slowly rebuilds. Thanks to a generous gift from the oil-rich United Arab Emirates (UAE), the mosque and related religious education complex will be rebuilt. The UAE's donation, along with United Nations support, will fund a five-year project to restore the beauty and cultural heritage of the famed mosque. Hopefully, by 2025, the scars of war will only be a bad memory, and the Grand Mosque will once again serve as Mosul's symbolic heart.

Defining Southwest Asia and North Africa

Climate, culture, and oil help define Southwest Asia and North Africa. Located where Europe, Asia, and Africa meet, the region includes thousands of square miles of parched deserts, rugged plateaus, and oasis-like river valleys (Figure 7.1).

> **Climate, culture, and oil help define Southwest Asia and North Africa.**

"Southwest Asia and North Africa" is both an awkward term and a complex region. The area is often called the Middle East, but some experts exclude the western parts of North Africa, as well as Turkey and Iran, from such a region. Moreover, "Middle East" suggests a European point of view, reflecting the states that colonized the region and still shape the names we give the world today.

Diverse languages, religions, and ethnic identities have molded land and life within the region, strongly wedding people and place. One traditional zone of conflict surrounds Israel, where Jewish, Christian, and Islamic peoples have yet to resolve long-standing differences. In the early 2010s, **Arab Spring** movements—a series of public protests, strikes, and rebellions calling for fundamental government and economic reforms—toppled several governments in the region and pressured others to accelerate political and economic change. Yet the uprisings failed to produce many democratic reforms or more stable political regimes, and some observers suggest that Arab Spring has been replaced by "Arab Winter."

At the same time, cycles of **sectarian violence**—conflicts dividing people along ethnic, religious, and sectarian lines—plague the region. For example, enduring differences have led to clashes between Jews and Muslims and among varying factions of Islam. Syria and Yemen have become violent settings for recent instability due to growing sectarian conflicts between various Sunni and Shiite factions. While still a destabilizing force, ISIS has been weakened since 2016, and its hold on territory in Iraq and Syria has been vastly diminished.

Islamic fundamentalism in the region more broadly advocates a return to traditional Islamic practices. A related political movement, termed **Islamism**, challenges the encroachment of global popular culture

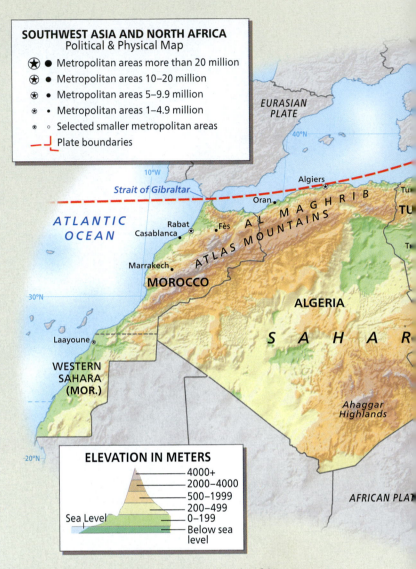

SOUTHWEST ASIA AND NORTH AFRICA
Political & Physical Map
- Metropolitan areas more than 20 million
- Metropolitan areas 10–20 million
- Metropolitan areas 5–9.9 million
- Metropolitan areas 1–4.9 million
- Selected smaller metropolitan areas
- Plate boundaries

ELEVATION IN METERS
- 4000+
- 2000–4000
- 500–1999
- 200–499
- 0–199
- Below sea level

Sea Level

▲ **Figure 7.1 Southwest Asia and North Africa** This vast region extends from the shores of the Atlantic Ocean to the Caspian Sea. Within its boundaries, major cultural differences and globally important petroleum reserves have contributed to political tensions.

and blames colonial, imperial, and Western elements for the region's political, economic, and social problems.

No world region better exemplifies globalization than Southwest Asia and North Africa. The region is a key **culture hearth**, producing religions and civilizations of global significance. Particularly within the past century, the region's strategic importance increasingly opened Southwest Asia and North Africa to outside influences. The 20th-century development of the petroleum industry, largely initiated by U.S. and European investment, had enormous consequences for the region. Today, many of the world's largest oil producers are found in the region, and they greatly influence global prices and production levels.

EURASIAN PLATE

Black Sea
Bosporus
Istanbul
Bursa
Dardanelles
Izmir
ANATOLIA
Konya
Antalya
Adana
TURKEY
Ankara
Kayseri
Gaziantep
Aleppo
Mosul
SYRIA
Hamah
Homs
Beirut
LEBANON
Damascus
ISRAEL
Haifa
West Bank
Tel Aviv
Jerusalem
Gaza Strip
Amman
JORDAN
Jordan River
Dead Sea
Nile Delta
Alexandria
Benghazi
Mediterranean Sea
Suez Canal
Cairo
Sinai
EGYPT
LIBYA
Libyan Desert
Nile River
Aswan
Aswan High Dam
Lake Nasser
Nubian Desert
Red Sea
SUDAN
Khartoum
White Nile R.
Blue Nile R.
Medina
Yanbu
Jeddah
Mecca
SAUDI ARABIA
KUWAIT
Kuwait
Riyadh
QATAR
BAHRAIN
Manama
Doha
Ad Dammam
Jubail
Sharjah
Dubai
Abu Dhabi
U.A.E.
OMAN
Gulf of Oman
Muscat
Tropic of Cancer
ARABIAN PENINSULA
Rub al Khali
OMAN
Tabriz
ELBURZ MTS.
Karaj
Tehran
Qom
Mashhad
DASHT-E KAVIR
ZAGROS MOUNTAINS
IRAN
Esfahan
Shiraz
Ahvaz
Al Basrah
Persian Gulf
Strait of Hormuz
Erbil
Sulaymaniyah
Kirkuk
Baghdad
IRAQ
Tigris R.
Euphrates R.
Caspian Sea

0   250   500 Miles
0   250   500 Kilometers

ARABIAN PLATE
20°N
Sanaa
YEMEN
INDIAN OCEAN
Gulf of Aden
Bab al-Mandeb
50°E
Socotra (YEMEN)
10°N
60°E
AFRICAN PLATE

LEARNING OBJECTIVES

After reading this chapter you should be able to:

7.1 Explain how latitude and topography produce the region's distinctive patterns of climate.

7.2 Connect the region's fragile, often arid setting to contemporary environmental challenges.

7.3 Identify four distinctive ways in which people have adapted their agricultural practices to the region's arid environment.

7.4 Summarize the major forces shaping recent migration patterns within the region.

7.5 List the major characteristics and patterns of diffusion of Islam.

7.6 Identify the key modern religions and language families that dominate the region.

7.7 Describe how cultural variables shape major conflicts in North Africa, Israel, Syria, Iraq, and the Arabian Peninsula.

7.8 Summarize the geography of oil and gas reserves in the region.

7.9 Describe traditional roles for Islamic women, and provide examples of recent changes.

Physical Geography and Environmental Issues: Life in a Fragile World

In the popular imagination, much of Southwest Asia and North Africa is a land of shifting sand dunes, searing heat, and scattered oases. Although examples of those stereotypes certainly exist, the actual physical setting is much more complex. One theme is dominant, however: A lengthy legacy of human settlement has left its mark on a fragile environment, and the entire region is facing increasingly difficult ecological problems.

Regional Landforms

A quick tour of Southwest Asia and North Africa reveals diverse environmental settings (see Figure 7.1). In North Africa, the **Maghreb** (meaning "western island") includes the nations of Morocco, Algeria, and Tunisia and is dominated near the Mediterranean coast by the Atlas Mountains. The rugged flanks of the Atlas rise like a series of islands above the narrow coastal plains to the north and the vast stretches of the lower Saharan deserts to the south (Figure 7.2). South and east of the Atlas Mountains, interior North Africa varies between rocky plateaus and extensive desert lowlands. In northeast Africa, the Nile River shapes regional drainage patterns as it flows north through Sudan and Egypt (Figure 7.3).

Southwest Asia is more mountainous than North Africa. In the **Levant**, or eastern Mediterranean region, mountains rise within 20 miles (30 km) of the sea, and the highlands of Lebanon reach heights of more than 10,000 feet (3000 meters). Farther south, the Arabian Peninsula forms a massive tilted plateau, with western highlands higher than 5000 feet (1500 meters) gradually sloping eastward to extensive lowlands in the Persian Gulf area. North and east of the Arabian Peninsula lie the two great upland areas of Southwest Asia: the Iranian and Anatolian plateaus (*Anatolia* refers to the large peninsula of Turkey, sometimes called Asia Minor; see Figure 7.1). Both plateaus, averaging 3000–5000 feet (1000–1500 meters) in elevation, are geologically active and prone to destructive earthquakes. In 2017, the world's deadliest earthquake that year struck near the Iraq border in western Iran, killing more than 600 people, injuring 8000, and leaving more than 70,000 homeless (Figure 7.4).

▲ **Figure 7.3 Nile Valley** This satellite image dramatically reveals the impact of water on the North African desert. Cairo lies at the southern end of the Nile Delta, where it begins to widen toward the Mediterranean Sea. The coastal city of Alexandria sits on the northwest edge of the low-lying delta. Lake Nasser is visible toward the bottom (center) of the image.

Smaller lowlands characterize other portions of Southwest Asia. Narrow coastal strips are common in the Levant, along the southern (Mediterranean Sea) and the northern (Black Sea) Turkish coastlines, and north of Iran's Elburz Mountains near the Caspian Sea.

▼ **Figure 7.4 Iranian Earthquake, 2017** Residents collect their useful household items from an earthquake-damaged apartment house in Kermanshah, Iran.

▼ **Figure 7.2 Atlas Mountains** The rugged Atlas Mountains dominate a broad area of interior Morocco. This view shows a landscape with cultivated fields and a small settlement in the Ait Bouguemez Valley.

▶ **Figure 7.5 Jordan Valley** This view of the fertile Jordan Valley shows a mix of irrigated vineyards and date palm plantations.

Explore the SIGHTS of the Jordan Valley
http://goo.gl/bTPRv5

Iraq contains the most extensive alluvial lowlands in Southwest Asia, dominated by the Tigris and Euphrates rivers, flowing southeast to empty into the Persian Gulf. The much smaller Jordan River Valley is a notable lowland that straddles the strategic borderlands of Israel, Jordan, and Syria and drains southward to the Dead Sea (Figure 7.5).

Patterns of Climate

Although the region of Southwest Asia and North Africa is often termed the "dry world," a closer look reveals more complex patterns (Figure 7.6). Both latitude and altitude come into play. Aridity

▶ **Figure 7.6**
Climate of Southwest Asia and North Africa Dry climates dominate from western Morocco to eastern Iran. Within these zones, persistent subtropical high-pressure systems offer only limited opportunities for precipitation. Elsewhere, mild midlatitude climates with wet winters are found near the Mediterranean Basin and Black Sea.

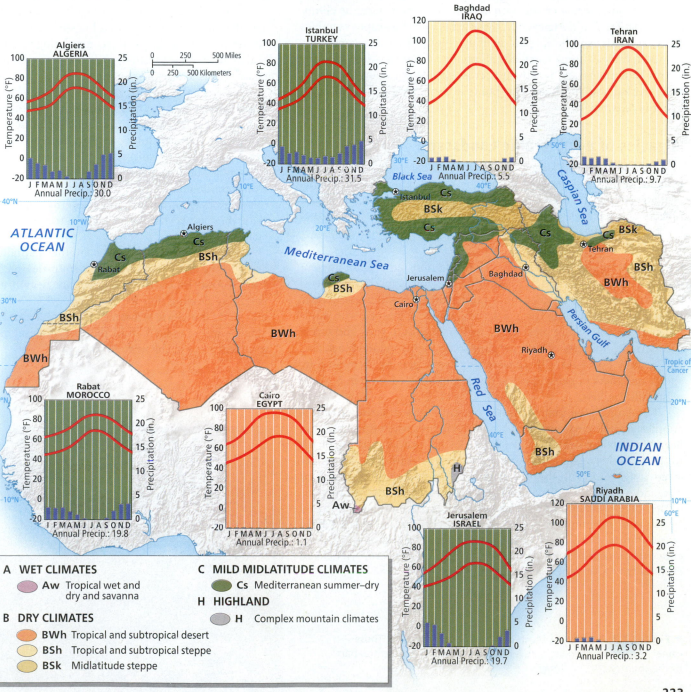

A WET CLIMATES

Aw Tropical wet and dry and savanna

B DRY CLIMATES

BWh Tropical and subtropical desert
BSh Tropical and subtropical steppe
BSk Midlatitude steppe

C MILD MIDLATITUDE CLIMATES

Cs Mediterranean summer–dry

H HIGHLAND

H Complex mountain climates

223

dominates large portions of the region (see the climographs for Cairo, Riyadh, Baghdad, and Tehran in Figure 7.6). A nearly continuous belt of desert land stretches eastward across interior North Africa, through the Arabian Peninsula, and into central and eastern Iran (Figure 7.7). Throughout this zone, plant and animal life adapts to extreme conditions. Deep or extensive root systems allow desert plants to benefit from the limited moisture they receive, and animals efficiently store water, hunt at night, or migrate seasonally to avoid the worst of the dry cycle. Intriguingly, the region's abundant sunshine and expansive desert landscapes also make it an ideal setting for solar energy. Recently, one of the world's largest and most modern solar power facilities opened in Morocco (see *Working Toward Sustainability: Noor 1 Shines Brightly in the North African Desert*).

Elsewhere, altitude and latitude produce a surprising variety of climates. The Atlas Mountains and nearby lowlands of northern Morocco, Algeria, and Tunisia experience a Mediterranean climate, in which dry summers alternate with cooler, wet winters (see the climographs for Rabat and Algiers). In these areas, the landscape resembles those in nearby southern Spain or Italy (Figure 7.8). A second zone of Mediterranean climate extends along the Levant coastline, into the nearby mountains, and northward across sizable portions of northern Syria, Turkey, and northwestern Iran (see the climographs for Jerusalem and Istanbul).

Legacies of a Vulnerable Landscape

The environmental history of Southwest Asia and North Africa reflects both the shortsighted and the resourceful practices of its human occupants. Littered with environmental problems, the region reveals the hazards of lengthy human settlement in a marginal land (Figure 7.9).

Deforestation and Salinization
Deforestation remains an ancient regional problem. Although much of the region is too dry for trees, the more humid and elevated lands bordering the Mediterranean once supported heavy forests. Human activities have combined with natural conditions to reduce most of the region's forests to grass and scrub. Mediterranean forests often grow slowly, are highly vulnerable to fire, and usually fare poorly if subjected to heavy grazing. Browsing by sheep and goats in particular has often been blamed for much of the region's forest loss.

The buildup of toxic salts in the soil, termed **salinization**, is another ancient environmental issue in a region where irrigation has been practiced for centuries (see Figure 7.9). The accumulation of salt in the topsoil is a common problem wherever desert lands are extensively irrigated. All freshwater contains a small amount of dissolved salt, and when water is diverted from streams into fields, salt remains in the soil after the moisture is absorbed by the plants and evaporated by the sun. In humid climates, accumulated salts are washed away by saturating rains, but in arid climates this rarely occurs. The salt concentrations build up over time, leading to lower crop yields and eventually to abandoned lands.

Water Management Southwest Asia and North Africa is the most water-stressed part of the world: about 5 percent of the world's population lives in a region possessing only about 1 percent of the planet's renewable freshwater resources. Water management will remain a critical problem because of continued population growth, widespread use of inefficient irrigation systems, overuse of depleted groundwater supplies, and the longer-term consequences of climate change projected for the region. Water has been manipulated and managed in the region for thousands of years, and traditional systems emphasized directing and conserving surface and groundwater resources at a local scale; but in the past half-century, the scope of environmental change has been greatly magnified.

One remarkable example is Egypt's Aswan High Dam, completed in 1970 on the Nile River south of Cairo (see Figure 7.9). Increased storage capacity in the upstream reservoir made more water available for agriculture and generates clean electricity. But irrigation has increased salinization because water is not rapidly flushed from fields. The dam has led to more use of costly fertilizers, the infilling of Lake Nasser behind the dam with accumulating sediments, and the collapse of the Mediterranean fishing industry near the Nile Delta, an area previously nourished by the river's silt.

Additional water-harvesting strategies have also proven useful. **Fossil water**, or water supplies stored underground during earlier and wetter climatic periods, is being mined. For example, Saudi Arabia has invested huge sums to develop deep-water wells, allowing it to greatly expand its food output. In another example, Libya's Great Manmade River project provides abundant groundwater from remote regions of the country to the northern coast. Unfortunately, underground supplies are being depleted much more rapidly than they are recharged in such settings, limiting their long-term sustainability.

Desalination of seawater is another partial solution to this water crisis in the dry world. Wealthier countries such as Saudi Arabia, the United Arab Emirates, and Israel are heavily investing in new and emerging desalination technologies, which can more affordably and efficiently transform seawater into potable drinking water. The region already accounts for half of the world's desalination plants, and that total will grow in the next 20 years.

▶ **Figure 7.7 Arid Iran** Much of Iran is arid or semiarid. These camel herders are trekking across Isfahan province in the central part of the country.

◀ **Figure 7.8** **Mediterranean Landscape, Northern Algeria**   The Mediterranean moisture in northern Algeria produces an agricultural landscape similar to that of southern Spain or Italy. Winter rains create a scene that contrasts sharply with deserts found elsewhere in the region.

huge environmental and economic consequences for a nation that depends on the Nile for 85 percent of its water.

In Southwest Asia, Turkey's growing development of the upper Tigris and Euphrates rivers (the Southeast Anatolia Project, or GAP), complete with 22 dams and 19 power plants, has raised issues with Iraq and Syria, who argue that capturing "their" water might be considered a provocative political act. Turkey has periodically withheld water from Syria by controlling flows along the Euphrates, provoking protests. Iraq remains highly vulnerable as well: many experts fear that once Turkey's huge Ilisu Dam (the largest in the country) is completed in 2019 and 2020, lower flows along the Tigris will impact broad areas downstream, even many of the valuable marshlands in southern Iraq near Basra (Figure 7.10).

Hydropolitics also has played into negotiations among Israel, the Palestinians, and other neighboring states, particularly in the Golan Heights (an important headwaters zone for multiple nations) and

Most dramatically, **hydropolitics**, or the interplay of water resource issues and politics, has raised tensions between countries that share drainage basins. For example, Ethiopia's Renaissance Dam project on the Nile is nearing completion and will be Africa's largest dam. Egypt, however, has protested, claiming that as the reservoir behind the dam fills up, downstream flows along the Nile could fall by 25 percent over a seven-year period. Such a disruption would have

▼ **Figure 7.9** **Environmental Issues in Southwest Asia and North Africa**   Growing populations, pressures for economic development, and pervasive aridity combine to create environmental hazards across the region. Long human occupancy has contributed to deforestation, irrigation-induced salinization, and expanding desertification. Q: Compare this map with Figure 7.6. What climate types are most strongly associated with desertification?

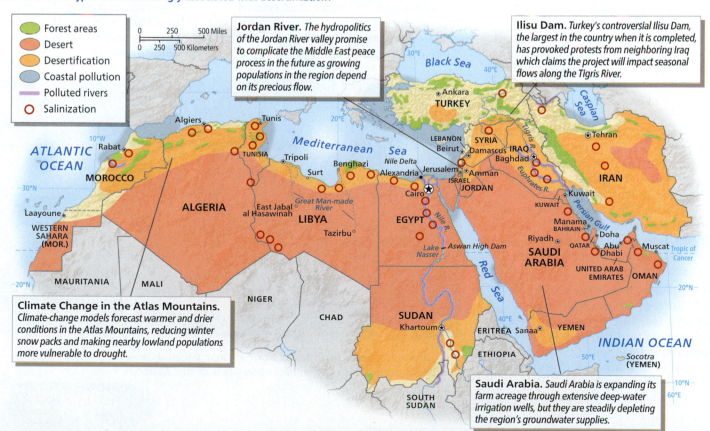

Noor 1 Shines Brightly in the North African Desert

Morocco's king performed the honors early in 2016 when he switched on the Noor 1 facility in the Moroccan desert near the town of Ouarzazate (Figure 7.1.1). The first of three phases, Noor 1 is a huge solar power plant that sprawls across more than 1100 acres (450 hectares) of the North African desert (Figure 7.1.2). It is already plainly visible from space. The final facility, completed in 2018, encompasses more than 7400 acres (3000 hectares), making the US$9 billion venture the world's largest concentrated solar power facility, designed to provide electricity for 1.1 million people. Initially, consumers will be in the local region, but ultimately Morocco's leaders and other investors hope that this and similar North African solar projects can light the living rooms of Europeans living far to the north in less sunny climes. The completed Noor project will have the capacity to generate 580 megawatts of electricity, saving hundreds of thousands of tons of carbon emissions annually and reducing Morocco's energy dependence by an equivalent of about 2.5 million tons of oil per year.

Shifting Solar Technologies? Noor 1, similar to several new facilities in the U.S. Southwest, relies on so-called concentrated solar technology, which is quite different from traditional (and still very competitive) photovoltaic technology. Photovoltaic solar cells generate power with sensitized panels that absorb sunlight and convert it directly to electricity (that typically must be transformed from DC [direct current] to AC [alternating current]), while concentrated solar technology systems use parabolic mirrors to capture and concentrate heat that can then be used to generate steam to power electricity-generating turbines. The Noor 1 facility contains an array of 500,000 solar mirrors that efficiently follow the sun along its diurnal journey.

New Regional Initiatives? With Noor 1's success, Morocco has taken a remarkable leap into the age of renewable energy. In fact, the country, poorly endowed with fossil fuels (it imports 97 percent of its traditional energy needs), hopes to generate more than half of its energy from homegrown renewables by 2030. Morocco's feat reflects a growing regional realization that solar energy's future shines bright in a part of the world endowed with some of the highest rates of annual sunshine on the planet. Other massive solar projects are gaining traction in Egypt, Saudi Arabia, the United Arab Emirates, and Kuwait. Even Dubai—the global epicenter of petroleum wealth—has developed its Clean Energy Strategy 2050 initiative that envisions the Gulf city generating 75 percent of its electricity with renewable sources by the middle of the century.

▲ **Figure 7.1.1 Noor 1, Morocco** The Noor 1 solar plant is located in central Morocco.

1. **Look at Figure 7.42a (Crude Petroleum Production) and identify other countries in the region, in addition to Morocco, that would logically be strong advocates for increasing their investment in renewable solar energy.**

2. **Find an online map of potential sunshine/solar potential for North America and describe the patterns you find. How much potential does your surrounding area have for solar energy?**

▲ **Figure 7.1.2 Solar Facilities at Noor 1** Solar panels shimmer in the desert sun near Ouarzazate at the newly opened Noor 1 facility in central Morocco.

GOOGLE EARTH
Virtual Tour Video
http://goo.gl/ODMV32

▲ Figure 7.10 Ilisu Dam, Southeast Anatolia Project, Turkey To make way for the Ilisu Dam, construction crews at work here recently destroyed an ancient Neolithic cave complex. When the dam is complete, portions of the nearby town of Hasankeyf will also be flooded.

within the Jordan River drainage basin (Figure 7.11). Israelis worry about Palestinian and Syrian pollution; nearby Jordanians argue for more water from Syria; and all regional residents must deal with the uncomfortable reality that, regardless of their political differences, they must drink from the same limited supplies of freshwater. A recent study by the International Institute of Sustainable Development is not encouraging; it estimated that the Jordan River may shrink by 80 percent by 2100, given increased demands and reduced flows. The Dead Sea is already shrinking rapidly (see Chapter 2), and Israel and Jordan still cannot agree on a joint "Red-Dead Project" or "Peace Corridor" that would pump desalinated seawater north from the Red Sea to the Dead Sea.

Water is not only a resource, but also a vital transportation link in the area. The region's physical geography creates enduring **choke points**, where narrow waterways are vulnerable to military blockade or disruption. For example, the Strait of Gibraltar (entrance to the Mediterranean), Turkey's Bosporus and Dardanelles, and the Suez Canal have all been key historical choke points within the region (see Figure 7.1). The narrow Bab-el-Mandeb, located at the southern entrance to the Red Sea, has gained strategic importance given Yemen's increasingly unstable political situation (Figure 7.12). Finally, Iran's periodic threat to close the Straits of Hormuz (at the eastern end of the Persian Gulf) to world oil shipments suggests the strategic role water continues to play in the region.

Climate Change in Southwest Asia and North Africa

Published in 2018, a special report by the Intergovernmental Panel on Climate Change (IPCC) suggested that 21st-century climate change in Southwest Asia and North Africa will aggravate already existing environmental issues. Temperature changes are predicted to have a greater impact on the region than changes in precipitation. The already arid and semiarid region will probably remain relatively dry, but warmer average temperatures are likely to have several major consequences:

- Higher overall evaporation rates and lower overall soil moisture across the region will stress crops, grasslands, and other vegetation. Semiarid lands in North Africa's Maghreb region are particularly vulnerable, especially dryland cropping systems that cannot

▲ Figure 7.11 Hydropolitics in the Jordan River Basin Many water-related issues complicate the geopolitical setting in the Middle East. The Jordan River system has been a particular focus of conflict.

▲ **Figure 7.12 Straits of Bab al-Mandeb** This satellite view shows the entrance to the Red Sea at the Straits of Bab al-Mandeb, one of the region's key choke points.

depend on irrigation. Desertification (see Chapter 6) has already struck broad swaths of Tunisia, for example, from the encroaching Sahara in the south to the degradation of overworked farm lands in the north.

• Warmer temperatures will diminish runoff into rivers, reducing hydroelectric potential and water available for the region's increasingly urban population. Less snow in the Atlas Mountains will hurt nearby farmers who depend on meltwater for irrigation, and reduced flows in the Tigris and Euphrates valleys will further stress an already marginal agricultural zone. Some experts argue that by the 2020s, the Euphrates will dry up completely during summer months, producing an environmental catastrophe (Figure 7.13).

• More extreme, record-setting summertime temperatures will lead to more heat-related deaths, particularly in cities where growing populations, sporadic electricity, and unreliable medical care may result in thousands more heat-related casualties.

Changes in sea level also pose special threats to the Nile Delta. This portion of northern Egypt is a vast, low-lying landscape of settlements, farms, and marshland. Studies that model sea-level changes suggest much of the delta could be lost to inundation, erosion, or salinization. Farmland losses of more than 250,000 acres (100,000 hectares) are quite possible with even modest sea-level changes. The IPCC estimates that a sea-level rise of 3.3 feet (1 meter) could affect 15 percent of Egypt's habitable land and displace 8 million Egyptians in coastal and delta settings.

Experts also estimate broader political and economic costs will accompany climate change. Given the region's political instability, even small changes in rainfall and water supplies, particularly where they might involve several nations, could add to potential conflicts. A study by the World Resources Institute, for example, suggests that Syria's civil war that began several years ago was at least partly triggered by drought, crop losses, and water shortages.

REVIEW

7.1 Describe the climatic changes you might experience as you travel on a line from the eastern Mediterranean coast at Beirut to the highlands of Yemen. What are some of the key climatic variables that explain these variations?

7.2 Discuss five important human modifications of the Southwest Asian and North African environment, and assess whether these changes have benefited the region.

KEY TERMS ISIS (Islamic State of Iraq and the Levant), Arab Spring, sectarian violence, Islamic fundamentalism, Islamism, culture hearth, Maghreb, Levant, salinization, fossil water, hydropolitics, choke point

Population and Settlement: Changing Rural And Urban Worlds

The human geography of Southwest Asia and North Africa demonstrates the intimate tie between water and life in this part of the world. The pattern is complex: Large areas of the population map remain almost devoid of permanent settlement, whereas lands with available moisture suffer increasingly from problems of crowding and overpopulation.

The Geography of Population

Today about 550 million people live in Southwest Asia and North Africa (Table 7.1). The distribution of that population is strikingly varied (Figure 7.14). In North Africa, the moist slopes of the Atlas Mountains and nearby better-watered coastal districts support dense populations, a stark contrast to thinly occupied lands southeast of the

◀ **Figure 7.13 Euphrates River, Syria** This portion of the Euphrates River flows through arid lands in eastern Syria.

Country	Population (millions) 2018	Population Density (per square kilometer)[1]	Rate of Natural Increase (RNI)	Total Fertility Rate	Life Expectancy Male	Life Expectancy Female	Percent Urban	Percent <15	Percent > 65	Net Migration (rate per 1000)
Algeria	42.7	17	2.2	3.1	77	78	73	29	6	0
Bahrain	1.5	1,936	1.1	1.9	76	78	89	20	3	23
Egypt	97.0	98	2.1	3.4	71	74	43	34	4	−1
Iran	81.6	50	1.4	2.0	75	77	74	24	6	−3
Iraq	40.2	88	2.7	4.1	68	72	70	40	3	2
Israel	8.5	403	1.6	3.1	81	84	91	28	11	3
Jordan	10.2	109	2.1	3.2	73	76	90	35	4	0
Kuwait	4.2	232	1.3	1.5	74	76	100	21	2	15
Lebanon	6.1	595	0.9	1.7	77	79	89	24	7	−1
Libya	6.5	4	1.4	2.3	69	75	80	28	4	−4
Morocco	35.2	80	1.2	2.2	75	78	62	24	8	−2
Oman	4.7	15	1.8	2.9	75	79	85	22	3	32
Palestinian Territories	4.8	778	2.7	4.1	72	75	76	39	3	−2
Qatar	2.7	227	0.9	1.9	77	80	100	14	1	28
Saudi Arabia	33.4	15	1.4	2.4	73	76	84	25	3	11
Sudan	41.7	23	2.7	4.7	63	66	36	41	4	−2
Syria	18.3	99	1.6	2.9	64	77	54	37	4	−21
Tunisia	11.6	74	1.4	2.3	75	78	68	24	8	0
Turkey	81.3	105	1.1	2.1	75	81	75	24	9	0
United Arab Emirates	9.5	112	0.8	1.8	77	79	87	14	1	8
Western Sahara	0.6	–	1.5	2.3	68	72	87	28	3	−7
Yemen	28.9	54	2.5	4.0	64	66	37	40	3	−1

Source: Population Reference Bureau, *World Population Data Sheet*, 2018.
[1] World Bank Open Data, 2018.

Log in to Mastering Geography & access MapMaster to explore these data!

1) Explain why the overall population densities of Egypt and Algeria are so low, while the population densities of Lebanon and the Palestinian Territories are so high. Why might these dramatic differences be misleading?

2) What are the three youngest populations (highest percentage <15) and the three oldest (highest percentage >65) in the region? What connections can you make to levels of economic development, if any?

mountains. Egypt's zones of almost empty desert differ dramatically from crowded, irrigated locations, such as those along the Nile River. In Southwest Asia, many residents live in well-watered coastal and highland settings and in desert localities where water is available from nearby rivers or subsurface aquifers. High population densities are found in better-watered portions of the eastern Mediterranean (Israel, Lebanon, and Syria), Turkey, and Iran. While overall population densities in such countries appear modest, the **physiological density**, which is the number of people per unit area of arable land, is quite high by global standards. Although many people in the region still live in rural settings, many states are now dominated by huge cities (for example, Cairo in Egypt, Istanbul in Turkey, and Tehran in Iran) and these suffer the same problems of urban crowding found elsewhere in the developing world (Figure 7.15).

Shifting Demographic Patterns

High population growth remains a critical issue throughout Southwest Asia and North Africa, but the demographic picture is shifting. Uniformly high growth rates in the 1960s have been replaced by more varied regional patterns (Figure 7.16). For example, women in Tunisia and Turkey now average fewer than three births, representing a large decline in total fertility rates (see Table 7.1). Various factors explain these changes. More urban, consumer-oriented populations opt for fewer children. Many Arab women now delay marriage into their middle 20s to early 30s. Family-planning initiatives are expanding in many countries; programs in Tunisia and Egypt have greatly increased access

to contraceptive pills, IUDs, and condoms, and even fundamentalist Iran has seen major declines in its birth rate since 1990.

Still, areas such as the West Bank, Gaza, and Yemen are growing much faster than the world average. Across the region, massive youth unemployment has been called a "ticking time bomb." Poverty and traditional ways of rural life contribute to large rates of population increase, and even in more urban Saudi Arabia, growth rates remain between 1.4 and 2 percent. The increases result from high birth rates combined with low death rates. In Egypt, even though birth rates may decline, the labor market will need to absorb more than 500,000 new workers annually over the next 10 to 15 years just to keep up with the country's large youthful population (see Figure 7.16).

Water and Life: Rural Settlement Patterns

Water and life are closely linked across rural settlement landscapes of Southwest Asia and North Africa (Figure 7.17). Indeed, Southwest Asia is one of the world's earliest hearths of **domestication**, where plants and animals were purposefully selected and bred for their desirable characteristics. Beginning around 10,000 years ago, increased experimentation with wild varieties of wheat and barley led to agricultural settlements that later included domesticated animals, such as cattle, sheep, and goats. Much of the early agricultural activity focused on the **Fertile Crescent**, an ecologically diverse zone stretching from the Levant inland through the fertile hill country of northern Syria into Iraq. Between 5000 and 6000 years ago, improved irrigation techniques and increasingly powerful political states encouraged the spread of

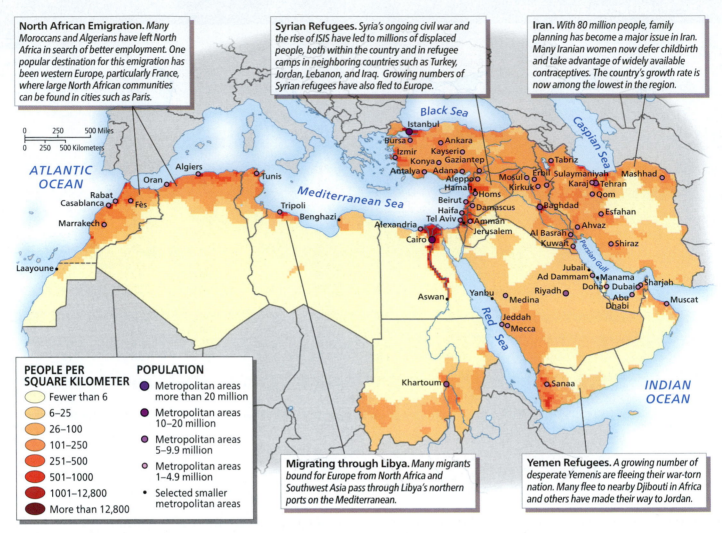

North African Emigration. *Many Moroccans and Algerians have left North Africa in search of better employment. One popular destination for this emigration has been western Europe, particularly France, where large North African communities can be found in cities such as Paris.*

Syrian Refugees. *Syria's ongoing civil war and the rise of ISIS have led to millions of displaced people, both within the country and in refugee camps in neighboring countries such as Turkey, Jordan, Lebanon, and Iraq. Growing numbers of Syrian refugees have also fled to Europe.*

Iran. *With 80 million people, family planning has become a major issue in Iran. Many Iranian women now defer childbirth and take advantage of widely available contraceptives. The country's growth rate is now among the lowest in the region.*

PEOPLE PER SQUARE KILOMETER
- Fewer than 6
- 6–25
- 26–100
- 101–250
- 251–500
- 501–1000
- 1001–12,800
- More than 12,800

POPULATION
- Metropolitan areas more than 20 million
- Metropolitan areas 10–20 million
- Metropolitan areas 5–9.9 million
- Metropolitan areas 1–4.9 million
- Selected smaller metropolitan areas

Migrating through Libya. *Many migrants bound for Europe from North Africa and Southwest Asia pass through Libya's northern ports on the Mediterranean.*

Yemen Refugees. *A growing number of desperate Yemenis are fleeing their war-torn nation. Many flee to nearby Djibouti in Africa and others have made their way to Jordan.*

▲ **Figure 7.14 Population of Southwest Asia and North Africa** The striking contrasts are clearly evident between large, sparsely occupied desert zones and much more densely settled regions where water is available. The Nile Valley and the Maghreb region contain most of North Africa's people, whereas Southwest Asian populations cluster in the highlands and along the better-watered shores of the Mediterranean.

▼ **Figure 7.15 Istanbul** Turkey's largest city is now home to more than 15 million people.

Explore the SIGHTS of Istanbul
http://goo.gl/LPKR26

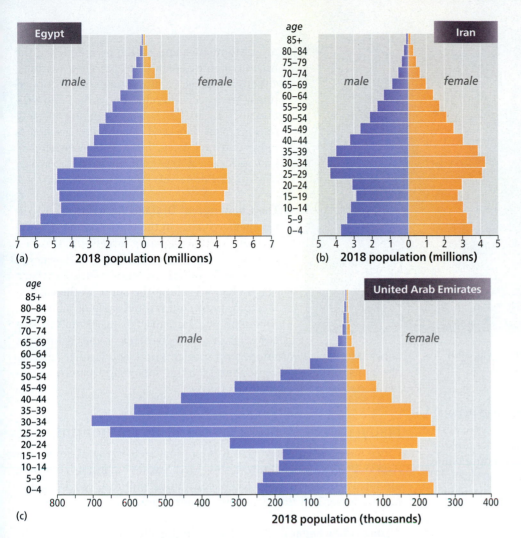

Three distinctive demographic snapshots highlight regional diversity: (a) Egypt's above-average growth rates differ sharply from those of (b) Iran, where a focused campaign on family planning has reduced recent family sizes. (c) Male immigrant laborers play a special role in skewing the pattern within the United Arab Emirates. **Q: For each example, cite a related demographic or cultural issue that you might potentially find in these countries.**

agriculture into nearby lowlands, such as the Tigris and Euphrates valleys (Mesopotamia; now modern Iraq) and North Africa's Nile Valley.

Pastoral Nomadism In the drier portions of the region, **pastoral nomadism**, where people move livestock seasonally, is a traditional form of subsistence agriculture. The settlement landscape of pastoral nomads reflects their need for mobility and flexibility as they move camels, sheep, and goats from place to place. Near highland zones such as the Atlas Mountains and the Anatolian Plateau, nomads practice **transhumance**—seasonally moving livestock to cooler, greener high-country pastures in the summer and returning them to valley and lowland settings for fall and winter grazing.

Oasis Life Permanent oases exist where high groundwater levels or modern deep-water wells provide reliable water (Figures 7.17 and 7.18). Tightly clustered, often walled villages sit next to small,

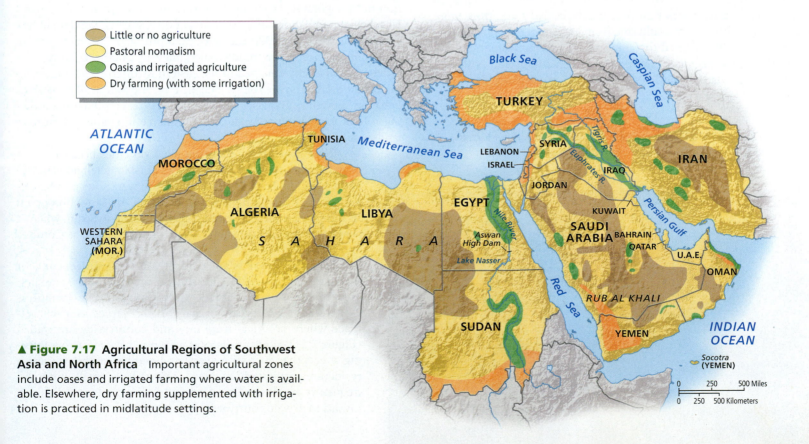

▲ **Figure 7.17** Agricultural Regions of Southwest Asia and North Africa Important agricultural zones include oases and irrigated farming where water is available. Elsewhere, dry farming supplemented with irrigation is practiced in midlatitude settings.

▲ **Figure 7.18 Oasis Settlement** This view of Morocco's Tinghir Oasis features small cultivated fields and date palms in the foreground.

intensely utilized fields where underground water is applied to tree and cereal crops. In newer oasis settlements, concrete blocks and prefabricated housing add a modern look. Traditional oasis settlements contain families that work their own irrigated plots or, more commonly, work for absentee landowners. While oases are usually small, eastern Saudi Arabia's Hofuf Oasis covers more than 30,000 acres (12,000 hectares).

Exotic Rivers   For centuries, the densest rural settlement of Southwest Asia and North Africa has been tied to its great irrigated river valleys and their seasonal floods of water and fertile nutrients. In such settings, **exotic rivers** transport much-needed water from more humid areas to drier regions suffering from long-term moisture deficits (Figure 7.19). The Nile and the combined Tigris and Euphrates rivers are the largest regional examples of such activity, and both systems have large, densely settled deltas. Similar settlements are found along the Jordan River in Israel and Jordan, in the foothills of the

▼ **Figure 7.19 Nile Valley Agriculture** Small, intensively cultivated irrigated fields are found along much of Egypt's lower Nile Valley.

Atlas Mountains, and on the peripheries of the Anatolian and Iranian plateaus. These settings, although capable of supporting sizable populations, are also vulnerable to overuse and salinization. Rural life is also changing in such settings. New dam- and canal-building schemes in Egypt, Israel, Syria, Turkey, and elsewhere are increasing the storage capacity of river systems, allowing for more year-round agriculture.

The Challenge of Dryland Agriculture   Mediterranean climates in the region permit dryland agriculture that depends largely on seasonal moisture. These zones include better-watered valleys and coastal lowlands of the northern Maghreb, lands along the shore of the eastern Mediterranean, and favored uplands across the Anatolian and Iranian plateaus. A mix of tree crops, grains, and livestock is raised in these zones. More mechanization, crop specialization, and fertilizer use are also transforming such agricultural settings, following a pattern set earlier in nearby areas of southern Europe.

Many-Layered Landscapes: The Urban Imprint

Cities have played a key role in the human geography of Southwest Asia and North Africa. Indeed, some of the world's oldest urban places are located in the region. Today continuing political, religious, and economic ties link the cities with the surrounding countryside.

A Long Urban Legacy   Cities have traditionally been centers of political and religious authority as well as focal points of trade. Urbanization in Mesopotamia began by 3500 BCE, and cities such as Eridu and Ur reached populations of 25,000 to 35,000 residents. Similar centers appeared in Egypt by 3000 BCE, with Memphis and Thebes assuming major importance in the middle Nile Valley. By 2000 BCE, however, a different kind of city emerged in the eastern Mediterranean and along important overland trade routes. Beirut, Tyre, and Sidon, all in modern Lebanon, as well as Damascus in nearby Syria, exemplified the growing role of long-distance trade and commerce in shaping the urban landscape.

Islam also left an enduring mark on cities because urban centers traditionally served as places of Islamic religious power and education. Both Baghdad and Cairo were seats of religious authority. Urban settlements from North Africa to Turkey felt its influence. Indeed, the Moors carried Islam to Spain, where it shaped the architecture and culture of centers such as Córdoba and Málaga.

The traditional Islamic city features a walled core, or **medina**, dominated by the central mosque and its associated religious, educational, and administrative functions (Figure 7.20). A nearby bazaar, or *suq*, serves as a marketplace where products from city and countryside are traded (Figure 7.21). Housing districts feature a maze of narrow, twisting streets that maximize shade and emphasize the privacy of residents, particularly women. Houses have small windows, frequently are situated on dead-end streets, and typically open inward to private courtyards.

▲ **Figure 7.20** **Mosque in Qom, Iran** *Qom's Hazrati Masumeh Shrine is visited annually by thousands of faithful Shiites in this sacred Iranian city south of Tehran.*

▲ **Figure 7.21** **Old Cairo** The Egyptian capital's narrow, twisting streets are often crowded with shoppers and pedestrians.

More recently, European colonialism has shaped selected cities. Particularly in North Africa, colonial builders added many architectural features from Great Britain and France. Victorian building blocks, French mansard roofs, suburban housing districts, and wide, European-style commercial boulevards remain landscape features in cities such as Fes, Algiers, and Cairo (Figure 7.22).

Signatures of Globalization Since 1950, cities in Southwest Asia and North Africa have become gateways to the global economy. Expanded airports, commercial and financial districts, industrial parks, and luxury tourist facilities all reveal the influence of globalization. Many cities, such as Algiers and Istanbul, have more than doubled in population in recent years. Crowded Cairo now has more than 15 million residents. Escalating demand for homes has produced ugly, cramped high-rise apartment houses, while extensive squatter settlements provide little in the way of quality housing or public services.

Undoubtedly, the most dramatic new urban landscapes are visible in the oil-rich states of the Persian Gulf. Before the 20th century, urban traditions were relatively weak in the area, and even as late as 1950 only 18 percent of Saudi Arabia's population lived in cities. All that has changed, however, and today Saudi Arabia is more urban than many industrialized nations, including the United States. Particularly

since 1970, cities such as Dubai (United Arab Emirates), Doha (Qatar), Manama (Bahrain), and Kuwait City (Kuwait) have mushroomed in size, often featuring central-city skylines showcasing futuristic architecture (Figure 7.23).

In addition, investments in petrochemical industries have fueled the creation of new urban centers, such as Jubail along Saudi Arabia's

▼ **Figure 7.22** **Map of Fes, Morocco** The tiny neighborhoods and twisting lanes of the old walled city reveal features of the traditional Islamic urban center. To the southwest, however, the rectangular street patterns, open spaces, and broad avenues suggest colonial European influences.

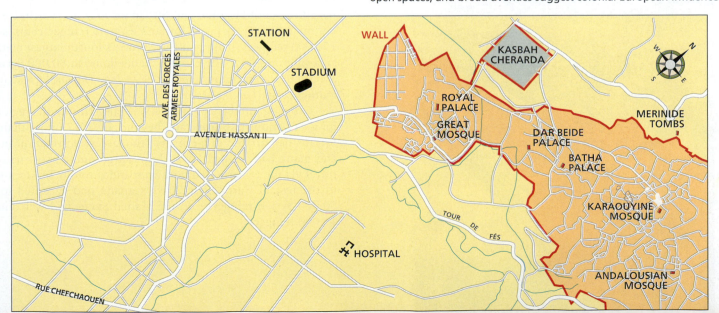

▲ **Figure 7.23 Burj Khalifa, United Arab Emirates** Soaring more than 2700 feet into the desert sky, Dubai's Burj Khalifa is home to more than 50 elevators that whisk visitors to a variety of hotel rooms, restaurants, residences, offices, and shops. It is the world's tallest building.

▼ **Figure 7.24 Masdar City, United Arab Emirates** Masdar City is being laid out near the Abu Dhabi Airport.

Persian Gulf coastline. Masdar City, one of the world's most carefully planned and energy-efficient cities, is also taking shape within the United Arab Emirates (Figure 7.24).

A Region on the Move

Although nomads have crisscrossed the region for ages, entirely new patterns of migration reflect the global economy and recent political events. The rural-to-urban shift seen widely in the less developed world is reworking population patterns across Southwest Asia and North Africa. Cities from Casablanca to Tehran are experiencing phenomenal growth, spurred by in-migration from rural areas.

Foreign workers have also migrated to areas within the region that have large labor demands. In particular, oil-rich countries of the Persian Gulf support immigrant workforces that often comprise large proportions of the overall population. Over 40 percent of the Gulf states' total population is made up of foreign-born workers. The influx has major economic, social, and demographic implications in nations such as the United Arab Emirates, where about 90 percent of the country's private workers are immigrants, often employed in construction or as domestic help. Source regions vary, but most immigrants come from South Asia and other Muslim countries within and beyond the region (Figure 7.25).

Other residents migrate to jobs elsewhere in the world. Because of its strong economy and close location, Europe is a powerful draw. More than 2 million Turkish "guest workers" live in Germany. Algeria and Morocco also have seen large out-migrations to western Europe, particularly France.

Political instability has also sparked migration. Wealthier residents, for example, have fled nations such as Lebanon, Syria, Iraq, and Iran since the 1980s and today live in cities such as Toronto, Los Angeles, and Paris. More recent political instability has provoked other refugee movements. Since 2003, huge numbers of people in western Sudan's unsettled Darfur region have moved to dozens of refugee camps in nearby Chad.

Syria's civil war and sectarian conflicts produced a massive refugee crisis that has displaced over half the country's population. In 2018, about 6.5 million internally displaced persons (IDPs) had left their homes and were living elsewhere in the country. About 6 million more Syrians were refugees and were living in other countries. Turkey is now a temporary home for more than 3.5 million Syrians (and hosts the world's largest refugee population) (Figure 7.26). To the south, both Lebanon and Jordan are also overwhelmed with Syrian migrants. At its peak, Jordan's Za'atari refugee camp supported 150,000 refugees, making it the country's fourth largest metropolitan area (Figure 7.27). The camp remains home to 80,000 Syrians, with about 60 percent under the age of 24 (see *Humanitarian Geography: Putting Za'atari on the Map*).

Growing numbers of Syrians were also fleeing the region, bound for destinations in Europe and beyond. By 2017, more than 600,000 Syrians had arrived in Germany (the largest destination in Europe) and many had applied for asylum. Large numbers of refugees also travel via the eastern Mediterranean and through Turkey and Greece to the rest of Europe. By 2018, Canada had accepted more than 50,000 Syrian refugees and the United States had welcomed about 18,000 refugees.

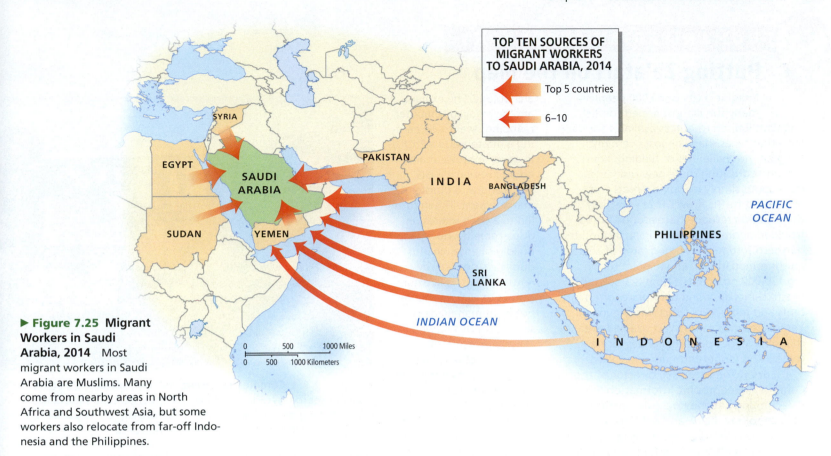

TOP TEN SOURCES OF MIGRANT WORKERS TO SAUDI ARABIA, 2014

Top 5 countries

6–10

► **Figure 7.25 Migrant Workers in Saudi Arabia, 2014** Most migrant workers in Saudi Arabia are Muslims. Many come from nearby areas in North Africa and Southwest Asia, but some workers also relocate from far-off Indonesia and the Philippines.

Elsewhere, North Africa has been another major focus for refugee populations. Many African and Southwest Asian migrants have traveled through war-torn Libya on their desperate search to find better lives in Europe and beyond (Figure 7.28). To the east, Yemen's bloody civil war is now generating its own growing refugee populations: in 2018, it was estimated that Djibouti (38,000) and Jordan (10,400) were two of the largest destinations for desperate, asylum-seeking Yemenis.

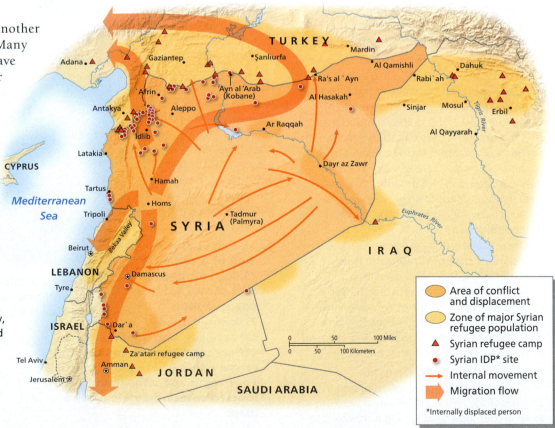

Area of conflict and displacement

Zone of major Syrian refugee population

▲ Syrian refugee camp

● Syrian IDP* site

→ Internal movement

➤ Migration flow

*Internally displaced person

► **Figure 7.26 Syrian Refugee Zones and Selected Camps** Neighboring areas of Turkey, Jordan, Iraq, and Lebanon have been inundated with refugees since 2012, and growing numbers have fled the region to Europe and beyond. Turkey is now home to the largest number of refugees. The large Za'atari refugee camp is located northeast of Amman, Jordan.

HUMANITARIAN GEOGRAPHY

Putting Za'atari on the Map

Imagine a city of 80,000 people—many survivors of armed conflict and with limited resources available for daily life—that takes shape almost overnight. Imagine the challenges of feeding, housing, and educating these new residents. How about simply mapping streets and organizing traffic flows? That's precisely the situation refugees face around Syria's troubled perimeter. Almost in an instant, the northern Jordanian refugee camp of Za'atari mushroomed in size, at one point reaching 150,000 occupants. Shops, schools, and health clinics all found their place, often in tents and temporary structures (Figure 7.2.1). Yet this refugee camp's vast size and ever-changing layout pose challenges for newcomers and longer-term residents alike.

Dr. Brian Tomaszewski, director of the Center for Geographic Information Science & Technology at the Rochester Institute of Technology, has made a difference for Za'atari's residents by teaching refugees to map the camp's thoroughfares and facilities. Partnering with the United Nations, several Jordanian universities, and Esri, a leading creator of GIS mapping software, Tomaszewski helped train a team of Syrian refugees in basic mapping and GIS skills.

Meeting Daily Challenges Tomaszewski first visited Za'atari in early 2015 as part of a team of experts on the scene to improve the camp's wireless and information infrastructure (Figure 7.2.2). He quickly realized that creating accurate, usable maps might be one of the most powerful tools he could provide to camp residents. Tomaszewski recalls that the geographer in him thought, "the refugees themselves could learn how to do GIS, because mapping and GIS are so fundamental to the refugee camp."

Once trained, the refugee team produced maps that facilitated the flow of information, goods, and people in the camp. This training benefitted the Syrian mappers as well: "A lot of refugees are looking for livelihoods," notes Tomaszewski. "They're smart, capable, and stressed as far as information management needs."

From Archeology to Disaster Management Tomaszewski majored in anthropology and used GIS while working as an archeologist. "Mapping is inherent to archeology," he explains, and he enjoyed computer mapping, eventually earning a masters and a PhD in geography. His main area of research is GIS applications to disaster management. This in turn led to a deep interest in displacement—how people build resilience when displaced, not only by natural disasters like hurricanes or climate change but by human-made crises such as conflict in Syria. "Displacement is really a global problem," he notes. "And it's not just in the developing world, it's everywhere."

Tomaszewski hopes his own experiences training refugees in mapping their new worlds can be transferred to settings around the globe. The maps will provide important everyday information for thousands of residents, and the GIS training can help refugees move on to productive lives elsewhere. "Geography matters," he explains. "We can't forget about real people, real places, and context. If you want to do something meaningful with your life, geography can open that up for you."

▲ **Figure 7.2.1 Commercial Street, Za'atari, Jordan** Za'atari's main market road is nicknamed the Champs-Élysées after the major urban thoroughfare in Paris, France. This merchant offers a wide variety of household goods to newly arriving refugees.

▲ **Figure 7.2.2 Brian Tomaszewski** Geographer Brian Tomaszewski's humanitarian work has contributed to bringing the world of GIS to the Jordanian refugee camp of Za'atari.

1. **Sketch your own map of a camp for 25,000 refugees that shows 1) basic street layout, 2) schools, medical services, and administrative offices, and 3) main commercial centers. In a brief essay, describe your camp and why activities are located where they are on the map.**

2. **As a recently arrived Syrian college student in Za'atari, write a letter to your uncle back home describing the chief challenges you face in the camp.**

GOOGLE EARTH
Virtual Tour Video
https://goo.gl/i7kRk1

▲ **Figure 7.27 Za'atari Refugee Camp, Jordan** Syria's civil war has created growing refugee populations. In 2018, Jordan's Zaatari camp was a temporary home to 80,000 refugees.

▼ **Figure 7.28 Libyan Highway to Europe** The map shows some of the overland routes across North Africa that converge on Libyan ports, as well as general routes across the Mediterranean that take desperate migrants to Europe.

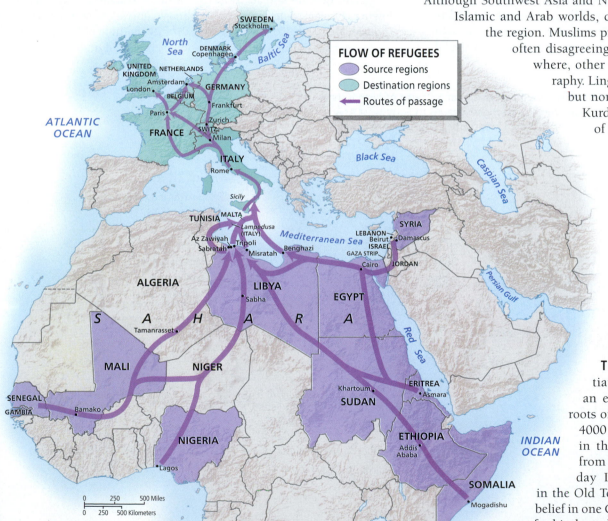

Cultural Coherence and Diversity: A Complex Cultural Mosaic

Although Southwest Asia and North Africa remain the heart of the Islamic and Arab worlds, cultural diversity also characterizes the region. Muslims practice their religion in varied ways, often disagreeing strongly on religious views. Elsewhere, other religions complicate cultural geography. Linguistically, Arabic languages are key; but non-Arab peoples, including Persians, Kurds, and Turks, also dominate portions of the region.

Patterns of Religion

Religion is an important part of the lives of most people in Southwest Asia and North Africa. Whether it is the quiet ritual of morning prayers or discussions about current political and social issues, religion remains part of the daily routine of residents from Casablanca to Tehran.

Hearth of the Judeo-Christian Tradition Both Jews and Christians trace their religious roots to an eastern Mediterranean hearth. The roots of Judaism lie deep in the past: Some 4000 years ago, Abraham, an early leader in the Jewish tradition, led his people from Mesopotamia to Canaan (modern-day Israel). Jewish history, recounted in the Old Testament of the Bible, focused on a belief in one God (or **monotheism**), a strong code of ethical conduct, and a powerful ethnic identity that continues to the present. During the Roman Empire, many Jews left the eastern Mediterranean to escape Roman

persecution. This forced migration, or *diaspora*, took Jews to
the far corners of Europe and
North Africa. Only in the past
century have many of the world's
far-flung Jewish people returned to Judaism's place of origin, a process
that gathered speed after the creation of Israel in 1948.

Christianity, an outgrowth of Judaism, was based on the teachings
of Jesus and his disciples, who lived and traveled in the eastern Mediterranean about 2000 years ago. Although many Christian traditions
became associated with European history, forms of early Christianity remain near the religion's hearth. For example, the Coptic Church
evolved in nearby Egypt. In Lebanon, Maronite Christians also retain
a separate cultural identity.

The Emergence of Islam   Islam originated in Southwest Asia in
622 CE, forming another cultural hearth of global significance. Muslims can be found from North America to the southern Philippines,
but the Islamic world remains centered on Southwest Asia. Most
Southwest Asian and North African peoples still follow its religious
teachings. Muhammad, the founder of Islam, was born in Makkah
(Mecca) in 570 CE and taught in nearby Medinah (Medina) (Figure 7.29). His beliefs parallel Judeo-Christian traditions. Muslims
believe both Moses and Jesus were prophets and that the Hebrew
Bible (or Old Testament) and the Christian New Testament, while
incomplete, are basically accurate. However, Muslims hold that the

Explore the SOUNDS of Jewish Folk Music
http://goo.gl/lgO9FS

Quran (or Koran), a book of teachings received by Muhammad from
Allah (God), represents God's highest religious and moral revelations
to humanity.

Islam offers a blueprint for leading an ethical and religious life.
Islam literally means "submission to the will of God," and its practice
rests on five essential activities: (1) repeating the basic creed ("There is
no god but God, and Muhammad is his prophet"); (2) praying facing
Makkah five times daily; (3) giving charitable contributions; (4) fast-
ing between sunup and sundown during the month of Ramadan; and
(5) making at least one religious pilgrimage, or **Hajj**, to Muhammad's
birthplace of Makkah (Figure 7.30). Islamic fundamentalists also argue
for a **theocratic state**, such as modern-day Iran, in which religious
leaders (ayatollahs) shape government policy.

A major religious division split Islam almost immediately after
the death of Muhammad in 632 CE and endures today. One group, now
called **Shiites**, favored passing on religious authority within Muhammad's family, specifically to Ali, his son-in-law. Most Muslims, later
known as **Sunnis**, advocated passing down power through established
clergy. This group was largely victorious. Ali was killed, and his Shiite
supporters went underground. Ever since, Sunni Islam has formed
the mainstream branch of the religion, to which Shiite Islam has presented a recurring and sometimes powerful challenge.

Islam quickly spread from the western Arabian Peninsula, following caravan routes and Arab military campaigns as it converted
thousands to its beliefs (see Figure 7.29). By the time of Muhammad's
death, peoples of the Arabian Peninsula were united under its banner.

▼ **Figure 7.29 Diffusion of Islam**   The rapid expansion of Islam that followed its birth is shown
here. From Spain to Southeast Asia, Islam's legacy remains strongest nearest its Southwest Asian
hearth. In some settings, its influence has ebbed or has come into conflict with other religions, such as
Christianity, Judaism, and Hinduism.

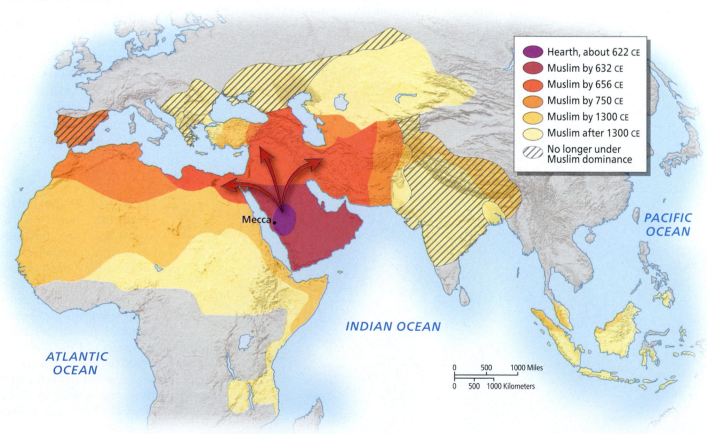

▲ **Figure 7.30 Makkah** Thousands of faithful Muslims gather at the Grand Mosque in central Makkah, part of the pilgrimage to this sacred place that draws several million visitors annually.

Explore the SIGHTS of Makkah
http://goo.gl/Kba8Ar

Shortly thereafter, the Persian Empire fell to Muslim forces, and the Eastern Roman (or Byzantine) Empire lost most of its territory to Islamic influences. By 750 CE, Arab armies swept across North Africa, conquered most of Spain and Portugal, and established footholds in Central and South Asia. Most people in the region were Muslims by the 13th century, and older religions such as Christianity and Judaism became minority faiths.

Between 1200 and 1500, Islamic influences expanded in some areas and contracted elsewhere. The Iberian Peninsula (Spain and Portugal) returned to Christianity in 1492, although Moorish (Islamic) cultural and architectural features remain today. At the same time, Muslims expanded southward and eastward into Africa, while Muslim Turks largely replaced Christian Greek influences in Southwest Asia after 1100. One group of Turks moved into the Anatolian Plateau and conquered the Byzantine Empire in 1453. These Turks soon created the huge **Ottoman Empire** (named after one of its leaders, Osman), which included southeastern Europe (including modern-day Albania, Bosnia, and Kosovo) and most of Southwest Asia and North Africa.

Modern Religious Diversity Today Muslims form the majority population in all the countries of Southwest Asia and North Africa except Israel, where Judaism dominates (Figure 7.31). Still, divisions within Islam create regional cultural differences. Many recent conflicts are defined along Sunni–Shiite lines, although specific issues focus more on power, politics, and economic policy than on theological differences.

Explore the SOUNDS of Islamic Call to Prayer
http://goo.gl/TDNnSs

The region is dominated by Sunni Muslims, but Shiites remain important elements in the contemporary cultural mix. In Iraq, for example, southern Shiites (around Najaf, Karbala, and Basra) asserted their cultural and political power following the fall of Saddam Hussein in 2003. Shiites also claim majorities in Iran and Bahrain; form substantial minorities in Lebanon, Yemen, and Egypt; and comprise about 10 percent of Saudi Arabia's population.

Since 1980, radicalized Shiite groups have pushed a cultural and political agenda of Islamic fundamentalism across the region. Tensions

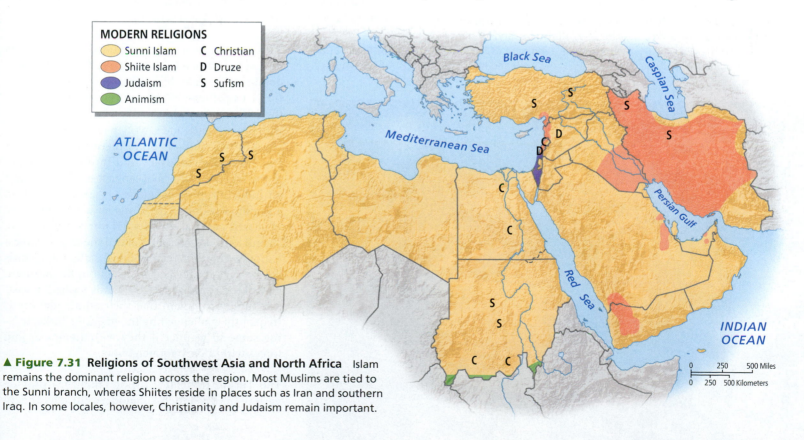

▲ **Figure 7.31 Religions of Southwest Asia and North Africa** Islam remains the dominant religion across the region. Most Muslims are tied to the Sunni branch, whereas Shiites reside in places such as Iran and southern Iraq. In some locales, however, Christianity and Judaism remain important.

are often evident within Shiite communities as fundamentalist elements clash with more secular Muslims. In Iran, for example, many younger, urban, and more affluent Muslims reject much of the extremist rhetoric of Shiite clerics who continue to reject Western values and cultural practices.

Sunni fundamentalism has also been on the rise, most evident in the presence of ISIS in Iraq, Syria, and beyond. More mainstream Sunnis still make up the vast majority of the region's population and reject its more radical cultural and political precepts, arguing for a modern Islam that accommodates some Western values and traditions.

While the Sunni–Shiite split is the Muslim world's great divide, other variations of Islam are also practiced in the region. One division separates the mystically inclined form of Islam known as *Sufism* from mainstream traditions. Sufism is prominent in the peripheries of the region, including the Atlas Mountains and across northwestern Iran and portions of Turkey. Elsewhere, the Salafists and Wahhabis, numerous in both Egypt and Saudi Arabia, are austere, conservative Sunnis who adhere to what they see as an earlier, purer form of Islamic doctrine. The Druze of Lebanon practice yet another variant of Islam.

Southwest Asia is also home to many non-Islamic communities. Israel has a Muslim minority (18 percent) dominated by that nation's Jewish population (75 percent). Even Israel's Jewish community is divided between Jewish fundamentalists and more reform-minded Jews. In neighboring Lebanon, a slight Christian (Maronite and Orthodox) majority was in evidence as recently as 1950, but Christian outmigration and higher Islamic birth rates created a nation that today is about 60 percent Muslim.

Jerusalem holds special religious significance for several faiths and also stands at the core of the region's political problems (Figure 7.32). When the United States moved its embassy to the

strategic city in 2018 (thus recognizing Jerusalem's symbolic status as Israel's capital), it was no surprise that Palestinians protested loudly and violently. Indeed, the sacred space of this ancient city remains deeply scarred and divided. Considering just the 220 acres of land within the Old City, Jews pray at the old Western Wall (the site of a Roman-era Jewish temple); Christians honor the Church of the Holy Sepulchre (the burial site of Jesus); and Muslims hold sacred rites in the city's eastern quarter (including the place from which Muhammad reputedly ascended to heaven).

Geographies of Language

Although the region is often termed the "Arab World," linguistic complexity creates important cultural divisions across Southwest Asia and North Africa (Figure 7.33).

Semites and Berbers   Afro-Asiatic languages dominate the region. Within that family is Arabic, spoken by Semitic peoples from the Persian Gulf to the Atlantic and southward into Sudan. Arabic has religious significance for Muslims because it was the sacred language in which God delivered his message to Muhammad. Although most of the world's Muslims do not speak Arabic, the faithful often memorize prayers in the language, and many Arabic words have entered the other important languages of the Islamic world.

Hebrew, another Semitic language, was reintroduced into the region with the creation of Israel. Hebrew originated in the Levant and was spoken by the ancient Israelites 3000 years ago. Today its modern version survives as the sacred tongue of the Jewish people and is the official language of Israel, although the country's non-Jewish population largely speaks Arabic.

Older Afro-Asiatic languages survive in more remote areas of North Africa. Collectively known as Berber, these languages are related, but not mutually intelligible. Berber speakers are spread across the North African interior, often in isolated rural settings.

Persians and Kurds   Although Arabic spread readily through portions of Southwest Asia, much of the Iranian Plateau and nearby mountains is dominated by older Indo-European languages. Here the principal tongue remains Persian, though since the 10th century, the language has been enriched with Arabic words and is written in the Arabic script.

Kurdish speakers of northern Iraq, northwest Iran, and eastern Turkey add further complexity to the regional pattern of languages. Kurdish, also an Indo-European language, is spoken by 10 to 15 million people. The Kurds have a strong sense of shared cultural identity (Figure 7.34). Indeed, "Kurdistan" is sometimes called the world's largest nation without its own political state. Iraqi Kurds have a well-established homeland in the northern part of the country. Nearby, Kurds located in eastern Turkey complain that their ethnic identity is frequently challenged by the majority Turks, leaving some to wonder if they will attempt to join forces with their Iraqi neighbors.

The Turkic Imprint   Turkic languages provide variety across much of modern Turkey and in

▼ **Figure 7.32 Old Jerusalem**   The historic center of Jerusalem reflects its varied religious legacy. Sacred sites for Jews, Christians, and Muslims are all located within the Old City. The Western Wall, a remnant of the ancient Jewish temple, stands at the base of the Dome of the Rock and Islam's al-Aqsa Mosque.

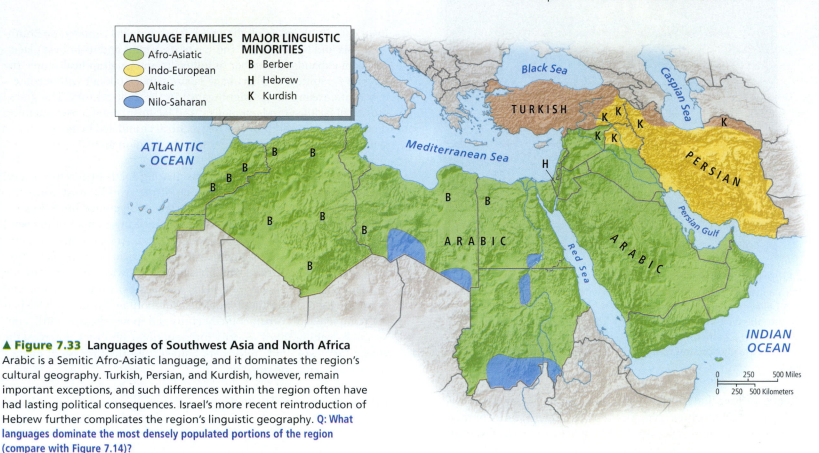

▲ **Figure 7.33 Languages of Southwest Asia and North Africa**
Arabic is a Semitic Afro-Asiatic language, and it dominates the region's cultural geography. Turkish, Persian, and Kurdish, however, remain important exceptions, and such differences within the region often have had lasting political consequences. Israel's more recent reintroduction of Hebrew further complicates the region's linguistic geography. **Q: What languages dominate the most densely populated portions of the region (compare with Figure 7.14)?**

▼ **Figure 7.34 Kurdish Women**   Eastern Turkey is home to these young Kurdish women. Many Kurds in this region face discrimination.

portions of far northern Iran. Turkic languages are a part of the larger Altaic language family that originated in Central Asia. Turkey remains the largest state in Southwest Asia dominated by this language family. Tens of millions of people in other countries of Southwest Asia and Central Asia speak related Altaic languages, such as Azeri, Uzbek, and Uyghur.

Regional Cultures in Global Context

Many cultural connections tie the region with the world beyond. These global connections are nuanced and complex and are expressed in fascinating, often unanticipated ways.

Religion and Dietary Preferences   For example, religion-based dietary preferences (known as *halal* in Islam and *kashrut* [kosher-style food preparation] in Judaism) shape food consumption patterns, both within the region and globally. From Indonesia to the United States, traditional Muslims and Jews incorporate dietary preferences based on holy teachings within the Quran and Torah. Both faiths share certain practices: pork and the consumption of animal blood are avoided, and foods are traditionally prepared using prescribed methods of slaughter (Figure 7.35). Alcohol consumption is prohibited in the Islamic world, and orthodox Jews carefully avoid consuming meat and dairy together.

In addition to traditional foodways, many younger, more urban residents in the region are growing more accustomed to American-style fast food, sushi, and other examples of changing global dietary preferences. It is also a two-way street: the growing popularity of Middle Eastern food in Europe, North America, and beyond represents another example of globalization (see *Globalization in Our World: Falafel Round the Globe*).

▲ **Figure 7.35 Ritual Food Preparation, Israel** An ultra-Orthodox Jew twirls a live chicken over the head of his wife as he recites a prayer. Religious Jews believe the practice transfers sins into the bird. The bird is then butchered in a kosher manner and the meat usually donated to the poor.

Explore the TASTES of of Kosher Meats
http://goo.gl/MkRk2M

Islamic Internationalism Islam is geographically and theologically divided, but all Muslims recognize the fundamental unity of their religion. Islamic communities are well established in such distant places as central China, European Russia, central Africa, Indonesia, and the southern Philippines. Today Muslim congregations also are expanding rapidly in the major urban areas of western Europe and North America.

Even with its global reach, however, Islam remains centered on Southwest Asia and North Africa, the site of its origins and its holiest places. As Islam expands in number of followers and geographical scope, the religion's tradition of pilgrimage ensures that Makkah will become a city of increasing global significance in the 21st century. The global growth of Islamist fundamentalism and Islamism also focuses attention on the region. In addition, the oil wealth accumulated by many Islamic states is used to sustain and promote the religion globally.

Changes in Saudi Arabia One fascinating focus of recent cultural change is Saudi Arabia, home to one of the world's most conservative Sunni Muslim societies. Crown Prince Mohammed bin Salman—the country's young, dynamic leader (son of King Salman in the Saud family)—has endorsed a more modern Saudi Arabia, both economically and culturally, packaging many of his ideas in the nation's "Vision 2030" plan. In cultural terms, this means allowing in more outside influences, loosening up long-held rules against women drivers, and permitting the flowering of a more cosmopolitan, Western-style society, complete with urban comedy clubs and movie theaters. Still, conservative-leaning Saudi security forces periodically crack down on the young "activists" and "reformers," suggesting the process of change will be slow and unpredictable.

Globalization and Technology Technology also shapes cultural and political change. Particularly among the young, millions within and beyond the region found themselves linked by the Internet, cell phones, and various forms of social media during the Arab Spring uprisings of 2011 and 2012. Texts, blogs, email, and tweets helped protesters plan events and coordinate strategies with their allies.

GLOBALIZATION IN OUR LIVES

Falafel Round the Globe

Falafel has arrived on the global stage. This ancient Middle Eastern food (perhaps originating among Coptic Christians in Egypt) is becoming a trendy, easy-to-prepare vegetarian alternative with worldwide appeal. From London to Los Angeles, these tasty deep-fried balls (or sometimes patties) are filled with some combination of ground chickpeas and fava beans, stuffed in a pita, and topped with green garnish, vegetables, and tasty tahini sauce (Figure 7.3.1). Long a national dish in both Egypt and Israel (falafel spans religious and political divides), it is well-designed as street food and as fast food (for a while, some McDonalds even marketed the McFalafel).

In 2016, London's inaugural Falafel Festival was marked by a fierce but nonviolent competition between Israeli, Lebanese, Palestinian, and Egyptian chefs. Judges finally awarded the Egyptian chef the top prize,

eking out a victory over Israel and prompting many lighthearted tweets celebrating Egypt's culinary conquest of its neighbor.

Falafel may be coming soon to a community near you. A global franchiser, JF Street Food, opened in 2007 in Abu Dhabi and wants to do for deep-fried chickpeas what Starbucks did for Frappuccinos. The chain has a presence in eight countries including Canada, the United States, and Australia.

1. **As a marketing director for your new local falafel fast-food restaurant, make a case for the food's appeal in your community.**

2. **Identify another authentic "foreign" food that is popular in your community, and explain why it is popular.**

▲ **Figure 7.3.1 Falafel** These deep-fried delicacies, usually made of ground chickpeas and fava beans, have become a popular Middle Eastern food choice around the world.

Explore the TASTES of Falafel
http://goo.gl/QydFqE

Intriguingly, ISIS has had a complex and unpredictable relationship with technology. Initially, the group claimed considerable success in recruiting new global converts to its cause (including those from Europe and the United States) via the digital world through the use of social media. Often, the group communicated to its adherents through a sophisticated digital communications strategy that involved online videos, Twitter, and Facebook. Recently, however, given greater online vigilance by social media companies and better tracking technologies by intelligence agencies, ISIS leaders have forbidden their members from using any online communications, surmising correctly that the technology was being successfully used to locate their whereabouts.

The Role of Sports Soccer plays a hugely important cultural role in everyday life within the region, both as a spectator sport and as an activity that many young people enjoy. Most countries have national football (soccer) associations that also participate in regional and global (FIFA) league competitions. The Union of Arab Football Associations (UAFA), headquartered in Riyadh, Saudi Arabia, offers many opportunities for regional competition, often carried on the Al Jazeera Sports network. Large soccer stadiums are a common part of modern urban landscapes, including Kuwait City's Sabah Al-Salem Stadium, which comfortably seats more than 28,000 fans. In the 2018 FIFA World Cup competition, teams from Morocco, Tunisia, Egypt, Saudi Arabia, and Iran all competed. Excitement is already growing about the potential for Qatar to host the 2022 FIFA World Cup event. As in other parts of the world, regional players can earn superstar status, such as Egypt's goal-shooting forward, Mohamed Salah (Figure 7.36).

REVIEW

7.6 Describe the key characteristics of Islam, and explain why distinctive Sunni and Shiite branches exist today.

7.7 Compare the modern maps of religion and language for the region, and identify three major non-Arabic-speaking areas where Islam dominates. Explain why that is the case.

▼ **Figure 7.36 Egyptian Soccer Star, Mohamed Salah** In the 2018 FIFA World Cup competition, Mohamed Salah played for his national team. He also plays as a forward for English club Liverpool.

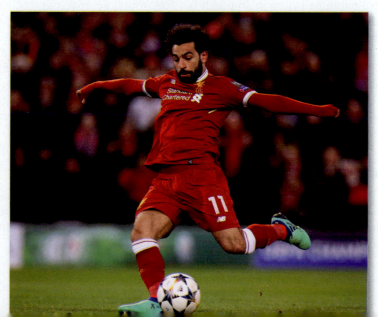

KEY TERMS monotheism, Quran, Hajj, theocratic state, Shiite, Sunni, Ottoman Empire

Geopolitical Framework: Never-Ending Tensions

Geopolitical tensions remain very high in Southwest Asia and North Africa (Figure 7.37). The Arab Spring rebellions of the early 2010s were a key turning point in marking this era of heightened regional tensions. Governments fell in Tunisia (where the regional movement began in late 2010), Egypt, Libya, and Yemen; widespread protests shook once-stable states such as Bahrain; and a more protracted civil war erupted in Syria, producing a huge and ongoing refugee crisis. Other countries witnessed shorter, more intermittent demonstrations against state authority in a region that has a long tradition of authoritarian political regimes. To varying degrees, these uprisings focused broadly on (1) charges of widespread government corruption; (2) limited opportunities for democracy and free elections; (3) rapidly rising food prices; and (4) the enduring reality of widespread poverty and high unemployment, especially for people under 30.

Sectarian conflicts between Sunnis and Shiites also dominate the geopolitical map. Iran (mainly Shiite) and Saudi Arabia (mainly Sunni) have each played major regional roles in bankrolling their respective supporters. For example, Iran supports the Shiite-dominated Iraq government and Syria's Assad regime (largely made up of Alawites, an offshoot of Shiite Islam). Iran also has close ties to Hezbollah, a well-armed group of militants located in Lebanon and to the Houthis, a Shiite faction leading a rebellion in Yemen. Saudi Arabia, from its perspective, has made it clear that it does not want to see nearby Yemen led by Shiite extremists, and began intervening directly in that conflict in 2015.

The Colonial Legacy

European colonialism arrived relatively late in Southwest Asia and North Africa, but the era left an important imprint on the region's modern political geography. Between 1550 and 1850, the region was dominated by the Ottoman Empire, which expanded from its Turkish hearth to engulf much of North Africa as well as nearby areas of the Levant, the western Arabian Peninsula, and modern-day Iraq. After 1850, Ottoman influences waned, and European colonial dominance grew after the dissolution of the Ottoman Empire in World War I (1918).

Both France and Great Britain were major colonial players within the region. French interests in North Africa included Tunisia and Morocco. French Algeria attracted large numbers of European immigrants. After World War I, France added more colonial territories in the Levant (Syria and Lebanon). The British loosely incorporated places such as Kuwait, Bahrain, Qatar, the United Arab Emirates, and Aden (in southern Yemen) into their empire to help control sea trade between Asia and Europe. Nearby Egypt also caught Britain's attention. Once the European-engineered **Suez Canal** linked the Mediterranean and Red seas in 1869, European banks and trading companies gained more influence over the Egyptian economy. In Southwest Asia, British and Arab forces joined to force out the Turks during World

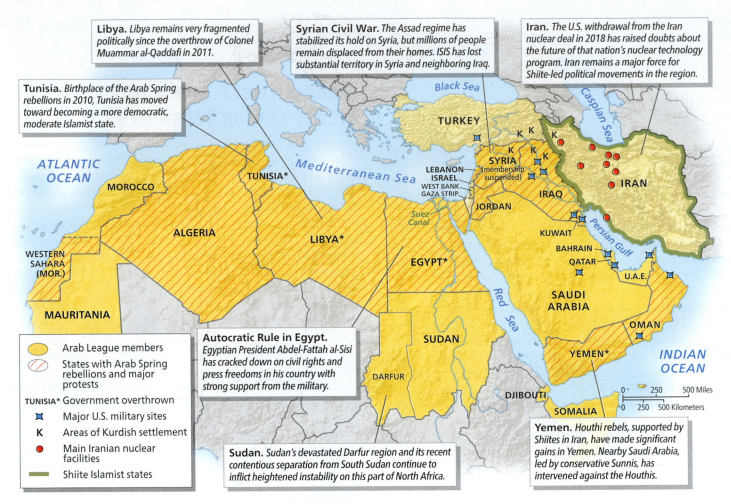

Libya. *Libya remains very fragmented politically since the overthrow of Colonel Muammar al-Qaddafi in 2011.*

Syrian Civil War. *The Assad regime has stabilized its hold on Syria, but millions of people remain displaced from their homes. ISIS has lost substantial territory in Syria and neighboring Iraq.*

Iran. *The U.S. withdrawal from the Iran nuclear deal in 2018 has raised doubts about the future of that nation's nuclear technology program. Iran remains a major force for Shiite-led political movements in the region.*

Tunisia. *Birthplace of the Arab Spring rebellions in 2010, Tunisia has moved toward becoming a more democratic, moderate Islamist state.*

Autocratic Rule in Egypt. *Egyptian President Abdel-Fattah al-Sisi has cracked down on civil rights and press freedoms in his country with strong support from the military.*

Sudan. *Sudan's devastated Darfur region and its recent contentious separation from South Sudan continue to inflict heightened instability on this part of North Africa.*

Yemen. *Houthi rebels, supported by Shiites in Iran, have made significant gains in Yemen. Nearby Saudi Arabia, led by conservative Sunnis, has intervened against the Houthis.*

Legend:
- Arab League members
- States with Arab Spring rebellions and major protests
- TUNISIA* Government overthrown
- ⚓ Major U.S. military sites
- K Areas of Kurdish settlement
- ● Main Iranian nuclear facilities
- Shiite Islamist states

▲ **Figure 7.37 Geopolitical Issues in Southwest Asia and North Africa** Political tensions continue across much of the region. The Arab Spring rebellions shaped subsequent political changes in several settings; and the rise of sectarian tensions within the Islamic World has disrupted life in Iraq, Syria, Yemen, and elsewhere. The Israeli–Palestinian conflict also remains pivotal.

War I. The Saud family convinced the British that a country should be established on the Arabian Peninsula, and Saudi Arabia became fully independent in 1932. Britain divided its other territories into three entities: Palestine (now Israel) along the Mediterranean coast; Transjordan to the east of the Jordan River (now Jordan); and a third zone that became Iraq.

Persia and Turkey were never directly occupied by European powers. In Persia, the British and Russians agreed to establish two spheres of economic influence in the region (the British in the south, the Russians in the north), while respecting Persian independence. In 1935, Persia's modernizing ruler, Reza Shah, changed the country's name to Iran. In Turkey, European powers attempted to divide up the Ottoman Empire following World War I. The successful Turkish resistance to European control was based on new leadership provided by Kemal Ataturk. Ataturk decided to imitate the European countries and establish a modern, culturally unified, secular state.

European colonial powers began withdrawing from Southwest Asian and North African colonies before World War II. By the 1950s, most countries in the region were independent. In North Africa, Britain withdrew troops from Sudan and Egypt in 1956. Libya (1951), Tunisia (1956), and Morocco (1956) achieved independence peacefully during the same era, but the French colony of Algeria became a major

problem. Several million French citizens resided there, and France had no intention of simply withdrawing. A bloody war for independence began in 1954, and France finally agreed to an independent Algeria in 1962.

Southwest Asia also lost its colonial status between 1930 and 1960. Iraq became independent from Britain in 1932, but its later instability resulted in part from its imposed borders, which never recognized its cultural diversity. Similarly, the French division of the Levant into Syria and Lebanon (1946) greatly angered local Arab populations and set the stage for future political instability. As a favor to the Lebanese Maronite Christian majority, France carved out a separate Lebanese state from largely Arab Syria, even guaranteeing the Maronites constitutional control of the government. The action created a culturally divided Lebanon as well as a Syrian state that has repeatedly asserted its influence over its Lebanese neighbors.

Modern Geopolitical Issues

The geopolitical instability in Southwest Asia and North Africa continues today. A quick regional transect from the shores of the Atlantic to the borders of Central Asia suggests how these forces are playing out in different settings early in the 21st century.

Across North Africa Varied North African settings have recently witnessed dramatic political changes (see Figure 7.37). In Tunisia, birthplace of the Arab Spring, a moderate Islamist government was elected to replace deposed dictator Zine el-Abidine Ben Ali, but the country has not been immune from terrorist attacks by extremists. In addition, in 2018, large-scale protests by many Tunisians frustrated by failed government economic policies resulted in hundreds of arrests and signaled the growing potential for future political instability.

In nearby Libya, while many cheered the end of Colonel Muammar al-Qaddafi's rule in 2011, rival militias split the nation into several dysfunctional fragments. The United Nations intervened, proposing a framework for elections and a return to a "government of national accord," but strong separatist factions, especially in the east, have shunned the process. For now, Libya is a country in name only, deeply divided internally into varied local and regional alliances.

Next door, Egypt remained politically unstable following the 2011 overthrow of Hosni Mubarak. Parliamentary and presidential elections in 2012 ushered in a brief period of rule by the Muslim Brotherhood. In 2013, the Egyptian military staged a coup and ousted the Brotherhood. Since then, some political stability has returned to the country within the shadows of a regime strongly influenced by the military. Egyptian president Abdel-Fattah al-Sisi has insisted that a strong authoritarian hand (including restrictions on demonstrations, social media, and a free press) is necessary to preserve the nation's relative stability. Egypt appears headed into another protracted period of virtual one-man rule with strong support from the military.

Sudan—a Sunni Islamic state—also faces daunting political issues. Civil war between the north and south produced more than 2 million casualties (mostly in the south) between 1988 and 2004. A tentative peace agreement was signed in 2005, opening the way for a successful vote on independence in South Sudan. Even though the two nations officially split in 2011, tensions remain, especially focused on an oil-rich and contested border zone between the two countries. Sudan's Darfur region also remains in shambles (see Figure 7.37). Ethnicity, race, and control of territory seem to be at the center of the struggle in the largely Muslim region as a well-armed, Arab-led militia group with links to the central government in Khartoum has attacked hundreds of black-populated villages. Since the beginning of the conflict in the early 2000s, more than 300,000 people have been killed (through violence, starvation, and disease) and 2.5 million people have been driven from their homes.

The Arab–Israeli Conflict The 1948 creation of the Jewish state of Israel produced another enduring zone of cultural and political tensions within the eastern Mediterranean (Figure 7.38). Jewish migration to Palestine increased after the defeat of the Ottoman Empire in World War I. In 1917, Britain issued the Balfour Declaration, a pledge to encourage the creation of a Jewish homeland in the region. After World War II, the UN divided the region into two states, one to be predominantly Jewish, the other primarily Muslim. Indigenous Arab Palestinians rejected the partition, and war erupted. Jewish forces proved victorious. Hundreds of thousands of Palestinian refugees fled from Israel to neighboring countries, where many of them remained in makeshift camps. Under these conditions, Palestinians nurtured the idea of creating their own state on land that had become part of Israel.

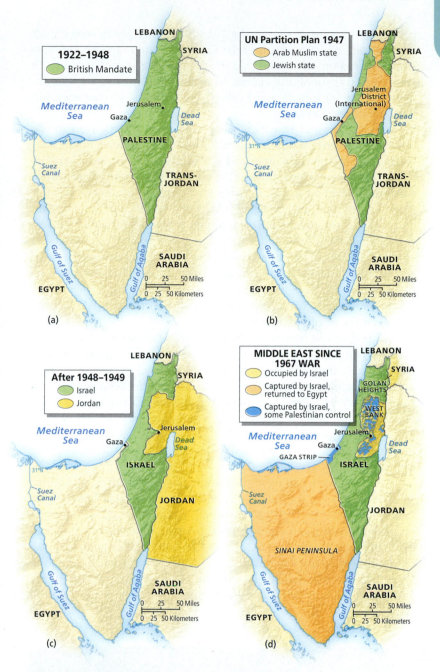

▲ **Figure 7.38 Evolution of Israel** Modern Israel's complex evolution began with (a) an earlier British colonial presence and (b) a UN partition plan in the late 1940s. (c) Thereafter, multiple wars with nearby Arab states produced Israeli territorial victories in Gaza, the West Bank, and the Golan Heights. (d) Each of these regions continues to be important in Israel's recent relations with nearby states and with resident Palestinian populations.

Israel's relations with neighboring countries remained poor. Supporters of Arab unity and Muslim solidarity sympathized with the Palestinians, and their antipathy toward Israel grew. Israel fought additional wars in 1956, 1967, and 1973. In territorial terms, the Six-Day War of 1967 was the most important conflict (see Figure 7.38). In this struggle against Egypt, Syria, and Jordan, Israel occupied substantial new territories in the Sinai Peninsula, the Gaza Strip, the West Bank, and the Golan Heights. Israel also annexed the eastern (Muslim) part of the formerly divided city of Jerusalem, arousing particular bitterness among the Palestinians. A peace treaty with Egypt resulted in the return of the Sinai Peninsula in 1982, but tensions focused on other occupied territories under Israeli control. To strengthen its geopolitical claims, Israel built additional Jewish settlements in the West Bank and in the Golan Heights, further angering Palestinian residents.

Palestinians and Israelis began to negotiate a settlement in the 1990s. Preliminary agreements called for a quasi-independent Palestinian state in the Gaza Strip and across much of the West Bank. A tentative agreement strengthened the control of the ruling **Palestinian Authority (PA)** in the Gaza Strip and portions of the West Bank (Figure 7.39). But new cycles of heightened violence have repeatedly erupted, and the Israelis have continued to build new settlements in the occupied West Bank. Indeed, the Israeli government has pledged to continue its support for new construction, including developments in and near Jerusalem and Bethlehem (Figure 7.40). Palestinian authorities continue to strongly protest these new West Bank Jewish settlements near Israel's capital, especially in an area (known as E1) that sits on hills just east of Jerusalem.

Adding to the friction is the ongoing construction of an Israeli security barrier, a partially completed series of concrete walls, electronic fences, trenches, and watchtowers designed to effectively separate the Israelis from Palestinians across much of the West Bank region (see Figure 7.39). Israeli supporters of the barrier (to be more than 400 miles long when completed) see it as the only way to protect their citizens from terrorist attacks. Palestinians see it as a land grab, an "apartheid wall" designed to isolate many of their settlements along the Israeli border.

Political fragmentation of the Palestinians adds further uncertainty. In 2006, control of the Palestinian government was split between the Fatah and Hamas political parties. Israelis have long regarded Hamas as an extremist political party, whereas Fatah has shown more willingness to work peacefully with Israel. Hamas gained effective control of the PA within Gaza, and Fatah maintains its greatest influence across the West Bank. Protests and attacks from Hamas-controlled Gaza continue and have repeatedly provoked Israeli counterattacks, decimating the Gaza economy.

(a)

(b)

▲ **Figure 7.39 West Bank**   (a) Portions of the West Bank were returned to Palestinian control in the 1990s, but Israel has partially reasserted its authority in some areas and has expanded the construction of its security barrier since 2000. New Israeli settlements are scattered throughout the West Bank in areas still under Israel's nominal control. (b) The photo shows a segment of the Israeli security barrier. **Q: Look carefully at the scale of the map. Measure the approximate distance between Jerusalem and Hebron, and find two local towns in your area that are a similar distance apart.**

One thing is certain: Geographical issues will remain at the center of the conflict. Israelis continue their search for secure borders to guarantee their political integrity. Most Palestinians still call for a "two-state solution," in which their autonomy is guaranteed; but a growing minority of Palestinians, frustrated with the stalemate, suggest

◄ **Figure 7.40 Jewish Settlement, West Bank**   This new housing is part of the Jewish settlement of Givat Ze'ev located just a few miles northwest of Jerusalem.

Explore the **SIGHTS** of **West Bank Settlement**
http://goo.gl/kPKByi

considering a "one-state solution," in which Israel would be compelled to recognize the Palestinians as equals.

Devastation in Syria   Elsewhere in the region, political instability in Syria erupted into civil war in 2011. Rebel (mostly Sunni Muslim) protests against the autocratic regime of President Bashar Hafez al-Assad reached a fever pitch, and government soldiers killed thousands of civilians and used chemical weapons in a series of violent confrontations. The larger regional Arab community reacted against Assad, suspending Syria from the **Arab League** (a regional political and economic organization focused on Arab unity and development; see Figure 7.37) and urging an international solution to the crisis. In 2014, ISIS also expanded its influence in the eastern portion of the country.

More recently, Assad has regained control over more of Syria, due in part to large-scale Russian assistance. There is also a growing Iranian presence in the country (allies of the Assad regime), which is seen by neighboring Israel as a direct threat to its own sovereignty. To the east, the influence of ISIS has declined, thanks to a mix of American, Kurdish, and Iranian forces which successfully reduced that extremist group's territorial control. By 2018, more than 400,000 deaths were directly related to the violence, and millions of people fled their homes (see Figure 7.26). In addition, many of Syria's great cultural landmarks—such as the ancient city of Palmyra—have been devastated in the conflict.

Iraq's Uncertain Path   Neighboring Iraq, another multinational state born during the colonial era, has yet to escape the consequences of its geopolitical origins. When the country was carved out of the British Empire in 1932, it contained the cultural seeds of its later troubles. Iraq remains culturally complex today (Figure 7.41). Most of the country's Shiites live in the lower Tigris and Euphrates river valleys south of Baghdad. Indeed, the region near Basra contains some of the world's holiest Shiite shrines. In northern Iraq, the Kurds have their own ethnic identity and political aspirations. A third major subregion, traditionally dominated by the Sunnis, encompasses part of the Baghdad area as well as territory to the north and west that includes strongholds such as Fallujah and Tikrit.

Since 2014, the Iraqi government has successfully reasserted its effective control over much of the country. With American and Iranian help, it has retaken areas north of Baghdad (in the vicinity of Mosul) that were controlled by ISIS. In addition, the central government (with Iran's assistance) has dislodged the Kurds from portions of the oil-rich north, shattering that group's short-term hopes for more autonomy even after they widely supported a referendum for Kurdish independence in 2017. Growing oil revenues have also added to the government's stability. In fact, thousands of Iraqi refugees have returned home, and many communities are rebuilding after more than two decades of political instability. Still, given the country's recent history, Iraq's future remains uncertain, although it seems likely to gravitate more closely into the orbit of neighboring Iran.

Iran Ascendant?   Iran increasingly attracts global notice. Islamic fundamentalism dramatically appeared on the political scene in 1978 as Shiite Muslim clerics overthrew Mohammad Reza Pahlavi, an authoritarian, pro-Western ruler friendly to U.S. interests. The new leaders proclaimed an Islamic republic in which religious officials ruled both clerical and political affairs. Today Iran's influence has grown across

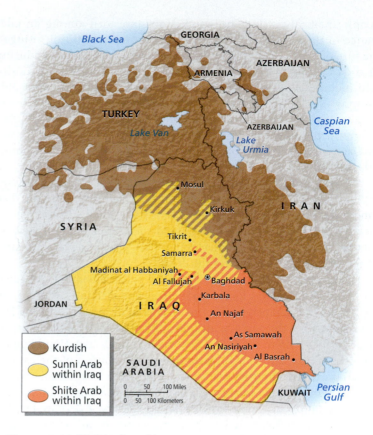

▲ **Figure 7.41 Multicultural Iraq**   Iraq's complex colonial origins produced a state with varying ethnic characteristics. Shiites dominate south of Baghdad; Sunnis hold sway in the western triangle zone; and Kurds are most numerous in the north, near oil-rich Kirkuk and Mosul.

the region, and the country has repeatedly threatened Israel. Israel and Arab states such as Saudi Arabia, the United Arab Emirates, and Egypt fear Iran's ascendance.

Adding uncertainty has been Iran's ongoing nuclear development program, an initiative its government claims is related to the peaceful construction of power plants (see Figure 7.37). Others in the West negotiated a settlement with Iran in 2015 (the U.S. withdrew from the agreement in 2018) that allows for more limited development of its nuclear capabilities and an end to many of the economic sanctions against the country.

Within Iran, varied political and cultural impulses are also evident. Many younger, wealthier, more cosmopolitan Iranians are hopeful that the country will become less isolated on the world stage. After all, more than 70 percent of the country's population was not even born at the time of the Iranian Revolution in the late 1970s. Popular interest in fundamentalism has waned, and many Iranians have actually moved toward a more secular lifestyle. Violence erupted across the country in late 2017, with critics of the government protesting high prices and corruption, and arguing for more domestic economic growth.

Politics in the Arabian Peninsula   Change has also rocked the Arabian Peninsula. In Saudi Arabia, the Al Saud royal family retains its control of the country, although the regime is gradually passing into the hands of younger family members who might be more inclined to democratize the nation's political structure. The Saudi people themselves, largely Sunni Arabs, are torn among an allegiance to their royal family (and the economic stability it brings); the lure of a more democratic, open Saudi society; and an enduring distrust of foreigners, particularly Westerners. The Sunni majority also includes Wahhabi sect members, whose radical Islamist philosophy has fostered anti-American sentiment. In addition, large numbers of foreign laborers and the persistent U.S. military and economic presence within the country (a chief complaint of former Al Qaeda leader Osama bin Laden) create a setting ripe for political instability.

In 2017 and 2018, Saudi Arabia also led a wider regional boycott and blockade of neighboring Qatar, a small but oil-rich Gulf state. The Saudis accused Qatar of supporting terrorist elements in the region, including Hezbollah in Lebanon. Qatar denied the claims, but the boycott has moved the Gulf nation closer to Iran, prompting new attempts by the United States to resolve the crisis.

Nearby Yemen has been devastated by political conflict. President Ali Abdullah Saleh was forced from office, and elections were held in 2012. Calls for democratic reforms were complicated by ongoing factionalism within the country, including the presence of Shiite militants (the Houthis) who maintain close connections with Iran. Houthi political gains in western Yemen (including taking control of the capital, Sanaa) provoked a military response in 2015 from Saudi Arabia, which fears growing Iranian interests in the region. The Saudis launched a bombing campaign against the Houthis and also have deployed troops (partnering with the United Arab Emirates and Sudan) on the ground to stabilize their hold in the war-ravaged nation. Added to the mix is Al Qaeda's influence (including terrorist training camps) in other portions of the country. Overall, the UN estimates that by 2018, the conflict led to more than 9,000 deaths and 50,000 injuries and displaced more than 2 million people from their homes. Widespread famine, malnutrition, and an outbreak of cholera have created one of the worst humanitarian crises in the world.

Tensions in Turkey   Turkey has also emerged as a geopolitical question mark, as it is strategically positioned between diverse, often contradictory geopolitical forces. Many pro-Westerners within Turkey have pushed for Turkey's admission into the EU. On the other hand, growing Islamist elements (mainly Sunni) within the country are wary of moving too close to Europe. In addition, in 2016, Turkey's president, Recip Erdogan, responded to a failed military coup within the country by cracking down on dissidents who were challenging his broad assertion of presidential powers. Erdogan's reelection to another five-year term in 2018 suggests he remains a popular political figure within his country. Increasingly, however, Erdogan's autocratic style, his support of an Islamist political agenda, and his growing alliance with Russia have clouded the country's democratic prospects.

Regional issues within the country also remain important. In the east, the Kurds (a key cultural minority in Turkey) have continued to press for more recognition and regional autonomy from the Turkish government (see Figure 7.41). In addition, Syria's political fragmentation has produced a huge refugee problem in the southern part of the country (see Figure 7.26). In response, Turkey has closed its border with its troubled neighbor and has moved its own troops into northern Syria to stabilize the situation and to limit the power of Syrian Kurds in that portion of the country.

REVIEW

7.8 Describe the role played by the French and British in shaping the modern political map of Southwest Asia and North Africa. Provide specific examples of their lasting legacy.

7.9 Discuss how the Sunni–Shiite split has recently played out in sectarian violence across the region.

7.10 Explain how ethnic differences have shaped Iraq's political conflicts in the past 50 years.

KEY TERMS Suez Canal, Palestinian Authority (PA), Arab League

Economic and Social Development: Lands of Wealth and Poverty

Southwest Asia and North Africa constitute a region of incredible wealth and discouraging poverty (Table 7.2). While some countries enjoy prosperity, due mainly to rich reserves of petroleum and natural gas, other nations are among the world's least developed. Continuing political instability contributes to the region's struggling economy. Civil wars within Syria, Libya, and Yemen have devastated their economies. Palestinians living in the Gaza and West Bank regions also suffer as political minorities within Israel. Petroleum will no doubt figure significantly in the region's future economy, but some countries in the area have also focused on increasing agricultural output, investing in new industries, and promoting tourism to broaden their economic base.

The Geography of Fossil Fuels

The striking global geographies of oil and natural gas reveal the region's continuing importance in the world economy as well as the extremely uneven distribution of these resources within the region (Figure 7.42). North African settings (especially Algeria and Libya), as well as the Persian Gulf region, have large sedimentary basins containing huge reserves of oil and gas, whereas other localities (for example, Israel, Jordan, and Lebanon) lie outside zones of major fossil fuel resources. Saudi Arabia, Iran, Iraq, Kuwait, and the United Arab Emirates hold large petroleum reserves, while Iran and Qatar possess the largest regional reserves of natural gas. In 2018, tiny Bahrain also announced huge new shale oil reserves in its offshore waters. The distribution of fossil fuel reserves suggests that regional supplies will not be exhausted anytime soon. Overall, with only 7 percent of the world's population, the region holds over half of the world's proven oil reserves.

Global Economic Relationships

Southwest Asia and North Africa share close economic ties with the rest of the world. This region is a major focus of the **Organization of the Petroleum Exporting Countries (OPEC)**, an international organization formed in 1960 that attempts to influence the global prices and supplies of oil. While oil and gas remain critical commodities that

TABLE 7.2   Development Indicators

Explore these data in MapMaster 2.0 https://goo.gl/TTQ6Ar

Country	GNI per Capita, PPP 2017[1]	GDP Average Annual Growth 2009-2015[2]	Human Development Index (2016)[3]	Percent Population Living Below $3.10 a Day[2]	Under Age 5 Mortality Rate (Per 1000 Live Births), 1990[1]	Under Age 5 Mortality Rate (per 1000 live births), 2016[1]	Secondary School Enrollment Ratios[4] Male (2009–2016)	Secondary School Enrollment Ratios[4] Female (2009–2016)	Gender Inequality Index (2016)[3,6]	Freedom Rating (2018)[5]
Algeria	15,275	3.3	0.745	–	49	25	98	102	0.429	5.5
Bahrain	47,527	3.9	0.824	–	23	8	102	102	0.233	6.5
Egypt	11,583	2.6	0.691	–	86	23	86	86	0.565	6.0
Iran	20,950	0.2	0.774	3.1	57	15	89	89	0.509	6.0
Iraq	17,197	6.7	0.649	–	54	31	–	–	0.525	5.5
Israel	38,413	3.8	0.899	–	12	4	102	103	0.103	2.0
Jordan	9153	2.7	0.741	–	37	18	80	85	0.478	5.0
Kuwait	71,943	3.3	0.800	–	18	8	88	103	0.335	5.0
Lebanon	14,676	2.3	0.763	–	33	8	61	61	0.381	5.0
Libya	19,631	–	0.716	–	42	13	–	–	0.167	6.5
Morocco	8218	3.9	0.647	25.7	80	27	74	64	0.494	5.0
Oman	41,675	4.3	0.796	–	39	11	101	108	0.281	5.5
Palestinian Territories	4885	4.7	0.684	–	45	19	79	87	–	6.5
Qatar	128,378	7.4	0.856	–	21	9	82	104	0.542	5.5
Saudi Arabia	53,845	5.0	0.847	–	45	13	123	94	0.257	7.0
Sudan	4904	2.0	0.490	–	131	65	44	41	0.575	7.0
Syria	2900*	–	0.536	–	37	18	50	51	0.554	7.0
Tunisia	11,911	2.1	0.725	13.3	57	14	90	94	0.289	2.5
Turkey	27,916	4.8	0.767	3.1	74	13	104	101	0.328	5.5
United Arab Emirates	73,879	4.5	0.840	–	17	8	–	–	0.232	6.5
Western Sahara*	2500*	–	–	–	–	–	–	–	–	7.0
Yemen	1595	–3.6	0.482	–	126	55	57	40	0.767	6.5

[1] World Bank Open Data, 2018. * Additional Data from CIA World Factbook.
[2] World Bank— *World Development Indicators*, 2017.
[3] United Nations, *Human Development Report*, 2016.
[4] Population Reference Bureau, *World Population Data Sheet*, 2017.
[5] Freedom House, Freedom in the World 2018. See Ch.1, pp. 33 –34, for more info. on this scale (1–7, with 7 representing states with the least freedom).
[6] See Ch. 1, p. 39, for more info. on this scale (0–1, with higher values representing less gender equality).

Log in to Mastering Geography & access MapMaster to explore these data!

1) Look at the HDI map and find the three countries with the *lowest* Human Development Index (HDI). What other development indicators correlate well with the HDI data? Why?

2) Where might you expect to live longest in this region? What are the top five countries in terms of life expectancy, and how do they measure up in terms of GNI per capita? Why might there be differences in these two measures?

dominate international economic linkages, the growth of manufacturing and tourism is also redefining the region's role in the world.

OPEC's Changing Fortunes

Although OPEC does not control global oil and gas prices, it still influences the cost and availability of these pivotal products. While the United States has made some progress toward greater energy independence from the Mideast (via its larger domestic oil and natural gas output), western Europe, Japan, China, and many less industrialized countries still depend on the region's fossil fuels.

Moderating energy prices, however, may be signaling changes for OPEC's fortunes. More competition from non-OPEC producers may limit the group's future influence. OPEC's move to keep production high indicates the organization (especially Saudi Arabia) is interested in driving high-cost producers (such as North American fracking operations; see Chapter 3) out of business. In the meantime, however, lower prices have already exerted pressure on the budgets of many OPEC nations within the region. Government programs are being cut, and many residents resent paying more for many services. Even the United Arab Emirates, one of the region's wealthiest and most diverse economies, operated with a deficit in 2016, the first in decades.

Other Global and Regional Linkages

Beyond key OPEC producers, other trade flows also contribute to global economic integration. Turkey, for example, ships textiles, food products, and manufactured goods

to its principal trading partners: Germany, the United States, Italy, France, and Russia. Tunisia sends more than 60 percent of its exports (mostly clothing, food products, and petroleum) to nearby France and Italy. Israeli exports emphasize the country's highly skilled workforce: Products such as cut diamonds, electronics, and machinery parts go to the United States, western Europe, and Japan. Localities such as Dubai (United Arab Emirates) have also become globally significant transportation hubs, attracting air travelers from around the world (see *Exploring Global Connections: Dubai's Role as a Global Travel Hub*).

Future interconnections between the global economy and Southwest Asia and North Africa may depend increasingly on cooperative economic initiatives far beyond OPEC. Relations with the European Union (EU) are critical. Euro-Med agreements have been signed between the EU and countries across the region that border the Mediterranean Sea.

Most Arab countries, however, are wary of too much European dominance. In 2005, 17 Arab League members established the **Greater Arab Free Trade Area (GAFTA)**, an organization designed to eliminate all intraregional trade barriers and spur economic cooperation. In addition, Saudi Arabia plays a pivotal role in regional economic development through organizations such as the Islamic Development Bank and the Arab Fund for Economic and Social Development. Many of these financial organizations offer services compliant with Islamic law (*sharia law*). In fact, these Islamic banking assets grew about 20 percent annually between 2015 and 2018.

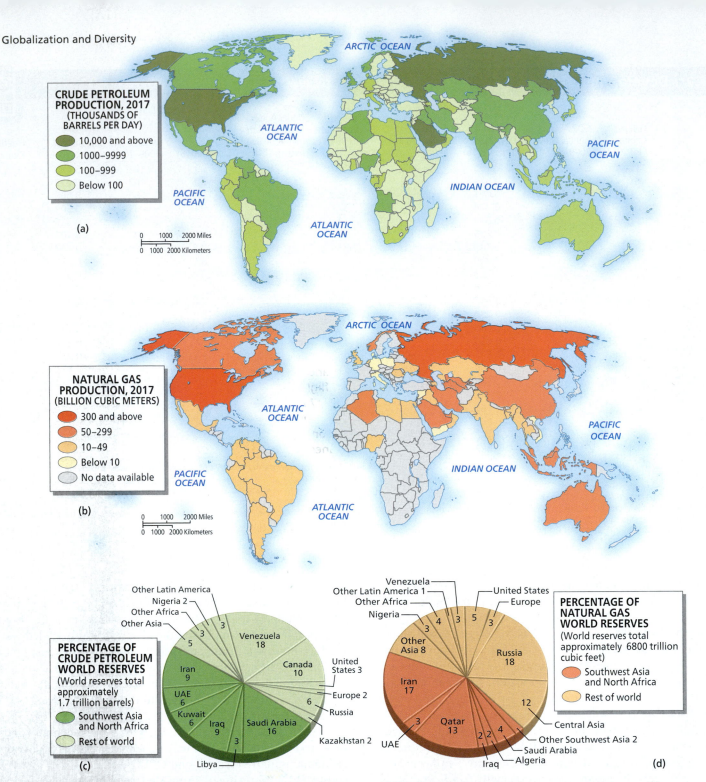

▲ **Figure 7.42 Crude Petroleum and Natural Gas Production and Reserves**   The region plays
a central role in the global geography of both (a) crude petroleum production and (b) natural gas
production. Abundant regional reserves of both (c) crude petroleum and (d) natural
gas suggest that the pattern will continue.

Regional Economic Patterns

Remarkable economic differences characterize Southwest Asia and North Africa (see Table 7.2). Some oil-rich countries have prospered greatly since the early 1970s; but in many cases, fluctuating oil prices, political disruptions, and rapidly growing populations threaten future economic growth. Regional levels of military spending have accelerated in many nations since 2010, suggesting that future conflicts may further disrupt the region's path to economic stability (Figure 7.43). Many countries in the region (Saudi Arabia and Israel, for example), on

a per capita basis, are some of the world's largest spenders on national defense, surpassing even the United States.

The richest countries of Southwest Asia and North Africa owe their wealth to massive oil reserves. Nations such as Saudi Arabia, Kuwait, Qatar, Bahrain, and the United Arab Emirates benefit from fossil fuel production as well as from their relatively small populations. Large investments in transportation networks, urban centers, and other petroleum-related industries have reshaped the cultural landscape (Figure 7.44). In 2016, Saudi Arabia released its Vision 2030 economic agenda aimed at diversifying its oil-based economy into sectors such as technology. It proposes (1) selling off a small part of its state-owned

Dubai's Role as a Global Travel Hub

The spidery web of the Air Emirates route map tells it all (Figure 7.4.1). Anchored strategically between Europe and Asia, Dubai's International Airport, the main route hub for Air Emirates, has blossomed into the planet's busiest center of global air travel. It is about six flying hours from Central Europe to the Middle East hub and another seven flying hours to East Asia.

A Growing Air Hub?

It wasn't so long ago that the airport claimed a single lonely runway in the desert, but in the past 25 years—thanks to the petroleum economy and the efficient management of the state-run facility and its flagship airline—Dubai has strikingly emerged to rule long-distance air traffic that connects many of the globe's most important places to one another. While Atlanta's Hartsfield airport still claims the most daily traffic, Dubai is the world's busiest international travel hub, surpassing London Heathrow in 2014.

In 2016, the Dubai Airport opened a massive new $1.2 billion concourse that boosted the facility's annual capacity from 75 to 90 million passengers. Its high-tech infrastructure and upscale passenger services has made it by far the busiest, most modern airport in the Middle East (Figure 7.4.2). Here, passengers from Hyderabad, Amsterdam, Cape Town, and Osaka mingle with one another as they grab a meal and pass time between flights.

Success in the Skies?

Air Emirates, based in Dubai, now boasts the world's largest fleet of Boeing 777s and Airbus A380s. In 2013, in a single staggering purchase, Air Emirates ordered $99 billion worth of new airplanes from the dominant commercial manufacturers (Boeing and Airbus), more wide-body jets than are flown by American Airlines and United Airlines combined. The airline's route map reveals its dominance of long-distance global travel with some nonstop flights to Latin America lasting more than 17 hours.

Part of Dubai's success as an aviation hub is simply its location. Long-distance travelers often change planes there on trips from Europe and eastern North America as they make their way to India, China, and other Asian destinations. Many competing U.S. carriers (such as Delta and United) also claim Dubai and its state-owned airline have unfair advantages because the government controls the facility, the airplanes, and all the regulations surrounding their integrated operation.

▲ **Figure 7.4.2 Dubai Airport** The Dubai Airport, located in the United Arab Emirates, is one of the world's busiest air traffic hubs.

In other words, the airline and related facilities can do the job cheaper and more efficiently, out-competing privately owned operators. Some see Dubai already emerging as a global *aerostate*, in which a country derives a significant portion of its economic muscle from controlling a key airport and associated air carriers. Indeed, it looks as if the airport and its associated airline will increasingly dominate long-distance travel in that part of the world.

1. **Looking at the global route map (Figure 7.4.1) for Air Emirates, which parts of the wealthier, more developed world appear to be most attractive targets for the airline's future expansion plans? Why?**

2. **Identify a nearby example of an airline "hub" in your area. What makes it a hub? What airlines dominate traffic?**

▲ **Figure 7.4.1 Air Emirates Route Map for Dubai** Dubai's striking global centrality is apparent as Air Emirates has successfully expanded to be a major international carrier.

Emirates routes, October 2017
Due to space limitations, not all locations are shown.

GOOGLE EARTH
Virtual Tour Video
http://goo.gl/JK4v8m

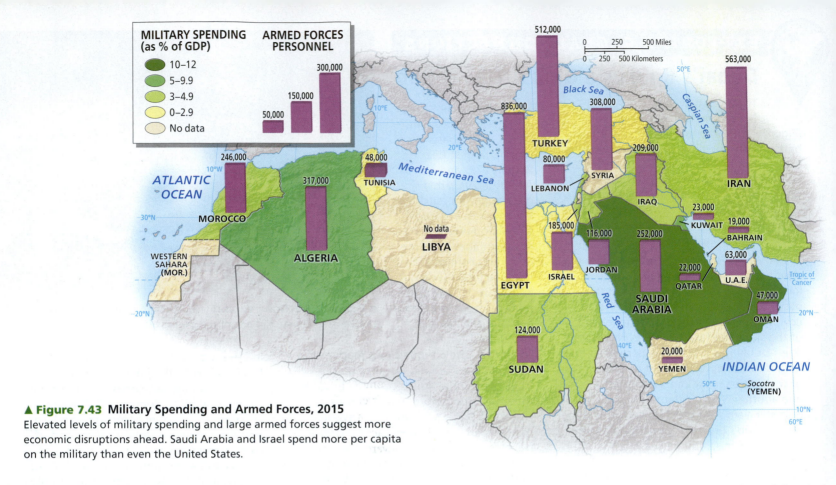

▲ Figure 7.43 Military Spending and Armed Forces, 2015
Elevated levels of military spending and large armed forces suggest more
economic disruptions ahead. Saudi Arabia and Israel spend more per capita
on the military than even the United States.

oil assets and then investing the proceeds in varied global investments;
(2) holding down domestic spending and making its economy more
competitive; and (3) generating more public revenues from fees, an
expanded tax base, and more tourism.

Some countries possess fossil fuel reserves, but political instability
has hampered sustained economic growth. For example, Libya's political disintegration has had profound economic consequences, sharply
reducing oil and gas output and causing severe economic disruptions.
Both Iraq and Iran are also blessed with rich reserves of fossil fuels, but

**▼ Figure 7.44 Saudi Arabian Oil
Refinery** These oil storage tanks are
a part of Saudi Arabia's sprawling Ras
Tanura refinery, located in the
eastern part of the country.

Explore the **SIGHTS** of
Yanbu Oil Refineries
http://goo.gl/tN8GWJ

output has often been limited or disrupted by internal instability and
conflicts with other states.

Other nations, while lacking petroleum resources, have still found
paths to economic prosperity. Israel, for example, supports one of the
highest standards of living in the region, even with its political challenges (see Table 7.2). The country is a global center for high-tech
computer and telecommunications products, known for its fast-paced
and highly entrepreneurial business culture that resembles California's
Silicon Valley. Israel also has daunting economic problems. Defense
spending absorbs a large share of total gross national income (GNI),
necessitating high tax rates. Poverty and unemployment among
Palestinians also remain unacceptably high, both in Gaza (recently
devastated by more violence) and in the West Bank.

Turkey has a diversified economy, even though incomes are
modest by regional standards. Lacking its own fossil fuels, Turkey has successfully attracted huge natural gas pipeline projects
through the country that have benefited infrastructure spending.
Gas began flowing through the Trans-Anatolian Natural Gas Pipeline in 2018, and the Trans-Adriatic Pipeline will be operational in
2020. About 20 percent of the workforce remains in agriculture;
and principal commercial products include cotton, tobacco, wheat,
and fruit. The industrial economy has grown since 1980—including exports of textiles, processed food, and chemicals. Turkey has
also gone high tech: About half the Turkish population uses the
Internet, and the country has been a fertile ground for dozens of
global Internet startup companies that connect well with younger
Turks (many are online or virtual gaming enterprises) as well as
with the global economy. Still, the country's growing political
instability already shows signs of contributing to more volatility
in the country's currency and in its overall economic performance.

Poorer countries of the region share the problems of the less
developed world. For example, Sudan, Egypt, Syria, and Yemen

each face unique economic challenges. Sudan's continuing political problems stand in the way of progress. Egypt's economic prospects are also unclear. While Egyptian leaders have been promoting the country's relatively stable political situation, many Egyptians still live in poverty, and the gap between rich and poor continues to widen. Costly state subsidies for food and fuel are rapidly growing the country's budget deficit. Illiteracy is widespread; and the country suffers from the **brain drain** phenomenon, as some of its brightest young people leave for better jobs in western Europe or the United States.

Syria once enjoyed both a growing economy and a stable political regime. Now it has neither, as its ongoing civil war and sectarian conflicts have decimated the economy. Millions of people who considered themselves part of the Syrian middle class have been thrown into utter poverty, often as desperate refugees. Similarly, Yemen's recent civil war has wreaked economic havoc and produced widespread declines in the country's already low standard of living (Figure 7.45).

Gender, Culture, and Politics: A Woman's Changing World

The role of women in largely Islamic Southwest Asia and North Africa remains a major social issue. Female participation rates in the workforce are among the world's lowest, and large gaps typically exist between levels of education for males and females. In conservative parts of the region, few women work outside the home.

More orthodox Islamic states impose legal restrictions on the activities of women. In Saudi Arabia, for example, women until recently have been prohibited from driving. In neighboring Qatar, women can vote (and hold public office), but they still need a husband's consent to obtain a driver's license. In Iran, full veiling remains mandatory in more conservative areas, but many wealthier Iranian women have adopted Western dress, reflecting a more secular outlook, especially among the young (Figure 7.46). Today, 60 percent of Iranian university students are female. Generally, Muslim women still lead more private lives than men: Much of their domestic space is shielded from the world by walls and shuttered windows, and their public appearances are filtered through the use of the *niqab* (face veil) or *chador* (full-body veil).

In some settings, women's lives are changing, even within norms of more conservative Islamist societies. From Tunisia to Yemen, women widely participated in the Arab Spring rebellions, asserting their new political visibility in public ways. Lebanon's Roadmap program, launched in 2018, partners with the UN and is dedicated to

▲ **Figure 7.46 Iranian Women** These fashionably dressed young women in Tehran suggest how Iran's more urban and affluent residents have embraced many elements of Western culture. **Q: What unique challenges confront educated women in Iran? Unique opportunities?**

empowering women in that nation's political process. Record numbers of women ran for election in Lebanon's recent parliamentary elections.

Even in conservative Saudi Arabia, women enjoy increased freedom. Some women have been encouraged to play more active roles in political affairs, including running for office. Many women also attend universities, making up 52 percent of the nation's student body. Women are pouring into the workplace, with female employment growth skyrocketing by 50 percent between 2010 and 2016. Two-income families are becoming the norm as women's roles change and as the country's economy struggles amid lower oil prices.

Even with these changes, many argue that the region still has a long way to go and that women, especially in the Arab World, are only in the early stages of a social transformation that will still take decades to unfold.

REVIEW

7.11 Describe the basic geography of oil reserves across the region, and compare the pattern with the geography of natural gas reserves.

7.12 Identify different strategies for economic development recently employed by nations such as Saudi Arabia, Turkey, Israel, and Egypt. How successful have they been, and how are they related to globalization?

KEY TERMS Organization of the Petroleum Exporting Countries (OPEC), Greater Arab Free Trade Area (GAFTA), brain drain

◄ **Figure 7.45 Sanaa, Yemen** Yemen's civil war has taken a huge economic toll on the country's capital city of Sanaa. This house was destroyed in a Saudi air raid in 2018.

Summary

- Many nations within Southwest Asia and North Africa suffer from significant environmental challenges. Twentieth-century population growth across the region was dramatic, but expanding the region's limited supplies of agricultural land and water resources is costly and difficult. The eroded soils of the Atlas Mountains to overworked garden plots along the Nile illustrate the environmental price paid when population growth outstrips the ability of the land to support it.

- The population geography of the region is strikingly uneven. Areas with higher rainfall or access to exotic water often have high physiological population densities, whereas nearby arid zones remain almost empty of settlement.

- Culturally, the region remains the hearth of Christianity, the spatial and spiritual core of Islam, and the political and territorial focus of modern Judaism. In addition, important sectarian divisions within religious traditions, as well as long-standing linguistic differences, continue to shape the area's local cultural geographies and regional identities.

- Political conflicts have disrupted economic development. Civil wars, sectarian violence, conflicts between states, and regional tensions hamper initiatives for greater cooperation and trade. Perhaps most important, the region must deal with the conflict between modernity and more fundamentalist interpretations of Islam.

- Abundant reserves of oil and natural gas, coupled with the world's continuing reliance on fossil fuels, ensure that the region will remain prominent in world petroleum markets. Yet economic prospects remain muted by fluctuating oil prices and never-ending political instability, which challenge progress in modernizing the region's social development, including expanded opportunities for women.

Review Questions

1. Discuss five important human modifications of the Southwest Asian and North African environment, and assess whether these changes have benefited the region.

2. Why are birth rates declining in many countries of the region? Despite cultural differences with North America, what common processes seem to be at work in both regions that have contributed to this demographic transition?

3. What are the special challenges in preserving traditional ways within conservative Islamic societies that are also major participants in the global oil economy? Give an example.

4. What economic changes could occur if Israel and the Palestinians achieved peace and if the Syrians could end their civil war? What kinds of general connections might be found between political conflict and economic conditions throughout the region?

5. Describe some of the similarities and differences in how lower oil prices have recently affected life in (a) France, (b) Saudi Arabia, and (c) Texas.

Image Analysis

1. This simple bar graph shows the percentage of female labor force participation in selected Middle Eastern and North African (MENA) countries for 2016. Why might rates be relatively high in Qatar and the United Arab Emirates (UAE) but so low in Jordan, the West Bank, and Gaza?

2. What underlying economic, cultural, and political variables are really being measured by female labor force participation rates?

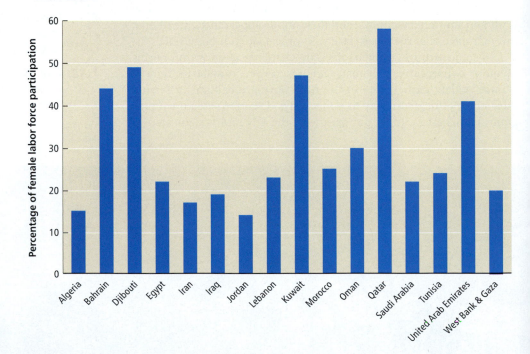

▲ **Figure IA7** Percentage of Female Labor Force Participation Rates in MENA Countries

▲ **Figure D7** Kurdish Women in Turkey

Join the Debate

Historically, Southwest Asia's Kurdish population has been a "nation without a state." An enduring "Kurdistan" almost came to fruition after World War I, but never became a reality. A recent referendum among Iraqi Kurds suggested widespread support for independence. Is the time now right for an independent Kurdish state?

The time is right to create a Kurdish Republic out of portions of Iraq, Turkey, Syria, and Iran.

- Kurdish-speaking peoples make up a large and distinctive cultural group in Southwest Asia, with a well-defined homeland that would define the limits of a new political entity within the region.

- If they were at peace and united, the Kurds would cease to be a constant irritant as a cultural and political minority in nations such as Turkey and Iraq, increasing stability in those countries.

- A Kurdish Republic would quickly become a real economic bright spot within the region, thanks to lessened hostilities, a solid economic base (including rich petroleum reserves), and a skilled, entrepreneurial population.

Although attractive in the abstract, a Kurdish nation would only disrupt an already unstable region further.

- Nations such as Turkey and Iraq would strongly resist giving up large, resource-rich portions of their national territory, and there would be endless arguments over boundaries that would define the new country.

- Massive, potentially violent migrations of both Kurds and non-Kurds could result from poorly drawn borders, similar to what happened with the partition of South Asia in the late 1940s.

- While not ideal, Kurds can work toward greater representation and increased autonomy within established nations such as Turkey and Iraq, and the past decade suggests they can make further progress in these efforts.

Key Terms

Arab League (p. 247)
Arab Spring (p. 220)
brain drain (p. 253)
choke point (p. 227)
culture hearth (p. 220)
domestication (p. 229)
exotic river (p. 232)
Fertile Crescent (p. 229)
fossil water (p. 224)
Greater Arab Free Trade Area (GAFTA) (p. 249)
Hajj (p. 238)

hydropolitics (p. 225)
ISIS (Islamic State of Iraq and the Levant; also known as ISIL, or IS) (p. 220)
Islamic fundamentalism (p. 220)
Islamism (p. 220)
Levant (p. 222)
Maghreb (p. 222)
medina (p. 232)
monotheism (p. 237)
Organization of the Petroleum Exporting Countries (OPEC) (p. 248)

Ottoman Empire (p. 239)
Palestinian Authority (PA) (p. 246)
pastoral nomadism (p. 231)
physiological density (p. 229)
Quran (p. 238)
salinization (p. 224)
sectarian violence (p. 220)
Shiite (p. 238)
Suez Canal (p. 243)
Sunni (p. 238)
theocratic state (p. 238)
transhumance (p. 231)

MasteringGeography

Looking for additional review and test prep materials? Visit the Study Area in Mastering Geography to enhance your geographic literacy, spatial reasoning skills, and understanding of this chapter's content by accessing a variety of resources, including MapMaster interactive maps, geoscience animations, videos, flashcards, web links, self-study quizzes, and an eText version of *Globalization and Diversity*.

Scan to read about Geographer at Work
Karen Culcasi and her examination of the challenges facing refugee women.
https://goo.gl/teXcji

GeoSpatial Data Analysis

Health Care Access Health care is often considered a basic human right in more developed portions of the world, but large parts of Southwest Asia and North Africa are poorly served by health care providers. The World Health Organization (WHO) gathers data on physicians per 1000 population, which can be used as a measure of access to health care as well as social development. According to recent data, the United States had about 2.5 physicians per 1000 and Germany about 3.9.

Open MapMaster 2.0 in the Mastering Geography Study Area. Go to the WHO website (www.who.int) and access the data/interactive atlas page on physicians per 1000 population (click on Health Topics and then click on Data to find Health Worker Density; then click on Density of Physicians and View Data). Download, import, and prepare the data in MapMaster.

WHO Data: Number of Physicians per 1000 Population
https://goo.gl/vBu7WR

1. Make your map showing the regional pattern of health care access across Southwest Asia and North Africa.

2. In a few sentences, summarize the general patterns and trends you see. How would you explain some of the major variations you observe across the region?

3. From Table 7.2, generate a map of HDI (Human Development Index) for the region and compare the two patterns you see in split-screen mode. What similarities and differences do you see? How might these two indicators be a good measure of future social development? How might they predict political stability?

Europe

Physical Geography and Environmental Issues

Diverse European environments range from subtropical Mediterranean lands to the arctic tundra. Europe is also one of the "greenest" world regions, with strong measures regarding pollution and renewable energy.

Population and Settlement

Europe has very low rates of natural growth and very high rates of internal mobility and international in-migration. Most current international migration consists of refugees from strife-torn Africa and Southwest Asia.

Cultural Coherence and Diversity

Europe has a long history of cultural tensions linked to internal differences in language and religion; however, today's tensions include those connected with immigration from other world regions, along with fracturing regional identities.

Geopolitical Framework

Two world wars and a lengthy Cold War divided 20th-century Europe, producing an ever-changing map. Although Europe is an integrated and peaceful region, geopolitical tensions linked to rising nationalism and devolution now dominate.

Economic and Social Development

For half a century, the European Union (EU) has worked successfully to integrate the region's diverse economies and political systems, making Europe a global superpower. Today, however, internal economic and social issues, including Brexit, challenge this unity.

▶ Catalan nationalists in Spain's northeastern autonomous region of Catalonia create castells, or human pyramids, at festivals and other occasions. Some reach 10 people high, with dozens more involved in creating each one. Castells have become popular in recent years as a way to celebrate Catalan identity and protest in favor of Catalan independence.

Catalonia, Spain

EUROPE

Europe faces a crisis of identity as forces of devolution and nationalism roil the continent and threaten its 50-year experiment in integration. On October 1, 2017, 93 percent of voters in Spain's wealthiest autonomous region, Catalonia, voted to become an independent country, although fewer than half the region's voters participated in this referendum. The Spanish government reacted by declaring the referendum invalid, and violence broke out between police and Catalan protesters across the picturesque region—which includes Spain's top tourist destination, the Mediterranean city of Barcelona. Images of Spanish police pulling elderly protesters out of polling places rocketed around the globe, challenging our perception of modern Europe as a stable and relatively strife-free place.

In the 1950s, after centuries of nationalistic and dynastic wars, competition, and conflict, Europe committed itself to an agenda of economic, political, and social integration through the **European Union (EU)**. The EU is a supranational organization made up of 28 countries, dropping to 27 once the United Kingdom completes its withdrawal, the so-called "**Brexit**," in April 2019. Although highly successful in uniting and integrating many aspects of European economic and political life, the EU's future is now being called into question due to tense geographies of separatism and political devolution.

The Catalonia example is the starkest, but the push for independence—or at least autonomy—is roiling countries across Europe. In 2014, Scotland narrowly voted to remain part of the United Kingdom but may push again for independence once Brexit is completed, as most Scots voted to stay in the EU. In many administrative regions of northern Italy, a growing movement for autonomy led the Veneto (home to Venice) to vote in late 2017 for greater control of its local affairs and taxes (Figure 8.1). Further votes for autonomy or independence may follow in other subregions of Europe, including Brittany in France, Wallonia in Belgium, or even for Northern Ireland to be united with the Republic of Ireland to its south. These votes raise questions about the survival of an integrated Europe.

> The EU's future is now being called into question due to tense geographies of separatism and political devolution.

▲ **Figure 8.1 Veneto Referendum** Ads urge voters in Italy's northeastern Veneto region, centered on the tourist mecca of Venice, to support a 2017 vote for additional autonomy from the central government. Wealthier northern Italians resent the tax money funneled by Rome to poorer regions in Italy's south. The referendum was supported by 98 percent of voters, but whether the central government will honor the results is unclear.

Defining Europe

Europe is one of the world's most diverse regions, encompassing a wide assortment of people and places in an area considerably smaller than North America. More than half a billion people reside in this region, living in 42 different countries ranging in size from France, Sweden, and Germany to microstates such as Andorra, and even Vatican City, the world's smallest independent state (Figure 8.2). Europe is commonly divided into five subregions: western, central, southern (or Mediterranean), southeastern (or Balkan), and northern (or Scandinavian and Baltic) Europe, terms we use throughout this chapter. A sixth subregion, eastern Europe, is primarily covered in Chapter 9. Europe is one of the wealthiest and most peaceful world regions, yet income and employment disparities, along with population movements both within and from outside Europe, have led to rising tensions.

ELEVATION IN METERS

4000+
2000–4000
500–1999
200–499
0–199
Sea Level
Below sea
level

EUROPE
Political & Physical Map

(★) ● Metropolitan areas
more than 20 million

(★) ● Metropolitan areas
10–20 million

(⊛) ● Metropolitan areas
5–9.9 million

(⊛) • Metropolitan areas
1–4.9 million

(⊛) ○ Selected smaller
metropolitan areas

▬ ▬ Plate boundaries

▲ **Figure 8.2 Europe**   Stretching east from Iceland in the Atlantic to Russia, Europe includes 42 countries, ranging in size from large states such as France and Sweden to the microstates of Liechtenstein, Andorra, Vatican City, and Monaco. Currently, the region's population is about 546 million. Europe is commonly divided into the five subregions of western, central, southern (or Mediterranean), southeastern (or Balkan), and northern (or Scandinavian) Europe. Tables 8.1 and 8.2 are organized by these subregions.

Physical Geography and Environmental Issues: Human Transformation of Diverse Landscapes

Despite Europe's small size, its environmental diversity is extraordinary. A startling array of landscapes is found within its borders, from the arctic tundra of northern Scandinavia to the semiarid hillsides of the Mediterranean islands, with explosive volcanoes in southern Italy and glaciated seacoasts in Norway and Iceland.

Three factors explain this impressive environmental diversity. The first is the complex geology of the western extension of the Eurasian land mass. Second is Europe's extensive latitudinal range from the Arctic to the Mediterranean subtropics, which influences climate, vegetation, and hydrology (Figure 8.3). However, its high latitudes are subject to the moderating influences of Europe's long coastlines along the Atlantic Ocean and its Gulf Stream, as well as the surrounding Baltic, Mediterranean, and Black seas. Finally, the history of human settlement, spanning thousands of years, has transformed and modified Europe's landscapes in fundamental ways.

Landform Regions

Europe can be organized into four general landform regions: the European Lowland, forming an arc from southern France to the northeast plains of Poland, but also including southeastern England; the Alpine mountain system, extending from the Pyrenees in the west to the Balkan mountain ranges of southeastern Europe; the Central

Uplands, positioned between the Alps and the European Lowland; and the Western Highlands, which include mountains in Spain and portions of the British Isles and the highlands of Scandinavia (see Figure 8.2). Iceland, unquestionably a part of Europe yet lying 900 miles (1500 km) west of Norway, has its own unique landforms, straddling the Mid-Atlantic Ridge.

The European Lowland　Also known as the North European Plain, the European Lowland is the unquestioned economic focus of western Europe, with its high population density, intensive agriculture, large cities, and major industrial regions. Though not completely flat, most of this lowland lies below 500 feet (150 meters) in elevation. Many of Europe's major rivers (the Rhine, Loire, Thames, and Elbe) meander across the lowland, forming broad estuaries before emptying into the Atlantic. Several of Europe's busiest ports, including London, Le Havre, Rotterdam, and Hamburg, are located in these lowland settings.

The Rhine River delta conveniently divides the unglaciated southern European Lowland from the glaciated plains to the north, which were covered by a Pleistocene ice sheet until about 15,000 years ago. This continental glacier renders the northern lowland, including the Netherlands, Germany, Denmark, and Poland, less fertile than the unglaciated portions of Belgium and France. Rocky clay materials in Scandinavia were eroded and transported south by glaciers. As the climate warmed and the glaciers retreated, piles of glacial debris were left on the plains of Germany and Poland.

The Alpine Mountain System　The Alpine system forms the topographic spine of Europe and consists of a series of ranges running west to east from the Atlantic to the Black Sea and the southeastern Mediterranean. These mountain ranges carry distinct regional names, such as the Pyrenees, Alps, Apennines, Carpathians, Dinaric Alps, and Rhodope Mountains, but share geologic traits.

The Pyrenees form the political border between Spain and France and include the microstate of Andorra. This rugged range extends almost 300 miles (480 km) from the Atlantic to the Mediterranean, containing glaciated peaks reaching to 11,000 feet (3350 meters) alternating with broad glacial valleys.

The centerpiece of this geologic region is the prototypical mountain range, the Alps, running more than 500 miles (800 km) from France to Slovenia (Figure 8.4). These impressive mountains are highest in the west, rising to more than 15,000 feet (4600 meters) at Mt. Blanc on the French–Italian border. In Austria and Slovenia, the Alps are more subdued, with few peaks exceeding 10,000 feet (3000 meters). Though easily crossed today by car or train through long tunnels and valley-spanning bridges, these mountains have historically formed an important cultural divide between the Mediterranean lands to the south and central and western Europe to the north.

The Apennine Mountains are located south of the Alps; the two ranges, however, are connected by the hilly coastline of the French and Italian Riviera. Forming Italy's spine, the Apennines are lower and lack the spectacular glaciated peaks and valleys of the Alps, but take on their own distinctive character farther south with the active volcanoes of Mt. Vesuvius (just over 4000 feet, or 1200 meters) outside of Naples and Mt. Etna (almost 11,000 feet, or 3350 meters) on the island of Sicily.

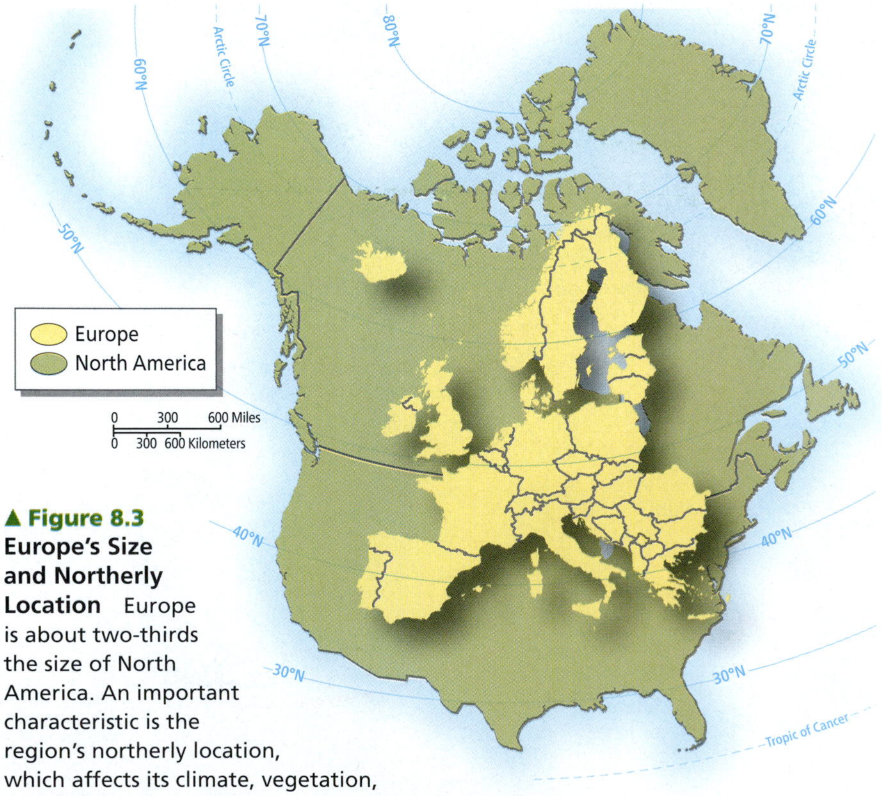

▲ **Figure 8.3**
Europe's Size and Northerly Location　Europe is about two-thirds the size of North America. An important characteristic is the region's northerly location, which affects its climate, vegetation, and agriculture. Much of Europe lies at the same latitude as Canada; note that even the Mediterranean lands are farther north than the United States–Mexico border. **Q: What parts of Europe are at the same latitude as your location?**

▲ **Figure 8.4 The Alps** Europe's most famous mountain range, the Alps, extend from France to Slovenia. The Triglav peak in Slovenia, pictured here, serves as the symbol of the country and shows evidence of glaciation from Europe's Pleistocene past.

The Carpathian Mountains define the eastern limits of the Alpine system. They are a plow-shaped upland area extending from Czechia to the Iron Gate gorge, a narrow passage where the Danube River cuts through along the Romanian–Serbian border.

Central Uplands A much older highland region forms an arc between the Alps and the European Lowland in France and Germany. These mountains are much lower in elevation than the Alpine system, with their highest peaks at 6000 feet (1800 meters). Much of this upland is characterized by rolling landscapes about 3000 feet (1000 meters) above sea level.

These uplands are important to western Europe because they contain the raw materials for Europe's industrial areas. In Germany and France, for example, they have provided the iron and coal central to each country's steel industry. To the east, mineral resources from the Bohemian Highlands have also fueled major industrial areas in Germany, Poland, and Czechia.

Western Highlands The ancient Western Highlands define the western edge of the European subcontinent, extending from Portugal in the south, through portions of the British Isles, to the highland backbone of Norway, Sweden, and Finland in the far north. As with other uplands that traverse many separate countries, the names for these mountains differ from country to country. A portion of the Western Highlands forms the spine of England, Wales, and Scotland, where picturesque glaciated landscapes are found at modest elevations of 4000 feet (1200 meters) or less. These U-shaped glaciated valleys also appear in Norway's uplands, where they produce a spectacular coastline of **fjords**, or flooded valley inlets, similar to the coastlines of Alaska and New Zealand (Figure 8.5).

Geologically, the far western edge of Europe is in Iceland, which, as mentioned, sits atop the Mid-Atlantic Ridge

separating the Eurasian and North American tectonic plates. Like other plate boundaries, Iceland has many active volcanoes that occasionally spew ash into the atmosphere, sometimes causing serious problems for the heavy airline traffic between Europe and North America.

Seas, Rivers, and Ports

In many ways, Europe is a maritime region with strong ties to its surrounding seas. Even landlocked countries such as Austria, Hungary, Serbia, and Czechia have access to the sea through extensive networks of navigable rivers and canals.

Europe's Ring of Seas Four major seas and the Atlantic Ocean encircle Europe. In the north, the Baltic Sea separates Scandinavia from north-central Europe. Denmark and Sweden have long controlled the narrow Skagerrak and Kattegat straits that connect the Baltic to the North Sea, both a major fishing ground and a principal source of Europe's oil and gas, mined from deep-sea drilling platforms.

The English Channel (in French, *La Manche*) separates the British Isles from continental Europe. At its narrowest point, the Dover Straits, the channel is only 20 miles (32 km) wide. Although England has long regarded the channel as a protective moat, it is primarily a symbolic barrier, for it deterred neither the French Normans from the continent nor the Viking raiders from the east. Only Nazi Germany found it a formidable barrier during World War II. Since 1993, the British Isles have been connected to France through the 31-mile (50-km) Channel Tunnel (or "Chunnel"), with its high-speed rail system carrying passengers, autos, and freight.

▼ **Figure 8.5 Fjord in Norway** During the Pleistocene epoch, continental ice sheets and glaciers carved deep U-shaped valleys along Norway's coastline. As the ice sheets melted and sea level rose, these valleys were flooded by Atlantic waters. Many fjord settlements are accessible only by boat, linked to the outside world by Norway's extensive ferry system.

Explore the **SIGHTS** of Norway's Fjords
http://goo.gl/zBlDX9

Gibraltar guards the narrow straits between Africa and Europe at the western entrance to the Mediterranean Sea, and Britain's stewardship of this passage remains an enduring symbol of a once great sea-based empire. On Europe's southeastern flanks are the straits of the Bosporus and Dardanelles, the narrows connecting the eastern Mediterranean with the Black Sea. Disputed for centuries, these pivotal waters are now controlled by Turkey. Though often considered the physical boundary between Europe and Asia, the straits are easily bridged in several places to facilitate truck and train transportation within Turkey and between Europe and Southwest Asia.

Rivers and Ports   Europe is a region of navigable rivers that allow inland barge travel from the Baltic and North Seas to the Mediterranean and between western Europe and the Black Sea. Many rivers on the European Lowland—namely the Loire, Seine, Rhine, Elbe, and Vistula—flow into Atlantic and Baltic waters. However, the Danube, Europe's longest river, rises in the Black Forest of Germany only a few miles from the Rhine River and runs southeastward to the Black Sea, offering a connecting artery between central and southeastern Europe (Figure 8.6). Similarly, the Rhône headwaters rise close to those of the Rhine in Switzerland, yet the Rhône flows southward into the Mediterranean. Both the Danube and the Rhône are connected by locks and canals to the rivers of the European Lowland, allowing barge traffic to travel between all of Europe's fringing seas and ocean.

Major ports found at the mouths of most western European rivers serve as transshipment points for inland waterways and rail and truck networks. From south to north, these ports include Bordeaux at the mouth of the Garonne, Le Havre on the Seine, London on the Thames, Rotterdam at the mouth of the Rhine, Hamburg on the Elbe and, in Poland, Szczecin on the Oder and Gdansk on the Vistula.

Europe's Climate

Three major climate types characterize Europe (Figure 8.7). Along the Atlantic coast, a moderate and moist **marine west coast climate**

modified by oceanic influences dominates. Farther inland, **continental climates** prevail, with hotter summers and colder winters. Finally, a dry-summer **Mediterranean climate** is found in southern Europe, from Portugal to Greece.

The Atlantic Ocean serves as an important climate control. Although much of Europe is at relatively high latitudes (see Figure 8.3; London, for example, is slightly farther north than Vancouver, British Columbia), the mild North Atlantic Current, which is a continuation of the warmer Atlantic Gulf Stream, moderates coastal temperatures from Iceland and Norway south to Portugal. This maritime influence gives western Europe a climate 5–10°F (3–6°C) warmer than regions at comparable latitudes that lack the moderating influence of a warm ocean current. As a result, in the marine west coast climate region, no winter months average below freezing, even though cold rain, sleet, and an occasional snowstorm are common winter events. Summers are often cloudy and overcast, with frequent drizzle and rain as moisture flows in from the ocean.

Inland, far removed from the ocean (or where a mountain chain limits the maritime influence, as in Scandinavia), landmass heating and cooling becomes a strong climatic control, producing hotter summers and colder winters. Indeed, all continental climates average at least one month below freezing during the winter. The transition between Europe's maritime and continental climates occurs close to the Rhine River border of France and Germany. Farther north, despite the moderating influence of the Baltic Sea, the higher latitude coupled with the blocking effect of the Norwegian mountains produces cold winter temperatures characteristic of true continental climates.

The Mediterranean climate has a distinct summer dry season resulting from the warm-season expansion of the Atlantic (or Azores) high-pressure area. As this warm air descends between latitudes of 30 and 40 degrees, it inhibits summer rainfall. This same phenomenon also produces the Mediterranean climates of California, western Australia, parts of South Africa, and central Chile. These rainless summers attract tourists from northern Europe, but the seasonal drought is problematic for agriculture. It is no coincidence that traditional Mediterranean cultures, such as the Egyptian, Moorish, Greek, and Roman civilizations, were major innovators of irrigation technology.

Environmental Issues: Local and Global

Because of its long history of agriculture, resource extraction, industrial manufacturing, and urbanization, Europe has its share of environmental issues (Figure 8.8). Compounding the situation, pollution rarely stays within political boundaries: air pollution from England creates serious acid rain problems in Sweden, and water pollution from Swiss factories on the upper Rhine River creates major problems downstream for the Netherlands, where Rhine water commonly supplies city water systems. These numerous trans-boundary issues have led the EU to address the region's environmental problems, and Europe is today probably the "greenest" of the major world regions.

Until recently, however, the countries of central and southeastern Europe were plagued by far more serious environmental problems than their western neighbors due to their history of communist domination, where economic planning emphasized short-term industrial output at the expense of environmental

▼ **Figure 8.6 Inland Barge Traffic**   Europe's river systems are connected by an extensive inland canal system so that barge traffic can readily move throughout the region, connecting ports on the Atlantic Ocean with those on the Baltic, Mediterranean, and Black Seas. This barge is traversing north Germany's Elbe Lateral Canal.

Explore the SIGHTS of the Elbe Canal
http://goo.gl/ryzGLv

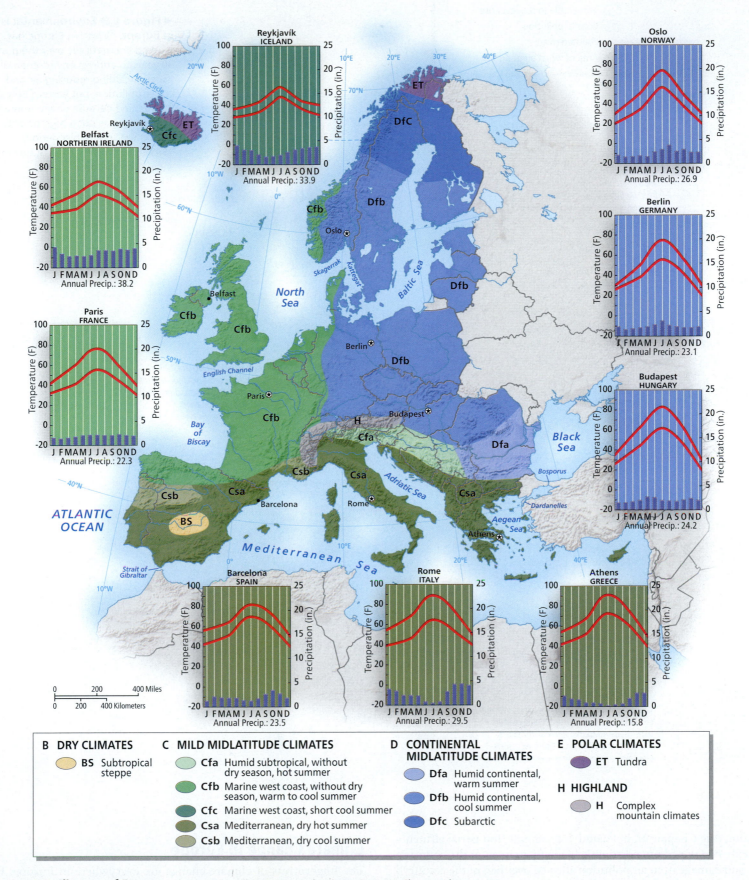

▲ Figure 8.7 Climates of Europe Three major climate zones dominate Europe. Close to the Atlantic Ocean, the marine west coast climate has cool seasons and steady rainfall throughout the year. Farther inland, continental climates have at least one month averaging below freezing, as well as hot summers, with a precipitation maximum occurring during the warm season. Southern Europe has a dry-summer Mediterranean climate.

Areas affected by acid precipitation
Vulnerable to sea-level rise
Coastal pollution
Polluted rivers
Area of worst air pollution

◄ **Figure 8.8** **Environmental Issues in Europe** Western Europe has worked energetically over the past 50 years to address environmental issues, including reducing air and water pollution, but eastern Europe lags behind because environmental protection was not a high priority during the communist period, 1945–1989. Current efforts, however, show great promise.

North Sea Coastline. *Low-lying coastal settlements and farmlands are threatened by sea-level rise from global warming.*

Acid Precipitation. *Half of Poland's forests and three-quarters of those in Czechia are damaged from acid precipitation.*

Climate Change in the Alps. *Warmer temperatures have caused Alpine glaciers to retreat, and sparse snowfall threatens the economic vitality of Alpine ski resorts.*

protection (see Chapter 9). In Poland, for example, industrial effluents reportedly had wiped out all aquatic life in 90 percent of the country's rivers, and damage from air pollution affected over half of the country's forests. Similar legacies from the communist period persist in Czechia, Romania, and Bulgaria. Today, however, most environmental issues in central and southeastern Europe have been resolved through EU funding and stronger national environmental laws.

Climate Change in Europe

The fingerprints of climate change are everywhere in Europe, from dwindling sea ice, melting glaciers, and sparse snow cover in arctic Scandinavia to more frequent droughts in the arid Mediterranean subregion. Furthermore, projections for future climate change are ominous: world-class Alpine ski resorts are forecast to have warmer winters with less snowpack, while in the lowlands, higher summer temperatures will probably produce more frequent heat waves like those in 2003 and 2015, affecting farmers and urban dwellers alike. In addition, rising sea levels from melting polar ice sheets will threaten the Netherlands, where much of the population lives in diked lands below sea level (Figure 8.9). Because of these threats, Europe has taken a strong stand in addressing climate change and, as a result, has implemented numerous policies and programs to reduce greenhouse gas (GHG) emissions.

The EU played a leading role in structuring the 2015 **Paris Agreement**, including the negotiations hosted in Paris, France. Almost 10 percent of global GHGs come from the European Economic Area (the EU and its neighbors in western Europe). The EU was the first major economy to submit its intended contribution to the Paris Agreement, committing to cutting GHG emissions by 40 percent from 1990 levels.

Energy and Emissions Greenhouse gas emissions are closely linked to a country's energy mix and population size. Not surprisingly, EU member countries having the largest populations and burning the most fossil fuels have the highest emissions. Germany, the largest European country by population with almost 83 million people, emits over 900 million metric tons of GHGs each year, although this amount is a significant decrease from 1990 levels. It is followed by the United Kingdom, France, and Italy, each with a population of about 60 million; all three countries emit just over half as much as Germany.

Europe has a diverse energy mix, with nuclear power, fossil fuels, and renewables all playing an important role. Although Europe's early industrialization was based on coal, those resources are now running thin in western Europe and, in fact, much of the EU's emission reductions came from shutting down British and German coal mines during

▼ **Figure 8.9 Protecting Low-Country Europe from Sea-Level Rise** Conceived more than half a century ago, the Dutch Delta Works were originally built to keep ocean storm surges and Rhine River flooding from the southwestern Netherlands. But given the forecasts for sea-level rise due to climate change, the Delta Works must be reengineered and made higher to protect the 50 percent of the Netherlands that lie below the current sea level.

▲ **Figure 8.10 French Nuclear Power** France relies on nuclear power for a greater portion of its electricity, 75 percent, than any other large country. In fact, the French produce so much electricity via nuclear that France ranks annually among the largest exporters of electricity, sending nuclear-derived power to many neighboring European countries.

the 1990s. To replace coal, Europe has shifted to a diverse array of renewable sources and a large nuclear power industry centered in France (Figure 8.10). The region also relies heavily on imported gas and oil, much of it from Russia, with the only local supplies coming from North Sea gas and oil wells developed by the UK, the Netherlands, and Norway.

Complementing EU emission reduction goals is a policy to increase the region's renewable energy resources so that the EU as a whole will generate 27 percent of its power from hydropower, wind, solar, and biofuels by 2030. This goal is already within reach given existing hydropower facilities in the Alpine and Scandinavian countries, coupled with the expansion of wind and solar power in the last two decades. Germany, Denmark, and Spain already generate well over 25 percent of their energy from renewable resources. Throughout Europe, wind power is the fastest-growing renewable energy segment, supplying 9 percent of the EU's power in 2015, with forecasts for twice that amount by 2020 (see *Working Toward Sustainability: Denmark's Offshore Wind Juggernaut*).

The EU's Emission-Trading Scheme As part of its emissions reduction strategy, the EU inaugurated the world's first carbon-trading scheme in 2005. Under this plan, specific yearly emission caps were set for the EU's largest GHG emitters. Exceeding those caps requires either purchasing carbon emission equivalences from a source below their own cap or, alternatively, buying credits from the EU carbon market. The goal of this cap-and-trade system was to make business more expensive for companies that pollute, while rewarding those that stay under their carbon quota, and has become the world's largest and most successful cap-and-trade plan.

REVIEW

8.1 Name and locate on a map the major lowland and mountainous areas of Europe, and identify major water routes.

8.2 Explain how Europe has taken a leading role globally in addressing climate issues including CO_2 emissions.

KEY TERMS European Union (EU), Brexit, fjord, marine west coast climate, continental climate, Mediterranean climate, Paris Agreement

WORKING TOWARD SUSTAINABILITY

Denmark's Offshore Wind Juggernaut

When the 1973 global energy crisis hit, Denmark relied on imported fossil fuels for almost all its energy needs. The Danish government immediately looked for ways to address this problem and started an energy revolution that has spread around the world—large-scale wind power development and, more specifically, offshore wind power.

Offshore Advantages Denmark was ideally suited to lead in wind power, as it sits in a particularly windy location. Offshore wind is especially suited to large-scale development, as the wind blows more uniformly and consistently over the sea than it does on land, and offshore development avoids some of the "not in my backyard" (NIMBY) concerns that energy development can bring. Offshore oil and gas had previously been developed in Denmark, so new offshore infrastructure was an easy sell to the Danish public.

Denmark built its first modern wind turbine in 1979, and in 1991, the first Danish offshore wind farm, Vindeby, with 11 turbines, was built. Since then, 12 additional offshore wind farms have been built in Denmark, with over 500 wind turbines now located off the coast. Its largest offshore wind farm, Anholt, is located in the Kattegat straights between Denmark and Sweden, with 111 wind turbines (Figure 8.1.1). This one wind farm can power over 400,000 homes!

Denmark now boasts one of the world's highest rates of "green" energy due to its massive offshore wind developments. In 2017, Denmark derived 43 percent of its total electricity from wind power and expects to derive over half of its electricity from wind by 2020, and 80 percent from renewable energy sources when solar and biomass are added to wind for electricity generation.

A Global Leader The Danish firm Vestas pioneered offshore wind technology through these sites and is one of the leading wind energy firms globally, having installed wind turbines in over 70 countries (Figure 8.1.2). Vestas has installed offshore wind farms in other European countries, including the UK and the Netherlands, as well as in countries farther away. In recent years, Vestas signed agreements to develop offshore wind sites in the United States, China, and India—three of the world's largest energy consumers. This places Denmark at the center of the ongoing energy revolution to clean energy and helps burnish its image as one of the world's "greenest" countries. In fact, the International Energy Agency points to Denmark as a world leader for decoupling economic growth from greenhouse gas emissions. Denmark's leading role in offshore wind and other "green" power has become an important engine of economic growth in the country.

1. **What makes Denmark particularly suited for offshore wind power? What geographic limitations make this particular technology less suited to coastal areas in other parts of the world?**

2. **Is there wind power where you live? Does it come from onshore or offshore sources?**

▲ **Figure 8.1.1 Royal Yacht at the Anholt Wind Farm** The Danish Royal Yacht sails through Denmark's largest offshore wind farm, Anholt, during its opening ceremonies in 2013, when Queen Margrethe officially opened the power plant. Anholt produces enough electricity to power over 400,000 homes.

▲ **Figure 8.1.2 Vestas Wind Turbine** Denmark's Vestas Wind Company is the world's largest manufacturer of wind power systems, which Vestas has built all over the world. Technological innovations, especially in deploying wind power systems offshore, make Denmark a global leader in alternative energy development.

Population and Settlement: Slow Growth and Migration Challenges

The major themes of Europe's population and settlement geography are its very low rates of natural growth, its aging population, widespread internal migration that is aggravating population loss in the Baltic states and southeastern Europe, and large streams of authorized and unauthorized international migration coming primarily from Africa and Southwest Asia. The highly urbanized, industrial, and relatively wealthy core of western and central Europe, which includes southern England, Belgium, the Netherlands, western Germany, and southern Sweden is the focus of most migration, both internal and international (Figure 8.11).

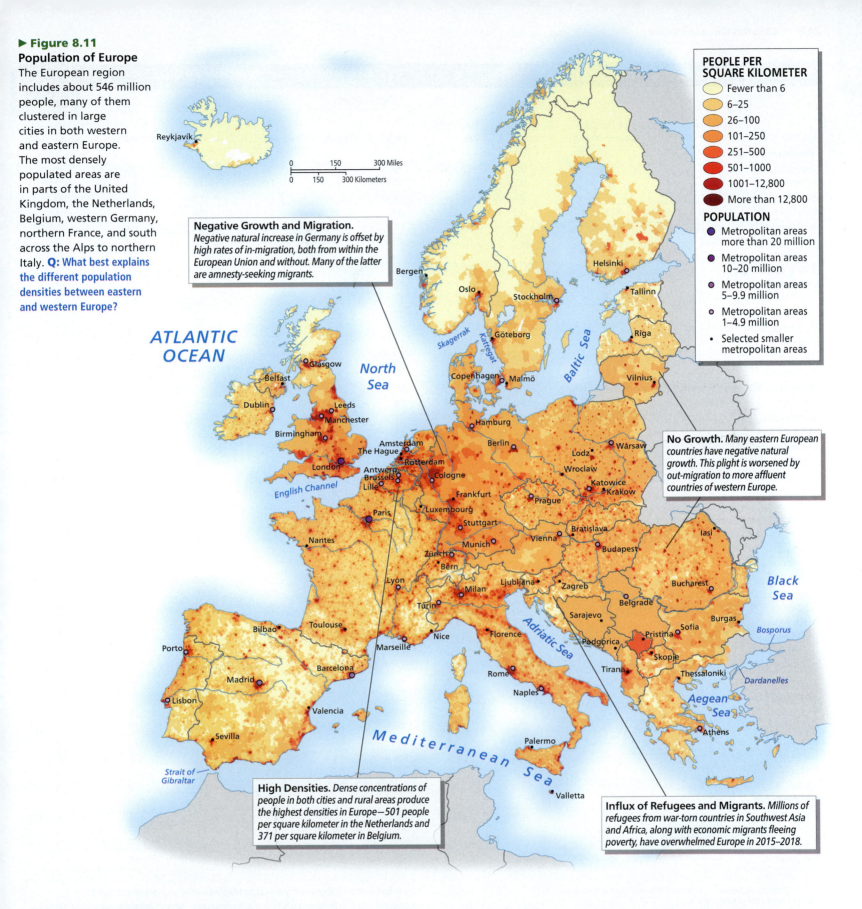

► **Figure 8.11**

Population of Europe
The European region includes about 546 million people, many of them clustered in large cities in both western and eastern Europe. The most densely populated areas are in parts of the United Kingdom, the Netherlands, Belgium, western Germany, northern France, and south across the Alps to northern Italy. **Q: What best explains the different population densities between eastern and western Europe?**

PEOPLE PER SQUARE KILOMETER

- Fewer than 6
- 6–25
- 26–100
- 101–250
- 251–500
- 501–1000
- 1001–12,800
- More than 12,800

POPULATION

- Metropolitan areas more than 20 million
- Metropolitan areas 10–20 million
- Metropolitan areas 5–9.9 million
- Metropolitan areas 1–4.9 million
- Selected smaller metropolitan areas

Negative Growth and Migration. *Negative natural increase in Germany is offset by high rates of in-migration, both from within the European Union and without. Many of the latter are amnesty-seeking migrants.*

No Growth. *Many eastern European countries have negative natural growth. This plight is worsened by out-migration to more affluent countries of western Europe.*

High Densities. *Dense concentrations of people in both cities and rural areas produce the highest densities in Europe—501 people per square kilometer in the Netherlands and 371 per square kilometer in Belgium.*

Influx of Refugees and Migrants. *Millions of refugees from war-torn countries in Southwest Asia and Africa, along with economic migrants fleeing poverty, have overwhelmed Europe in 2015–2018.*

Low (or No) Natural Growth

Perhaps the most striking characteristic of Europe's demography is the lack of natural growth, as death rates exceed birth rates (Table 8.1). Several large countries, notably Germany and Italy, actually have negative natural growth, and their populations could decrease over the next decades unless enough immigrants offset this loss. Numerous smaller European countries (such as Latvia, Spain, Bulgaria, Romania, Serbia, and Portugal) are projected to have smaller populations in the coming decades.

Europe's population, like that of Japan and even the United States, is characterized by the fifth, or *postindustrial*, stage of the demographic

TABLE 8.1   Population Indicators

Explore these data in MapMaster 2.0 https://goo.gl/54jnC9

Country	Population (millions) 2018	Population Density (per square kilometer)[1]	Rate of Natural Increase (RNI)	Total Fertility Rate	Life Expectancy Male	Life Expectancy Female	Percent Urban	Percent <15	Percent >65	Net Migration (rate per 1000)
Western Europe										
Belgium	11.4	376	0.1	1.6	79	84	98	17	19	4
France	65.1	123	0.2	1.9	80	85	80	18	20	1
Ireland	4.9	70	0.7	1.8	78	83	63	21	14	4
Luxembourg	0.6	231	0.3	1.6	80	84	91	16	14	5
Netherlands	17.2	509	0.1	1.6	19	21	92	16	18	5
United Kingdom	66.4	273	0.3	1.8	79	83	83	18	18	6
Central Europe										
Austria	8.8	107	0.0	1.5	79	84	58	14	19	5
Czechia	10.6	137	0.0	1.7	76	82	74	16	19	3
Germany	82.8	237	−0.2	1.6	78	83	77	13	21	6
Hungary	9.8	108	−0.4	1.5	72	79	71	15	19	1
Poland	38.4	124	0.0	1.4	74	82	60	15	17	0
Slovakia	5.4	113	0.1	1.5	74	80	54	16	16	1
Switzerland	8.5	214	0.2	1.5	20	23	85	15	18	6
Northern Europe (Scandinavia and the Baltics)										
Denmark	5.8	137	0.2	1.8	79	83	88	17	19	4
Estonia	1.3	31	−0.1	1.6	73	82	69	16	20	4
Finland	5.5	18	−0.1	1.5	79	84	71	16	21	3
Iceland	0.4	3	0.5	1.7	81	84	94	19	14	24
Latvia	1.9	31	−0.4	1.7	70	79	68	16	20	−8
Lithuania	2.8	45	−0.4	1.6	69	80	67	15	19	−10
Norway	5.3	14	0.3	1.6	81	84	81	18	17	4
Sweden	10.2	25	0.2	1.8	81	84	87	18	20	10
Southeastern Europe/the Balkans										
Albania	2.9	105	0.3	1.6	77	80	60	18	14	−5
Bosnia & Herzegovina	3.5	68	−0.2	1.3	74	79	48	14	16	0
Bulgaria	7.0	65	−0.7	1.6	71	78	73	14	21	−1
Croatia	4.1	74	−0.3	1.4	74	80	57	15	19	−5
Kosovo	1.8	168	0.8	1.7	74	79	38	24	8	0
Macedonia	2.1	83	0.1	1.4	73	78	58	17	14	1
Montenegro	0.6	46	0.1	1.8	74	79	67	18	15	−1
Romania	19.5	85	−0.4	1.4	72	79	54	16	18	−2
Serbia	7.0	80	−0.5	1.5	73	78	60	14	19	3
Slovenia	2.1	103	0.0	1.6	78	84	55	15	19	1
Southern Europe										
Cyprus	1.2	128	0.5	1.4	80	85	68	16	16	3
Greece	10.6	83	−0.2	1.3	78	84	79	14	21	−4
Italy	60.6	206	−0.3	1.3	81	85	70	13	23	3
Malta	0.5	1454	0.3	1.4	81	84	95	14	19	19
Portugal	10.3	112	−0.2	1.4	78	83	65	14	21	0
Spain	46.1	93	−0.1	1.3	80	86	80	15	19	4
Microstates										
Andorra	2.9	164	0.4	1.1	–	–	88	14	14	24
Liechtenstein	0.04	237	0.3	1.6	81	84	14	15	17	5
Monaco	0.04	19,348	0.1	1.5	–	–	100	13	26	20
San Marino	0.03	557	−0.2	1.0	82	87	97	15	18	5
Vatican City	–	–	–	–	–	–	–	–	–	–

Source: Population Reference Bureau, *World Population Data Sheet,* 2018.
[1] World Bank Open Data, 2018.

Log in to Mastering Geography & access MapMaster to explore these data!

1) Map the rate of natural increase for European countries. Which five countries have the highest rates of increase? Which have the lowest rates? Is there an identifiable pattern by subregion?
2) Use the percentage of the population over age 65 data to make a map showing which European countries have the highest number of older people. Suggest reasons for this pattern.

transition (see Chapter 1), in which fertility falls below the replacement level. The consequences of shrinking national populations—labor shortages, smaller internal markets, and reduced tax revenues to support social services (such as retirement pensions) essential for their aging populations—could be significant (Figure 8.12).

Pro-growth Policies To address concerns about population loss, many European countries try to promote growth through various programs and policies. These range from bans on abortion and the sale of certain contraceptives (Hungary, Poland) to what are commonly called **family-friendly policies** (Germany, France, and Scandinavia). In these countries, pro-growth policies include full-pay maternity and paternity leaves for both parents, guarantees of continued employment once these leaves conclude, extensive child-care facilities for working parents, outright cash subsidies for having children, and free or low-cost public education and job training for their offspring. However, even with these family-friendly policies, no European country has a total fertility rate above the replacement level of 2.1. Therefore, any population growth would come solely through in-migration.

Migration Within Europe The EU has worked toward the goal of free movement of both people and goods within the larger European community. Consequently, residents of the EU member countries and certain neighbors can generally move about as they please. And they are doing just that. In the past decade, for example, thousands of young people have left Lithuania and Latvia for Germany. Recent net migration figures show significant out-migration from the Baltics, Spain, Greece, and Croatia, the newest EU member. Germany leads the way in terms of in-migration within Europe, along with Norway, Belgium, and Austria.

The Schengen Agreement Underlying this new intra-Europe mobility is a treaty that eroded Europe's historical national borders: the **Schengen Agreement**, named after the city in Luxembourg where it was signed in 1985. This agreement covers much of the EU, as well as several neighboring countries such as Switzerland and Norway that have opted to remain outside the EU.

Before Schengen, crossing a European border always involved showing passports and auto insurance papers, car inspection records, and so on at every European border. Today, however, there are either no border stations or only the most cursory formalities for those traveling between Schengen countries (Figure 8.13). To older Europeans who traveled before Schengen, it's a remarkable experience to freely cross a national border.

Yet today the Schengen Agreement has become increasingly controversial due to a combination of fears over increased terrorism and unauthorized international migration. Once inside a peripheral Schengen country, such as Italy or Greece, an undocumented migrant can theoretically move freely across borders, just like EU citizens. Even without formal border-crossing points, most countries still maintain some sort of border-policing organization that attempts to limit unauthorized migration, but these efforts are often ineffective. This has led some Schengen countries to seriously consider returning to the pre-Schengen era of formal border controls. Supporting this cautious mentality is the fact that the UK opted out of Schengen in 1985, citing its historical insular location as the rationale. Current conversations about renewed border

▼ **Figure 8.12 Retirees in Portugal** This couple, one from the United Kingdom and the other from Belgium, have chosen to retire to Portugal's south coast, which has become a mecca for retirees from northern Europe. This boom in elderly migrants has led to English-language radio and TV stations, along with Anglican churches popping up in small Portuguese towns. The combination of low taxes and warm weather make this phenomenon similar to American retirees flocking to states such as Florida or Arizona.

▼ **Figure 8.13 Schengen Border** A truck whizzes across the border between Belgium and the Netherlands. The Schengen Agreement removed border checkpoints across much of Europe, but fears of unauthorized migration and terrorism are leading to random stops along these borders. Many wonder if the border crossings of old will return to fortress Europe.

controls seem to conflate legitimate concerns about illegal activity (terrorism and organized crime) with equal amounts of nationalism, cultural xenophobia, and political opportunism, topics discussed later in the chapter.

Authorized Migration to Europe   Western Europe has long accepted international migrants, particularly in the former colonial powers of Spain, the Netherlands, France, and the United Kingdom, countries that willingly provided their overseas citizens with visas and residential permits. Thus, historically, South Asians came to the UK, Indonesians to the Netherlands, Africans and Vietnamese to France, and Latin Americans to Spain.

Europe's doors opened wider to foreigners to solve postwar labor shortages as cities and factories rebuilt from the destruction of World War II. The former West Germany, for example, drew heavily on workers from Europe's rural and poorer periphery—Italy, the former Yugoslavia, Greece, and even Turkey—to fill industrial, construction, and service jobs. Later, with the 1991 collapse of the Soviet Union, emigrants from former satellite countries poured into western Europe, seeking relief from the economic chaos in Russia and central and eastern European countries. This post–Cold War migration also included refugees from war-torn areas of the former Yugoslavia, particularly Bosnia and Kosovo.

As a result of these different migration streams, foreigners now make up about 5 percent of the EU population; Germany, the largest country, has the highest percentage of foreigners (10 percent), with France and the United Kingdom having about 5 percent each. (For comparison, 11.7 percent of the U.S. population are foreign-born.)

Unauthorized Migration, Leaky Borders, and "Fortress Europe"

While at one level the distinction between authorized and unauthorized migration may seem clear-cut—a migrant either does or does not have the proper entry papers—the situation today is much more complex because of the region's **asylum laws** that protect refugees from global political and ethnic persecution. Current asylum laws stem from Europe's post–World War II humanitarian efforts to care for the region's refugees displaced by the war as well as those displaced by the resulting Cold War that divided the continent politically from 1945 to 1991.

Today, however, Europe is inundated with asylum seekers from afar, primarily from Africa, Southwest Asia, and Afghanistan. Although it is theoretically possible to apply for asylum in Europe from one's home country, the very nature of persecution often prevents this, leading most asylum seekers to try to enter Europe without documentation. Once (or if) they reach European soil, refugees can then ask authorities for political asylum. In 2015, the number of refugees and migrants reaching Europe peaked at almost one million and, although it has since slowed, Europe faces its largest inflow of migrants in decades. This increase is primarily caused by worsening conditions in war-torn countries, and many migrants reach Europe via organized crime groups engaged in human trafficking (Figure 8.14).

◄ **Figure 8.14 Migration into and Within Europe**   The migrant crisis peaked in 2015, with almost one million migrants coming into Europe from Southwest Asia and Africa. Despite a drop-off in subsequent years, the challenges of migration are impacting European politics and threaten to close the open borders of "Schengenland."

Getting to European soil is not simple, cheap, or safe. Refugees reportedly pay thousands of dollars to smugglers for a risky land or sea journey to Europe (see *Humanitarian Geography: Mediterranean Rescues and the Migrant Crisis*). Many die in the process. Those that reach Europe face a long legal process to determine if they may stay, since separating those with legitimate claims of persecution from the so-called economic migrants is a difficult and time-consuming task. If the refugees are granted asylum, they will be sent to an accepting European country, where they will face the additional challenges of adapting to a new culture and environment; if they are denied asylum, they will be summarily deported to their home country.

On a per capita basis (of the native population), Sweden and Norway have accepted the largest number of migrants granted asylum. Germany's 2014–2017 intake of asylum seekers was five times that of the previous five years, and more than any other European country (over 250,000). France and the UK have also accepted large numbers. Currently, the EU is developing a quota system that will require all member states to accept a specific number of refugees annually, but some countries in the former communist east are pushing back on taking even a token number of asylum seekers. To break the logjam of migrants crowded into relocation camps in Italy, Greece, and Spain, the EU is also working on plans to send migrants to non-Mediterranean countries where they can file their asylum requests. This does not mean that these refugees must stay in those countries; if their asylum claims are validated, they will be able to move freely among the Schengen countries.

To help the EU perimeter countries of Greece, Italy, Malta, and Spain police their borders, the EU has provided funds for more guards and, in some places, for physical border barriers to inhibit unauthorized entry. For those with longer memories, these fortifications are disturbingly reminiscent of the Cold War's Iron Curtain that divided Europe into west and east (Figure 8.15).

Today some observers describe Europe as a geographical system having a perimeter of hard borders—a "Fortress Europe," as critics (and anti-immigrant groups) call the plan—while its internal borders are deliberately soft and porous due to the Schengen Agreement. However, until the unauthorized migration issue is resolved, the Schengen borders will become increasingly controversial, challenging earlier political and economic goals of a "Europe without borders."

Landscapes of Urban Europe

Europe is highly urbanized, with almost three-quarters of its population living in cities. In fact, several European microstates are essentially city-states: Monaco, Malta, and Vatican City. But urbanization data can produce different landscapes. Belgium, for example, is 98 percent urban, but this results from a landscape of connected mid-sized towns rather than huge mega-cities. And no traveler in Iceland would believe it to be 95 percent urban—beyond its only city, Reykjavik, the landscape is rural and wild. At the other end of the scale is Bosnia–Herzegovina, the only European country less than 50 percent urban.

Europe's largest countries—Germany, France, the United Kingdom, and Italy—are more typical of the region: three-quarters urban, with numerous large cities scattered among expansive rural landscapes.

The Past in the Present   North American visitors often find European cities far more interesting than their own because of the mosaic of historical and modern landscapes, featuring medieval churches and squares interspersed with high-rise buildings and modernist apartment blocks. The imprints of three historical periods are visible: the medieval (900–1500), Renaissance–Baroque (1500–1800), and industrial/modern (1800–present) periods have left traces on the European urban scene (Figure 8.16).

▼ **Figure 8.16 Old and New Architecture**   The main square in Wroclaw, Poland, features Renaissance and Baroque buildings that exhibit architecture of the German style, as this city was then part of the Kingdom of Saxony, only joining Poland after World War II. Today, modern features, like this fountain, sit side-by-side with buildings of the past in Wroclaw and many cities throughout Europe.

Explore the SIGHTS of Wroclaw
http://goo.gl/T9jPQZ

▼ **Figure 8.15 Hungarian Border Fence**   Syrian refugee families are trapped by a newly erected border fence between Hungary and Serbia in 2016. Hungarian authorities sealed the border quickly in reaction to nationalist calls to limit the flow of refugees coming from the conflicts in Syria, Libya, and Afghanistan.

Mediterranean Rescues and the Migrant Crisis

GOOGLE EARTH
Virtual Tour Video
http://goo.gl/aJaZCO

Although the refugee crisis peaked in 2015, massive flows of refugees and other migrants continue to seek passage into the EU, many by unauthorized means. Some of the most dramatic scenes have played out in the waters of the Mediterranean Sea, Europe's southern divide with the volatile regions of North Africa and Southwest Asia.

Human traffickers in Turkey, Libya, Morocco, Tunisia, and Algeria offer passage to desperate people via the sea, often on rickety small boats or even inflatable rafts (Figure 8.2.1). The crisis has become especially pronounced in the failed state of Libya, where terrorist groups and militias control vast swaths of the coastline. Refugees from the Libyan and Syrian wars, as well as those fleeing violence and extreme poverty in Sub-Saharan Africa, reach the Libyan coast and pay to be smuggled to Europe, risking drowning if their craft fails to make the 184-mile trip to Italy's small island of Lampedusa, which has become ground zero in humanitarian efforts to rescue these vulnerable migrants before their boats fail.

Geography Matters Various humanitarian organizations and Italian Coast Guard ships scour the international waters between Libya and Italy in search of migrant boats. Rescuers involved with humanitarian groups bring these desperate migrants onward to Italian shores (Figure 8.2.2), while government ships may instead return them to Libya. Geography plays a critical role in these operations, as boats still in Libyan waters are beyond the legal reach of European-flagged vessels, and those that can make it to dry EU land have the right to request political asylum, which may or may not be granted. These refugees are indeed victims of their geography, and of the stark difference across borders between their volatile home countries and the prosperity and peace of western Europe.

No Easy Solutions Unlike land borders, where many European countries have built fences to keep out migrant flows (like the walls and fences along parts of the U.S.–Mexico border), at sea there is no easy way to monitor vast expanses of international waters that lie between the territorial waters of Europe and North Africa. Non-governmental humanitarian organizations and European coast guards plying these waters have different goals in mind. In fact, many anti-migrant political parties in Europe are calling for an end to humanitarian patrols and a more militant response wherein boatloads of refugees are all brought back to Libya and other countries, rather than allowed in the EU and a chance at asylum.

As these debates rage, the humanitarian crisis on the Mediterranean Sea continues. In 2017, an estimated 3100 migrants drowned in the Mediterranean, and another 150,000 reached Europe. Although a dramatic drop from 2015, when almost 150,000 migrants a month reached Europe (primarily from Turkey to Greek islands at that time), the numbers remain high, and the costs—both monetary and human—weigh heavily on the EU.

1. **How can Europe best address the humanitarian issue of migrants arriving via the sea? List the consequences of granting asylum and of returning migrants.**

2. **How might the debate over migrants in Europe, especially refugees, be similar or different from the debate on migration in the United States?**

▲ **Figure 8.2.1 African Migrants** Small makeshift boats, overloaded with refugees and migrants, head toward Europe from Libya. Lucky boats are interdicted by rescue groups or the coast guard. Unlucky boats often capsize, and thousands of migrants drown as they attempt to escape to Europe.

▲ **Figure 8.2.2 Humanitarian Rescue Ship** Many humanitarian organizations scour the Mediterranean Sea trying to find migrant boats before they encounter trouble, and then bring migrants to Italy, Greece, or Spain for initial processing and assistance.

The **medieval landscape** is one of narrow, winding streets, crowded with three- or four-story masonry buildings with little setback from the street. This is a dense landscape with few open spaces, except around churches and town halls, where public squares once housed open-air marketplaces. These picturesque medieval-era districts present challenges to contemporary inhabitants because of their narrow, congested streets and old housing.

Many cities have enacted legislation to restore and protect their historic central cities. This movement began in the late 1960s in the Marais area of central Paris and has become increasingly popular throughout Europe as cultures work to preserve the unique sense of place provided by their urban medieval sections. High restoration costs have led to a demographic change where low- and fixed-income people are displaced by those able to pay higher rents. Further, historical areas often attract tourists, and with increased foot traffic the array of street-level shops also often changes from neighborhood-serving stores to those catering to tourists. Urban planners use the term *gentrification* to describe these changes to historic districts.

In contrast to the cramped medieval landscape, those areas of the city built during the **Renaissance–Baroque period** are much more open and spacious, with expansive ceremonial buildings and squares, monuments, ornamental gardens, and wide boulevards lined with palatial residences. During this period (1500–1800), a new artistic sense of urban planning arose in Europe, restructuring many European cities, particularly large capitals such as Paris and Vienna where grand boulevards replaced older, more densely settled quarters. These changes were primarily for the benefit of the new urban elite—the royalty and rich merchants.

The advent of assault artillery during the Renaissance–Baroque era led to the building of extensive systems of defensive walls and, once encircled by these walls, the cities could not expand outward. Instead, as the demand for space increased within the cities, a common solution was to add several new stories to the medieval houses.

Industrialization dramatically altered the landscape of European cities. Beginning in the early 19th century, factories clustered in cities, drawn by their large markets and labor force and supplied by raw materials shipped via barge and railroad. Industrial districts of factories and worker tenements grew up around these transportation lines. In continental Europe, where many cities retained their defensive walls until the late 19th century, the new industrial districts were often located outside the former city walls, removed from the historic city center. When cities removed these defensive walls, these spaces were commonly converted into ring roads, as seen in Vienna, Austria, and Toulouse, France.

Not to be overlooked are the post–World War II changes to European cities as they rebuilt and adapted to the political and economic demands of the postwar era. As in North America, suburban sprawl has become an issue in many European countries as people seek lower-density housing in nearby rural environments. Unlike most North American cities, skyscrapers are primarily located on the fringes of city centers beyond the medieval and Renaissance-Baroque districts. Another major difference is the well-developed public transportation systems that offer attractive alternatives to commuting by car in major European cities.

REVIEW

8.3 What is the Schengen Agreement, and how is it related to population movement?

8.4 Name three stages of historical urban development still commonly found in European urban landscapes and discuss the characteristics of each stage.

KEY TERMS

family-friendly policies, Schengen Agreement, asylum laws, medieval landscape, Renaissance–Baroque period

Cultural Coherence and Diversity: A Mosaic of Differences

The rich cultural geography of Europe demands our attention for several reasons. First, the highly varied mosaic of languages, customs, religions, and ways of life that characterize Europe not only strongly shaped regional identities, but also often stoked the fires of conflict. Embers from those historical conflicts still smolder today in several areas.

Second, European cultures played leading roles in globalization as European colonialism brought about changes in languages, religions, political systems, economies, and social values in every corner of the globe. Examples include bullfights in Mexico, high tea in India, Dutch architecture in South Africa, and French bread in Vietnamese sandwiches (*banh mi*). European cultural exports broadly include classical music, various popular couple dance forms, and many types of pop music (see *Globalization in Our Lives: The Eurovision Song Contest*).

Explore the **SOUNDS** of Polish Folk Dance
https://goo.gl/Sh3BxM

Today, however, waves of global culture are spreading back into Europe, and while some Europeans embrace these changes, others actively resist. France, for example, often struggles with both global popular culture and the multicultural influences of its large Muslim migrant population (Figure 8.17).

Geographies of Language

Language has always been an important component of nationalism and group identity in Europe. While some small ethnic groups such as the Irish and the Bretons work hard to preserve their local language, millions of other Europeans are busy learning multiple languages—primarily English—so they can better communicate across cultural and national boundaries.

GLOBALIZATION IN OUR LIVES

The *Eurovision Song Contest*

American Idol was not the first major televised program featuring singers and call-in votes for favorites. The *Eurovision Song Contest* has been running every year since 1973 and stands today as the longest-running TV song competition (Figure 8.3.1). Each participating country nominates one act to perform a popular song in a preliminary broadcast, and the top vote getters come back for a grand finale. *Eurovision* has become one of the most popular TV broadcasts across Europe and, indeed, around the world.

In the early years, around 20 countries from western and southern Europe, along with Israel (which also competes in European soccer) entered *Eurovision*. With the fall of the Soviet Union, the contest spread eastward to include former communist countries and newly independent states in Europe and Eurasia. It has now spread to include even Australia, which first participated in 2015, and *Eurovision* first aired on American live television in 2017.

Eurovision is best known for its campy entries featuring outrageous costumes and energetic backup dancers. It has, however, produced winners that have gone on to global fame, such as the Swedish band ABBA in 1974 and Switzerland's 1988 winner, Celine Dion. Contestants can perform any style of music, and entrants have performed pop, traditional folk music, turbofolk (a modern techno take on traditional folk music), country/western, and even death metal and punk. While some styles are unique to the country entering, many come from musical traditions that began in North America.

The *Eurovision* spectacle continues to entertain millions in a sort of song Olympics, where nationality and showmanship often plays as much a role as talent in determining the winner.

1. **Do contests like *Eurovision* help build a common European identity or reinforce existing national divides?**

2. **Do you know or like any of the bands that have played or won at *Eurovision?***

▲ **Figure 8.3.1** *Eurovision* **Performance**   Ukrainian drag performer Verka Serduchka performs at the 2007 *Eurovision Song Contest,* where Serduchka finished in second place. *Eurovision*'s campy performances and spirited international competition have led to the contest spreading far beyond its original western European countries to encompass much of the former Soviet bloc, and even Australia.

At the broadest scale, most Europeans speak an **Indo-European language** (Figure 8.18). As their mother tongue, 90 percent of Europe's population speaks a Germanic, Romance, or Slavic language, which are linguistic groups within the Indo-European family. Germanic and Romance speakers each number almost 200 million in the European region. Although Slavic languages are spoken by 400 million when Russia and its immediate neighbors are included, there are only 80 million Slavic speakers within Europe proper.

Germanic Languages   Germanic languages dominate Europe north of the Alps. Today German, claimed by about 90 million people as their mother tongue, is spoken in Germany, Austria, Liechtenstein, Luxembourg, eastern Switzerland, and far northern Italy.

English is the second-largest Germanic language, with about 60 million speaking it as their first language. In addition, a large number of Europeans learn English as a second language, and many are fluent. Linguistically, English is closest to the Low German spoken along the North Sea, which reinforces the theory that an early form of English evolved in the British Isles through contact with the coastal peoples of northern Europe. One distinctive trait of English that sets it apart from German, however, is that almost one-third of the English vocabulary is made up of Romance words brought to England during the Norman French conquest of the 11th century.

◀ **Figure 8.17 Muslims in Europe** Europe has long had a small Muslim population, historically in the Balkans and Spain but more recently in western Europe because of Europe's colonial ties to Asia and Africa. The postwar guest-worker program led to the creation of Turkish communities in many German cities. Currently, nationalistic, anti-migrant groups have voiced concerns about the "Islamization" of Europe from the large numbers of migrants from Southwest Asia and North Africa.

▼ **Figure 8.18 Languages of Europe** Ninety percent of Europeans speak a Germanic, Romance, or Slavic language, all members of the Indo-European language family. Ninety million Europeans speak German as a first language, which places it ahead of the 60 million who list English as their mother tongue. However, given the large number of Europeans who speak fluent English as a second language, one could make the case that English is the dominant language of modern Europe.

Elsewhere in the Germanic linguistic region, Dutch (in the Netherlands) and Flemish (in northern Belgium) together account for another 20 million speakers, with roughly the same number of Scandinavians speaking the closely related languages of Danish, Norwegian, and Swedish. Icelandic is a more distinctive Scandinavian language because of that country's geographic isolation.

Romance Languages
Romance languages, including French, Spanish, Italian, and Romanian, evolved from the vulgar (or everyday) Latin spoken within the Roman Empire. Today Italian is the most widely used of these Romance languages in Europe, with about 60 million speaking it as their first language. In addition to being spoken in Italy, Italian is an official language of Switzerland and is also spoken on the French island of Corsica.

French is spoken in France, western Switzerland, and southern Belgium. Today there are about 55 million native French speakers in Europe. As with other languages, French has very strong regional dialects.

Spanish also has very strong regional variations. About 25 million people speak Castilian Spanish, Spain's official language, which dominates the interior and northern areas of that large country. However, the related language of Catalan is spoken along the eastern coastal fringe, centered on Barcelona. Other dialects of Spanish dominate Spain's former colonies in Latin America, making it one of the world's major languages.

Portuguese is spoken by 12 million in Portugal and in the northwestern corner of Spain, although considerably more people speak the language in former Portuguese colonies including Brazil in Latin America and several African countries. Finally, Romanian represents an eastern outlier of the Romance language family, spoken by 24 million people in Romania. Though unquestionably a Romance language, Romanian also contains many Slavic words.

Slavic Languages
Slavic speakers are traditionally separated into northern and southern groups, divided by the non-Slavic speakers of Hungary and Romania. To the north, Polish has 35 million speakers, with Czech and Slovak speakers totaling about 15 million. As noted earlier, these numbers pale in comparison to the number of northern Slav speakers in nearby Ukraine, Belarus, and Russia, which easily total around 200 million. Southern Slav languages include the 14 million speakers of Serbian and Croatian (primarily differentiated by which alphabet they use), 11 million Bulgarian and Macedonian speakers, and 2 million Slovenian speakers.

The use of two distinct alphabets further complicates the geography of Slavic languages. In countries with a strong Roman Catholic heritage, such as Poland, Czechia and Croatia, the **Latin alphabet** is used. In contrast, countries with close ties to the Orthodox Church—Bulgaria, Macedonia, parts of Bosnia–Herzegovina, and Serbia—use the Greek-derived **Cyrillic alphabet** (Figure 8.19).

Outlier Linguistic Groups
Europe is also home to several linguistic/cultural groups that lie outside the major Indo-European language families described above. Europe's most dispersed minority, the Roma (commonly known by the term Gypsy, which most Roma reject as racist) speak a group of related languages called Romani that are more closely related to Hindi, India's most common language, than they are to the Indo-European languages spoken in Europe. Hungarian

(or Magyar), Finnish, and Estonian all come from another language family, the Finno-Ugric Languages, which come from Siberia. The Basque of France and Spain speak Euskara, a language unrelated to any other major world language.

Geographies of Religion, Past and Present
Religion is a critical part of the geography of cultural coherence and diversity in Europe because many of today's ethnic tensions result from historical religious events (Figure 8.20). To illustrate, significant cultural borders in the Balkans are based on the 11th-century split of Christianity into eastern and western churches as well as on the division between Christianity and Islam. Much of the genocide perpetrated in the former Yugoslavia during the 1990s was based on these religious differences. In western Europe, blood is still occasionally shed in Northern Ireland over the tensions resulting from the 17th-century split of western Christianity into Catholicism and Protestantism.

The Schism Between Western and Eastern Christianity
In southeastern Europe, early Greek missionaries spread Christianity throughout the Balkans and into the lower reaches of the Danube. Because these missionaries refused to accept the control of Roman Catholic bishops in western Europe, there was a formal split with western Christianity in 1054 CE. This eastern church subsequently splintered into Orthodox sects closely linked to specific nationalities. Today we find Greek Orthodox, Bulgarian Orthodox, and Romanian Orthodox churches, among others, all with slightly different rites and rituals.

The Protestant Revolt
The other great split within Christianity occurred between Catholicism and Protestantism during the 16th century, and has divided the region ever since. However, with the exception of "the Troubles" in Northern Ireland, tensions today between these two major groups are far less problematic than in the distant past.

▲ **Figure 8.19 Cyrillic Alphabet** A road sign in Belgrade, Serbia, uses both the Cyrillic and the Latin alphabets to guide locals and visitors to tourist sites in the city.

▲ **Figure 8.20 Religions of Europe** This map shows the divide in western Europe between the Protestant north and the Roman Catholic south. This distinction was much more important historically than it is today. Note also the location of the former Jewish Pale, the area devastated by the Nazis during World War II. Today ethnic tensions with religious overtones are found primarily in the Balkans, where adherents to Roman Catholicism, Eastern Orthodoxy, and Islam live in close proximity. **Q: After comparing this map to the one of Europe's languages (Figure 8.18), list those areas where language families and religion appear to be related.**

Historical Conflicts with Islam Both the eastern and western Christian churches struggled with challenges from the Islamic empires to Europe's south and east. Even though historical Islam was reasonably tolerant of Christianity in its conquered lands, Christian Europe was far less accepting of Muslim imperialism. The first crusade to reclaim Jerusalem from the Turks took place in 1095. After the Ottoman Turks conquered Constantinople in 1453 and gained control over the Bosporus Strait and the Black Sea, they moved rapidly to spread a Muslim empire throughout the Balkans, arriving at the gates of Vienna in the middle of the 16th century. There Christian Europe stood firm militarily and stopped Islam from expanding into western Europe.

Ottoman control of southeastern Europe, however, lasted until the empire's demise in the early 20th century. This historical presence of Islam explains the current geography of religions in the Balkans, with intermixed areas of Muslims, Orthodox Christians, and Roman Catholics.

Islam was also the dominant religion in Portugal and much of Spain from the 8th to the 15th century, when the Catholic kingdoms in Spain's northeast conquered most of the Iberian Peninsula (Figure 8.21).

A Geography of Judaism Europe has long been a difficult home for the Jews forced to leave Palestine during the Roman Empire. At that time, small Jewish settlements were located in cities throughout the Mediterranean. By 900 CE, about 20 percent of the Jewish population was clustered in the Muslim lands of the Iberian Peninsula, where Islam showed greater tolerance for Judaism than did Christianity. After the Christian reconquest of Iberia, however, Jews once more faced severe persecution and fled from Spain to more tolerant countries in western and central Europe.

One focus for this exodus was the area in eastern Europe that became known as the Jewish Pale. In the late Middle Ages, at the invitation of the Kingdom of Poland, Jews settled in cities and small villages in what are now eastern Poland, Belarus, western Ukraine, and northern Romania, in the hope of establishing a true European homeland.

▼ **Figure 8.21 Mosque Cathedral of Cordoba, Spain** Following the Christian reconquest of Spain, Islamic buildings were repurposed for Christian use, including the massive mosque in Cordoba. A Gothic cathedral and Catholic chapels were built into the middle of the mosque, which still maintains numerous Islamic architectural features, such as the arches seen surrounding this chapel.

Until emigration to North America began in the 1890s, 90 percent of the world's Jewish population lived in Europe, and most were clustered in the Pale region. In 1939, on the eve of World War II, 9.5 million Jews lived in Europe. During the war, German Nazis murdered some 6 million Jews in the horror of the Holocaust. Today fewer than 2 million Jews, about 10 percent of the world's Jewish population, live in Europe, primarily in the west.

Patterns of Contemporary Religion Estimates of religious adherence in contemporary Europe suggest there are 250 million Roman Catholics, fewer than 100 million Protestants, and 13 million Muslims. Generally, Catholics live in the southern half of the region, except for the significant numbers in Ireland and Poland. Protestantism is most widespread in northern Germany (with Catholicism stronger in southern Germany), the Scandinavian countries, and the United Kingdom, and it is intermixed with Catholicism in the Netherlands, Belgium, and Switzerland. Muslims historically are found in Albania, Kosovo, Bosnia–Herzegovina, and Bulgaria. Adding to Europe's Muslim population are postwar migrants from Turkey and northern Africa, with 4.8 million Muslims today in Germany and 4.7 million in France. Between migration and a generally higher natural birth rate, combined with a stagnating number of practicing Christians, Islam is Europe's fastest-growing religion.

Explore the **TASTES** of **Turkish Food in Germany**
https://goo.gl/Lb9rJB

Because of Europe's long history of religious wars and tensions, the EU's agenda of European unity is explicitly secular, a position that causes its own cultural tensions. For example, the euro, the EU's common currency, has purged all national symbols of Christian crosses and saints, much to the chagrin of countries like Hungary and Poland whose sense of nationalism is inseparable from Catholicism. This secularism is particularly difficult to accept in those former Soviet satellites where churches were closed during the communist period and have only recently reopened as places of gathering and worship.

European Food and Culture

Along with European colonialism, European food spread across the planet, yet it also maintains a central place in the region's cultural identity. From pasta in Italy to Polish kielbasa sausages and French wine, various food products have become closely associated with Europe's countries and regions.

Varied climates and long coastlines allow for a vast array of agricultural production across the continent. In the Mediterranean south, olives, grapes, and other heat- and drought-tolerant crops dominate. In western Europe, cereal crops and a wide variety of vegetables are grown, while in northern Europe where harsh winters limit agriculture, fish is a central part of the traditional Scandinavian diet.

European integration has also impacted Europe's agricultural economy and food culture. The EU has developed policies to protect European farmers, many of whom receive large crop subsidies, as well as protections for special foods that traditionally come from one region. Known colloquially as *terroir*—French for earth or soil—and officially as "Protected Geographical Status," certain EU products may only use the official name if they come from that specific place. Thus champagne wines must come from the French region of Champagne, Parmesan cheese can only be produced in Emilia-Romagna in Italy, and palinka brandy is limited to Hungary. These special protections link

European food production to the specific climate, soil, and culture of the places where that food was invented (Figure 8.22).

Explore the TASTES of Bologna, Italy
https://goo.gl/hHXfqf

Migrants and Culture

New migration streams from Africa and Asia are profoundly influencing the dynamic cultural geography of Europe, but in some areas the products of this recent cultural exchange are highly troubling.

Immigrant clustering, leading to the formation of ethnic neighborhoods and even ghettos, is common in the cities and towns of western Europe. The high-density apartment buildings of suburban Paris and Brussels, for example, are home to large numbers of French-speaking Africans and Arab Muslims caught in a web of high unemployment, poverty, and racial and religious discrimination. As a result, cultural struggles, both on the streets and in the courtrooms, are now common in many European countries. For example, in 2004, French leaders drew on the country's constitutional separation of state and religion to ban a key symbol of conservative Muslim life—the head scarf (*hijab*)—for female students in public schools because, officials argued, it interfered with the educational process. And in 2010, full-face veils were banned in public places. Another rationale for the legislation was that traditional Muslim dress inhibited the assimilation process and immigrants should blend into contemporary French society. Other European countries have followed suit.

Since the outbreak of the migrant crisis in which Syrian Muslims make up the largest group, anti-Muslim animus is especially prevalent among Europeans who support far-right political parties in several EU member states. This is not surprising, since far-right, neonationalistic political parties throughout Europe share anti-migrant, anti-immigration, anti-asylum positions.

Another group, the Roma, is treated by many as a migrant group due to their nomadic lifestyle, although they have lived in Europe for centuries after arriving from India. Millions of Roma are spread across Europe, often living in desperately poor conditions and facing severe discrimination from majority populations (Figure 8.23).

Sports in Europe

Soccer (which Europeans call football) is unquestionably Europe's most popular sport, played everywhere from sandlots to stadiums, by both women and men, at all levels from family picnics to multiple-level professional leagues. At the highest pro level, soccer teams draw crowds into stadiums holding 100,000 people. Smaller stadiums seating 30,000–40,000 are common in every European town.

Like many sports throughout the world, soccer is irrevocably linked to globalized culture, with fanatical fans rooting for place-based teams constituted largely of international players lacking any local allegiance. But this contradiction doesn't keep fans from taking their local fandom across Europe's borders to rival towns and cities, where team loyalties sometimes turn violent. Soccer hooliganism, unfortunately, has become a common outlet for Europe's anti-migrant racism and xenophobia.

Aside from homegrown field sports like soccer and rugby, Europe has shown some interest in North American sports. Basketball is unquestionably the favorite American sport, with hoops and courts found across the region's gyms and playgrounds. Pro leagues at all levels abound for both men and women, with most European cities supporting at least one pro team. The linkages go across the Atlantic; many NBA players come from the Balkans, and U.S. Women's National Basketball Association (WNBA) players commonly spend their off-season playing for a European pro team to augment their modest WNBA salaries.

Tennis is popular, with both Wimbledon (London) and the French Open (Paris) serving as two of the four Grand Slam tournaments. In

▲ **Figure 8.22 Champagne Region**   In the EU and many other parts of the world, the only sparkling wine that can be called champagne comes from the region of the same name in France where the wine was first developed. Protected *terroir* for products like champagne often lead to higher prices for the specialty products that many European producers claim cannot be made the same, as they depend on the climate, soils, and other geographic phenomenon of specific places.

▲ **Figure 8.23 Roma in Romania**   Despite widespread discrimination, the Roma people have managed to maintain their unique culture and way of life across Europe. Roma often live in harsh conditions, even in the wealthiest countries of Europe.

Explore the SOUNDS of Roma Music
https://goo.gl/wxhbtN

the Balkans and Hungary, water polo is another dominant sport, with most of the top Olympic teams coming from this region.

Europe is also home to many popular winter sports—both downhill and cross-country skiing were invented on the continent. In the Winter Olympic Games, the countries of the Alps and Scandinavia bring home medals in skiing, while the Netherlands dominates speed skating, first developed on the frozen canals of that country (Figure 8.24).

REVIEW

8.5 Describe the general location within Europe of the three major language groups: Germanic, Romance, and Slavic.

8.6 Summarize the historical distribution within Europe of Catholicism, Protestantism, Judaism, and Islam.

8.7 Discuss the overlap between languages and religion in Europe.

KEY TERMS Indo-European languages, Latin alphabet, Cyrillic alphabet

◀ **Figure 8.24 Ski Jumping at the Olympics** A ski jumper flies high over the Olympic stadium during the 1994 Winter Olympics in Lillehammer, Norway. Legend has it that ski jumping was invented in Norway by people skiing off their barn roofs. Norway and its Scandinavian neighbors dominate in various winter sports including ski jump, cross-country skiing (also developed in Norway), and downhill skiing.

Geopolitical Framework: A Dynamic Map

One of Europe's unique characteristics is its dense fabric of 42 independent states within a relatively small area. The ideal of democratic **nation-states** (see Chapter 1 for a discussion of the nation-state concept) arose in Europe and, over time, replaced the fiefdoms and empires ruled by autocratic royal families. France, Italy, and Germany are major examples of this phenomenon. This royal legacy remains, as many European countries are now **constitutional monarchies**, where a royal serves as the symbol of the state while democratically elected officials run the country. The UK, Spain, and Norway are examples.

The shift to a fully democratic Europe involved tremendous violence and upheaval, and conflict continued to shape the drawing of new borders in Europe through the end of the 20th century (Figure 8.25).

The Violent Redrawing of Europe's Map

Two world wars radically reshaped the geopolitical map of 20th-century Europe (Figure 8.26). Although World War I was referred to as the "war to end all wars," it fell far short of solving Europe's geopolitical problems. The peace treaty instead made another world war virtually unavoidable.

When Germany and Austria–Hungary surrendered in 1918, the Paris treaties (known colloquially as the Treaty of Versailles) set about redrawing the map of Europe with two goals in mind: first, to punish the losers through loss of territory and severe financial reparations and, second, to recognize the aspirations of unrepresented nationalities by creating several new states. As a result, the new states of Czechoslovakia, Yugoslavia, and the Baltic states of Estonia and Latvia were created. In addition, Poland and Lithuania were reestablished (see Figure 8.28b). Though the goals of the treaties were admirable, few European states were satisfied with the resulting map. New states were resentful when some of their conationals were left outside the redrawn borders. This created an epidemic of **irredentism**, state policies directed toward reclaiming lost territory and peoples.

These imperfect geopolitical solutions were greatly aggravated by the global economic depression of the 1930s, which brought high unemployment, inflation, food shortages, and even more political unrest to Europe. Three competing ideologies promoted their own solutions to Europe's pressing problems: Western democracy (and capitalism); communism from the Soviet revolution to the east; and a fascist totalitarianism promoted by Mussolini in Italy and Hitler in Germany (see Chapter 1 for descriptions of these political ideologies). With unemployment at record highs in western Europe, public opinion fluctuated wildly between far-right fascism and far-left communism and socialism. In 1936, Italy and Germany joined forces through the Rome–Berlin "axis" agreement. As in World War I, this alignment was countered with mutual protection treaties among France, Poland, and Britain. When an imperialist Japan signed a pact with Germany, the scene was set for a second global war.

◄ Figure 8.25 Geopolitical Issues in Europe
Although the major geopolitical issue of the early 21st century remains the integration of eastern and western Europe into the EU, numerous issues of micro- and ethnic nationalism also engender geopolitical fragmentation. In other parts of Europe, such as Spain, France, and Great Britain, questions of local ethnic autonomy within the nation-state structure challenge central governments.

Map legend:
- North Atlantic Treaty Organization (NATO) member
- Former Warsaw Pact member
- ⊕ NATO headquarters

Note: The United States and Canada are also members of NATO.

Scotland. *In 2014 Scots narrowly rejected a referendum on independence from the United Kingdom; however, separatist sentiment remains strong.*

United Kingdom. *June 2016 vote to leave the European Union destabilizes Europe's agenda for political and economic cooperation.*

Catalonia. *Separatists held a referendum voting to leave Spain in 2017. Spain's government rejected the result, leading to flaring tensions in the subregion.*

A New Cold War? *Tensions between NATO and Russia have increased due to Russia's aggressive actions in Ukraine and Georgia, along with provocative military activities in international waters and airspace.*

By early 1944, the Soviet army had recovered most of its territorial losses and moved against the Germans in central Europe, reaching Berlin in April 1945 and beginning the long communist domination of most of central and southeastern Europe. At that time, Allied forces crossed the Rhine River and began their occupation of Germany. Germany's surrender in May 1945 ended the war in Europe, but with Soviet forces firmly entrenched in much of central and eastern Europe, the military battles of World War II were immediately replaced by an ideological **Cold War** between communism and democracy that lasted until 1991.

A Divided Europe, East and West

From 1945 until 1991, Europe was divided into two geopolitical and economic blocs, east and west, separated by the infamous **Iron Curtain** that descended shortly after the peace agreement ending World War II. East of the Iron Curtain boundary, the Soviet Union imposed the heavy imprint of communism throughout all facets of life—political, economic, military, and cultural. To the west, as Europe rebuilt from the destruction of the war, new alliances and institutions were created to counter the Soviet presence in Europe.

Cold War Geography   The seeds of the Cold War were planted at the 1945 Yalta Conference, when the leaders of Britain, the Soviet Union, and the United States met to plan the shape of postwar Europe (see Figure 8.26c). Because the Soviet army was already in central Europe and moving quickly on Berlin, Britain and the United States agreed that the Soviet Union would occupy eastern parts of Europe and the Western allies would occupy parts of Germany.

The larger geopolitical issue, though, was the Soviet desire for a **buffer zone** between its own territory and western Europe. This buffer consisted of an extensive bloc of satellite countries, dominated politically and economically by the Soviet Union, to cushion the Soviet heartland against possible attack from western Europe. In the east, the Soviet Union annexed the Baltic states while imposing communist puppet governments in Poland, Czechoslovakia, Hungary, Bulgaria, Romania, and Albania. Austria and Germany were divided into occupied sectors by the four former Allied powers. In both cases, the Soviet Union dominated the eastern portion of the country containing the capital cities of Berlin and Vienna, which in turn were divided into French, British, U.S., and Soviet sectors.

In 1955, with the creation of an independent and neutral Austria, the Soviets withdrew from their sector, effectively moving the Iron

Nazi Germany tested western European resolve in 1938 by annexing Austria, the country of Hitler's birth, and then Czechoslovakia, under the pretense of providing protection for ethnic Germans living there. After Germany signed a nonaggression pact with the Soviet Union, Hitler's armies invaded Poland on September 1, 1939. Two days later, France and Britain declared war on Germany. Within a month, the Soviet Union moved into eastern Poland, the Baltic states, and Finland to reclaim territories lost through the peace treaties following World War I. Nazi Germany then moved north and west and occupied Denmark, Norway, the Netherlands, Belgium, and France, after which it began preparations to invade Britain.

In 1941, the war took several startling new turns. In June, Hitler broke the nonaggression pact with the Soviet Union and, catching its Red Army by surprise, took the Baltic states and then drove deep into Soviet territory. When Japan attacked the American naval fleet in December 1941, the United States entered the war in both the Pacific and Europe.

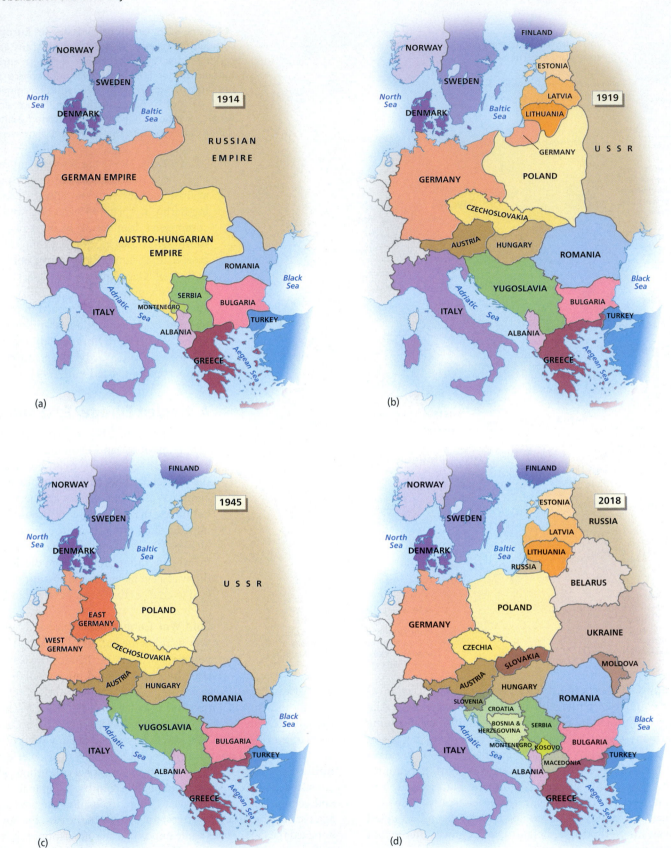

▲ **Figure 8.26 A Century of Geopolitical Change**   (a) At the outset of the 20th century, central Europe was dominated by the German, Austro-Hungarian (or Hapsburg), and Russian empires. (b) Following World War I, these empires were largely replaced by a mosaic of nation-states. (c) More border changes followed World War II, largely as a result of the Soviet Union turning the area into a buffer zone between itself and western Europe. (d) With the demise of Soviet hegemony in 1990, further political change took place. **Q: Where are the strongest relationships between political change and cultural factors such as language and religion?**